Neutrons and Synchrotron Radiation in Engineering Materials Science

Neutrons and Synchrotron Radiation in Engineering Materials Science

From Fundamentals to Applications

Edited by Peter Staron, Andreas Schreyer, Helmut Clemens, and Svea Mayer

Second Edition

WILEY-VCH
Verlag GmbH & Co. KGaA

The Editors

Dr. Peter Staron
Helmholtz-Zentrum Geesthacht
Institute of Materials Research
Max Planck-Str. 1
21502 Geesthacht
Germany

Prof. Dr. Andreas Schreyer
European Spallation Source ERIC
P.O. Box 176
22100 Lund
Sweden

Prof. Helmut Clemens
Montanuniversität Leoben
Department of Physical Metallurgy
and Materials Testing
Franz-Josef-Straße 18
8700 Leoben
Austria

Dr. Svea Mayer
Montanuniversität Leoben
Department of Physical Metallurgy
and Materials Testing
Franz-Josef-Straße 18
8700 Leoben
Austria

Cover
Airplane: fotolia/Gudellaphoto
Car: Porsche

Library of Congress Card No.: applied for

British Library Cataloguing-in-Publication Data
A catalogue record for this book is available from the British Library.

Bibliographic information published by the Deutsche Nationalbibliothek
The Deutsche Nationalbibliothek lists this publication in the Deutsche Nationalbibliografie; detailed bibliographic data are available on the Internet at <http://dnb.d-nb.de>.

Print ISBN: 978-3-527-33592-3
ePDF ISBN: 978-3-527-68451-9
ePub ISBN: 978-3-527-68450-2
Mobi ISBN: 978-3-527-68449-6
oBook ISBN: 978-3-527-68448-9

Cover Design Schulz Grafik-Design, Fußgönheim, Germany
Typesetting SPi Global, Chennai, India
Printed in the UK

Contents

List of Contributors

Volker Abetz
Helmholtz-Zentrum Geesthacht
Institute of Polymer Research
Max Planck-Str. 1
21502 Geesthacht
Germany

Felix Beckmann
Helmholtz-Zentrum Geesthacht
Institute of Materials Research
Max Planck-Str. 1
21502 Geesthacht
Germany

Michael Bischof
Hilti AG
Feldkircher Strasse 100
9494 Schaan
Liechtenstein

Heinz G. Brokmeier
Clausthal University of Technology
Institute of Material Science
and Engineering
Agricolastrasse 6
38678 Clausthal-Zellerfeld
Germany

Bettina Camin
TU Berlin
Institut für Werkstoffwissenschaften
und -technologien
Metallische Werkstoffe
Ernst Reuter-Platz 1
10587 Berlin
Germany

Helmut Clemens
Montanuniversität Leoben
Department of Physical Metallurgy and
Materials Testing
Franz-Josef-Straße 18
8700 Leoben
Austria

Shahrokh Daneshpour
Sogeti Deutschland GmbH
Lübecker Str. 128
22087 Hamburg
Germany

Gerhard Dehm
Max-Planck-Institut für Eisenforschung
GmbH
Max-Planck-Str. 1
40237 Düsseldorf
Germany

Elisabeth Eidenberger-Schober
Plansee Group Service GmbH
Metallwerk-Plansee-Str. 71
6600 Reutte
Austria

Jörg Eßlinger
MTU Aero Engines AG
Dachauer Straße 665
80995 München
Germany

Torben Fischer
Helmholtz-Zentrum Geesthacht
Institute of Materials Research
Max Planck-Str. 1
21502 Geesthacht
Germany

Weimin Gan
Helmholtz-Zentrum Geesthacht
Institute of Materials Research
Outstation at Heinz Maier-Leibnitz
Zentrum (MLZ)
Lichtenbergstr. 1
85747 Garching
Germany

Christoph Genzel
Helmholtz-Zentrum Berlin für
Materialien und Energie GmbH
Abteilung für Mikrostruktur- und
Eigenspannungsanalyse
Albert-Einstein-Str. 15
12489 Berlin
Germany

Robert Gerstenberger
TU Berlin
Institute for Machine Tools and Factory
Management
Pascalstr. 8 -9
10587 Berlin
Germany

Astrid Haibel
Beuth Hochschule für Technik Berlin
Luxemburger Straße 10
13353 Berlin
Germany

André Heinemann
Helmholtz-Zentrum Geesthacht
Institute of Materials Research
Outstation at Heinz Maier-Leibnitz
Zentrum (MLZ)
Lichtenbergstr. 1
85748 Garching
Germany

Julia Herzen
Technical University of Munich
Department of Physics
Chair of Biomedical Physics
James-Franck-Str. 1
85748 Garching
Germany

Norbert Huber
Helmholtz-Zentrum Geesthacht
Institute of Materials Research
Max Planck-Str. 1
21502 Geesthacht
Germany

Gene E. Ice
Oak Ridge National Laboratory
Materials Science and Technology
Division
P.O. Box 2008
Oak Ridge
TN 37831-6132
USA

Karl U. Kainer
Helmholtz-Zentrum Geesthacht
Institute of Materials Research
Max Planck-Str. 1
21502 Geesthacht
Germany

Wolfgang Kaysser
Helmholtz-Zentrum Geesthacht
Max Planck-Str. 1
21502 Geesthacht
Germany

Jozef Keckes
Montanuniversität Leoben
Department of Materials Physics
Franz-Josef-Str. 18
8700 Leoben
Austria

Christoph Kirchlechner
Montanuniversität Leoben
Department of Materials Physics
Franz-Josef-Str. 18
8700 Leoben
Austria

Thomas Klassen
Helmholtz-Zentrum Geesthacht
Institute of Materials Research
Max Planck-Str. 1
21502 Geesthacht
Germany

Manuela Klaus
Helmholtz-Zentrum Berlin für
Materialien und Energie GmbH
Abteilung für Mikrostruktur- und
Eigenspannungsanalyse
Albert-Einstein-Str. 15
12489 Berlin
Germany

Wolfgang Knop
Helmholtz-Zentrum Geesthacht
Research Reactor
Max Planck-Str. 1
21502 Geesthacht
Germany

Harald Leitner
Böhler Edelstahl GmbH &Co KG
Mariazellerstraße 25
8605 Kapfenberg
Austria

Thomas Lippmann
Helmholtz-Zentrum Geesthacht
Institute of Materials Research
Max Planck-Str. 1
21502 Geesthacht
Germany

Klaus-Dieter Liss
Australian Nuclear Science and
Technology Organization
The Bragg Institute
New Illawarra Road
Lucas Heights, NSW 2234
Australia

Ina Lommatzsch
Heinz Maier-Leibnitz Zentrum (MLZ)
Technische Universität München
Lichtenbergstr. 1
85748 Garching
Germany

Wolfgang Ludwig
INSA Lyon
25 av. J. Capelle
69621 Villeurbann
France

René V. Martins
European Commission
Directorate-General for Research &
Innovation
1049 Brussels
Belgium

Svea Mayer
Montanuniversität Leoben
Department of Physical Metallurgy and
Materials Testing
Franz-Josef-Straße 18
8700 Leoben
Austria

Jean S. Micha
CEA-Grenoble
Institute Nanosciences et Cryogénie
B.P. 220
38043 Grenoble Cedex 9
France

Henning F. Poulsen
Technical University of Denmark
Department of Physics
Anker Engelundsvej 1
2800 Kgs. Lyngby
Denmark

Philipp K. Pranzas
Helmholtz-Zentrum Geesthacht
Institute of Materials Research
Max Planck-Str. 1
21502 Geesthacht
Germany

Florian Pyczak
Helmholtz-Zentrum Geesthacht
Institute of Materials Research
Max Planck-Str. 1
21502 Geesthacht
Germany

Anke Kaysser-Pyzalla
Helmholtz-Zentrum Berlin
für Materialien und Energie
Hahn-Meitner-Platz 1
14109 Berlin
Germany

Walter Reimers
TU Berlin
Institut für Werkstoffwissenschaften
und -technologien
Metallische Werkstoffe
Ernst Reuter-Platz 1
10587 Berlin
Germany

Javier Santisteban
Comisión Nacional de Energía Atómica
CONICET and Universidad Nacional de
Cuyo
Neutron Physics Department
Centro Atómico Bariloche
Av Bustillo 9.500
San Carlos de Bariloche 8400
Argentina

Christina Scheu
Max-Planck-Institut für Eisenforschung
GmbH
Max-Planck-Straße 1
40237 Düsseldorf
Germany

Søren Schmidt
Technical University of Denmark
Department of Physics
Anker Engelundsvej 1
2800 Kgs. Lyngby
Denmark

Thomas Schmoelzer
ABB Corporate Research Center
Segelhofstraße 1K
5405 Baden-Dättwil
Switzerland

Ronald Schnitzer
Montanuniversität Leoben
Department of Physical Metallurgy and
Materials Testing
Franz-Josef-Str. 18
8700 Leoben
Austria

Dirk Schnubel
Nemak Europe
Product Development Center
Marie Curie-Strasse
66763 Dillingen
Germany

Michael Eidenberger-Schober
Plansee SE
Metallwerk Plansee Straße 71
6600 Reutte
Austria

Peter Schreiner
Helmholtz-Zentrum Geesthacht
Research Reactor
Max Planck-Str. 1
21502 Geesthacht
Germany

Andreas Schreyer
European Spallation Source ESS ERIC
P.O. Box 176
22100 Lund
Sweden

Emanuel Schwaighofer
Böhler Edelstahl GmbH & Co KG
Mariazellerstraße 25
8605 Kapfenberg
Austria

Peter Staron
Helmholtz-Zentrum Geesthacht
Institute of Materials Science
Max-Planck-Str. 1
21502 Geesthacht
Germany

Andreas Stark
Helmholtz-Zentrum Geesthacht
Institute of Materials Research
Max Planck-Str. 1
21502 Geesthacht
Germany

Wolfgang Treimer
University of Applied Sciences Berlin
Department II, Physics, Mathematics & Chemistry
Luxemburgerstr. 10
13353 Berlin
Germany

Rolf Treusch
Deutsches Elektronen-Synchrotron DESY
Notkestr. 85
22607 Hamburg
Germany

Eckart Uhlmann
TU Berlin
Institute for Machine Tools and Factory Management
Pascalstr. 8–9
10587 Berlin
Germany

Ulla Vainio
Aalto University
Department of Applied Physics
Puumiehenkuja 2
02150 Espoo
Finland

Sangbong Yi
Helmholtz-Zentrum Geesthacht
Institute of Materials Research
Max Planck-Str. 1
21502 Geesthacht
Germany

Gerald A. Zickler
Montanuniversität Leoben
Institute of Mechanics
Franz-Josef-Str. 18
8700 Leoben
Austria

Preface to Second Edition

Since the first autumn school in 2005, four further very successful autumn schools on "Engineering Materials Science with Neutrons and Synchrotron Radiation" with changing topics have taken place in 2007, 2009, 2011, and 2013. At all schools the book has been very valuable for the attendees for deepening their knowledge and as a reference for further reading. Now the time has come for a second edition. All contributions published in the first edition have been updated and in some cases even replaced, for example, when the instruments described therein are no longer in operation. Furthermore, Part IV of the new edition has been significantly extended incorporating revised contributions from the special issue "Application of Photons and Neutrons for the Innovation of Engineering Materials" published in Advanced Engineering Materials Vol. 13, No. 8 (2011).

Since the publication of the first edition, the field of research covered in this book has continued to grow as shown in the figure displayed below which is an update of the

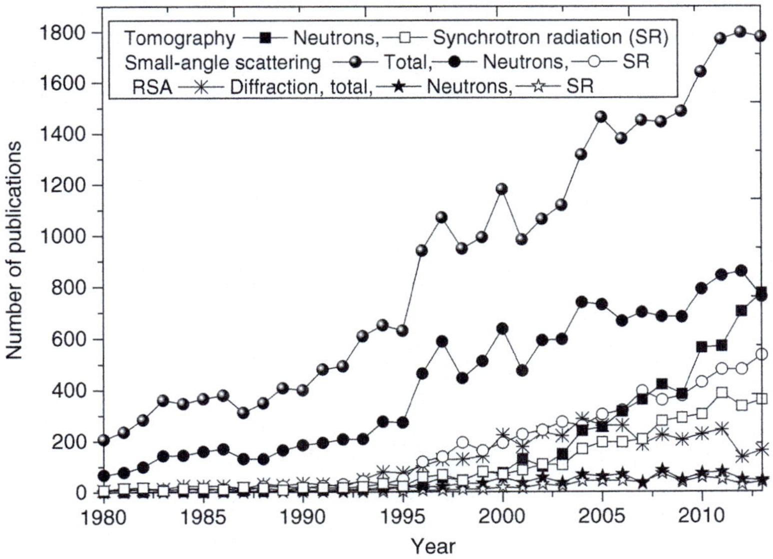

Updated number of publications where X-rays, synchrotron radiation, and neutrons were employed in engineering materials science. As in the preface of the first edition, the publication numbers were determined using the search engine SCOPUS (www.scopus.com) in May 2014 with the keywords residual stress analyses + diffraction, small-angle scattering, and tomography combined with the keywords synchrotron and neutron.

compilation given in the preface of the first edition. The fact that the absolute numbers of relevant publications found in the search engine SCOPUS using the same key words is somewhat higher than those in 2006 indicates that currently SCOPUS includes more journals in its database.

For the second edition, the team of editors has changed with Svea Mayer and Peter Staron replacing Anke Rita Kaysser-Pyzalla and Walter Reimers. The editors would like to thank all authors for their contributions to the present edition. We are very grateful to Nicola Kampner (HZG) for her help in organizing the editorial process and to Thomas Klein (MU Leoben) for preparing the revised publication statistics. Finally, the editors like to thank Dr Martin Preuß of Wiley-VCH for his substantial and enduring support during writing and editing this book.

Geesthacht, Lund and Leoben
October 2016

Peter Staron
Andreas Schreyer
Helmut Clemens
Svea Mayer

Part I

General

1

Microstructure and Properties of Engineering Materials

Helmut Clemens, Svea Mayer, and Christina Scheu

1.1 Introduction

In general, engineering materials are grouped into four basic classifications: metals, ceramics, polymers, and semiconductors. While semiconductors represent exclusively functional materials, the remaining three – depending on their application – can be assigned to the group of either structural or functional materials. Independent of the group they belong to, the important properties of solid materials depend on the geometrical atomic arrangement and also the type of bonding that exists between the constituent atoms. The three types of primary or chemical bonds that are found in engineering materials – covalent, ionic, and metallic – and the main contributions to the individual groups are shown in Figure 1.1. Metals and their alloys possess primarily metallic bonding; semiconductors have mainly covalent bonds, whereas many ceramics exhibit a mixture of covalent and ionic bonding. In engineering polymers, weak secondary forces of attraction (van der Waals forces) exist between the extended covalently bound hydrocarbon chains (Figure 1.1). In general, the nature of bonding depends on the electronic structure of the constituent atoms forming the solid and arises from the tendency of atoms to obtain stable electron configurations.

The structure of engineering materials relates to the arrangement of its internal components. On an atomic level, a structure is understood as the organization of atoms relative to each other. In crystalline materials, the atoms are arranged in periodically repeating arrays which are termed crystal or lattice structures. Metals, for instance, have particularly simple crystal structures: (i) face-centered cubic (fcc), (ii) body-centered cubic (bcc), (iii) hexagonal closed-packed (hcp), and (iv) tetragonal. Many metals and their alloys exist in more than one crystal structure depending on the temperature and composition, but, in most cases, transitions are between these four crystal structures. In contrast, semiconductors usually crystallize either in the diamond structure (silicon, germanium) or often in the zincblende structure (e.g., gallium arsenide).

The next larger structural level is the microscopic level. Here, large groups of atomic arrangements are considered as components of the microstructure, which determines most of the properties of the material. The microstructure of engineering materials is described by the grain size, types of phases present, and description of their structure, shape, and size distributions. In addition, two-dimensional defects such as grain boundaries and heterophase interfaces, one-dimensional defects such as dislocations,

Neutrons and Synchrotron Radiation in Engineering Materials Science: From Fundamentals to Applications, Second Edition. Edited by Peter Staron, Andreas Schreyer, Helmut Clemens, and Svea Mayer.

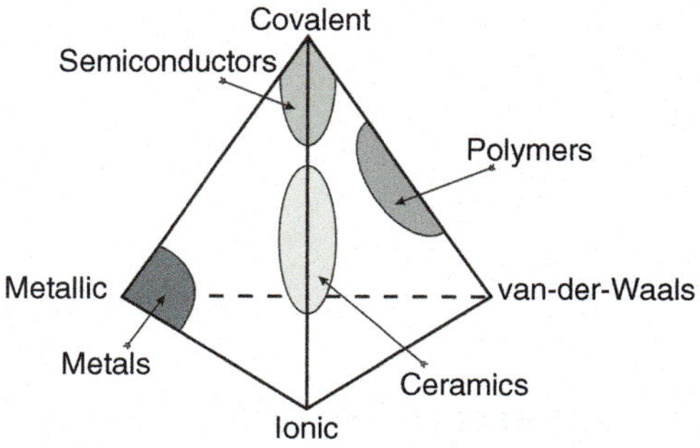

Figure 1.1 Bonding behavior present in different groups of engineering materials. (After [1].) Covalent, ionic, and metallic bonds represent strong primary bonds, whereas van der Waals attraction is a weak secondary bond.

Table 1.1 Influence of atomic arrangement and microstructure on the properties of engineering metallic materials.

Property	Influence of atomic arrangements and atomic defects	Influence of microstructure
Mechanical (e.g., strength and ductility)	Strong	Strong
Electrical, magnetic, and thermal (e.g., resistivity, magnetization, conductivity)	Moderate to strong	Slight to strong
Chemical (e.g., corrosive resistance, catalytic potential)	Slight	Slight to moderate

Control of the atomic arrangement and the microstructure is possible through processes such as casting, powder metallurgy, working, and heat treatment.
Source: Verhoeven 1994 [2]. Reproduced with permission of Wiley.

and zero-dimensional defects such as point defects are important microstructural features that often control the resulting properties.

In this introductory chapter, the microstructure of engineering materials is explained with focus on structural metallic materials, showing a polycrystalline multiphase assembly. The most important microstructure parameters are presented and their influence on mechanical properties is briefly discussed. Table 1.1 roughly summarizes the influence of atomic arrangement, atomic defects, and microstructure upon the properties of metallic materials. In addition, the most important methods for microstructural characterization on a nanometer and micrometer scale will be outlined in this chapter with emphasis on analytical electron microscopy. At the end of the chapter, a selection of textbooks and journal articles is listed, which might be helpful for the reader to deepen his/her understanding of the microstructure and properties of engineering materials [1–15] as well as of methods used for microstructural characterization [7, 16–23].

1.2 Microstructure

Figure 1.2 shows schematically the microstructure of a polycrystalline multiphase metallic material. For a comprehensive and better understanding of the following explanations, Table 1.2 lists the typical mole fractions and size ranges of the individual

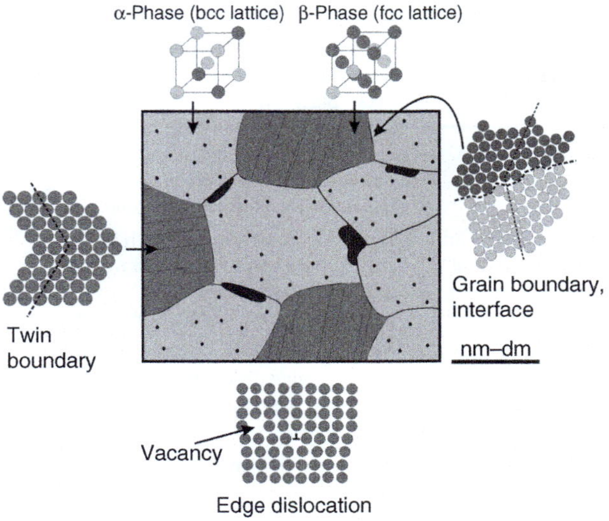

Figure 1.2 Schematic microstructure of a polycrystalline multiphase metallic material. The microstructure consists of α and β grains showing bcc and fcc crystal structures. Within the grains the existence of atomic defects is indicated (vacancies, dislocations, twin boundaries). The grains are separated by grain (phase) boundaries. On some grain boundaries large precipitates are visible. Within one type of grains nanometer-sized particles are present. For further explanations see text and refer to Tables 1.2 and 1.3.

Table 1.2 Microstructure of engineering metallic materials: constituents and their concentration and size ranges.

Microstructural constituents	**Range**
Vacancies	Equilibrium concentration (mole fraction): 10^{-15} (room temperature) to 10^{-4} (near the melting point)
Dislocations	Density[a]: 10^{10} (annealed) to 10^{16} m^{-2} (heavily cold-worked)
Grains	Size: nm – dm
Subgrains/domains	Size: nm – μm
Alloying elements	Concentration (mole fraction): ppm – 50%
Second phases	Volume fraction: 0 – 70%
Particles (precipitates, dispersoids)	Size: nm – μm; volume fraction: 0 – 70%

a) Dislocation density: the number of dislocations that intersect a unit area of a random surface section; alternatively: the total dislocation length per unit volume of material.

microstructural features and Table 1.3 describes their most important characteristics and their influence on various properties.

The schematic microstructure shown in Figure 1.2 consists of grains of two different phases. The phases differ in their crystal structures (fcc and bcc) and their chemical compositions. As indicated in the depicted crystal structures, each phase forms a solid solution. A solid solution represents a homogeneous crystalline phase that contains two

Table 1.3 The role of microstructural constituents in engineering metallic materials.

Microstructural constituents	Dependent on/characteristics (selection)	Responsible for (examples)
Vacancies	Temperature, deformation	Hardening at low temperatures; diffusion processes at elevated temperatures; diffusional creep
Dislocations	Deformation, temperature, recovery and recrystallization processes; at elevated temperatures edge dislocations may climb, and leave their slip planes	Plastic deformation; strength is controlled by their number and motion; driving force for recrystallization; dislocation creep
Stacking faults	Crystal structure, alloying	Mobility of dislocations, for example, climb of edge dislocations and cross-slip of screw dislocations is hampered
Mechanical twins	Stacking fault energy, deformation, temperature	Additional deformation mechanism at low temperatures and/or high strain rates
Subgrains/domains	Deformation, temperature, stacking fault energy/ordered crystal structure; antiphase boundary energy	Work hardening, creep, creation of antiphase boundaries
Grain boundaries	Lattice orientation between neighboring grains; subdivision in small-angle, medium-angle and high-angle grain boundaries	Work hardening by acting as barriers to slip from one grain to the next; segregation site of impurity atoms
Phase boundaries	Alloy system, composition, phase stability at elevated temperatures	Strengthening effects, for example, in duplex or multiphase steels
Grains	Alloy system, type of nucleation, processing, deformation, heat treatment, recrystallization	Strengthening (see grain boundaries) but ductility is maintained; grain boundary sliding at elevated temperatures (creep, superplasticity)
Annealing twins	Stacking fault energy; characteristic of face-centered cubic materials exhibiting a low stacking fault energy	Lowering of total boundary energy during grain growth
Precipitates/ dispersoids	Alloy system, composition, heat treatment, processing; the interface between particle and matrix can be coherent, semicoherent, or incoherent	Increase in strength by the interaction of moving dislocation; dislocations can loop, cut through or cross-slip the particles at ambient temperatures; at elevated temperatures the dislocations can surmount the particles by climb processes
Phase arrangement (e.g., eutectics, duplex, dual phase)	Alloy system, composition, processing, heat treatment	Positive: control of mechanical and thermo-physical properties; negative: embrittlement in case of brittle phases situated at grain boundaries

or more chemical species. Both substitutional and interstitial solid solutions are possible, such as nickel and chromium in iron (e.g., austenitic steels) and carbon in iron (e.g., heat treatable steels), respectively. The solubility of a metal i.e, its alloying behavior, depends on the atomic size factor (difference in size between solute and solvent atom), the electrochemical effect (the higher the difference in electronegativity, the higher the tendency for the elements to form intermetallic phases rather than extensive solid solutions), and the relative electron valency (a metal of higher electron valency is more likely to dissolve to a large extent in one of lower electron valency than vice versa).

1.2.1 Crystal Defects

The grains in a microstructure represent individual crystals within the polycrystalline material (Figure 1.2). Within each grain, atoms are regularly arranged according to the basic crystal structure but a variety of imperfections, termed crystal defects, may also occur. These defects are point defects (vacancies, interstitial atoms), line defects (dislocations), planar defects (stacking faults, twin boundaries), and volume defects (voids, cavities). Of particular interest are dislocations, because plastic deformation mainly corresponds to the motion of dislocation in response to an applied shear stress (see Chapters 17 and 18). In contrast, hindering of dislocation movement is the basic concept of all strengthening mechanisms (see Section 1.3). Dislocations are subdivided into edge and screw dislocations. At temperatures where no thermally activated diffusion processes take place edge dislocations are confined to their slip planes, whereas screw dislocations can change their slip planes rather easily by cross-slip processes. A schematic drawing of an edge dislocation is shown in Figure 1.2. An edge dislocation is a linear crystalline defect associated with the lattice distortion produced in the vicinity of the end of an extra half-plane of atoms within a crystal. Depending on processing history and/or mechanical loading, subgrains or cell structure, separated by dislocation networks or tangles, can be formed within the grains. In grains showing an ordered crystal structure (e.g., intermetallic phases) domain structures may appear. The individual domains are separated by antiphase boundaries. The corresponding energy is referred to as antiphase boundary energy.

1.2.2 Grain (Phase) Boundaries and Twins

The size of the grains depends on materials processing and heat treatments and can be adjusted in a wide range. In most technically relevant structural metallic materials, such as steels, aluminum alloys, and titanium alloys, the grain size is in the range of several ten micrometers. In contrast, in nanostructured functional materials, for example, superhard coatings with high wear resistance, a grain size in the range of few nanometers is required. The grains as shown in Figure 1.2 are separated by grain (phase) boundaries. In general, grain (phase) boundaries are interfaces that separate two adjoining grains (phases) having different crystallographic orientations and, in the case of phases, different crystal structures and/or chemical compositions. Within the boundary region, which can have a width of one to several atomic distances, an atomic mismatch due to the transition from the crystalline orientation of one grain to that of an adjacent one can occur. Depending on the structure, one can distinguish between high-angle grain boundaries, small-angle grain boundaries, and so on. Since the atoms are differently coordinated and/or bonded along grain boundaries, there is an interfacial or grain boundary energy associated with them. The magnitude of this energy is a function of the degree of misorientation between the grains, being larger for high-angle grain boundaries although some energy cusps can occur for special grain boundaries. Simple small-angle grain boundaries can be described by dislocation arrangements. A twin boundary as shown in Figure 1.2 is a special type of grain boundary. Atoms of one side of the boundary are located in mirror image positions of the atoms of the other side. Twins result from atomic displacements that are produced from an external stress state (mechanical or deformation twins) and also during annealing heat treatments subsequent to deformation (annealing twins). The formation of twins is closely related to the stacking fault energy of the material. In general, low stacking fault energy facilitates twinning as can be seen in the high density of annealing twins in fcc metals and their alloys, such as copper, α-brass, and austenitic steels. The positive effect of deformation twinning on strain hardening and deformability, for example, is exploited in TWIP (twinning-induced plasticity) steels.

1.2.3 Precipitates and Dispersions

In many structural engineering metal materials precipitates occur. In Figure 1.2 two types of precipitates are drawn schematically: few large ones at grain boundaries and a large number of small particles homogeneously dispersed within individual grains. In many alloys, for example, steels or nickel-based alloys, these precipitates are carbides or intermetallic phases. Their influence on mechanical properties primarily depends on volume fraction, size, distribution, type of precipitate, and arrangement in the microstructure. Large precipitates along grain boundaries as shown in Figure 1.2 can either have a positive or negative effect on the properties. For example, in nickel-based superalloys, precipitates are generated at grain boundaries by means of a special heat treatment in order to minimize grain boundary sliding at high service temperatures. However, such a phase arrangement can also lead to serious embrittlement as observed in steels containing nonmetallic inclusions or cementite films along grain boundaries. Nanometer-sized particles of a second phase, which are uniformly dispersed within the grains, provide the most versatile strengthening mechanism for metallic materials in addition to solid solution strengthening (see Section 1.3). There are different ways to produce extremely fine particles in a metallic matrix: a variety of metallic alloy systems have been developed for which so-called precipitation heat treatments are employed to precipitate a new phase from a supersaturated solid solution. Examples of engineering alloys that are hardened by precipitation treatments include aluminum–copper (e.g., Duraluminum or Dural), nickel–aluminum (e.g., nickel-based superalloys), and some ferrous alloys (e.g., maraging and tool steels). A common feature of these nanometer-sized particles, which usually precipitate in the form of metastable phases, is their coherency with the matrix in the early stages of precipitation. However, during exposure at service temperatures these particles may change their chemistry and are prone to coarsening. Very often this process is accompanied by loss of coherency; thus, a semicoherent or an incoherent interface between particle and matrix is formed. As a consequence the initial hardening mechanism is altered, leading to a decrease in strength.

Another way to strengthen metals and their alloys is to produce a uniform dispersion of several volume fractions of extremely small particles of a very hard and inert material. The dispersed phase may be either metallic or nonmetallic and they usually do not show coherency with the matrix. Examples are oxide dispersion strengthened (ODS) superalloys: hard nanometer-sized Y_2O_3 particles are mechanically alloyed into the matrix powder and consolidated and processed by powder metallurgical techniques. The dispersion-strengthening effect is often technologically more difficult to realize, however, the strengthening effect is retained at elevated temperatures and for extended service times. This is a direct effect of the inertness of the extremely fine particles, leading to a high resistance against particle growth and re-dissolution effects.

The previous explanation was focused on the various microstructural constituents that range from atomic dimensions to the mesoscopic scale. In engineering metallic materials these constituents appear in a great variety of arrangements that in turn determine many of their properties (Tables 1.1–1.3). In Figure 1.3 a schematic drawing of different microstructures is given along with references to structural metallic materials that are widely used. For completeness it should be mentioned that metals and their alloys that have undergone a severe amount of deformation, as in rolling, forging, or wire drawing, will develop a preferred orientation or deformation texture, in which certain crystallographic planes within the deformed grains tend to align themselves in a preferred manner with respect to the direction of maximum strain. A recrystallization heat treatment, conducted on a cold-worked metal, can produce a preferred orientation

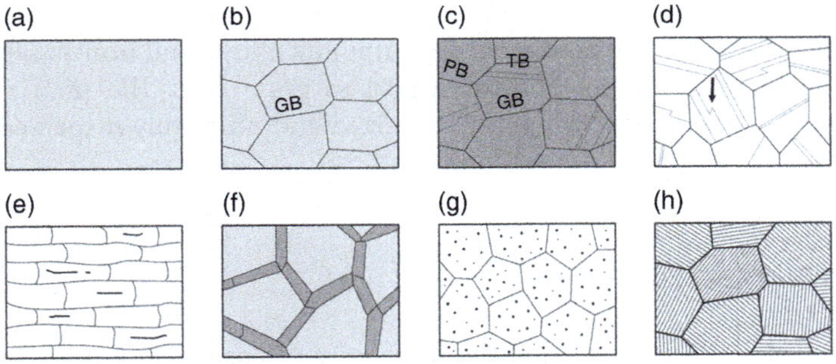

Figure 1.3 Schematic drawings of different microstructures: (a) single crystal: crystalline solid for which the periodic and repeated atomic pattern extends throughout the entire sample without interruption. The properties depend strongly on the orientation of the crystal. Example: single-crystal made of nickel-based superalloys. (b) Polycrystalline single-phase material. The individual grains differ in their crystallographic orientations and are separated by grain boundaries (GBs). Example: α-iron (ferrite) with body-centered cubic (bcc) lattice structure. (c) Two-phase material. The phases differ in chemical composition and crystal structure. The grains are separated either by phase boundaries (PBs) or GBs. One phase, most probably a phase possessing a face-centered cubic (fcc) lattice structure, shows the appearance of annealing twins. TB denotes a coherent twin boundary. Example: $\alpha + \beta$-brass, consisting of α-grains (fcc) and β-grains (bcc). (d) Single-phase material exhibiting a large number of annealing twins; arrow: incoherent TB. Example: α-brass (fcc), austenitic stainless steel. (e) Deformed grains with elongated inclusions. Example: ferritic steel with nonmetallic inclusions after rolling to sheet. Due to rolling, the sheet can exhibit a marked texture that may be reflected in anisotropic mechanical properties. (f) Two-phase material, where one of the phases is situated along GBs. Example: pearlitic steel with proeutectoid ferrite on GBs. (g) Polycrystalline material with precipitates. Example: nickel-based superalloy containing γ'-Ni_3Al precipitates. (h) Two-phase material after eutectoid transformation which represents the outcome of a diffusion-controlled reaction. The grains consist of alternating layers (or lamellae) of the constituting phases. The mechanical properties, for example, the yield strength, depend primarily on the lamellae spacing. Example: pearlitic steels.

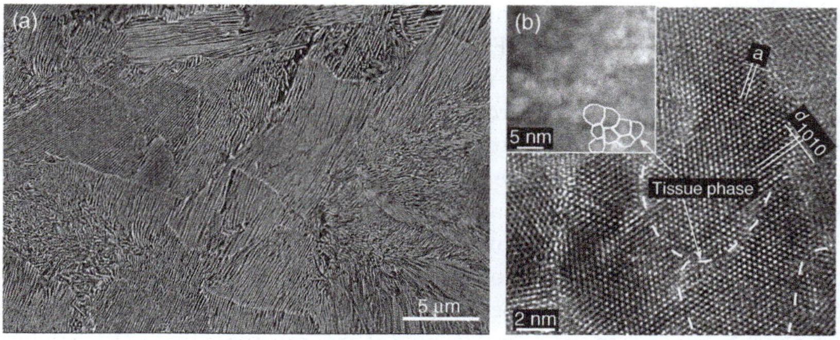

Figure 1.4 (a) "Conventional microstructure" of a pearlitic steel (scanning electron microscope image) and (b) "advanced nanostructure" of a superhard TiB_2 coating (high-resolution transmission electron microscope image). (Mayrhofer 2005 [15]. Reproduced with permission of Wiley.) The grain size of the pearlitic steel is about 10 µm, whereas the grain size of the TiB_2 coating is below 5 nm.

that is different to that existing in the deformed material. This type of texture is termed annealing or recrystallization texture (see Chapters 3, 10, and 18).

As examples for the described microstructures, Figure 1.4 displays images of a pearlitic steel and the nanostructure of a superhard TiB_2 coating. The grain size of the pearlitic steel is about 10 µm, whereas the grain size of the TiB_2 coating is below 5 nm. Today's advanced engineering metallic materials represent a combination of both features. For example, nickel-based superalloys, some aluminum alloys, and iron-based tool steels possess a "conventional" matrix with regard to grain size. The matrix, however, is hardened and strengthened by nanometer-sized and uniformly dispersed particles that precipitate from a supersaturated solid solution.

1.3 Microstructure and Properties

In the previous section it was pointed out that the properties of engineering metallic materials depend on the atomic arrangement, the prevailing crystal defects as well as the arrangement and morphology of the constituting phases/particles (see Figure 1.2 and Tables 1.1–1.3). In the following text, the influence of microstructural parameters on mechanical strength will be discussed. In general, the strength of a metal is controlled by the number and motion of dislocations. The stress required to move dislocations, the Peierls–Nabarro stress, is relatively low in pure metals. Consequently, in order to strengthen metals one must restrict the motion of dislocations by either generating internal stresses that oppose their motion, or by placing particles in their path that require them to cut or to loop the particles. Figure 1.5a,b summarizes the basic strengthening mechanisms for metallic materials at low ($T < 0.3T_M$) and high ($T > 0.3T_M$) temperatures. T_M is the melting point (in Kelvin) of the metal or alloy under consideration. Practically, there are four major strengthening mechanisms that will be outlined in the following: (1) work (dislocation density) hardening, (2) strengthening by grain size reduction, (3) solid solution strengthening, and (4) strengthening by particles.

1) The work hardening phenomenon can be explained on the basis of dislocation–dislocation strain field interactions. Plastic deformation during cold working

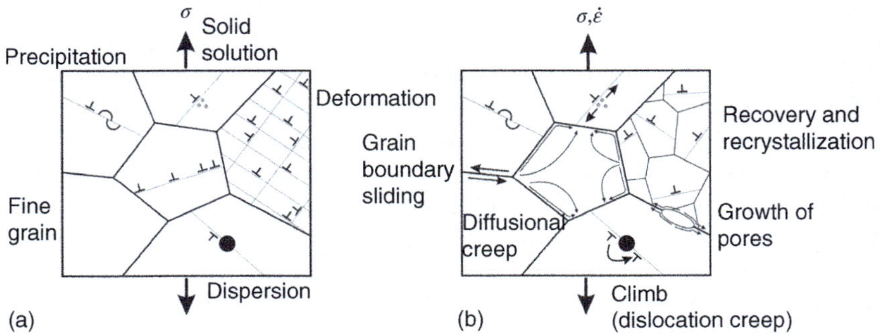

Figure 1.5 Basic strengthening mechanisms for metallic materials at low ($T < 0.3T_M$) and high ($T > 0.3T_M$) temperatures. T_M is the melting point in Kelvin. (a) At low temperatures the (yield) strength of a material is controlled by dislocation density (work hardening or strain hardening), grain size (grain boundary strengthening), concentration and size of alloying atoms (solid solution strengthening), and size and volume fraction of particles (precipitation or dispersion strengthening). (b) At high temperatures thermally activated processes and creep determine the occurring strength. For example, a high dislocation density is reduced by recovery and recrystallization. Fine-grained microstructures lead to high diffusion creep rates and pronounced grain boundary sliding. Particles that are effective barriers to dislocations at low temperatures are surmounted by climb processes (dislocation creep). Depending on the loading conditions, pores nucleate and grow at grain boundaries leading to micro- and macrostructural damage and consequently to a reduced lifetime.

produces an increase in the number of dislocations (Table 1.2). As a consequence, the mean distance between individual dislocations decreases. On average, dislocation–dislocation strain interactions are repulsive. The net result is that the motion of a dislocation is hindered by the presence of other dislocations. As the dislocation density increases with increasing degree of deformation, the resistance to dislocation motion becomes more pronounced.

2) The yield strength of a metal is almost universally observed to increase as the grain size decreases. The experimental data virtually always show a linear relationship between yield strength and the reciprocal value of the square root of the grain diameter. The strengthening effect produced by grain size reduction results from the blockage of dislocations by grain boundaries. Therefore, a fine-grained material is stronger than one that is coarse grained, since the former has a greater total grain boundary area to obstruct dislocation motion. Two reasons can be given why grain boundaries act as barriers to dislocation motion during plastic deformation: firstly, grains are of different crystallographic orientations. If a dislocation passes from one grain to another it will have to change its direction of motion. This process becomes more difficult as the misorientation between the grains increases. Secondly, the atomic disorder within a grain boundary region results in a discontinuity of slip planes from one grain into the other. Boundaries between two different phases are also barriers to dislocations. Such a behavior is utilized in the strengthening of complex multi-phase metallic materials.
3) Solid solution hardening is another effective technique to strengthen and harden metals. When a solute atom (alloying atom) dissolves in a solid metal it may act as an atomic-sized obstacle to dislocation motion. The strengthening effect depends on the nature of the interaction of the dislocation with the solute atoms. Usually, two general interactions are considered, one of a chemical nature and the other of an

elastic nature. The difference in chemical bonding between solute atoms and solvent atoms is reflected in the difference in their elastic shear moduli. This difference gives rise to a change in the dislocation–atom interaction. If the solute atom has a different size than that of the matrix atoms, then a misfit strain field is produced around the solute atom that interacts with the strain field of the dislocations.

4) Particles may be introduced into the matrix either by precipitation or powder metallurgical approaches (see Section 1.2). These particles will interact with the dislocations causing the dislocations either to cut through the particles or to loop them. It should be noted that particle cutting is restricted to particles that are coherent or at least semicoherent to the matrix. The degree of strengthening resulting from nanometer-sized particles depends on their distribution in the matrix (see Chapter 12). In addition to the shape, the second-phase dispersion can be described by specifying volume fraction, average particle diameter, and mean interparticle spacing.

At elevated temperatures ($T > 0.3T_M$), the microstructure may become thermally unstable (Figure 1.5b) and thermally activated processes such as diffusion and creep start determining the strength of the material. For example, a hardened cold-worked material can lose its strength due to recovery and recrystallization (see Chapter 17). Fine-grained materials that show good strength properties at ambient temperatures are prone to diffusional creep and pronounced grain boundary sliding. Furthermore, particles that are effective barriers to dislocations at low temperatures can be surmounted by diffusion-assisted climb processes (dislocation creep).

As a conclusion, Table 1.4 summarizes the discussed basic strengthening mechanisms and assesses their effect at low and high temperatures. It is worth mentioning here that the mechanical properties might not show the same size dependence when the grain sizes or the material dimensions reach the nanometer regime. This is most likely related to the difficulty to generate and move dislocations in these materials and ongoing research works are addressing this problem. The interested reader might find an introduction to this research field in [24–28].

1.4 Microstructural Characterization

As has been outlined in the previous sections, the microstructure has a major influence on the properties of engineering materials, and the most relevant microstructural features are summarized in Table 1.5. Usually, a combination of different characterization methods has to be applied to obtain the necessary information, and this section is devoted to this topic. However, only a rough guideline can be provided as to which methods can be applied to assess a specific microstructural feature, which may not be exhaustive. For a detailed description of the operating modes of the different methods the reader is referred to literature [7, 16–23] and to specific chapters of this book.

The most frequently used characterization techniques for studying the microstructure of engineering materials are light optical microscopy, electron and ion beam microscopy and corresponding analytical measurements as well as X-ray, neutron, and electron diffraction experiments. All of these methods are based on the elastic or inelastic interaction of a probe (visible light, electrons, ions, X-rays, neutrons) with the material under investigation giving rise to a scattered intensity of the initial beam and to the generation of secondary signals (photons, electrons, ions). Each method

Table 1.4 Basic strengthening mechanisms for engineering metallic materials and their assessment.

	Dislocations	Grain boundaries	Solute atoms	Particles	Transformation	Anisotropy	
Strengthening mechanism	Work hardening	Grain boundary hardening	Solid-solution hardening	Precipitation/ dispersion hardening	Transformation hardening, for example, martensitic transformation	Crystal (structure, texture)	Microstructure (fiber composites; directed grains; duplex microstructure)
Scaling parameters	Dislocation density	Grain size	Concentration of solute atoms	Particle size and volume fraction	Quenching rate, alloy composition	Crystal orientation, intensity of texture	Strength, orientation and volume fraction of fibers; grain orientation; deformation behavior of constituent phases
Assessment (after [14])							
Low-temperature strength	++	++	+	++	++	+	+
High-temperature strength (creep)	±[a)]	–	+	++	±[b)]	+	+

Impact on yield strength: ++ increases strongly
+ increases
– decreases.

a) + if dislocations are pinned by stable particles; – if recrystallization takes place during high-temperature application (creep).

b) + if martensitic structure is maintained.

Table 1.5 Information on microstructure needed.

- Grain/subgrain/domain size
- Crystal structure and chemistry of grains and particles
- Preferred grain orientation (texture)
- Three-dimensional arrangement of phases
- Phase transitions (onset, temperatures …)
- Size, shape, and volume fraction of particles (precipitates, dispersoids)
- Structure and type of appearing interfaces, segregation to interfaces
- Types of defects and defect density (pores, cracks …)
- Vacancy concentration and dislocation density
- Local/residual stresses
- Microstructural evolution during deformation and/or thermal treatment
- Nucleation and growth processes

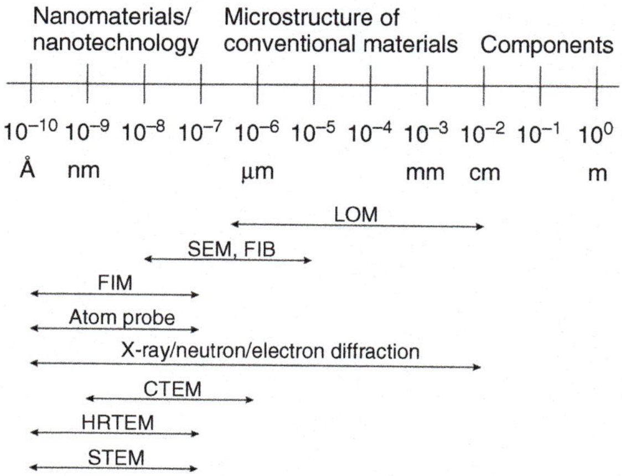

Figure 1.6 Length scale, which is covered in engineering materials, ranging from the atomic/nano scale to the dimensions of large components. Several characterization methods are listed as well as their resolution limits. LOM: light optical microscopy; SEM: scanning electron microscopy; FIB: focused ion beam; IM: ion microscopy; FIM: field ion microscopy; CTEM: conventional transmission electron microscopy; HRTEM: high resolution transmission electron microscopy; and STEM: scanning transmission electron microscopy.

allows the access of microstructural features on different length scales as indicated in Figure 1.6 and, except for X-ray and neutron diffraction techniques that will be addressed in the remaining chapters of this book, the others will be shortly described in the following text.

Light optical microscopy (LOM) is the common method that is employed to determine the grain size of engineering materials. In addition, the size and distribution of larger inclusions and precipitates can be investigated. However, due to a resolution limit in the order of the wavelength of light (i.e., around 500 nm) it is not suitable to investigate nanocrystalline materials or sub-micrometer precipitates. In addition, no information on the chemical composition or crystal structure of the individual phases can be obtained.

Scanning electron microscopy (SEM) and focused ion beam microscopy (FIB) enable us to study grain and precipitate sizes as well as their arrangement with a spatial resolution in the order of several ten nanometers. The resolution in the image is thereby mainly governed by the beam size. Often these microscopes are equipped with analytical tools to perform energy-dispersive X-ray spectroscopy (EDX) and wavelength dispersive X-ray spectroscopy (WDX) (in the SEM) or secondary ion mass spectroscopy (SIMS) (in the FIB). These analytical methods can be used to determine the chemical composition of different phases. Since the interaction volume of the incident electron beam within the sample is much larger than the beam size (which can be as small as a few nanometers), semiquantitative EDX or WDX measurements can only be done with a resolution of about 1 μm. In principle, for SIMS a sub-micrometer lateral resolution can be achieved; however, in practice this resolution is often not obtained due to insufficient counting statistics. Thus, even if the size distribution of small particles can be determined, the classification of the corresponding particle types (regarding, e.g., chemical composition) is not possible and requires the use of an additional characterization method. Modern SEM are often equipped with an electron back scatter diffraction (EBSD) detector that allows to investigate the crystal structure of the occurring phases and their preferred orientation (texture) within the sample surface. The spatial resolution is in the range of 50 nm. An example of FIB/SEM tomography conducted on an intermetallic Ti-44 at.%Al-7 at.%Mo-0.1 at.%B alloy is depicted in Figure 1.7. The investigated alloy consists of two ordered phases, γ-TiAl ($L1_0$ structure) and β_0-TiAl (B_2 structure), which exhibit nearly the same volume fraction. Figure 1.7a,b shows the reconstruction of the γ- and β_0-phase, respectively. For both phases almost the whole volume is interconnected and forms a network. Specifics concerning the shown FIB/SEM tomography as well as 3D image analysis are reported in [29] and the references cited therein.

The crystal structure of sub-micrometer-sized particles and precipitates can be studied by transmission electron microscopy (TEM) using electron diffraction experiments. These studies can also be conducted to determine the orientation relationship between different phases or to show the presence of special grain boundaries such as twin boundaries. With the help of conventional TEM images (bright-field, dark-field, weak-beam) microstructural features such as dislocation densities, antiphase boundaries, grain/subgrain/domain sizes, particle shape, size, and distribution can be addressed. The spatial resolution for conventional TEM investigations is in the order of some nanometers. Figure 1.8a shows the presence of dislocations and mechanical twins

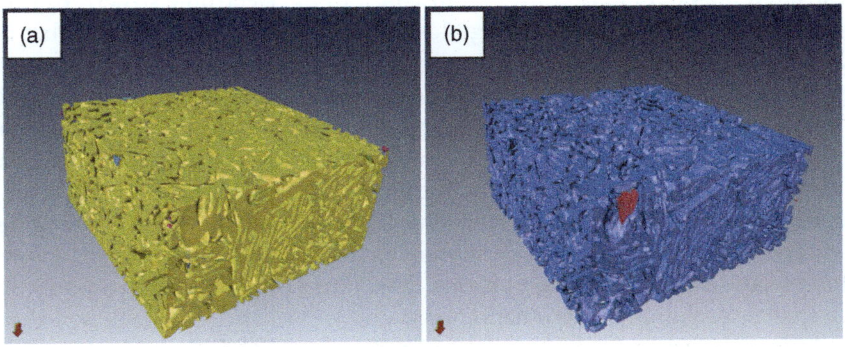

Figure 1.7 3D reconstruction of the constituting phases of an intermetallic Ti-44 at%Al-7 at.%Mo-0.1 at.%B alloy: (a) γ-TiAl and (b) β_0-TiAl. For both phases almost the whole volume is interconnected, except small particles at the border which are marked with different colors. The region of interest has an approximate size of 35 × 17 × 35 μm^3. (Engstler 2013 [29]. Reproduced with permission of Wiley.)

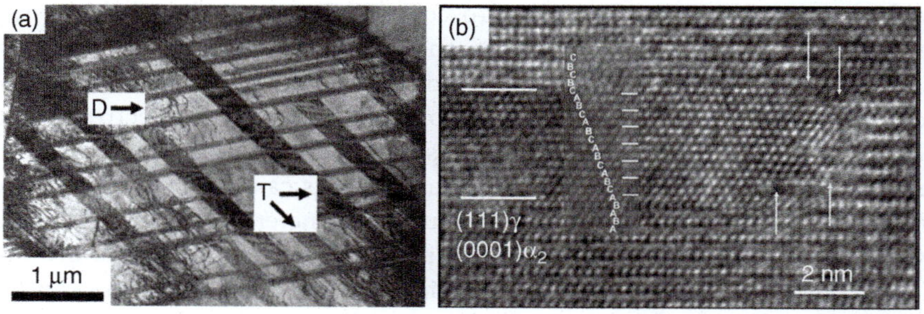

Figure 1.8 (a) TEM image of a Ti-46.5 at.%Al-4 at.% (Cr, Nb, Ta, B) sample deformed in compression up to 5% at room temperature reveals two different deformation mechanisms acting in γ-grains, mechanical (deformation) twinning (T) and dislocation glide (D). Note that cross-twinning with the primary twinning system limits the extension of the second twinning system. (Kauffmann 2000 [30]. Reproduced with permission of Elsevier.) (b) HRTEM image of a γ-TiAl lamella that terminates within the supersaturated α_2-Ti_3Al matrix in a Ti-45 at.%Al-7.5 at.%Nb specimen (Fischer 2010 [32]. Reproduced with permission of Elsevier.). The stacking sequence of the close packed planes is indicated. The $(111)\gamma//(0001)\alpha_2$ habit planes are marked by full lines. The arrows point to misfit dislocations. Beam direction = $[\bar{1}10]\gamma||[\bar{1}\bar{2}\bar{1}0]\alpha_2$.

in a deformed γ-TiAl grain within a Ti-46.5 at.%Al-4 at.%(Cr, Nb, Ta, B) alloy. From the TEM image it is evident that in the observed γ-grain more than one twin system has been activated [30, 31].

Analytical TEM measurements such as EDX and electron energy-loss spectroscopy (EELS) allow to determine the chemical composition of individual phases, particles, or at interfaces. The spatial resolution of these methods depends strongly on the beam size, and for modern TEM with a scanning unit (scanning transmission electron microscopy (STEM)) a resolution of about 1 nm is achieved for EDX and ≥0.1 nm for EELS measurements. The main reason for the differences in resolution is attributed to the larger specimen thickness (and thus stronger effect of beam broadening within the sample) for EDX measurements (to obtain a better signal-to-noise ratio in the data) and due to the detection geometries.

The EELS measurements can be used not only to determine the chemical composition of the investigated region, but also to get an insight into the electronic structure. This is obtained by analyzing the electron energy-loss near-edge structure (ELNES) that is associated with each element-specific ionization edge and which contains information on, for example, bonding characteristics and nominal oxidation states of the probed atoms. In addition, studying the extended energy-loss fine structure (EXELFS), which occurs around 50 eV above the ionization edge onset, allows to obtain information on the radial distribution function of the atoms. The valence loss region with its characteristic plasmon features at an energy loss of around 15–25 eV can be used to investigate the optical properties of the materials by a subsequent Kramers–Kronig analysis. However, due to the nature of the excitation process these latter measurements can only be done with a spatial resolution of a few nanometers.

High-resolution transmission electron microscopy (HRTEM) and so-called Z-contrast images (Z stands for the atomic number) using a STEM allow to study the atomic structure of interfaces or the crystal structure of nanometer-sized precipitates.

The HRTEM image formation can be described with the help of Abbe's theory, and the image can be understood as an interference pattern of different diffracted beams. For the imaging a parallel beam is used, and the whole interference pattern is detected simultaneously. In contrast, for a Z-contrast image a convergent electron beam is used and scanned over the sample. At each position of the beam, the intensity of electrons scattered in large scattering angles is detected and the image is formed serial point by point. The detected signal is roughly proportional to the square of the atomic number. The Z-contrast image can be understood as a convolution of the specimen function (atomic columns) with the electron beam function. The resolution limit of both methods is mainly governed by the spherical aberration (C_s) of the electron lenses, that is, of the objective lens, which is most important for the imaging in HRTEM and of the condenser lens, which is responsible for the electron beam size in the case for Z-contrast imaging in the STEM. Recent developments of C_s correctors allow obtaining, for both methods, a spatial resolution of $\leq$0.1 nm. Figure 1.8b shows a high-resolution HRTEM image of an intermetallic Ti-45 at.%Al-7.5 at.%Nb alloy [32]. At first, the specimen was oil quenched from the single α-phase field region. In a second step, the sample, which consists of supersaturated α_2-Ti_3Al grains, was continuously heated to 790 °C at a rate of 20 °C min^{-1}, immediately followed by oil quenching. The image shows a γ-TiAl lath that terminates within a α_2-Ti_3Al grain. The beam direction is $[\bar{1}10]\gamma||[\bar{1}2\bar{1}0]\alpha_2$ and, therefore, the $(111)\gamma||(0001)\alpha_2$ interfaces, indicated by solid lines, are edge on. The stacking sequence of the close packed planes is indicated. Some steps exist along this interface with a height corresponding to a $(111)\gamma$ lattice plane distance. At the terminating end of the γ-TiAl lath, the lattice mismatch between the γ-TiAl and the α_2-Ti_3Al phase can be compensated by transformation dislocations that have the Burgers vector of a Shockley partial dislocation and occur on every other close packed plane [32].

It is important to mention that all TEM images reveal a two-dimensional projection of the three-dimensional sample. This can cause problems, for example, if particle distributions are investigated, and thus complementary methods have to be applied. In addition, problems can occur if the sample thickness is too large since then small particles embedded in the matrix might become invisible. Dislocation densities can only be estimated up to 10^{15} m^{-2}, and other methods have then to be applied. In addition, the TEM specimen preparation has to be taken into account, and care has to be taken not to change the original structure or at least to minimize possible damaging effects. In addition, only a limited specimen volume is analyzed in TEM and, statistically, evaluations of microstructural features are time consuming. Therefore, integral methods that probe the features over a large sample volume and that are nondestructive (regarding the sample preparation) such as X-ray and neutron scattering should be performed in addition.

Another method to image lattice defects such as dislocations and grain boundaries at an atomic level is the field-ion microscope (FIM). A positive voltage is applied to a fine tip of the material of interest, which leads to the ionization of an imaging gas (e.g., He, Ne). The ions of the imaging gas are then radially accelerated to a fluorescent screen, which is at a negative potential. The image represents the geometry of the atomic arrangement of the terraces of the tip. Particles of a second phase might lead to a different ionization behavior of the imaging gas and thus appear differently. If the applied electrical field is

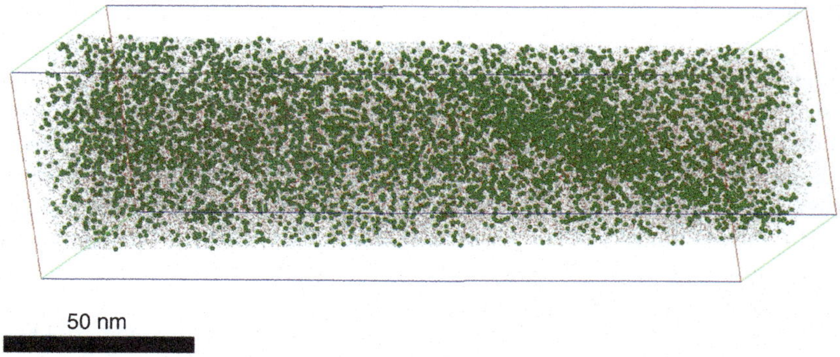

Figure 1.9 Atom probe image of C enrichment at a dislocation in the γ-TiAl-phase of an intermetallic Ti-45 at.%Al-5 at.%Nb-0.5 at.%C alloy. Every dot corresponds to a C atom. (Scheu 2009 [33]. Reproduced with permission of Elsevier.)

high enough, the atoms of the tip can be ionized themselves and leave the tip in radial directions. The ions can be identified with the help of a time-of-flight mass spectrometer, which is the basic working principle of an atom probe. A three-dimensional image of the tip can be obtained with suitable detectors. This method is especially suitable to study the initial stages of precipitations or to determine the chemical composition of impurity elements at defects such as dislocations or interfaces. As an example, Figure 1.9 illustrates the three-dimensional elemental distribution of C atoms within the γ-TiAl phase of a Ti-45 at.%Al-5 at.%Nb-0.5 at%C alloy. In the investigated Nb-containing alloy, C was found to be homogeneously distributed within the γ-TiAl-phase. Locally C enriched areas in the γ-TiAl-phase were found only at dislocations, forming so-called Cottrell atmospheres as described in [33]. More information with regard to intermetallic TiAl alloys and the contribution of synchrotron radiation and neutrons to their characterization and development is given in Chapter 22.

The tip preparation of samples containing defects can be rather time consuming, but using an FIB can help to produce a needle from the area of interest. However, not all materials can withstand the high electrical field and, as for the TEM investigations, the analyzed sample volume is rather small. Again, complementary methods are required to access the microstructural features governing the properties of engineering materials.

Despite the methods described so far, a variety of other imaging characterization techniques exist such as scanning probe microscopes, for example, scanning tunneling microscope and atomic force microscope [23]. These methods are helpful to get an insight into the surface structure of engineering materials down to the atomic level, but information on, for example, surface stresses on a larger scale are not easy to address. Since material scientists are generally interested to obtain all the information listed in Table 1.5 with a high statistical relevance, diffraction techniques are the right choice for microstructural characterization – if possible always linked to complementary methods such as the ones mentioned in this chapter. The following chapters will provide the basic background in the underlying physics of X-ray and neutron diffraction. In addition, the experimental setups used for the measurements are explained and fundamental descriptions of data treatment and analysis are given.

References

1 Shackelford, J.F. (2009) *Introduction to Materials Science for Engineers*, Pearson Education, New Jersey.
2 Verhoeven, J.D. (1994) *Fundamentals of Physical Metallurgy*, Wiley, Inc., New York.
3 Callister, W.D. (2011) *Materials Science and Engineering – An Introduction*, Wiley, New York, Weinheim.
4 Tetelman, A., Barrett, C.R., and Nix, W.D. (2005) *The Principles of Engineering Materials*, Prentice-Hall, Englewood Cliffs, NJ.
5 Weidmann, G., Lewis, P., and Reid, N. (1990) *Structural Materials*, Butterworth, London.
6 Gottstein, G. (2007) *Physikalische Grundlagen der Materialkunde*, Springer, Berlin.
7 Haasen, P. (2013) *Physikalische Metallkunde*, Springer, Berlin.
8 Cahn, R.W., Haasen, P., and Kramer, E.J. (eds) (2005) *Materials Science and Technology*, vol. **2a/2b, 6/7, 15/16**, Weinheim, Wiley-VCH Verlag GmbH.
9 Porter, D.A. and Easterling, K.E. (2009) *Transformations in Metals and Alloys*, Nelson Thornes, Cheltenham.
10 Hull, D. and Bacon, D.J. (2011) *Introduction to Dislocations*, Butterworth-Heinemann, Oxford.
11 Dieter, G.E. (1988) *Mechanical Metallurgy*, McGraw-Hill, London.
12 Smallman, R.E. and Bishop, R.J. (2014) *Modern Physical Metallurgy & Materials Engineering*, Butterworth-Heinemann, Oxford.
13 Courtney, T.H. (2013) *Mechanical Behavior of Materials*, McGraw-Hill, London.
14 Hornbogen, E. (1974) in *High-Temperature Materials in Gas Turbines* (eds P.R. Sahm and M.O. Speidel), Elsevier, Amsterdam, pp. 187–205.
15 Mayrhofer, P.H., Mitterer, C., and Clemens, H. (2005) Self-organized nanostructures in hard ceramic coatings. *Adv. Eng. Mater.*, 7, 1071–1082.
16 Brandon, D. and Kaplan, W.D. (2008) *Microstructural Characterization of Materials*, John Wiley & Sons, Ltd., West Sussex.
17 Reimer, L. (1998) *Scanning Electron Microscopy*, Springer Series in Optical Sciences, 2nd edn, Springer, Berlin.
18 Williams, D.B. and Carter, C.B. (2011) *Transmission Electron Microscopy*, vol. **1–4**, Plenum Press, New York.
19 Fultz, B. and Howe, J.M. (2008) *Transmission Electron Microscopy and Diffractometry of Materials*, Springer, Berlin.
20 Reimer, L. (1997) *Transmission Electron Microscopy*, Springer Series in Optical Sciences, 4th edn, Springer, Berlin.
21 Miller, M.K., Cerezo, A., Hetherington, M.G., and Smith, G.D.W. (2006) *Atom Probe Field Ion Microscopy*, Oxford University Press, Oxford.
22 Hono, K. (2002) Nanoscale microstructural analysis of metallic materials by atom probe field ion microscopy. *Prog. Mater Sci.*, **46** (**6**), 621–729.
23 Brundle, C.R., Evans, C.A., and Wilson, S. (1992) *Encyclopedia of Materials Characterization – Surfaces, Interfaces, Thin Films*, Butterworth-Heinemann, Stoneham, MA.

24 Nix, W.D. (1989) Mechanical properties of thin films. *Metall. Trans. A*, **20** (**11**), 2217–2245.

25 Gleiter, H. (2000) Nanostructured materials: basic concepts and microstructure. *Acta Mater.*, **48** (**1**), 1–29.

26 Arzt, E., Dehm, G., Gumbsch, P., Kraft, O., and Weiss, D. (2001) Interface controlled plasticity in metals: dispersion hardening and thin film deformation. *Prog. Mater Sci.*, **46** (**3–4**), 283–307.

27 Freund, L.B. and Suresh, S. (2004) *Thin Film Materials: Stress, Defect Formation and Surface Evolution*, Cambridge University Press, Cambridge.

28 Dehm, G., Motz, C., Scheu, C., Clemens, H., Mayrhofer, P., and Mitterer, C. (2006) Mechanical size-effects in miniaturized and bulk materials. *Adv. Eng. Mater.*, **8**, 1033–1045.

29 Engstler, M., Mayer, S., Pauly, C., Clemens, H., and Mücklich, F. (2013) 3D characterization of an intermetallic β/γ-titanium aluminide alloy. *Adv. Eng. Mater.*, **15**, 1125–1128.

30 Kauffmann, F., Bidlingmaier, T., Dehm, G., Wanner, A., and Clemens, H. (2000) On the origin of acoustic emission during room temperature compressive deformation of a γ-TiAl based alloy. *Intermetallics*, **8**, 823–830.

31 Clemens, H. and Mayer, S. (2013) Design, processing, microstructure, properties, and applications of advanced intermetallic TiAl alloys 2013. *Adv. Eng. Mater.*, **15**, 191–215.

32 Fischer, F.D., Waitz, T., Scheu, C., Cha, L., Dehm, G., Antretter, T., and Clemens, H. (2010) Study of nanometer-scaled lamellar microstructure in a Ti-45Al-7.5Nb alloy – experiments and modelling. *Intermetallics*, **18**, 509–517.

33 Scheu, C., Stergar, E., Schober, M., Cha, L., Clemens, H., Bartels, A., Schimansky, F.-P., and Cerezo, A. (2009) High carbon solubility of a γ-TiAl based Ti-45Al-5Nb-0.5C alloy and its effect on hardening. *Acta Mater.*, **57**, 1504–1511.

2

Internal Stresses in Engineering Materials

Anke Kaysser-Pyzalla

2.1 Definition

2.1.1 Stress Tensor, Strain Tensor, and Elasticity Tensor

2.1.1.1 Stress Tensor

In all materials, workpieces, and components, every (infinitesimal) small volume element is subject to forces exerted by its surroundings onto the surface of the volume element. The surface of the volume element can be described by surface areas, mathematically equivalent to planes, whose spatial orientation is given by their normal vector $\vec{n}$

$$\vec{n} = \begin{pmatrix} n_1 \\ n_2 \\ n_3 \end{pmatrix} \tag{2.1}$$

The force $\vec{F}$ acting on the volume element is

$$\vec{F} = \begin{pmatrix} F_1 \\ F_2 \\ F_3 \end{pmatrix} \tag{2.2}$$

The stress tensor $\tilde{\sigma}$, which has the components σ_{ij} acting on a cubic volume element results from the forces exerted on each orthogonal section plane A of the volume element:

$$\tilde{\sigma} \cdot \vec{n} = \frac{\vec{F}}{A}, \quad \sigma_{ij} n_j = \frac{F_i}{A} = t_i \tag{2.3}$$

where $\vec{t}$ is called the *stress vector* (Figure 2.1, [1]).

The diagonal components σ_{kk} of the stress tensor, which act in the normal direction onto the cubic volume element as shown in Figure 2.1, are often referred to as *normal stresses*. The stress tensor components σ_{ij}, $i \neq j$, which act parallel to the surface elements are referred to as *shear stresses*. The stress tensor $\tilde{\sigma}$ is a symmetric second-order tensor

$$\tilde{\sigma} = \begin{pmatrix} \sigma_{11} & \sigma_{12} & \sigma_{13} \\ \sigma_{12} & \sigma_{22} & \sigma_{23} \\ \sigma_{13} & \sigma_{23} & \sigma_{33} \end{pmatrix} \tag{2.4}$$

Neutrons and Synchrotron Radiation in Engineering Materials Science: From Fundamentals to Applications, Second Edition. Edited by Peter Staron, Andreas Schreyer, Helmut Clemens, and Svea Mayer.

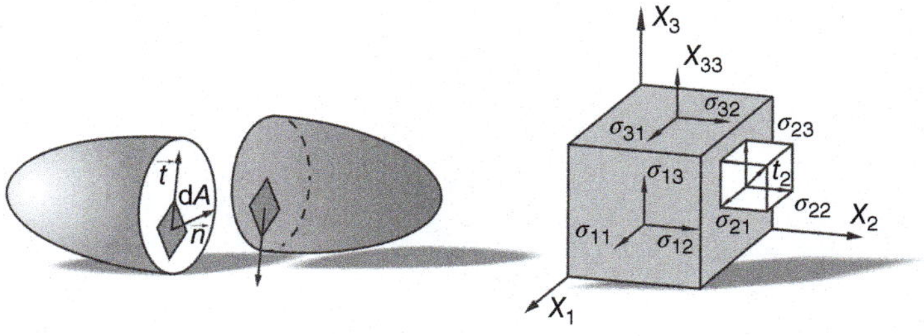

Figure 2.1 Stress vector and stress tensor. (Gross 1992 [1]. Reproduced with permission of Springer.)

The stress tensor can be transposed to any other reference system using the tensor transformation laws. It can, therefore, be diagonalized ($\sigma_{ij} = 0$ if $i \neq j$) by a principal axis tensor transformation. In the principal axis system only normal stresses act in the direction of the principal axes and there are no shear stresses between them.

2.1.1.2 Strain Tensor

The external forces exerted on a material, workpiece, or the component result in displacements $\vec{u}(\vec{x})$. The components of the displacement vector $\vec{u}$ in the direction of the coordinates x_1, x_2, and x_3 (for simplicity we assume a Cartesian coordinate system in the following paragraphs) are defined by:

$$\vec{u} = \begin{pmatrix} u_1 \\ u_2 \\ u_3 \end{pmatrix} \tag{2.5}$$

The deformation associated with the displacement is described by the symmetric second-order strain tensor $\tilde{\varepsilon}$:

$$\tilde{\varepsilon} = \begin{pmatrix} \varepsilon_{11} & \varepsilon_{12} & \varepsilon_{13} \\ \varepsilon_{12} & \varepsilon_{22} & \varepsilon_{23} \\ \varepsilon_{13} & \varepsilon_{23} & \varepsilon_{33} \end{pmatrix} \tag{2.6}$$

Within the limit of "small" deformations (this is nearly always the case for metals and other engineering materials but, not for example rubber-like materials) the relation between deformations and strain tensor components is given by

$$\varepsilon_{ij} = \frac{1}{2}\left(\frac{\partial u_i}{\partial x_j} + \frac{\partial u_j}{\partial x_i}\right) = \frac{1}{2}(u_{i,j} + u_{j,i}) \tag{2.7}$$

In the case of pure elastic deformation the relation between strain and stress tensor is given by Hooke's law:

$$\tilde{\sigma} = \tilde{C}\tilde{\varepsilon} \tag{2.8}$$

The elasticity tensor $\tilde{C}$ is a fourth-order tensor and its components are C_{ijkl}. The symmetry of the strain and the stress tensor results in, for example,

$$C_{ijkl} = C_{klij}$$

$\tilde{C}$ in general has 21 different components. $\tilde{C}$ is also referred to as the *stiffness tensor*; its inverse tensor $\tilde{S}$ is the compliance tensor.

For isotropic crystals, the elasticity tensor has only two independent components, c_{12} and c_{44}:

$$\tilde{C} = \begin{pmatrix} c_{12}+2c_{44} & c_{12} & c_{12} & 0 & 0 & 0 \\ c_{12} & c_{12}+2c_{44} & c_{12} & 0 & 0 & 0 \\ c_{12} & c_{12} & c_{12}+2c_{44} & 0 & 0 & 0 \\ 0 & 0 & 0 & c_{44} & 0 & 0 \\ 0 & 0 & 0 & 0 & c_{44} & 0 \\ 0 & 0 & 0 & 0 & 0 & c_{44} \end{pmatrix} \tag{2.9}$$

In the case of isotropy, Hooke's law can be simplified to the expressions

$$\sigma_{ij} = \frac{E}{1+\nu}\varepsilon_{ij} - \frac{\nu E}{(1+\nu)(1-2\nu)}\delta_{ij}\varepsilon_{kk} \tag{2.10}$$

$$\varepsilon_{ij} = \frac{1+\nu}{E}\sigma_{ij} - \delta_{ij}\frac{\nu}{E}\sigma_{kk} \tag{2.11}$$

δ is the Kronecker δ-function ($\delta = 1$ for $i = j$, $\delta = 0$ for $i \neq j$), E the Young's modulus, and ν the Poisson's ratio of the material.

2.1.2 Definitions, Residual Stresses

2.1.2.1 Stress Equilibrium

Residual stress is the stress, which remains in a body that is stationary and at equilibrium with its surroundings [2]. Residual stresses, thus, are mechanical stresses present in a workpiece or a component, which is not subject to external forces, a momentum, or temperature gradient. Residual stresses $\tilde{\sigma}$ are by definition self-equilibrating stresses. At the surface of the material, in particular, of the component, the residual stresses must fulfill the boundary conditions

$$\tilde{\sigma}\cdot\vec{n} = 0, \quad \sigma_{ij}n_j = 0 \tag{2.12}$$

In each point of the material, respectively, the component equilibrium means

$$\frac{\partial \sigma_{ij}}{\partial x_j} = 0 \tag{2.13}$$

In a Cartesian coordinate system, the equilibrium conditions thus are [3]

$$\begin{aligned} \frac{\partial \sigma_{11}}{\partial x_1} + \frac{\partial \sigma_{12}}{\partial x_2} + \frac{\partial \sigma_{13}}{\partial x_3} &= 0 \\ \frac{\partial \sigma_{12}}{\partial x_1} + \frac{\partial \sigma_{22}}{\partial x_2} + \frac{\partial \sigma_{23}}{\partial x_3} &= 0 \\ \frac{\partial \sigma_{13}}{\partial x_1} + \frac{\partial \sigma_{23}}{\partial x_2} + \frac{\sigma_{33}}{\partial x_3} &= 0 \end{aligned} \tag{2.14}$$

In a cylinder coordinate system, Eq. (2.13) becomes [3]

$$\begin{aligned} \frac{\partial \sigma_{rr}}{\partial r} + \frac{1}{r}\frac{\partial \sigma_{r\phi}}{\partial \phi} + \frac{\partial \sigma_{rz}}{\partial z} + \frac{1}{r}(\sigma_{rr} - \sigma_{\phi\phi}) &= 0 \\ \frac{\partial \sigma_{r\phi}}{\partial r} + \frac{1}{r}\frac{\partial \sigma_{\phi\phi}}{\partial \phi} + \frac{\partial \sigma_{\phi z}}{\partial z} + \frac{2}{r}\sigma_{r\phi} &= 0 \end{aligned} \tag{2.15}$$

$$\frac{\partial \sigma_{rz}}{\partial r} + \frac{1}{r}\frac{\partial \sigma_{\phi z}}{\partial \phi} + \frac{\partial \sigma_{zz}}{\partial z} + \frac{1}{r}\sigma_{rz} = 0$$

where r is the cylinder radius, ϕ the azimuth, and z is the coordinate in the direction of the cylinder longitudinal axis.

In case of flat parts such as sheets or, for example, in coatings due to their small expansion in one direction (e.g., the x_3-direction) as compared to the other two directions (x_1, x_2), often stresses in the x_3-direction can be assumed to be negligible ($\sigma_{13} = \sigma_{23} = \sigma_{33} = 0$); this is called the *plane stress* condition. In the case of "plane stress" Eq. (2.13) simplifies to [3]

$$\frac{\partial \sigma_{xx}}{\partial x} + \frac{\partial \sigma_{xy}}{\partial y} = 0 \quad \text{and} \quad \frac{\partial \sigma_{xy}}{\partial x} + \frac{\partial \sigma_{yy}}{\partial y} = 0 \tag{2.16}$$

Hooke's law can be expressed simply by

$$\sigma_{11} = \frac{E}{1-\nu^2}(\varepsilon_{11} + \nu\varepsilon_{22}) \quad \text{and} \quad \sigma_{22} = \frac{E}{1-\nu^2}(\varepsilon_{22} + \nu\varepsilon_{11}) \tag{2.17}$$

The stress equilibrium conditions imply that tensile residual stresses in a certain direction within one part of a body are always balanced by matching compressive residual stresses in another part. Thus, the residual stress state of a workpiece or a component can never be expressed by a single residual stress tensor (or even a single residual stress value), but only by a three-dimensional residual stress distribution [4]. This also implies the presence of residual stress gradients, since the residual stress distribution has to be (mathematically) continuous. Strong residual stress gradients are often present in the near-surface area of components, because the residual stress component normal to the surface due to the equilibrium condition (2.12) needs to vanish but stress continuity has to be observed in the bulk material (Figure 2.2).

2.1.2.2 Residual Macro- and Microstresses

Residual stresses can be categorized by various schemes, for example, as per the manufacturing method producing the residual stresses (e.g., welding residual stresses, shot-peening residual stresses), the method by which they are measured, and their cause (e.g., temperature or deformation inhomogeneities). The most common categorization of residual stresses introduced by Macherauch *et al.* [5, 6] when dealing with diffraction

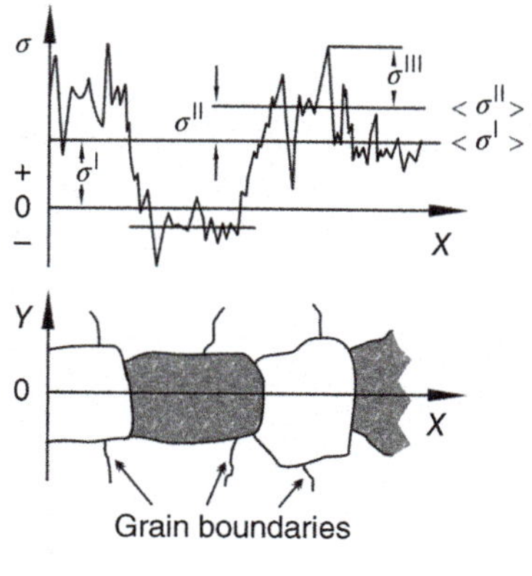

Figure 2.2 Definition of type I, type II, and type III residual stresses in a single phase material. (Macherauch 1973 [5]. Reproduced with permission of Härterei Technische Mitteilungen.)

methods is based on the length scale. This categorization distinguishes between type I, type II, and type III residual stresses (Figure 2.2). Type I residual stresses are also referred to as *residual macrostresses*, type II and type III residual stresses are commonly denominated as residual microstresses, see [7].

The residual stress distribution in a material, respectively, a component is the sum of type I, type II, and type III residual stresses:

$$\tilde{\sigma}(\vec{x}) = \tilde{\sigma}^{\mathrm{I}}(\vec{x}) + \tilde{\sigma}^{\mathrm{II}}(\vec{x}) + \tilde{\sigma}^{\mathrm{III}}(\vec{x}), \quad \int_V \tilde{\sigma}(\vec{x})\,\mathrm{d}V = 0 \tag{2.18}$$

Type I residual stresses $\tilde{\sigma}^{\mathrm{I}}(\vec{x})$ represent the average residual stresses $\tilde{\sigma}(\vec{x})$ acting within all phases and crystallites in the gauge volume V.

$$\tilde{\sigma}^{\mathrm{I}}(\vec{x}) = \frac{1}{V}\int \tilde{\sigma}(\vec{x})\,\mathrm{d}V \tag{2.19}$$

The gauge volume must be large enough to represent the macroscopic material (e.g., contain a sufficient number of crystallites and all phases present). Releasing $\tilde{\sigma}^{\mathrm{I}}(\vec{x})$ causes macroscopic shape changes of the material, respectively, the component.

Type II residual stresses $\tilde{\sigma}^{\mathrm{II}}(\vec{x})$ describe the mean deviation from the macroscopic residual stress level $\tilde{\sigma}^{\mathrm{I}}(\vec{x})$ of an individual crystallite (single phase material)

$$\tilde{\sigma}^{\mathrm{II}}(\vec{x}) = \frac{1}{V}\int [\tilde{\sigma}(\vec{x}) - \tilde{\sigma}^{\mathrm{I}}(\vec{x})]\,\mathrm{d}V \tag{2.20}$$

In a multiphase material type II, residual stresses are taken as the average residual stresses $\langle\tilde{\sigma}^{\mathrm{II}}(\vec{x})\rangle^{\alpha}$ of the crystallites belonging to a phase α or as the average residual stresses of those crystallites of the phase α contributing to the measurement:

$$\langle\tilde{\sigma}^{\mathrm{II}}(\vec{x})\rangle^{\alpha} = \frac{1}{V^{\alpha}}\int [\tilde{\sigma}(\vec{x}) - \tilde{\sigma}^{\mathrm{I}}(\vec{x})]\,\mathrm{d}V^{\alpha} \tag{2.21}$$

Release of the type II residual stresses may result in macroscopic distortions.

Type III residual stresses, $\tilde{\sigma}^{\mathrm{III}}(\vec{x})$ represent the local deviation of the residual stresses within an individual crystallite from its average residual stress (variation on the atomic scale). Thus, the average type III residual stress $\tilde{\sigma}^{\mathrm{III}}(\vec{x})$ of a crystallite is zero by definition. Release of type III residual stresses does not result in macroscopic distortions. Since the categorization of type I, type II, and type III residual stresses is based on their length scale and not on their magnitude, for example, type III residual stresses may in some cases be as detrimental to a component's lifetime as type I residual stresses (Figure 2.3, [5]).

2.2 Origin of Residual Macro- and Microstresses

Residual stresses originate from the misfit between different regions [2]. These misfits result from gradients in deformation or temperature or both deformation and temperature across the component. The most common sources of residual stresses in a component are manufacturing processes, but, residual stresses can also result from temperature or deformation gradients exerted under service conditions.

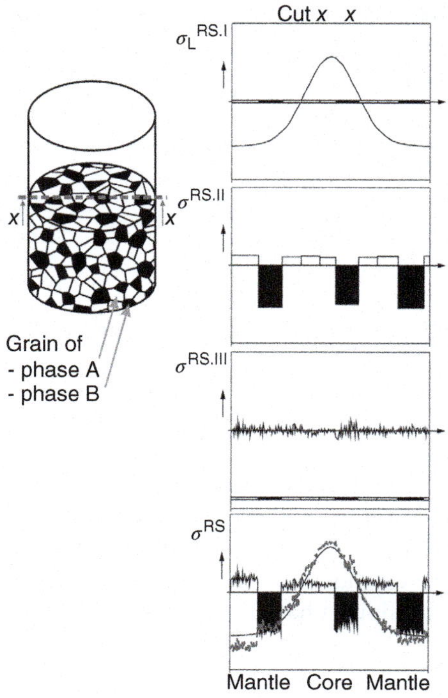

Figure 2.3 Definition of type I, type II, and type III residual stresses in a multiphase material. (Macherauch 1973 [5]. Reproduced with permission of Härterei Technische Mitteilungen.)

Type I residual stresses $\tilde{\sigma}^{I}(\vec{x})$ result from long-range strain incompatibilities introduced, for example, by strain or temperature gradients in a manufacturing process. These strain inhomogeneities arise, for example, due to friction between the semifinished component and the die for instance in cold rolling or extrusion [8]. In welding processes, temperature gradients across the welds cause misfits in thermal expansion and strength and, thus, residual macrostresses, see [9].

While the distribution and magnitude of type I residual stresses often can be controlled by modifying the process parameters of a manufacturing process, this in general is not possible for type II and type III residual stresses. Type II residual stresses arise for instance due to deformation misfits between neighboring grains and due to temperature- or deformation-induced misfits between the phases of a multiphase material. Type III residual stresses are caused, for example, by voids, solute atoms, or dislocations in the crystal lattice (Figure 2.4, [5]).

For about two decades, research efforts have revealed the elementary processes of residual stress formation in manufacturing processes, some of the results will be outlined in the next sections.

2.2.1 Residual Stress Formation in Primary Forming Processes

Primary forming by processes, such as casting, sintering, and isostatic pressing, usually occurs at elevated temperatures, the ingot or semifinished product is then cooled down to room temperature. In large cast or sintered parts, temperature and density gradients occur, which can lead to gradients in the thermal expansion and, thus, to residual strains

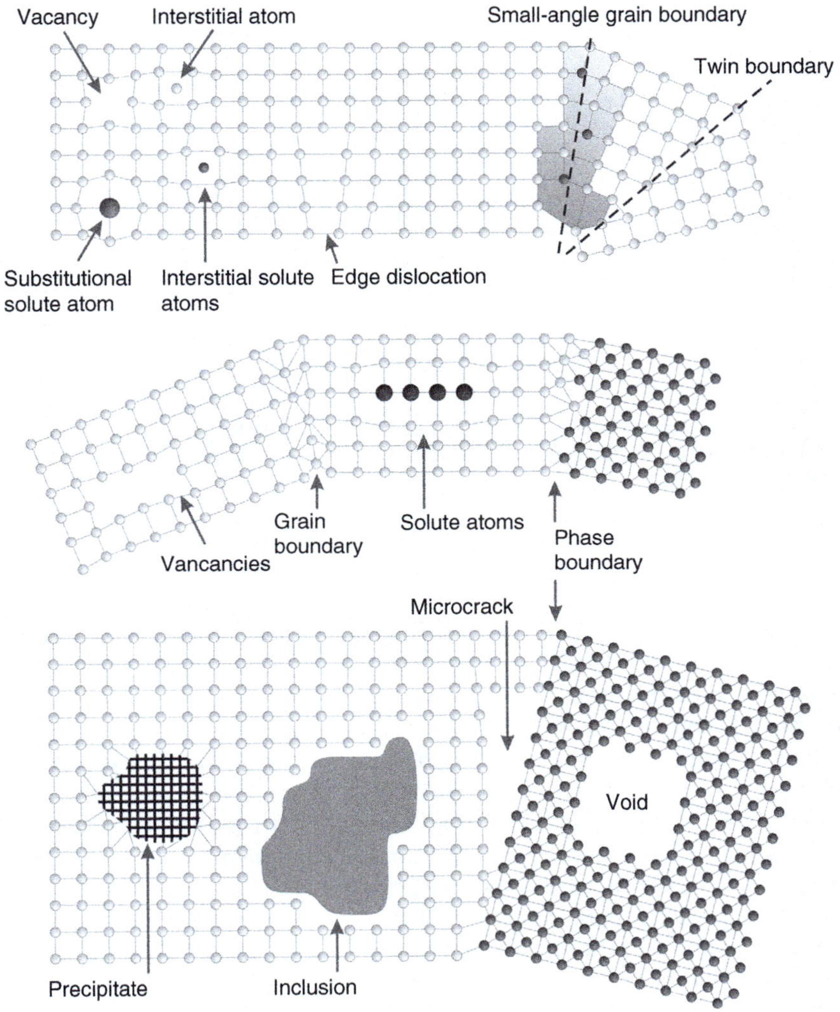

Figure 2.4 Origin of type III residual stresses. (Macherauch 1973 [5]. Reproduced with permission of Härterei Technische Mitteilungen.)

and stresses. The formation of residual stresses during primary forming by hot isostatic pressing (HIP) of a two-phase material is a well-defined problem that can be understood relatively easily and, thus, will be used as an example here:

"When a child squeezes snow to make a snowball, he is using the technique of hot-isostatic pressing (inelegantly, but universally known as *HIP*ing)" [10] is a very descriptive explanation of the HIP process. HIP is used to produce precisely shaped – often near net-shape – bodies from metal and/or ceramic powders. The powder is packed into a sheet metal pre-form, which is then evacuated and inserted into a pressure vessel. Within the pressure vessel the preform is heated and simultaneously subjected to a high pressure, often by argon gas. Depending on the process

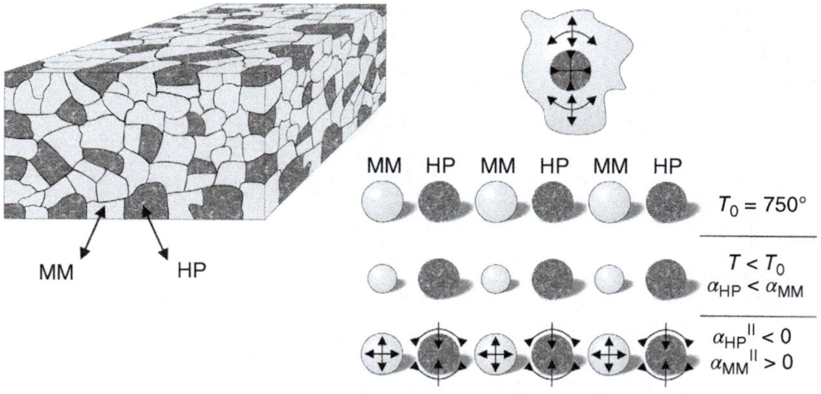

Figure 2.5 Residual stress formation during cooling of a composite. The dark particles could be, for example, ceramic hard particles (HPs), which have a lower thermal expansion coefficient than the metal matrix (MM, "white" grains).

parameters, several mechanisms contribute to the densification of the powders, they involve diffusional redistribution of matter, plastic flow, and/or power law creep, see, [10].

HIP is one of the predominant techniques for producing particle reinforced wear resistant metal matrix composites (PMMCs). These PMMCs consist of a metal matrix with up to 30% coarse (several micrometer diameter) ceramic hard phase particles. During cooling from the HIP temperature (about 1100 °C) residual stresses emerge due to differences in the thermal expansion of the metal matrix and the ceramic hard phase particles (Figure 2.5).

A very simple model, using a parallel or a series connection between metal matrix and ceramic particles after cooling yields compressive residual microstresses in the ceramic particles (due to their lower thermal expansion), while tensile residual microstresses develop in the metal matrix. A more detailed analysis of the residual stresses in the metal matrix using a model of a spherical ceramic particle embedded in an infinite metal matrix reveals tensile residual microstresses that are parallel to the interface to the ceramic particle, whereas compressive residual microstresses are perpendicular.

2.2.2 Residual Stress Formation in Heat Treatment Processes

Heat treatment processes usually lead to residual stress formation during cooling the component to room temperature because of temperature gradients and/or volume changes induced by phase transformations. Among the technically most important heat treatment processes are hardening and annealing. The following sections therefore will concentrate on residual stress formation during these processes, starting with the less complex residual stress formation during annealing and simplifying the complex geometry of technical components, for example, camshafts or crankshafts by a cylinder model (Figure 2.6, [12]).

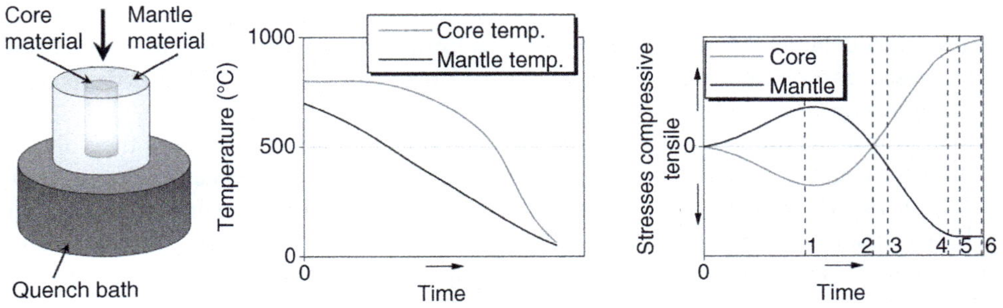

Figure 2.6 Temperature and stress development of a material without phase transformation during water cooling [11].

2.2.2.1 Residual Stresses in a Material without Phase Transformation (Pure Cooling Residual Stresses)

When immersing a cylinder lengthwise into a cooling medium, for example, water, oil, or a salt bath, during the first seconds of cooling the temperature in the mantle decreases faster than in the core of the cylinder. Thus, volume shrinkage (due to thermal expansion) is stronger in the mantle than in the cylinder core. Because the mantle aims at shrinking more than the core permits, tensile stresses evolve in the mantle, which are balanced by compressive stresses in the core. The differences in the volume changes and, thus, the stresses increase with increasing temperature difference of mantle and core (until point 1 in Figure 2.6 is reached).

After reaching the maximum temperature difference between mantle and core, where tensile stresses exist in the mantle, respectively, compressive stresses in the core, further cooling results in a decrease of the temperature difference in the mantle and core. The macroscopic stresses of the cylinder therefore decrease. Once the temperature decrease in the core becomes faster than that in the cylinder mantle (point 2 in Figure 2.6), the material in the core aims at shrinking faster than that in the mantle. Thus, tensile stresses evolve in the core and compressive stresses develop in the cylinder mantle. The magnitude of the stresses increases with decreasing temperature. The residual stress level after cooling to room temperature is limited by the strength of the material; once the stresses reach the yield point (point 3 in Figure 2.6), further stress increase is limited by hardening of the material and the increase of yield strength with decreasing temperature (points 4 and 5 in Figure 2.6).

The stresses evolving due to temperature gradients between mantle and core during cooling the cylinder are macroscopic stresses. If the stresses formed exceed the yield stress of the cylinder and inhomogeneous plastic deformations occur, residual stresses remain after cooling. The distribution and magnitude of these residual stresses depend strongly on the cooling process (e.g., medium, immersion speed), the geometry of the cylinder (diameter, length-to-diameter ratio) and (the temperature dependent!) material properties (e.g., thermal diffusivity, thermal expansion coefficient, yield strength) (Figure 2.6).

2.2.2.2 Residual Stresses in a Material with Phase Transformation

In the particular case of steels, residual stress formation processes have to take volume changes due to phase transformations into account. These volume changes depend on the phases formed and the carbon content of the steel (Table 2.1, [13]).

The transformation of pearlite into austenite during heating decreases the volume, while, the transformation of austenite into martensite, bainite, and pearlite during cooling increases the volume. The strains induced by the volume increase during martensitic phase transformation, in most cases, are larger than the thermal strains generated during cooling a component. Thus, phase transformations significantly alter the residual stress state after thermal treatments.

In the following text, a long axle-like component is again simplified by a cylinder model; first, a through-hardened cylinder, and then a surface-hardened cylinder will be considered:

Through-hardening component: Hardening of the steel cylinder necessitates rapid cooling of high-temperature austenite. During cooling, the cylinder mantle cools faster than the core, and thus, as long as both the mantle and the core remain austenitic, tensile stresses develop in the mantle, which are balanced by compressive stresses in the cylinder core (Figure 2.7, [11, 14]). At a time t_1, the mantle reaches the martensite start temperature M_s and the austenite → martensite transformation starts. Due to martensite formation, the volume of the cylinder mantle increases and,

Table 2.1 Phase transformation induced volume changes during heat treatment processes of steels, for example, phase transformation from austenite into martensite of a steel C45 with 0.45%C, results in a volume increase of $(4.64 - 0.53 \times 0.45) = 4.4\%$.

Phase transformation	Volume change (%C)
Pearlite → austenite	$-4.64 + 2.21$
Austenite → martensite	$4.64 - 0.53$
Austenite → lower bainite	$4.64 - 1.43$
Austenite → upper bainite or pearlite	$4.64 - 2.21$

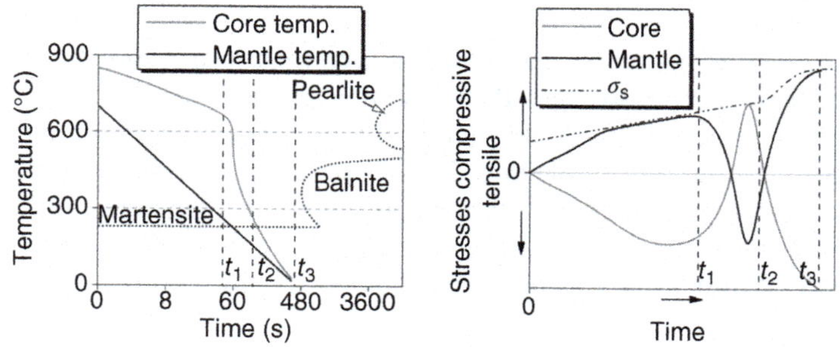

Figure 2.7 Residual stresses in mantle and core of a through-hardening ⌀100 mm steel cylinder after water cooling [11, 14].

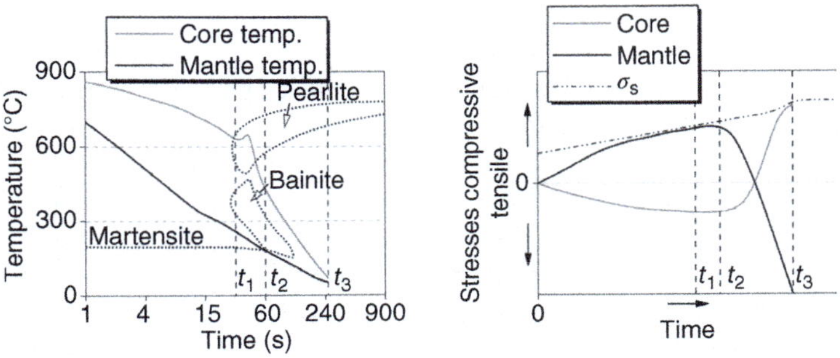

Figure 2.8 Residual stresses in mantle and core of a surface hardening ⌀100 mm steel cylinder after water cooling, [11, 14].

consequently, the tensile stresses in the mantle decrease. As soon as the difference of strains due to temperature differences are outweighed by the volume increase in the mantle, compressive stresses develop in the mantle and tensile stresses form in the core of the cylinder. At time t_2, the temperature in the cylinder core reaches M_s and martensite formation leads to a volume increase within the cylinder core. Thus, the tensile stresses in the core and the balancing compressive stresses in the mantle decrease. Since at this time due to the low temperature, the mantle is rigid (yield strength increases with decreasing temperature), expansion of the core is limited, leading to compressive stresses in the core and tensile stresses in the cylinder mantle. In through-hardening components, thus, the residual stress state induced by martensitic hardening usually is the contrary of the residual stress state resulting solely from temperature gradients in the component.

Partially hardened component: In the case of a nonthrough hardening, component stress formation during cooling starts again with the development of tensile stresses in the mantle and compressive stresses in the cylinder core (Figure 2.8, [11, 14]). Reaching the time t_1 austenite → pearlite transformation starts within the cylinder core. The compressive strain induced by the volume increase caused by phase transformation and the compressive stresses due to cooling add to each other. As soon as the time t_2 is reached, martensite formation of the cylinder mantle starts and the volume of the cylinder mantle increases. As a consequence of the volume increase, the tensile stresses in the mantle and the compressive stresses in the cylinder core decrease and after cooling (t_3) the cylinder mantle is under compressive residual stresses, the core contains the balancing tensile residual stresses. Thus, in surface-hardening components the residual stress state after cooling is similar to the residual stress state of a component experiencing only thermal stresses as against the residual stress state of a through-hardened component.

The residual stress state of components, which are neither through-hardened nor only hardened within a thin surface layer, depends on a number of factors, such as the component dimensions, the phase transformation behavior of the material, the (temperature dependent) mechanical strength of the material, its martensite start temperature, and the cooling medium.

2.2.2.3 Residual Stress Formation in Surface-Hardening Processes (Nitriding, Carbo-Nitriding, and Case Hardening)

In order to increase strength and wear resistance in the near-surface region, steel parts can be subjected to nitriding, carbo-nitriding, and case hardening treatments. In principle, the elementary processes of residual stress formation in case of hardened components are similar to those discussed for the surface-hardened components. However, due to its different chemical composition, the near-surface region must be regarded as a material different from that of the core. In the case of hardening, the carbon content of the cylinder mantle is higher as compared to the core; thus, the phase transformation behavior of mantle and core differ significantly (Figure 2.9, [11, 15]). The M_s temperature of the mantle of case-hardened components is, therefore, substantially lower than in the core. During cooling of case hardening workpieces martensite transformation and volume increase in the cylinder mantle, thus, occurs usually later than the phase transformation in the core, resulting in compressive residual stresses in the cylinder mantle and tensile residual stresses in the cylinder core. In addition to decreasing the M_s temperature the high carbon content in the cylinder mantle implies that M_f is not reached and, therefore, the cylinder mantle contains martensite apart from the retained austenite. The retained austenite volume fraction due to the gradient in carbon content is highest and, thus, the martensite volume fraction is lowest at the cylinder surface and decreases toward the interface between the case-hardened zone and the parent material. Because of this gradient in phase composition, maximum compressive residual stresses are present in case-hardened components at the interface between the hardened zone and the parent material.

Residual stresses formed during nitriding are in most cases comparatively small compressive residual stresses in the near-surface zone balanced by tensile residual stresses in the bulk [11, 16, 17], due to nitriding temperatures being lower than case-hardening ones. Besides temperature gradients the volume increase due to nitrogen diffusion and nitride precipitation in the near-surface zone also contributes to the compressive stresses. Because the steels are hardened and annealed previous to nitriding, phase transformations usually do not occur [18].

2.2.3 Residual Stress Formation in Forming Processes

Forming processes necessitate plastic deformation of a workpiece or semifinished product. Due to the unavoidable inhomogeneity of plastic deformation, forming processes

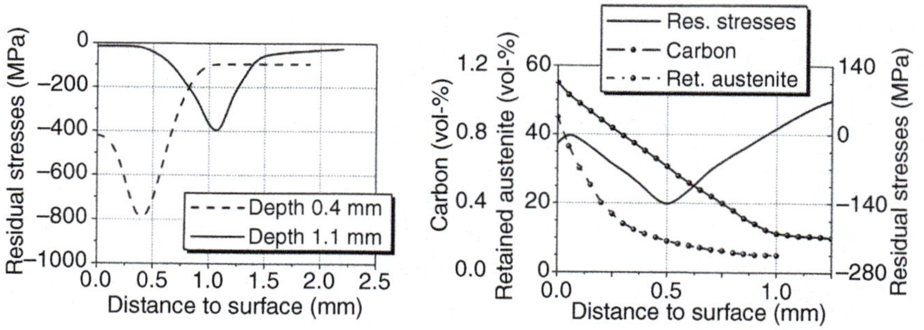

Figure 2.9 Relation between carbon, retained austenite volume fraction, and residual stresses in a case-hardened component (distance to surface) [11, 15].

create residual stresses. In addition to residual macrostresses in forming processes with strong plastic deformation, residual microstresses of the same magnitude also evolve due to locally very high defect densities or due to texture formation. In order to introduce the processes of residual stress formation in forming processes, in the following section, deep-rolling and cold extrusion are taken as examples.

2.2.3.1 Deep-Rolling Residual Stresses

Deep-rolling is a well-established industrial process for deformation hardening of axial symmetric components such as axles, bolts, and threads. Deep-rolling substantially improves the fatigue strength of dynamically loaded parts [19–22], for example, by deep-rolling of the shoulder transient between jowl and bearing (Figure 2.10) the bending fatigue strength of crankshafts could be improved by up to 200% [23] (Figure 2.10).

During deep-rolling, the rolls exert Hertzian pressure onto the surface of the workpiece. The Hertzian pressure and friction forces lead to a triaxial stress state in the near-surface area (Figure 2.11, [24]). In those regions of the near-surface zone, where the equivalent stress exceed the yield stress, the material deforms plastically. In deep-rolling processes the near-surface zone is usually stretched during the process. If the yield strength is exceeded compressive residual stresses remain. Thus, after loading compressive residual stresses remain in the near-surface zone, which are balanced by tensile residual stresses in the bulk material.

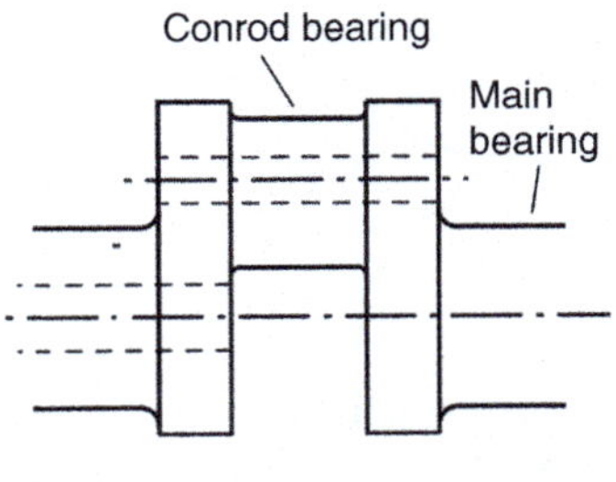

Figure 2.10 Crankshaft.

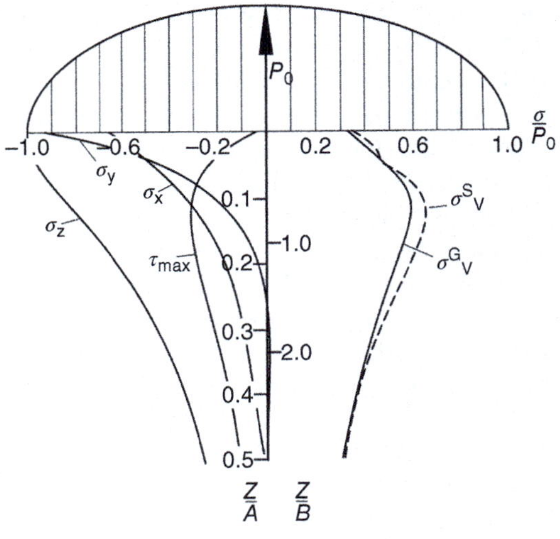

Figure 2.11 Stresses due to Hertzian pressure [24].

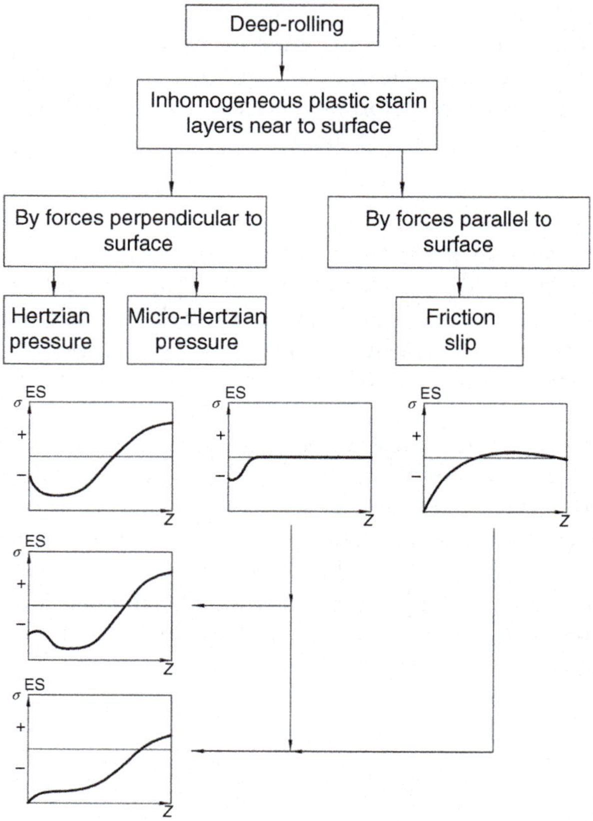

Figure 2.12 Residual stresses introduced by deep-rolling. (Adapted [24].)

While deformation due to tangential forces, for example, due to friction between the tool and the workpiece, leads to compressive stress maxima at the surface of the workpiece, the maximum stress and deformation caused by Hertzian pressure is not directly at, but, below the surface (Figure 2.12).

The depth of the residual stress maximum after deep-rolling, thus, depends on the process parameters, material behavior, and geometry of tool and workpiece (Figure 2.12): for instance, an increase in rolling force increases the distance between the depth of the residual stress maximum and the surface of the workpiece [24]. Besides the depth and the magnitude of the residual stresses also surface roughness and hardness need to be taken into account as criteria for choosing an optimum rolling force. An excess in rolling force results in an increase of surface roughness and, thus, a decrease of fatigue strength. The hardness distribution in the near-surface zone correlates with the residual stress distribution because of increasing lattice defect formation with increasing plastic deformation (Figure 2.13).

2.2.3.2 Cold Extrusion Residual Stresses

Cold forward extrusion (Figure 2.14) is industrially used for manufacturing bolts, which will be further processed resulting, for example, in screws or thread rods (Figure 2.14).

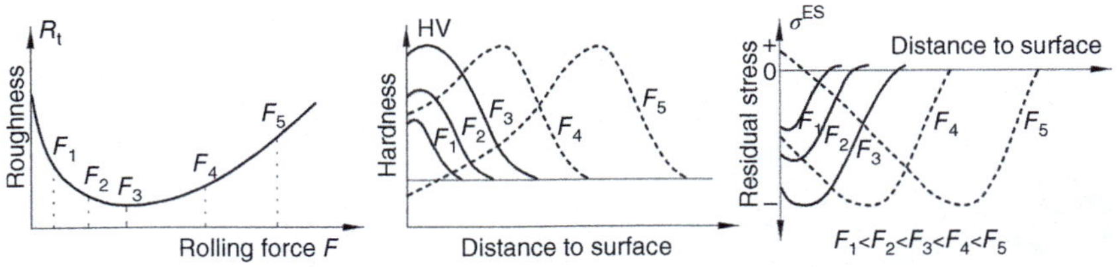

Figure 2.13 Relation between rolling force and surface roughness, hardness, and residual stresses in deep-rolling. (Adapted [24].)

Figure 2.14 Scheme of a forward extrusion process of a full and a hollow workpiece.

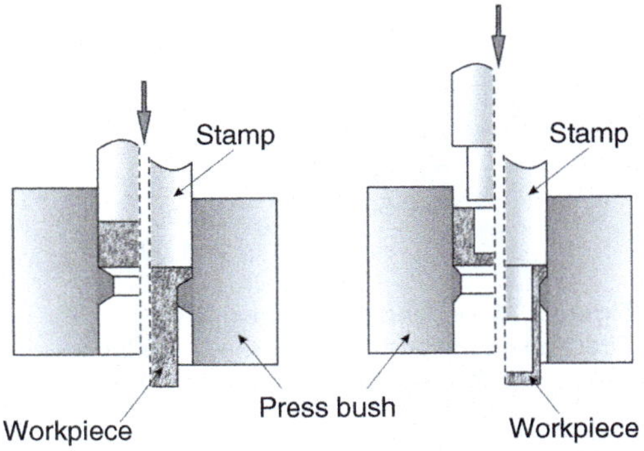

Figure 2.15 Axial residual macrostress in hollow forward extruded steel samples. (Pyzalla [27] http://www.hindawi.com/archive/1999/362913/abs/. Used under CC BY 3.0 license http://creativecommons.org/licenses/by/3.0/.)

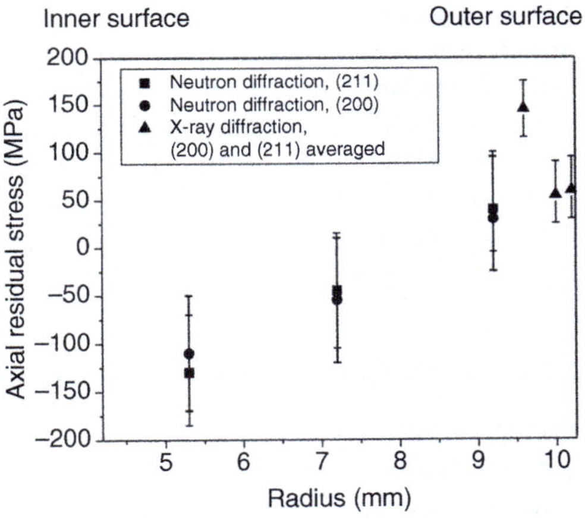

Both in forward extrusion processes of full (Figure 2.14) and hollow workpieces, the workpiece is heavily plastically deformed. Due to deformation obstruction at the shoulder of the die, the material flow at the outer surface is slower and more inhomogeneous than in the bulk of the workpiece. Therefore, within the bulk of the workpiece the grains are homogeneously stretched, whereas at the outer surface of the samples the grains are first compressed and later stretched, while passing the transient radius of the die [25–27]. Thus, tensile residual stresses remain in the near-surface zone, while the bulk of the workpiece is under compressive residual stresses (Figure 2.15, [27]).

The heavy plastic deformation of the workpiece during the cold extrusion process leads to fiber texture formation. The changes of grain orientation in fiber texture

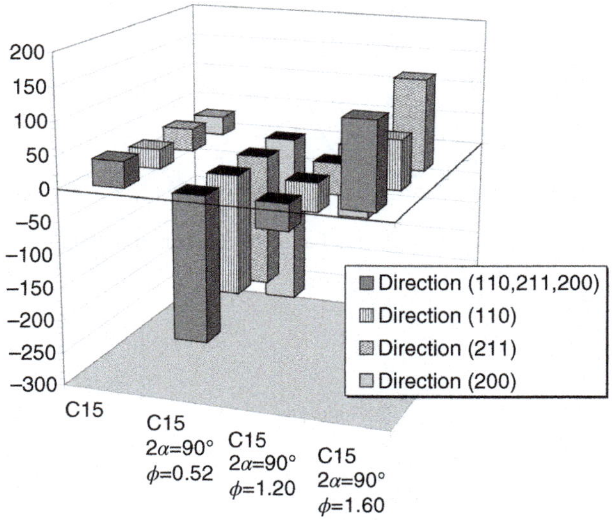

Figure 2.16 Dependence of the difference of the principal residual stresses in axial and radial direction on the orientation of the crystallite group. (Pyzalla [27] http://www.hindawi.com/archive/1999/362913/abs/. Used under CC BY 3.0 license http://creativecommons.org/licenses/by/3.0/.)

evolution result in strain incompatibilities between neighboring crystallites. As a consequence, the residual stress state of a crystallite depends on its crystallographic orientation (Figure 2.16). These residual microstresses between different oriented crystallites usually are called *intergranular stresses*. The intergranular microstresses in different crystallite groups (crystallites of similar orientation) depend on the amount of plastic deformation. Since these intergranular residual stresses can reach the magnitude of the residual macrostress, the formation of the intergranular stresses during plastic deformation has been extensively investigated for uniaxial [28, 29], for example, Holden, Daymond, and multiaxial deformation [30] (Figure 2.16, [27]).

2.2.4 Residual Stress Formation in Metal Cutting Manufacturing

In chip-forming metal cutting processes, such as drilling, turning, machining, and grinding, the dimensions and the surface topography of the workpiece and, therefore, its residual macrostress state are altered. The main mechanisms of residual stress formation in metal cutting processes depend on the material, its heat treatment (e.g., hardened or annealed steel), the machining process, and the process parameters. During chip formation the near-surface zone of the workpiece experiences elastic and plastic deformations, thermal loading, and may undergo phase transformations [31]. Elastic and plastic deformations are caused by the mechanical forces the die exerts on the workpiece. Thermal loading is the result of plastic deformations and of friction, for example, between the die and the workpiece. Because of the decrease of yield strength with temperature, additional plastic deformation can result from heat production. Phase magnitude of transformations may occur during heating as well as during rapid cooling because of the heat flow from the rather thin near-surface zone into the cold surrounding workpiece (Figure 2.17, [31]).

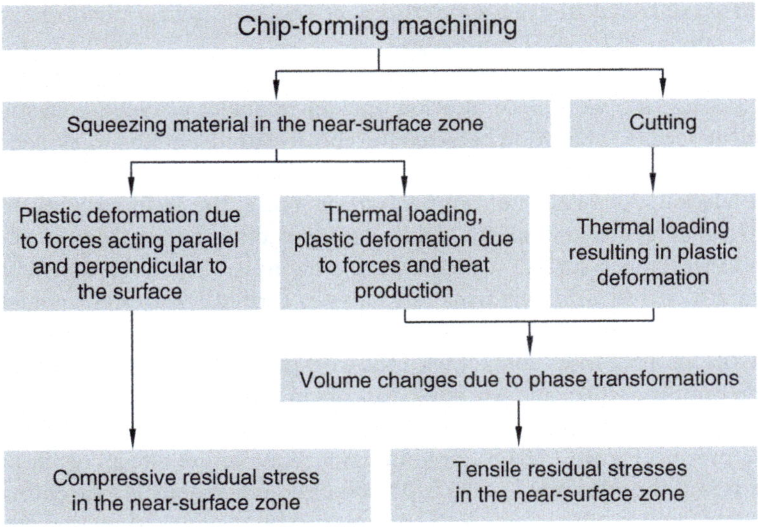

Figure 2.17 Mechanisms of residual stress formation in chip-forming processes. (Adapted from [31].)

Plastic deformation introduced by squeezing the material in the near-surface zone will extend the surface area, thus, introducing compressive residual stresses in the near-surface zone and tensile residual stresses in the bulk material. In contrast, thermal loading results in a hot near-surface zone compared to the bulk, thus, in materials without phase transformations tensile residual stresses form in the near-surface zone, which are balanced by compressive residual stresses in the bulk. The resulting residual stress state after machining is, therefore, determined by the superposition of the residual stresses introduced by deformation and by temperature. In the near-surface region predominant deformation causes compressive residual stresses, whereas a predominant thermal loading results in tensile residual stresses (Figure 2.17). In materials undergoing phase transformations, for example, in steels, strains induced by the phase transformation also need to be considered. Since only the surface zone of the workpiece hardens, in analogy to surface-hardening processes (see Section 2.2.2), compressive residual stresses will form in the near-surface region due to the volume increase caused by martensitic transformation.

2.2.4.1 Grinding Residual Stresses

Grinding often is the last step in the manufacturing process of a component. Surface integrity after grinding, thus, has a strong influence on component behavior under service conditions, for example, low cycle fatigue (LCF) and high cycle fatigue (HCF) behavior, for example, [24].

Abrasive particles on the surface of a grinding wheel perform the cutting process. Depending on the workpiece material (e.g., steel or ceramics), the grinding wheel (e.g., type of abrasive and binder, dressing), and the parameters of the grinding process (e.g., process type, depth of cut, feed rate, cooling medium) strong plastic deformations and high temperatures may occur [24]. Because the surface zone influenced by the grinding process usually is rather thin, the residual stress state after grinding is a plane stress state (see Section 2.1.2); however, particularly in the case of ground ceramics and ground hardened steels also triaxial residual stress states and steep residual stress gradients have been observed, for example, [32–34] (Figure 2.18).

A detailed view on chip formation in grinding shows that first the cutting edge of an abrasive grain touches the workpiece at a low incidence angle, producing elastic deformations and heat because of friction between the abrasive grain and the workpiece (Figure 2.18). Increasing forces result in plastic deformation and temperature increase due to increase of friction. As a result of temperature increase, the yield strength of the workpiece material decreases and plastic deformation becomes easier. As a consequence of changes in workpiece surface topography, for example, local bulges form at the workpiece surface. Local shear band formation starts, contributing strongly to heat formation and preparing chip cutting. The cutting process ends by elastic or elastoplastic deformation, while the abrasive loses contact with the workpiece [24].The macroscopic interaction between grinding wheel and workpiece occurs in a very similar way [24].

The residual stress state after grinding depends on whether plastic deformation or temperature effects prevail (Figure 2.19a,b). Predominant temperature effects result in tensile residual stresses at the workpiece surface [24, 35]. Inhomogeneous plastic deformations and negligible temperature effects lead to compressive residual stresses at the surface and tensile residual stresses beneath the surface. A combination of both temperature effects and inhomogeneous plastic deformation often results in compressive residual stresses at the surface and a wide near-surface zone containing lower tensile residual stresses (Figure 2.19c).

An assessment of the influence of the grinding process on the workpiece residual stress state revealed that the material removal rate (workpiece speed × depth of cut) does not relate to the residual stress state of the material in a unique way. Instead, individual process parameters of the grinding process need to be taken into account. For instance, increasing workpiece speed diminishes local heating and, thus, favors compressive residual stresses at the surface (Figure 2.20). Increasing the depth of cut on the other hand increases heat input and, therefore, induces surface tensile residual stresses.

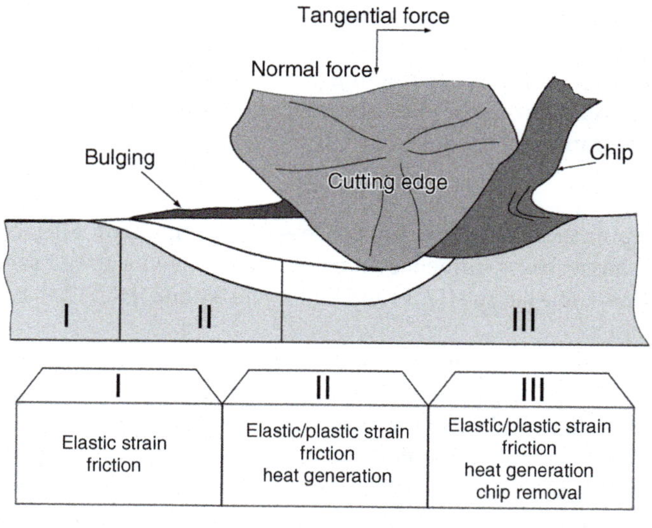

Figure 2.18 Elementary processes in chip formation in grinding processes [24].

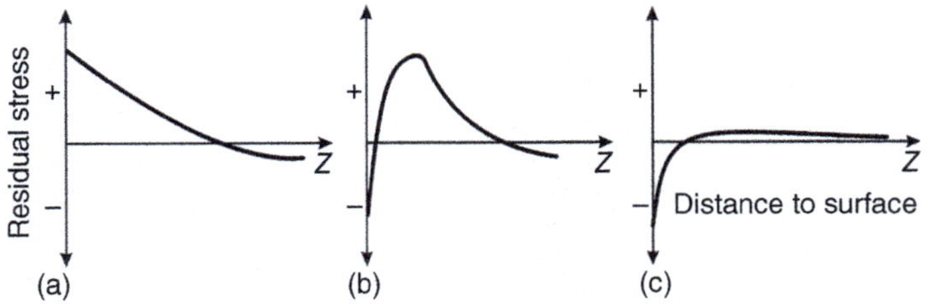

Figure 2.19 Typical residual stress distributions in the near-surface zone of ground materials.

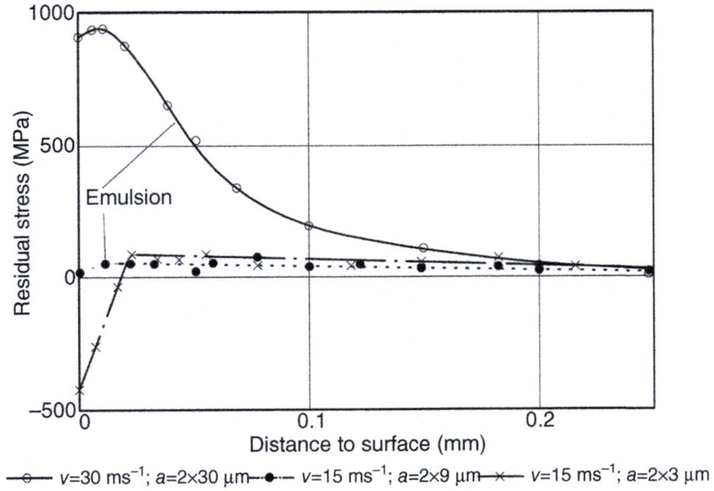

Figure 2.20 Dependence of near-surface residual stresses on grinding conditions. (Adapted [24].)

The selection of grinding fluids is of particular importance both with respect to their cooling and lubrication capacity, details are given, for example, in [31, 36].

2.2.5 Residual Stress Formation in Joining Processes

Joining by substance-to-substance-bonding includes the process groups welding, brazing, soldering, and gluing. Differentiation between welding and brazing, respectively, soldering, is usually based on the melting of the parent material during the welding process, while only the filler melts in brazing and soldering processes. Due to the high industrial and commercial relevance of welding, the number of welding processes has been steeply increasing for the past 100 years. In the following section residual stress formation in two fusion welding techniques using plasma or electric arcs is considered in detail.

In fusion welding, residual stresses are caused by inhomogeneous temperature fields. In the weld seam, temperatures will be significantly higher than that in the parent material, thus, the weld seam during cooling shrinks, but, is hindered by the surrounding colder parent material (Figure 2.21, [37, 38]).

Residual stress distribution in welds of materials without phase transformation are, therefore, characterized in the longitudinal joint direction (i.e., along the weld line) by

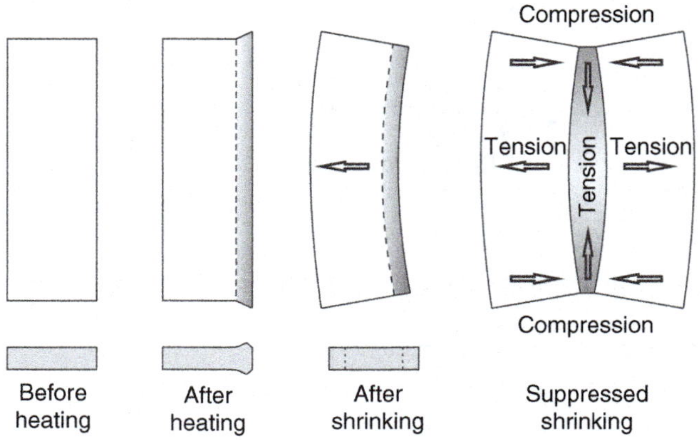

Figure 2.21 Residual stress formation in thin sheet fusion welds [37, 38].

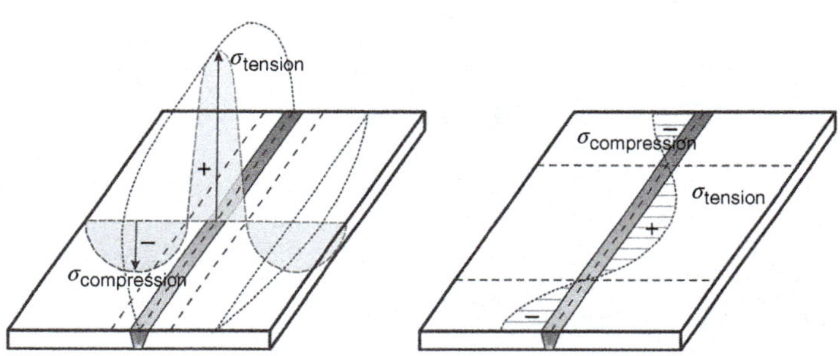

Figure 2.22 Residual stresses in thin sheet welds of a material without phase transformation.

tensile residual stresses in the weld seam and heat affected zone (HAZ) of the parent material (next to the weld seam) and balancing compressive residual stresses in the parent material (Figure 2.22, [38]).

In the case of thin sheets and single pass fusion welding, the welding residual stress state often is a plane stress state, whereas in the case of thick sheets or profiles the welding residual stress state usually is triaxial. With respect to the weld behavior under load, for example, fatigue loading, the biaxial, respectively, triaxial tensile residual stresses in the center of the weld are critical (Figure 2.22).

In the case of fusion welding of steel with austenite → martensite (or bainite) phase transformation, martensite forms in the weld seam and the HAZ. As a consequence, the volume in the weld seam and the HAZ increases and usually at room temperature compressive residual stresses are present in the weld seam (Figure 2.23) and the HAZ (except for materials with very high M_s temperature). These compressive residual stresses will be higher, the lower the phase transformation temperature of the material is, because the yield strength increases with increasing bainite/martensite volume fraction and decreasing temperature [39].

Besides martensitic phase transformation, other physical processes, such as recovery and recrystallization processes in the case of cold rolled sheets or precipitation aging

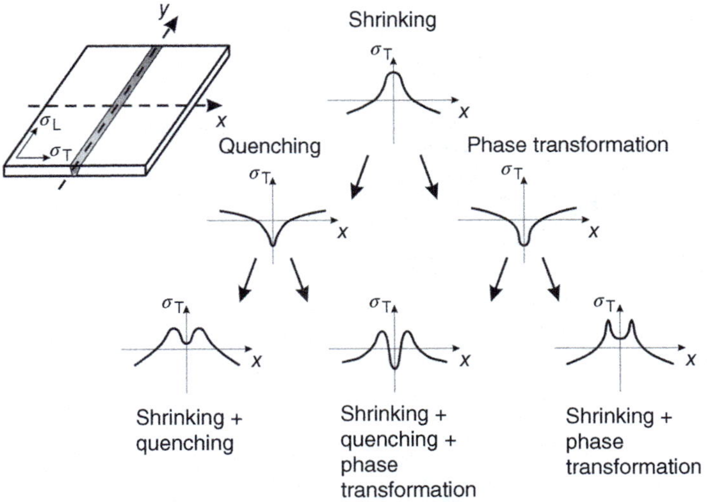

Figure 2.23 Residual stress formation in a thin sheet weld of a steel with phase transformation.

Figure 2.24 Relation between temperature distribution and physical processes in a steel weld with phase transformation.

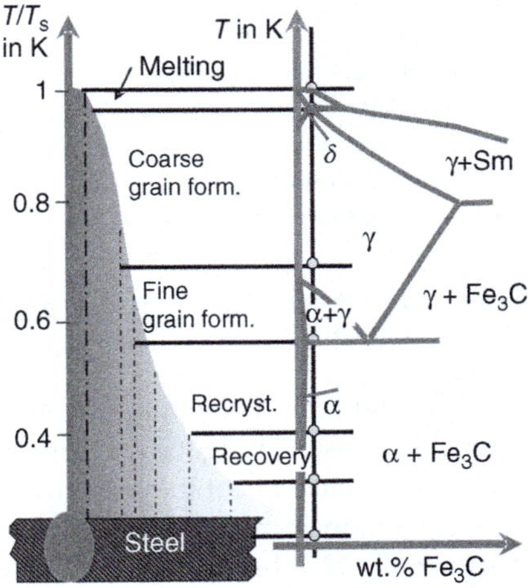

in the HAZ, affect the residual stress state and the mechanical properties of the weld (Figure 2.24).

The distribution and the magnitude of the welding residual stresses in both materials with and without phase transformation depend on the material (e.g., temperature-dependent thermal expansion coefficient and mechanical properties) and the process parameters (e.g., heat input, heat source, welding speed, weld geometry, environment temperature) of the welding process (Figure 2.25, [39]).

The restraint within the weld can be described by the ratio between the extension of the hot zone and the cold parent material and (in the case of materials with phase transformation) by the ratio between the transforming and the nontransforming zone [39]. Thus, comparing the residual stress distribution of a weld produced by conventional gas

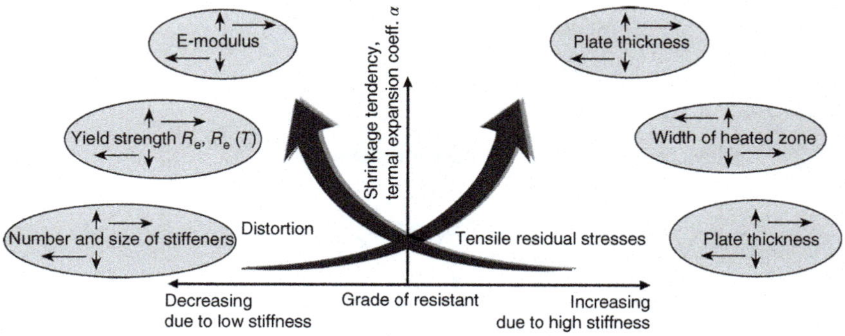

Figure 2.25 Relation between grade of restraint, shrinkage tendency, stiffness, and residual stresses in welds [39].

Constructive	Manufacturing
➡ **Minimum structure stiffness** - plate thickness as low as possible - no external restraint if possible - number of stiffeners as low as possible ➡ **Small number of weld seams** - utilization of high extension lengths - avoidance of long, parallel weld seams - avoidance of weld seam crossings - avoidance of weld seams in zones with high load stresses ➡ **Optimized welding sequence**	- Welding of parts of a structure considering symmetry aspects and flame straightening in the different phases - Heat input as small as possible - Elastic prestrain - Warm hammering of the weld seams - Heat sink welding of austenitic steels - No overmatching ➡ **Consequently follow the planned welding sequence** ➡ **Transverse welds have to be welded first** (welding sequence should consider possibilities of free or less hindered shrinkage)

Figure 2.26 Preventive possibilities for avoiding high tensile residual stresses and their consequences in welded structures [39].

welding and a weld of similar geometry produced by tungsten inert gas (TIG) welding, residual stresses of higher magnitude are expected in the TIG weld due to the higher and more focused heat input during TIG welding.

A reduction in magnitude of high-tensile residual stresses in welded joints can be performed either by preventive means (Figure 2.26, [39]) or by postweld heat treatments. The most effective method of reducing the magnitude of welding residual stresses is the reduction in stiffness of the construction [39]. A lower stiffness, however, may result in unacceptable distortions of the welded structure.

Most effective in reducing the magnitude of the residual stresses in a welded structure is a postweld heat treatment (stress relief annealing), which is more effective the higher the annealing temperature [39]. The annealing temperature, however, is limited by phase transformations or other significant microstructure changes, such as grain coarsening. In the case of dissimilar material welds, postweld heat treatments may even be detrimental due to the thermal expansion mismatch of the dissimilar parent materials [35].

Besides construction welds, repair welds are also introduced into components and structures by routine inspection either to remedy initial fabrication defects found in castings or welds or to rectify in-service degradation of components and thereby extend their lifetime [40]. Omitting postweld heat treatments, repair welds can play an important role in many subsequent component failures due to the high residual stresses associated with the repair process. For instance, in case of steam leak at a nonstress relieved pipe-work repair weld, both the magnitude and multiaxial nature of the residual stress field was instrumental in driving creep damage leading to reheat cracking [40, 41].

2.2.6 Residual Stress Formation in Coatings

Thin films and coatings are applied to increase, for example, the wear resistance [42], high temperature corrosion resistance [43], resistance against wet corrosion, and biocompatibility [44] of a substrate. Multilayer thin films are extensively used in the microelectronics industry because of their superior magnetic properties [45] and are increasingly used in X-ray diffraction [46].

The methods for applying coatings are manifold, including, for example, thermal spraying, thermally activated chemical vapor deposition (CVD), and plasma-assisted physical vapor deposition (PVD) techniques. In coatings produced by CVD, residual stresses form during cooling from deposition temperature due to the difference in thermal expansion of the coating and the substrate. In addition to these extrinsic residual stresses, PVD coatings undergo so-called intrinsic residual stresses due to the ion bombardment during the PVD process [47] (Figure 2.27, [42, 48–50]).

These incoming ions or knock-on atoms will remove atoms from their lattice positions (Figure 2.27). Collision cascades result in a rearrangement of the lattice atoms. The impinging particles may also be physically implanted and surface species may be recoil-implanted into the subsurface lattice. Residual interstitials and vacancies are formed, which can, in turn, lead to an increased density of extended defects and form one-, two- (e.g., grain and sub-grain boundaries, twins), or three-dimensional defects (e.g., voids, cracks or particles) [42].

The microstructure of PVD coatings depends strongly on deposition parameters. Thornton [51] developed a "zone model," which describes the dependence of the microstructure on argon partial pressure, respectively, temperature for thinner PVD deposited layers based on a correlation of the sputter particle direction, shadowing effects and surface, respectively, volume diffusion. As a consequence of the deposition process, the layers possess a microstructure gradient, the maximum of defect density usually is at the interface between substrate and coating and as a consequence residual macro- and microstress gradients are present (Figure 2.28, [52]).

The magnitude of compressive residual macrostresses in hard coatings can reach up to several gigapascal [26]. The steepness of the residual stress gradients within the coating depends on the process parameters. Often, steep residual stress gradients of up to several hundred megapascal per micrometer coating thickness are encountered (Figure 2.29). Residual microstresses in thin films also strongly depend on the deposition process, because it determines the defect density and distribution (e.g., Dieter). By annealing thin films relaxation processes, such as plastic flow, creep, and microcracking can substantially reduce the residual macro- and microstresses [53].

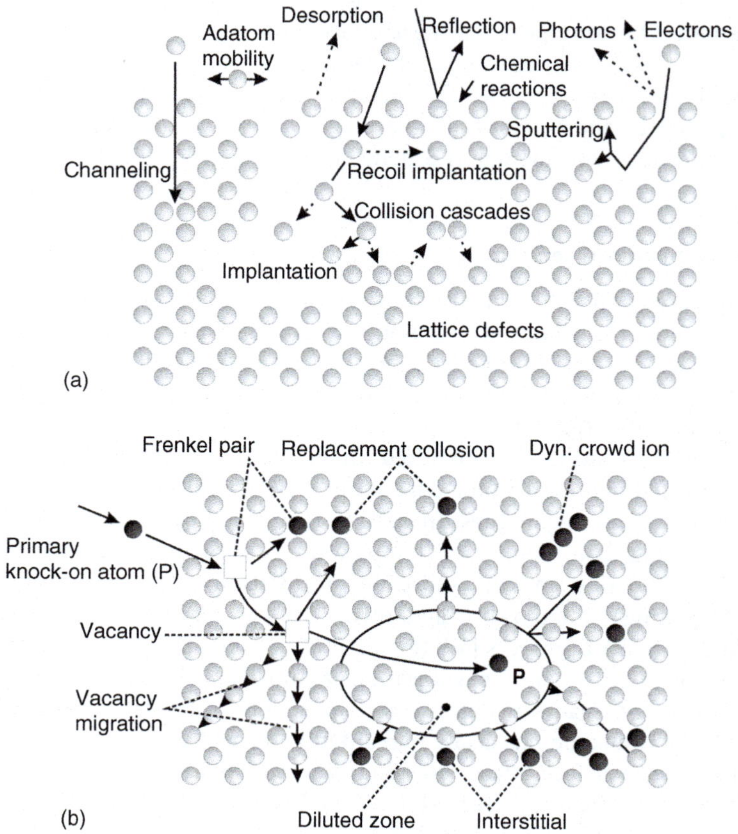

Figure 2.27 (a) Effects of an ion bombardment on a growing film [48, 49] and in the 10–1000 eV range commonly employed in thin film processing. (b) Schematic view of possible lattice defects created by an impinging energetic atom, primary knock-on atom (P) [42, 50].

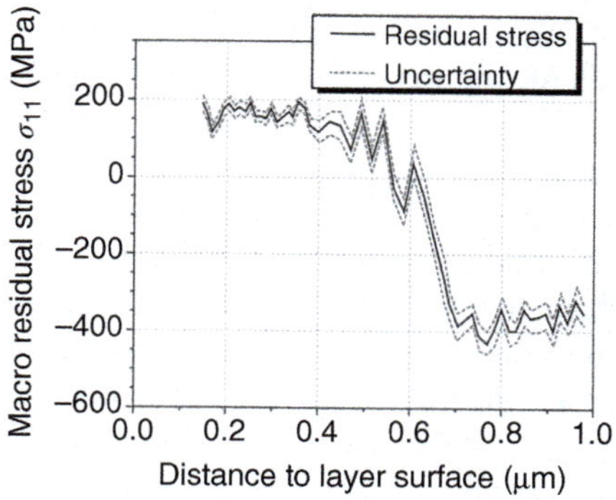

Figure 2.28 Residual macrostress distribution within a PVD sputtered CoFe single layer [52].

Figure 2.29 Dependence on macro-internal stress gradients in CoFe-single layers on sputter power and argon partial pressure [52].

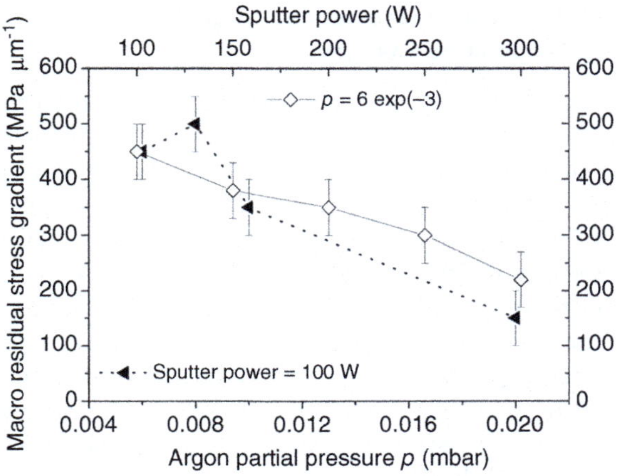

2.3 Relevance

The basis for assessing the influence of residual stresses on the strength and lifetime of a component is the superposition principle, which means that in a first approximation a linear superposition (addition) of load and residual stress distributions represent the total stress distribution within a workpiece or a component. Residual stresses, therefore, have influence on the strength, the fatigue behavior, corrosion resistance (in particular resistance against stress corrosion cracking), magnetic properties, and so on. Workpiece or component failure in most cases occurs due to the onset of plastic deformation or fracture when subjected to tensile loads. The near-surface area is critical, for example, because of notch effects. Therefore, usually compressive residual stresses in the near-surface zone are beneficial with respect to workpiece or component lifetime. Thus, in large cast or forged technical components, such as generator shafts and pressure vessels, terms of delivery often include specifications regarding maximum residual stress values and permissible residual stress distributions [54]. In component design, known residual stress distributions can be included in the design criteria (e.g., the effective stress), whereas unknown residual stress distributions are incorporated into lifetime predictions by sweeping assumptions, conservative application of statistics, or by an increase in safety factors [54, 55].

The relevance of residual stress distributions measured or calculated can only be assessed taking its context into account [54], for example, material, microstructure, heat treatment, mechanical properties (brittle, respectively, ductile state), surface roughness, loading condition, environment (corrosive media, high temperature, extreme low temperature, etc.), and the method used for measurement or calculation. For instance, in the case of a brittle steel workpiece, the residual stress state needs to be considered both regarding the onset of plastic deformation and fracture (at not much higher plastic strains). In contrast, in case of a ductile workpiece, residual stresses need to be considered with respect to the onset of plastic deformation, but, their influence on fracture is not pronounced [56].

In the following sections the relevance of residual stresses on workpiece and component behavior under external loading will be explained in some more detail using technically relevant examples.

2.3.1 Failure due to Residual Stress Formation or Residual Stress Relief Induced by Temperature Changes

Depending on the yield strength of a material stresses formed during cooling will either result in residual stresses or distortion after cooling (see Section 2.2). Once thermal stresses are higher than the yield strength, distortion induced by plastic deformation results in size and shape alterations and, thus, may reduce the component to scrap. Chatterjee-Fischer [11] compared the distortions of two 200 mm × 200 mm × 20 mm-sized sheets of a plain carbon steel (0.1% C) and an austenitic stainless steel X5CrNi18-8 after cooling from 920 °C in water (Figure 2.30). The sheet cross-section after cooling depicts well the tendency of the component to reach a spherical shape. Due to the higher yield strength of the stainless steel compared to the plain carbon steel, the distortions introduced are significantly smaller, thus, residual stresses in the X5CrNi18-8 steel sheet will be significantly higher than in the plain steel sheet.

In the case of materials with phase transformation, the residual stress state of long components, such as shafts or axles (see Section 2.2), formed after cooling usually consists of compressive residual stresses in the core and tensile residual stresses in the mantle. When the tensile residual stresses in the mantle exceed the strength of the material, cracks usually occur in longitudinal direction of shafts or axles [57].

In order to prevent crack formation, large surface-hardening tools are usually cooled in air before quenching, while through-hardening tools are instantly quenched. Thus, a further increase of tensile residual stresses in the near-surface zone due to the martensitic transformation can be avoided [57].

Failure due to stress formation may occur also during rapid heating. Due to the faster temperature increase in the mantle compared to the core of the cylindrical component, the mantle will try to expand faster than the core, but, its expansion is hindered by

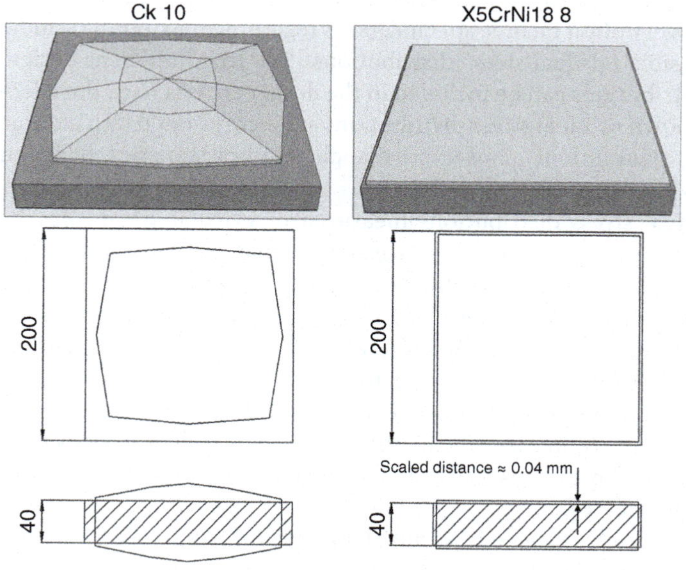

Figure 2.30 Distortions introduced by quenching of steel sheets [11].

the more rigid core. Thus, the mantle experiences compressive stresses, while tensile stresses evolve in the core of the component. Particularly in the case of large dies used in continuous casting, tensile stresses introduced by fast heating processes may cause crack formation in the core of dies, for example, in high alloyed steels with low thermal conductivity and low toughness [11].

2.3.2 Influence of Residual Stresses on Component Failure under Static and Dynamic Mechanical Loads

Compressive residual stresses can stop fatigue cracks. In a fatigue experiment using a notched specimen a defined compressive residual stress field was produced in front of the growing crack by a hardness indentation (Figure 2.31). During cyclic loading the crack growth rate initially increases with increasing crack length. As soon as the crack tip reaches the compressive residual stress field, the crack growth rate decreases and the crack is finally arrested [58].

A very simple but impressive example of the influence of residual stresses on fatigue lifetime gives the SN (Wöhler-) curves obtained on hardened and ground low-carbon (C45) steels (Figure 2.32). Tensile residual stresses introduced by the grinding process

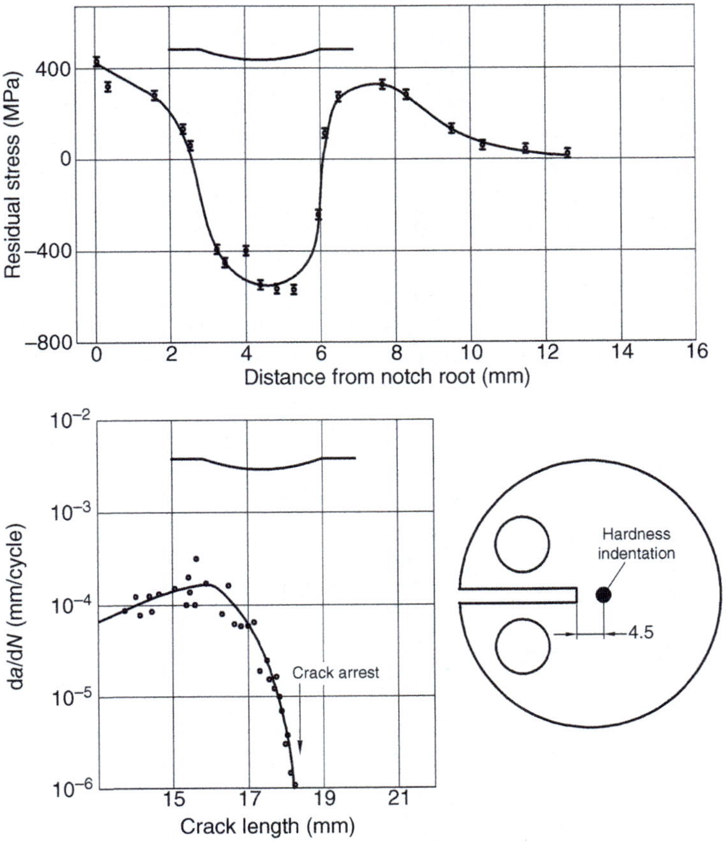

Figure 2.31 Interaction of crack growth and residual stress field.

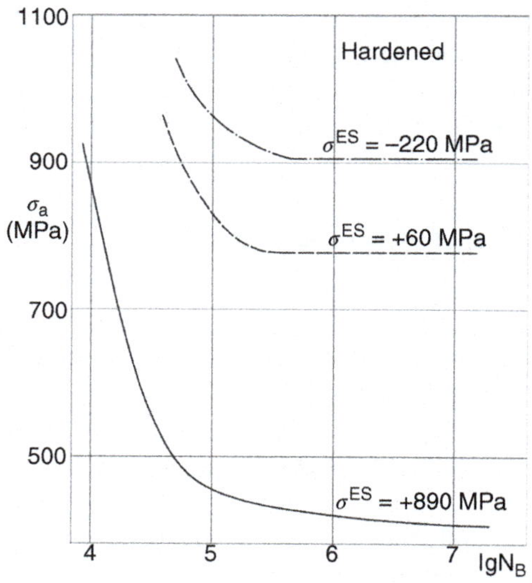

Figure 2.32 Influence of surface residual stresses on fatigue strength (Quelle), N_B is the number of loading cycles, σ_a is the stress amplitude.

yield a significantly lower fatigue strength and endurance limit compared to an almost stress-free sample, while compressive residual stresses in the near-surface zone increased both the fatigue strength and the endurance limit [58].

The highest benefit of compressive residual stresses appears for hardened steel, also quenched and tempered steel shows an increase of bending fatigue strength with increasing compressive residual stresses. In contrast, the bending fatigue strength of the same steel in normalized condition benefits only very slightly from compressive surface stresses due to its higher ductility. In the opposite case of tensile residual stresses, the hardened steel is the most susceptible to premature failure under fatigue loading (Figure 2.33).

Based on many of these experiments, Starker *et al.* [59] and Kloos and Braisch [60] proposed to consider the influence of residual stresses in the design of components exposed to fatigue loading by adding in the Haigh diagram; the residual stress value

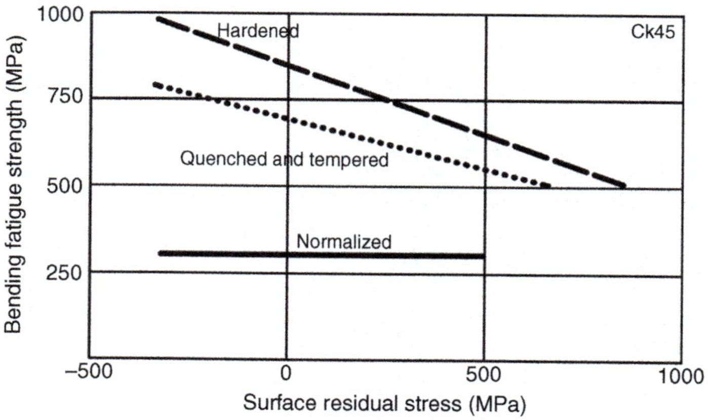

Figure 2.33 Influence of heat treatment and surface residual stresses on bending fatigue strength [31].

measured to the mean fatigue strength. The local failure condition for single step cyclically loaded components then is given by

$$\sigma_{eq,a} = R_{tf}\left[1 - \frac{(\sigma_{1,m} + \sigma_1^{RS}) + (\sigma_{2,m} + \sigma_2^{RS})}{R_m}\right] \tag{2.22}$$

$\sigma_{eq,a}$: equivalent stress amplitude according to the Von Mises failure criterion
$\sigma_{1,m}, \sigma_{2,m}$: principal mean stresses
$\sigma_1^{RS}, \sigma_2^{RS}$: principal residual stresses
R_{tf} : tension compression fatigue strength
R_m : tensile strength.

The increase in component lifetime due to the introduction of compressive residual stresses is one of the main motivations of mechanical surface treatments, such as deep-rolling and shot-peening.

2.3.3 Influence of Residual Stresses on Component Failure in Corrosive Environments

Although the influence of residual stresses on component failure in corrosive environments is in general affirmed, detailed and quantitative experimental investigations are still sparse. This is probably due to the interdependence of corrosion resistance, microstructure, and, for example, heat treatment and the residual stress state, which complicates the derivation of a simple relation between residual stresses and characteristic values that describe corrosion sensitivity.

For duplex stainless steels, recent investigations revealed that the local stress gradient was the key parameter in pit initiation and that the local average stress was the parameter governing the transition from metastable to stable pitting [61]. In the case of uniform corrosion of martensitic tool steels, tensile residual surface stresses introduced local maxima into the current-density potential curve [62]. In addition, electrochemical measurements for quantifying intergranular corrosion of an aluminum alloy revealed a significantly higher breakdown potential in zones with residual compressive stress [63].

Residual stresses, thus, can have a strong influence on corrosion processes; however, corrosion processes can also exert a strong influence on the residual stress state of a component. Amirat *et al.* [64] showed by experimental and numerical investigations the redistribution of residual stresses in a pipeline tube caused by corrosion and the effect of the residual stress redistribution on the reliability and lifetime of the tube.

2.3.4 Influence of Residual Stresses on Wear

Residual stresses can both directly and indirectly affect the wear rate of components (Figure 2.34).

Until recently consensus indicate that because of the very high contact stresses in wear processes, the effect of the residual stress state of the first and the second body (counterbody) in a tribo-system on the wear process appeared negligible. The magnitude of residual stresses experimentally determined on worn surfaces, for example, railway rails [65] and laser-hardened steels [62, 66, 67], however, are in the same order of magnitude as the stresses generated by Hertzian pressure and friction in wear processes (Figure 2.34).

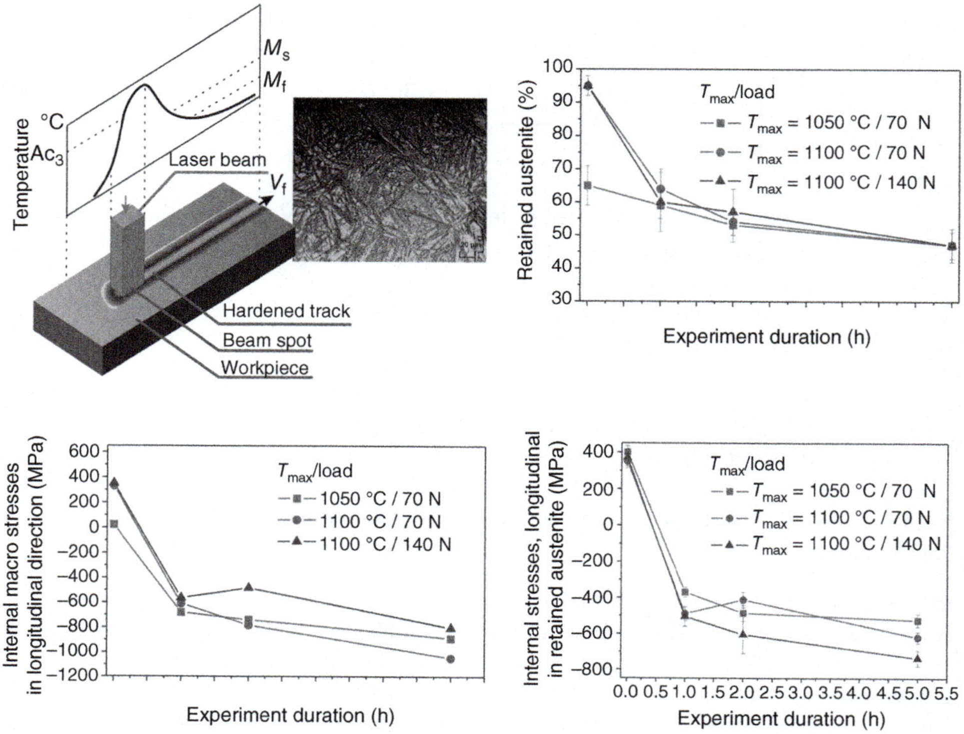

Figure 2.34 Development of retained austenite volume fraction and residual stresses during wear of a laser surface-hardened martensitic steel.

These experimental results indicate that in some tribo-systems significant interaction between residual stresses and wear rate exists. This will be explained in more detail on the example of a nitrogen alloyed tool steel X30CrMoV15-1+0.3%N that was laser surface hardened in order to increase its wear resistance, while maintaining corrosion resistance. The amount of martensite and retained austenite generated in the near-surface region depends on the maximum temperature reached during the laser-hardening process and the heating ratio. High maximum surface temperatures result in high retained austenite volume fractions in the near-surface zone due to nitride dissolution in the steel matrix. During the wear test, the surface-hardened zone is subjected to Hertzian pressure and friction, which lead to an expansion of the near-surface layer. Thus, compressive residual stresses develop during unloading (similar to the mechanism of residual stress formation discusses for deep-rolling). In addition, compressive residual stresses form due to the transformation of part of the retained austenite into martensite. As a consequence, the near-surface zone contains compressive residual macrostresses and the phase-specific residual stresses in the retained austenite are also compressive. The high compressive macro- and microstresses stabilize the retained austenite, because its transformation into martensite would generate additional compressive residual stresses due to the volume expansion.

An indirect influence of residual stresses on wear rate occurs due to their interaction with the damage induced by the wear process. James *et al.* [55] present an example of a cutter head gear box casing, where a fatigue crack is initiated at local fretting damage and preferentially follows the local residual stress field.

References

1 Gross, D. (1992) *Bruchmechanik 1*, Springer, Berlin.

2 Withers, P.J. and Badeshia, H.D.K.H. (2001) *Mater. Sci. Technol.*, **17**, 355–365.

3 Eschenauer, H. and Schnell, W. (1986) *Elastizitätstheorie I*, BI Wissenschaftsverlag, Zürich.

4 Pyzalla, A. and Reimers, W. (2003) in *Analysis of Residual Stress by Diffraction Using Neutron and Synchrotron Radiation* (eds M. Fitzpatrick and A. Lodini), Taylor & Francis, London, pp. 219–232.

5 Macherauch, E., Wohlfahrt, H., and Wolfstieg, U. (1973) *HTM, Haerterei Tech. Mit.*, **28**, 201–211.

6 Behnken, H. (1997) in *Structural and Residual Stress Analysis by Nondestructive Methods* (ed. V. Hauk), Elsevier, Amsterdam, pp. 39–65.

7 Noyan, I.C. and Cohen, J.B. (1987) *Residual Stress–Measurement by Diffraction and Interpretation*, Springer, New York.

8 Pyzalla, A. and Reimers, W. (1999) *Textures Microstruct.*, **33**, 291–301.

9 Prime, M.B., Gnäupel-Herold, T., Baumann, J.A., Lederich, R.J., Bowden, D.M., and Sebring, R.J. (2006) *Acta Mater.*, **54**, 4013–4021.

10 Helle, A.S., Easterling, K.E., and Ashby, M.F. (1985) *Acta Metall.*, **33**, 2163–2174.

11 Chatterjee-Fischer, R. (1973) *HTM, Haerterei Tech. Mit.*, **28**, 276–288.

12 Buchholtz, H. and Bühler, H. (1933) *Arch. Eisenhüttenwesen*, **6**, 335–340.

13 Lement, B. (1959) American Society for Metals, in *Distortion in Tool Steels*, Metals Park, OH.

14 Rose, A. (1966) *HTM, Haerterei Tech. Mit.*, **21**, 1–6.

15 Koistinen, D.P. (1958) *Trans. Am. Soc. Met.*, **50**, 227–241.

16 Glocker, R. and Hasenmaier, H. (1949) *Z. Metallkunde*, **40**, 182–186.

17 Koch, M. (1973) *Z Werkstofftech.*, **4**, 81–85.

18 da Rocha, A. and Hirsch, T. (2005) *Mater. Sci. Eng., A*, **395** (**1–2**), 195–207.

19 Roettger, K., Wilcke, G., and Mader, S. (2005) *Materialwiss. Werkstofftech.*, **36** (**6**), 270–274.

20 Wagner, L. (1999) *Mater. Sci. Eng., A*, **263** (**2**), 210–216.

21 Altenberger, I., Noster, U., Boyce, B.L., Peters, J.O., Scholtes, B., and Ritchie, R.O. (2002) *Mater. Sci. Forum*, **404–407**, 457–462.

22 Zinn, W. and Scholtes, B. (1999) *J. Mater. Eng. Perform.*, **8** (**2**), 145–151.

23 Mayer, H.M., Achmus, C., Pyzalla, A., and Reimers, W. (2000) *Mater. Sci. Forum*, **347**, 340–345.

24 Scholtes, B. (1991) *Eigenspannungen in mechanisch randschicht-verformten Werkstoffzuständen*, DGM Informationsgesellschaft Verlag, Oberurseland Habilitationsschrift, TU Karlsruhe (1990).

25 Tekkaya, A.E. (1986) Ermittlung von Eigenspannungen in der Kaltmassivumformung. Doctoral Thesis. Universität Stuttgart.
26 Genzel, C., Reimers, W., Malek, R., and Pöhlandt, K. (1996) *Mater. Sci. Eng., A*, **205**, 79.
27 Pyzalla, A., Reimers, W., and Pöhlandt, K. (1997) Residual stresses and texture evolution in cold extrusion of full and hollow steel bodies, in *Proceedings of the ICRS 5, Linkoeping, Schweden, June 16–18, 1997* (eds T. Ericsson, M. Odén, and A. Andersson), Institute of Technology, Linköping University, pp. 58–63.
28 Pang, J.W.L., Holden, T.M., Wright, J.S., and Mason, T.E. (2000) *Acta Mater.*, **48**, 1131–1140.
29 Korsunsky, A.M., James, K.E., and Daymond, M.R. (2004) *Eng. Fract. Mech.*, **71**, 805–812.
30 Martins, R.V., Lienert, U., Margulies, L., and Pyzalla, A. (2005) *Mater. Sci. Eng., A*, **402**, 278–287.
31 Scholtes, B. (1987) *Residual Stresses Technology, Applications, Effects*, Advances in Surface Treatments, vol. **4**, Pergamon Press, pp. 59–71.
32 Immelmann, S., Welle, E., and Reimers, W. (1997) *Mater. Sci. Eng., A*, **238**, 287–292.
33 Eigenmann, B., Scholtes, B., and Macherauch, E. (1990) *Materialwiss. Werkstofftech.*, **21**, 257–265.
34 Ruppersberg, H. and Detemple, I. (1993) *Mater. Sci. Eng., A*, **161**, 41–44.
35 Balart, M.J., Bouzina, A., Edwards, L., and Fitzpatrick, M.E. (2004) *Mater. Sci. Eng., A*, **367**, 132–142.
36 Brinksmeier, E., Schneider, E., Theiner, W.A., and Tönshoff, H.K. (1984) *CIRP Ann.*, **33** (**2**), 489–509.
37 Malisius, R. (1977) *Schrumpfungen, Spannungen und Risse beim Schweißen*, DVS-Verlag, Düsseldorf.
38 Boerse, U., Werner, D., and Wirtz, H. (1995) *Das Verhalten der Stähle beim Schweißen*, Tl.1,2 Grundlagen, DVS-Verlag, Duisburg, p. 2000.
39 Nitschke-Pagel, T. and Wohlfahrt, H. (2002) *Mater. Sci. Forum*, **404–407**, 215–224.
40 Bouchard, P.J., George, D., Santisteban, J.R., Bruno, G., Dutta, M., Edwards, L., Kingston, E., and Smith, D.J. (2005) *Int. J. Press. Vessels Pip.*, **82** (**4**), 299–310.
41 Dunn, J., MacGuigan, J., McLean, R.L., Miles, L., and Stevens, R.A. (1998) Investigation and repair of a leak at a high temperature stainless steel butt weld. Proceedings of International Conference on Integrity of High Temperature Welds, Nottingham, pp. 241–258.
42 Mayrhofer, P.H., Mitterer, C., Hultman, L., and Clemens, H. (2006) *Prog. Mater Sci.*, **51**, 1032–1114.
43 Kuroda, S. and Clyne, T.W. (1991) *Thin Solid Films*, **200** (**1**), 49–66.
44 Tadano, S., Todoh, M., Shibano, J.-I., and Ukai, T. (1997) *JSME Int. J., Ser. A: Mech. Mater. Eng.*, **40** (**3**), 328–335.
45 Grünberg, P. and Brodsky, M.B. (1986) *Phys. Rev. Lett.*, **57** (**19**), 2442–2445.
46 Schuster, M., Goebel, H., Bruegemann, L., Bahr, D., Burgaezy, F., Michaelsen, C., Stoermer, M., Ricardo, P., Dietsch, R., Holz, T., and Maid, H. (1999) *Proc. SPIE Int. Soc. Opt. Eng.*, **3767**, 183–193.
47 Perry, A.J., Sue, J.A., and Martin, P.J. (1996) Practical measurement of the residual stress in coatings. Surface and Coatings Technology, First Australian-USA Workshop

on Critical Issues in High Performance Wear Resistant Films, vol. **81**, Issue **1**, pp. 17–28.

48 Mattox, D.M. (1989) *J. Vac. Sci. Technol., A*, **7** (**3**), 1105–1114.

49 Ensinger, W. (1997) *Nucl. Instrum. Methods Phys. Res., Sect. B*, **127–128**, 796–808.

50 Cahn, R.W. and Haasen, P. (1996) *Physical Metallurgy*, 4th edn, University Press, Cambridge.

51 Thornton, J.A. (1986) *J. Vac. Sci. Technol., A*, **4** (**6**), 3059–3065.

52 Dieter, S., Pyzalla, A., Bauer, A., Schell, N., McCord, J., Seemann, K., Wanderka, N., and Reimers, W. (2004) *Z. Metallkde*, **95**, 163–175.

53 Withers, P.J. and Badeshia, H.D.K.H. (2001) *Mater. Sci. Technol.*, **17**, 366–375.

54 Hauk, V. (1983) Eigenspannungen, ihre Bedeutung für Wissenschaft und Technik, in *Eigenspannungen–Entstehung-Messung-Bewertung-* (eds E. Macherauch and V. Hauk), Deutsche Gesellschaft für Metallkunde e.V, Oberursel, pp. 9–48.

55 James, M.N., Hughes, D.J., Chen, Z., Lombard, H., Hattingh, D.G., Asquith, D., Yates, J.R., and Webster, P.J. (2004) *Eng. Fail. Anal.*, **14**, 384–395.

56 Macherauch, E. and Kloos, K.H. (1982) *HTM Beiheft Last- und Eigenspannungen*, Carl Hanser Verlag, München, Wien, pp. 175–190.

57 Bühler, H. and Scheil, E. (1933) *Arch. Eisenhüttenwesen*, **6**, 283–288.

58 Macherauch, E. and Kloos, K.H. (1986) Origin, measurement and evaluation of residual stresses, in *Eigenspannungen, Entstehung, Messung, Bewertung* (eds E. Macherauch and V. Hauk), Deutsche Gesellschaft für Metallkunde e.V, Oberursel, pp. 3–26.

59 Starker, P., Wohlfahrt, H., and Macherauch, E. (1980) *Arch. Eisenhüttenwesen*, **51**, 439–445.

60 Kloos, K.H. and Braisch, P.K. (1982) *HTM*, **37**, 83–86.

61 Vignal, V., Mary, N., Oltra, R., and Peultier, J. (2006) *J. Electrochem. Soc.*, **153**, B352–B357.

62 Pyzalla, A., Bohne, C., Heitkemper, M., and Fischer, A. (2001) *Mater. Corros.*, **52**, 99–105.

63 Liu, X. and Frankel, G.S. (2006) *Corros. Sci.*, **48**, 3309–3329.

64 Amirat, A., Mohamed-Chateauneuf, A., and Chaoui, K. (2006) *Int. J. Press. Vessels Pip.*, **83**, 107–117.

65 Wild, E., Wang, L., Hasse, B., Wroblewski, T., Goerigk, G., and Pyzalla, A. (2003) *Wear*, **254**, 876–883.

66 Heitkemper, M., Fischer, A., Bohne, C., and Pyzalla, A. (2001) *Wear*, **250**, 477–484.

67 Heitkemper, M., Bohne, C., Pyzalla, A., and Fischer, A. (2003) *Int. J. Fatigue*, **25**, 101–106.

3

Textures in Engineering Materials

Heinz G. Brokmeier and Sangbong Yi

3.1 Introduction

In the past decades, there have been significant advances in texture measuring techniques. These originate not only from the technical progress, for example, instrumentation, computation, and detector technologies, but also from the increasing requests for understanding the texture evolution, which appears unavoidably during thermo-mechanical processes of almost all polycrystalline materials. A textured material in which the composed crystallites align in a preferred orientation is anisotropic regarding their physical and mechanical properties. This means in turn that a material property can be designed by a proper control of the crystallographic texture. Furthermore, the metal-physical mechanisms of thermo-mechanical treatments, such as deformation, recrystallization, and phase transition, can be studied more effectively with the aid of texture analysis, since those processes are orientation-dependent and the resulting texture is strongly influenced by the activated mechanisms.

The crystallographic texture describes the orientation distribution of all crystallites in a polycrystalline sample with respect to a sample orientation. Together with the microstructure, it is one of the basic parameters to characterize polycrystalline materials. General introductions to texture analysis can be found in the textbooks by Wassermann and Grewen [1], Bunge [2], Kocks *et al.* [3], Bunge and Esling [4], and Engler and Randle [5]. Texture analysis is a statistical method, so that a sufficient number of crystallites, given by their crystal coordinate system K_A (Figure 3.1), are needed. Firstly, the orientation distribution has to be related to a sample coordinate system K_B using the orientation g of all grains. The sample coordinate system is determined by the deformation mode or by the shape of the sample. In Figure 3.1 the sample coordinate system is given for a rolling process (RD – rolling direction, TD – transverse direction, and ND – normal direction). Similar to the rolling geometry, geologists define a sample coordinate system by the foliation (equivalent to the rolling plane) and the lineation (equivalent to RD). The resulting orientation distribution function (ODF) $f(g)$ is given either by the volume fraction $\mathrm{d}V$ of grains or by the number of grains in an orientation increment $\mathrm{d}g$ (Eq. 3.1). For a detailed description of different ways to define the rotation g by Euler angles (φ_1, Φ, φ_2), by $(hkl)[uvw]$-indices, or by the orientation matrix, see Bunge [2].

Neutrons and Synchrotron Radiation in Engineering Materials Science: From Fundamentals to Applications, Second Edition. Edited by Peter Staron, Andreas Schreyer, Helmut Clemens, and Svea Mayer.

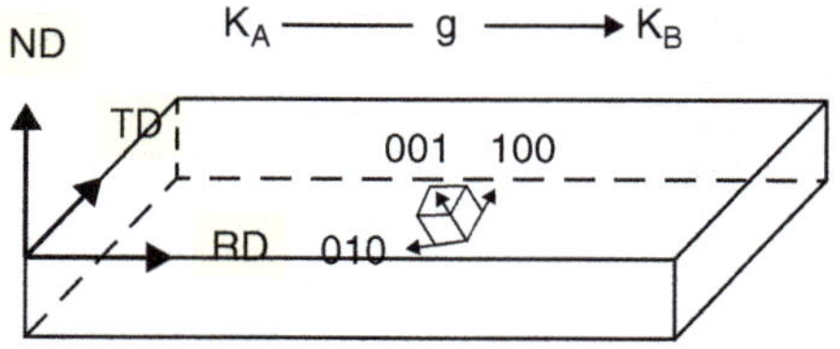

Figure 3.1 Definition of the orientation g of an individual grain. K_A is the coordinate system of a crystallite, K_B is the sample coordinate system with the labeled directions RD (rolling direction), TD (transverse direction), and ND (normal direction).

$$f^i(g) = \frac{\mathrm{d}V^i/V^i}{\mathrm{d}g} \quad f^i(g) = \frac{\mathrm{d}N^i/N^i}{\mathrm{d}g} \tag{3.1}$$

Secondly, the orientation distribution may depend on the location of the volume element inside a compact sample, so that a texture gradient is observed. Typical examples are more or less all semi-finished products, pipes, and severe plastic deformed materials. Consequently, the ODF has to be analyzed not only in relation to the sample coordinate system but also to the position x, y, z in the sample, which is called *orientation stereology* [6] or *texture mapping* [7].

Thirdly, the crystallographic texture documents the present state of the material. Any thermomechanical treatment or aging can change the texture, so that the texture dynamically develops during the corresponding treatment as a function of time and process parameters. For a detailed understanding of such processes it is of great importance to trace the texture changes during the process. Therefore, recently developed methods for *in situ* experiments [8–10] need fast scanning capabilities that can adapt to the reaction speed of, for example, recrystallization or loading.

Fourthly, also the orientation stereology is a function of the external conditions acting on the material that can change with time, see for instance nucleation processes. However, automatic scanning of crystallite orientations over a sufficient number of individual grains requires a certain amount of time and, thus, the use of this method for the investigation of time-dependent processes where, for example, the temperature changes with time, is limited. In Table 3.1, a summary of multi-dimensional texture analyses with up to nine dimensions is given.

Due to the crystallographic structure, material properties of single crystals are anisotropic, so that the crystallographic texture also describes the anisotropy of the material properties of a polycrystalline product. Consequently, the anisotropy in the

Table 3.1 Summary of different texture definitions.

Dimension	Texture presentation
Two-dimensional texture	Pole figure, inverse pole figure
Three-dimensional texture	Orientation distribution function (ODF)
Six-dimensional texture	Orientation stereology (crystal orientation as function of x, y, z)
Six-dimensional texture	Orientation distribution function (ODF) as function of temperature, time, and mechanical loading
Nine-dimensional texture	Orientation stereology as function of temperature, time, and pressure

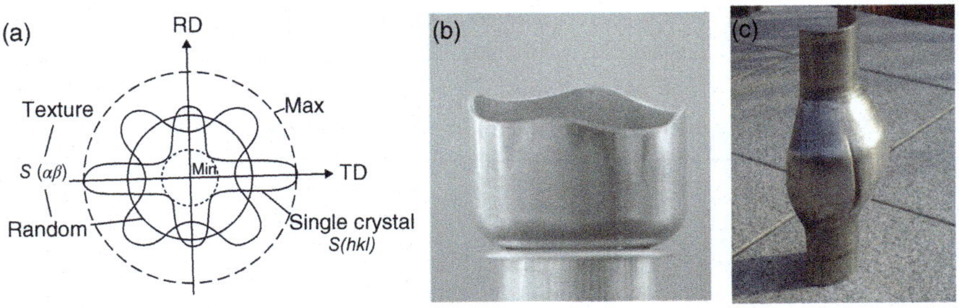

Figure 3.2 (a) Single crystal and polycrystal anisotropy [4]; (b) deep drawn cup made of Al showing the earing effect; and (c) failure after hydroforming due to exceeded deep drawing limit [11].

material properties of a textured polycrystalline material (like, e.g., Young's modulus, electric conductivity, thermal expansion, plastic anisotropy, or magnetic behavior) is between the single crystal anisotropy and the isotropic behavior of a texture-free powder (Figure 3.2a). Thus, the knowledge of texture is essential for understanding the properties of technical polycrystalline materials. Here, one has to distinguish on the one hand the texture type, which is the qualitative information, and on the other hand, the texture sharpness, which gives also quantitative information. A well-known example of the influence of crystallographic texture on the deformation behavior is shown in Figure 3.2b, where deep drawing produces an uneven rim of an Al cup showing the appearance of four ears. The visible failure in Figure 3.2c results from anisotropic flow during hydroforming caused by texture anisotropy around the circumference.

In the past decades, there have been significant advances in measuring techniques for texture analysis. These originate not only from technical progresses in instrumentation, computer, and detector technologies, but also from the increasing request for understanding the mechanism of texture evolution, which appears unavoidably in thermo-mechanical processing of almost all polycrystalline materials such as metals, ceramics, or rocks. The need for efficient use of materials in industrial processes increased the interest in quantitative texture analysis as an effective tool for understanding materials behavior. In this sense, the directional anisotropy of physical and mechanical properties of textured materials is the driving force for advances in quantitative texture analysis. These advances go hand in hand with progress in numerical methods and software development [12–14] based, for example, on the series expansion method [2, 15] or the WIMV algorithms [16] for the calculation of ODF. Current topics in texture analysis are, for example, prediction of anisotropic behavior, optimization of textures for industrial applications, comparison of texture, and residual strain mapping as well as new aspects in texture research, like modeling and simulation of polycrystalline plasticity.

In this chapter, the basics of experimental texture research that nowadays are widely used will be introduced and compared to each other. By comparing each method, the necessity for large-scale facilities in texture analysis and the importance of selecting the adequate equipment for user demands will be emphasized. An important issue is the penetration depth because it largely determines the experimental technique to be used for a specific problem. In Table 3.2, penetration depths of different types of radiation are collected.

Table 3.2 Penetration power of different radiations given as half value thickness $I/I_0 = 50\%$.

Metal	X-ray CuKα 1.54 Å (cm)	High-energy X-ray 0.124 Å (100 keV) (cm)	High-energy X-ray 0.062 Å (200 keV) (cm)	Neutrons 1.00 Å (cm)
Mg	0.0140	3.40	4.20	6.10
Al	0.0053	1.50	2.70	7.67
Cu	0.0015	0.20	0.50	0.85
Ti	0.0011	0.60	1.30	1.61
Pb	0.0003	0.03	0.04	2.10

3.2 Measurement of Preferred Orientations

Regarding the microstructure of a polycrystalline material, one can apply two different groups of techniques to determine preferred orientation. The first group of techniques is summarized as single crystal methods:

- Optical microscopy using the universal stage method; for details of the arrangement see [17]
- Micro-Laue technique using an energy-dispersive X-ray beam
- Orientation imaging method (OIM) by electron backscatter diffraction (EBSD)
- 3DXRD (three-dimensional X-ray diffraction) at the synchrotron (see Chapter 17).

The beam is optimized to focus on individual grains to get the orientation g directly in Euler coordinates φ_1, Φ, φ_2, see Figure 3.3a. Depending on the grain size of the material in interest, a measuring method is selected from the four methods listed above. One has to analyze the sufficient number of crystals measured one after the other (Figure 3.3b), which is performed for all techniques by automatic scanning routines. In the case of sufficient statistics the true ODF can be calculated (Figure 3.3c) in such a way that a quantitative description of the texture is available. The standard is the EBSD-method using well-indexed Kikuchi-patterns to obtain the grain orientation, see more details in [5, 6].

Single grain methods have the great advantage in that one obtains simultaneously the ODF, the misorientation distribution function, the grain size distribution, as well as the qualitative and quantitative phase composition. With EBSD and 3DXRD [7] one can

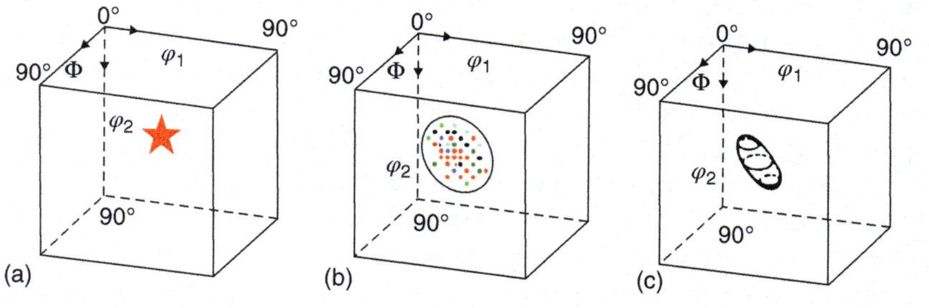

Figure 3.3 Orientation of (a) one single crystal, (b) a number of crystals, and (c) the calculated orientation distribution function (ODF).

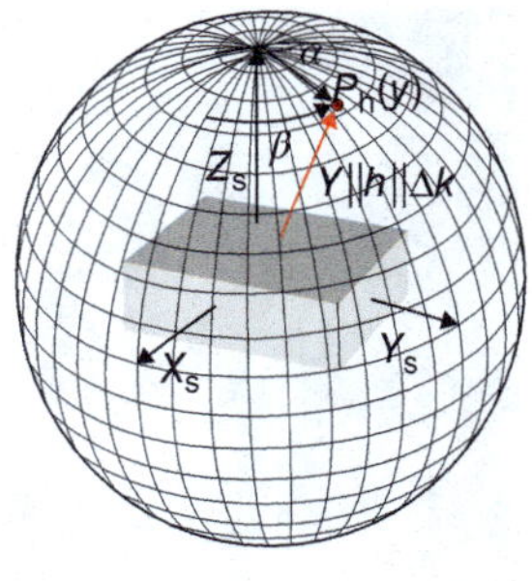

Figure 3.4 Polar sphere where one sample orientation $P_h(y)$ is indicated.

obtain the three-dimensional microstructure. The main disadvantage is that comparably small sample volumes are investigated causing sometimes problems to obtain the true ODF. Besides, nondestructive measurements are impossible.

The alternatives are the so-called volume methods, which are based on comparably large beams so that many crystals are measured simultaneously. Similarly to single crystal methods one needs a sufficient number of crystals for statistical reasons:

- XTA, X-ray texture analysis (working horse)
- NTA, Neutron texture analysis (large volume, multi-phase systems)
- STA, Synchrotron texture analysis (fast, small volumes, mapping).

The main difference between these methods is the applied penetration power. For many materials neutrons have the highest penetration power (Table 3.2), which makes them ideal for bulk investigations. Laboratory X-rays with about 9 keV have the lowest penetration power used for surface textures. Synchrotron offers a wide range of energies with much higher photon flux, which makes them ideal for fast *in situ* investigations as well as for texture gradient studies.

A direct determination of the orientation g and the ODF is impossible by using the volume methods listed above. The information of the crystal orientation of a polycrystalline sample is related to the integral intensity of the scattered beam I (α,β) with respect to the sample position α and β. The point $P_h(y)$ on the polar sphere, see Figure 3.4, shows the information of one sample orientation α, β.

The higher the intensity I (α,β) the more crystallites fulfill the diffraction condition at the corresponding sample position. Intensity variation for different α and β indicates the presence of preferred orientation. In case of no variation of the diffraction intensity, the sample has a random orientation distribution. In practice, orientation determination using volume methods means to collect integral intensities I (α,β) at a sufficient number of sample orientations α and β. The sample manipulation to move α from 0° to 90° and β from 0° to 360° can be carried out by an Eulerian cradle or a robot (see Figure 3.5) and is called *pole figure measurement* [18]. For a large sample, texture mapping using an Euler cradle is strongly limited due to spatial and weight restriction.

The result is the intensity distribution I (α, β) of a lattice plane (hkl) on a polar sphere as shown in Figure 3.6 for (110) plane of Fe-2.6%Si.

3.3 Presentation of Preferred Orientations

There are several options for the graphical representation of textures depending on the information one wants to present.

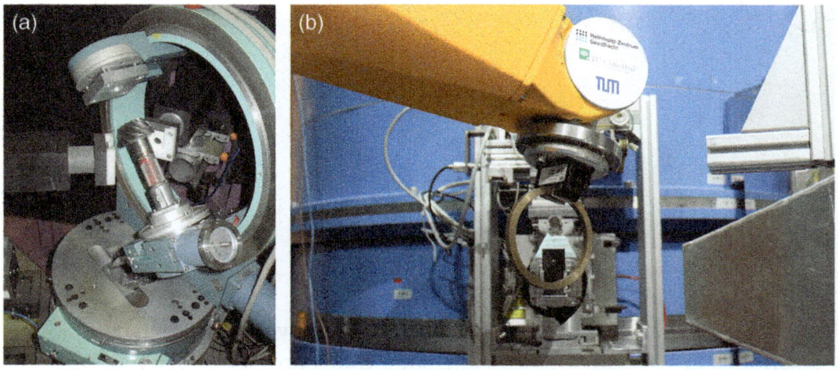

Figure 3.5 Eulerian cradle (a) and robot (b) to collect the intensities on the polar sphere.

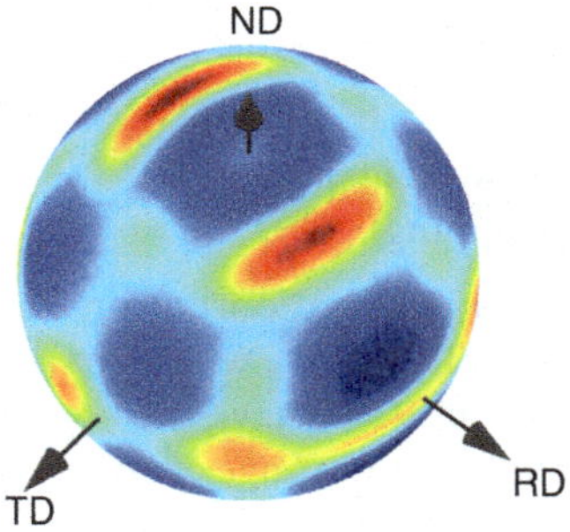

Figure 3.6 Polar sphere intensity distribution (Fe-110) of rolled Fe-2.6%Si.

- pole figures, measured and recalculated from ODF
- complete ODF
- selected information from complete ODF (sections or orientation bands)
- inverse pole figures (IPFs)
- graphical representation of materials properties calculated from single crystal properties and the ODF-based preferred orientations.

3.3.1 Pole Figure

A pole figure (*hkl*) is a projection of the intensity distribution of a polar sphere and one option of a two-dimensional projection of the three-dimensional ODF. In materials science often the stereographic projection, a kind of equal angular projection, is used, which is shown in Figure 3.7a. A direction is defined by a line from the center of the polar sphere to a point on the surface of the polar sphere (A). The direction can be, for example, the normal of a set of lattice planes. When the point A is observed from point P, the point of intersection B marks the stereographic projection of direction $\overline{OA}$ on a plane. Directions determined in this way can be drawn into a polar grid (Figure 3.7b). When the intensity distribution gained from the measurement of a lattice reflection with indices (*hkl*), where the sample is rotated around two axes (angles α and β), is drawn into such a polar grid using contour lines, it is called a *pole figure*. The diffracted intensity

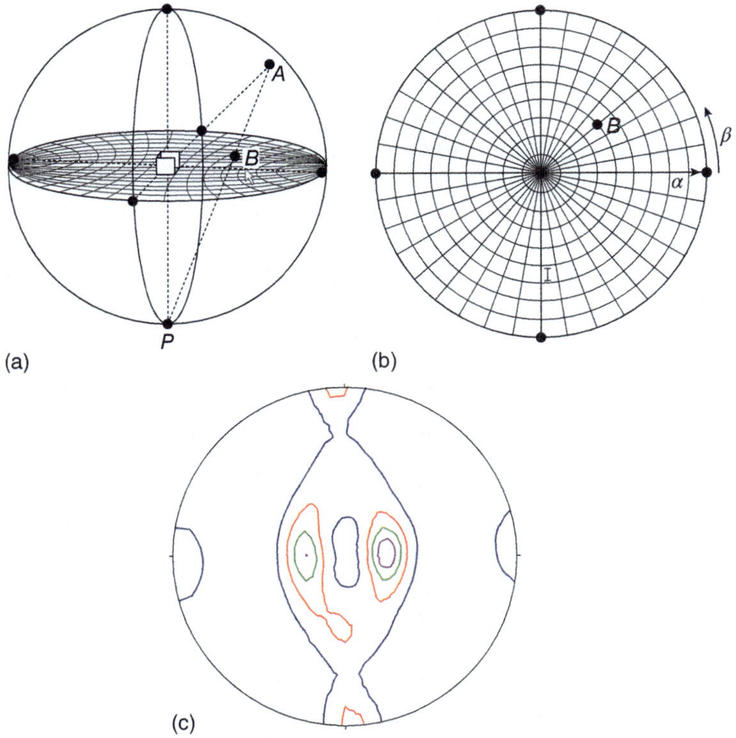

Figure 3.7 (a) Stereographic projection (see text), (b) polar grid in which pole figures are drawn, and (c) pole figure plot of Ti (0002) using 1368 $I_{(0002)}(\alpha, \beta)$ values ($\Delta\alpha = 5°$, $\Delta\beta = 5°$).

is proportional to the fraction of crystallites within the tested volume oriented in the specific direction. The pole figure is drawn as contour plot by connecting point (α, β) with same intensities, as shown in Figure 3.7c.

For comparing different pole figures, the intensities have to be normalized because scattered intensities depend on lattice plane reflectivity, gauge volumes, experimental conditions, and different methods (X-ray, high-energy X-ray, neutrons). Theoretically, normalization requires the intensity of a random sample of identical material (Eq. 3.2). Practically, one uses the integration over the whole polar sphere (Eqs. 3.3 and 3.4). The result is the orientation $P_{hkl}(\alpha,\beta)$ as multiples of the random distribution (mrd).

$$P_{hkl}(\alpha, \beta) = I_{hkl}(\alpha, \beta)/I_{hkl}\,\text{random} \tag{3.2}$$

$$P_{hkl}(\alpha, \beta) = I_{hkl}(\alpha, \beta)/N_{hkl} \tag{3.3}$$

$$N_{hkl} = 1/2\pi \int I_{hkl}(\alpha, \beta) \sin\alpha \, d\alpha \, d\beta \tag{3.4}$$

In the case of incomplete pole figures, the ODF is needed to normalize measured intensities.

The main use of pole figures is to interpret qualitative textures giving the texture type.

3.3.2 Orientation Distribution Function

Since the diffracted intensity does not depend on the rotation angle around the scattering vector, two or more pole figures have to be determined for a calculation of the ODF, where the required number of pole figures and suitable choices of (*hkl*) depend on the crystal symmetry, the sample symmetry, and the sharpness of texture. Three angles are required for defining the orientation of a crystallite in relation to the sample axes, where different sets of angles are possible. A common choice is the set of Euler angles φ_1, ϕ, and φ_2 for the consecutive rotations around z-axis, x-axis, and again z-axis [2]. Mathematically, texture can be described by the ODF $f(g)$ with the orientation $g(\varphi_1, \phi, \varphi_2)$, see an ODF-plot in Figure 3.8a. Each point in the orientation space is described on the one hand by the Euler coordinates φ_1, ϕ, φ_2 and on the other hand by ideal crystal orientations (*hkl*)//rolling plane and [*uvw*]//RD. For practical reasons, the ODF is usually plotted as φ_1 or φ_2 sections in steps of 5° (e.g., every 5° in φ_2 with φ_1 from 0° → 90° and, $\phi = 0° \rightarrow 90°$, Figure 3.8b). Details of the ODF and its representation can be found in [2, 14, 16, 20–22].

The $\varphi_2 = 45°$ section is selected for cubic materials, revealing the main information of the orientation distribution as shown in Figure 3.9a. Another option is to concentrate on an ODF-band as shown in Figure 3.9b. ODF bands are defined as lines in the orientation space fixing one {*hkl*} or ⟨*uvw*⟩. Examples are the γ-fiber ranging with ⟨111⟩//ND for $\Phi = 55°$, $\varphi_2 = 45°$, $\varphi_1 = 0° \rightarrow 90°$. This fiber is essential for the deep drawing behavior of metals and their alloys.

The ODF is needed firstly for quantitative texture description to include texture sharpness and volume fractions of texture components and secondly to incorporate the quantitative texture into simulation and property calculations of polycrystalline materials.

3.3.3 Inverse Pole Figures

The IPF is another type of two-dimensional projection of the ODF with a fixed sample orientation and the normalized intensity distribution of all possible crystal orientations (*hkl*). Due to the symmetry of the texture one needs a minimum of two IPFs after uniaxial deformation (IPF//to RD and IPF ⊥ to RD) and three IPFs after plane strain deformation (IPF//RD, IPF//TD, IPF//ND). Each crystal system (cubic, hexagonal, tetragonal, trigonal, etc.) has its own shapes of the IPF (Figure 3.10). The IPF includes all crystal directions allowed in the corresponding crystal system. For example, all cubic space groups have an identical IPF (Figure 3.10a) due to the definition $a = b = c$, $\alpha = \beta = \gamma = 90°$, while this is different for tetragonal systems ($a = b \neq c$, $\alpha = \beta = \gamma = 90°$). It has to be noticed that any sample orientation has its own inverse pole figure and the information of IPFs are limited.

IPFs are used to correlate the texture with materials properties in defined sample directions, which is one option to explain the anisotropy of the properties. In EBSD presentation, the IPF is routinely used as a legend to define the color code of the plot. In general, IPFs do not give a full description of the texture of a sample.

3.4 Interpretation of Textures

Texture is the statistical information of the grain orientation in polycrystalline materials. Thus, five relevant issues for the interpretation of textures will be discussed in the following text.

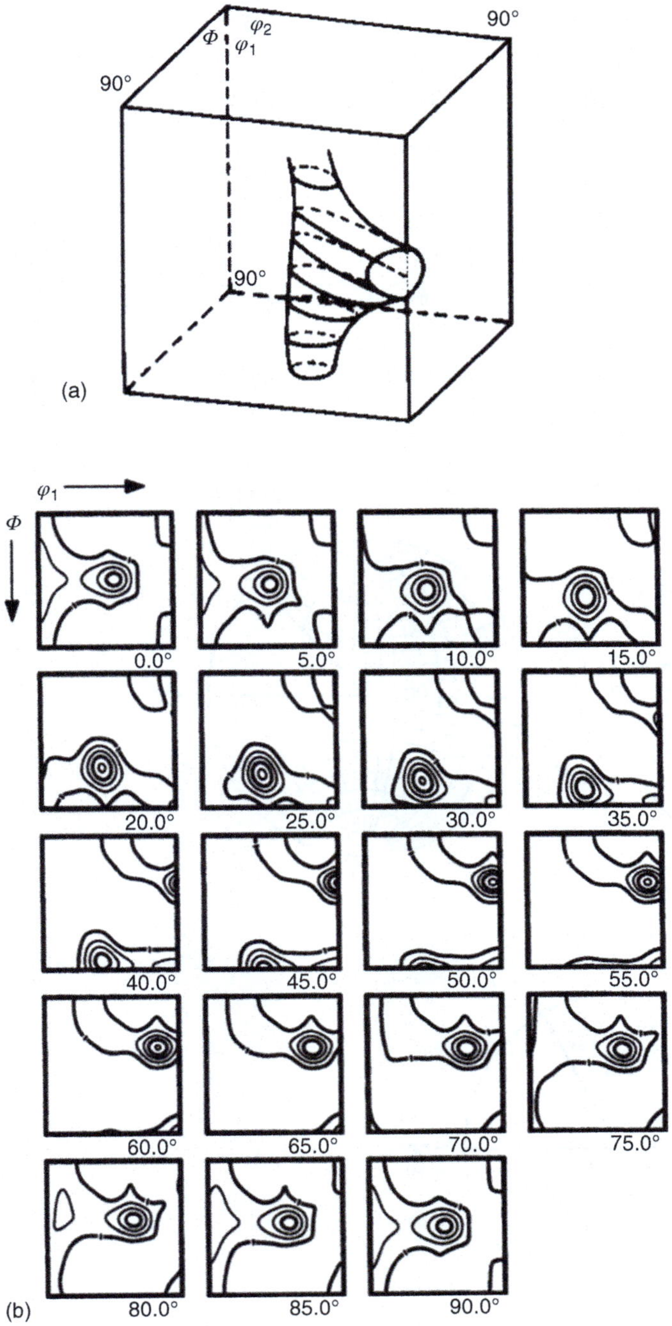

Figure 3.8 (a) Example of a 3D-ODF [4]. (b) φ_2 sections with φ_1 from 0° → 90°, $\phi = 0° \rightarrow 90°$ [19].

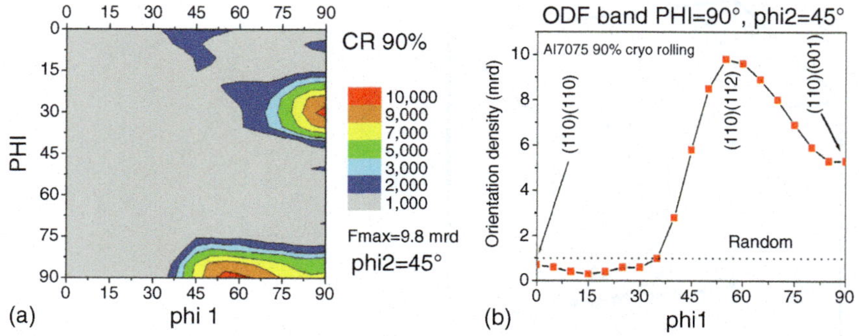

Figure 3.9 (a) Example of ODF φ_2 sections of rolled Al and (b) the ODF band $\varphi_2 = 45°$. (Jayaganthan 2010 [23]. Reproduced with permission of Elsevier.)

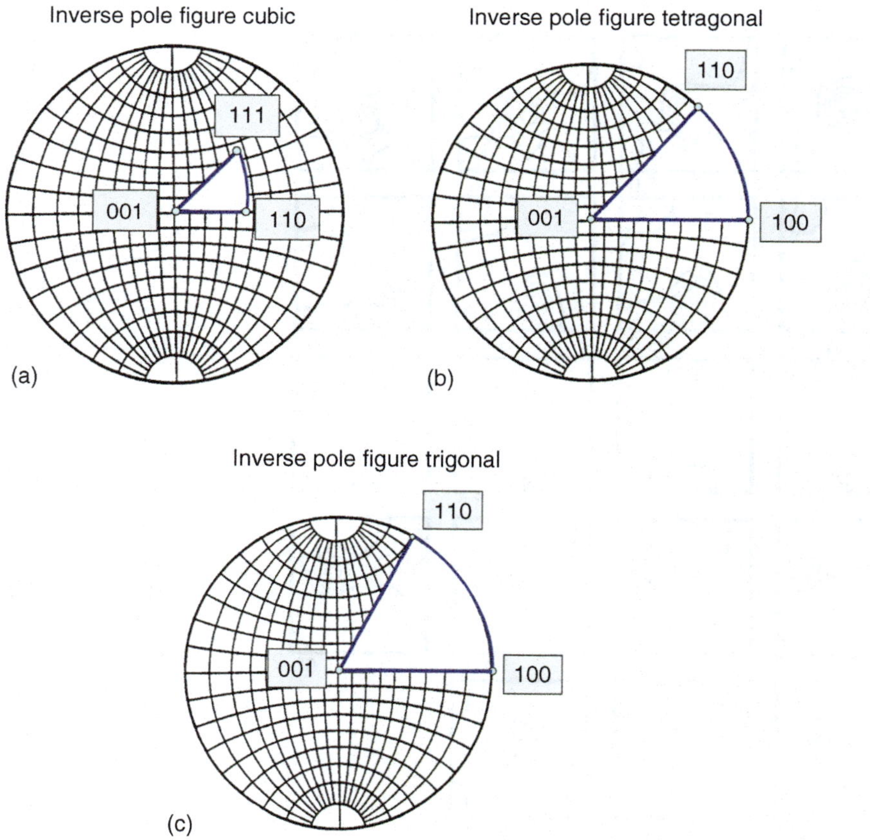

Figure 3.10 Inverse pole figure for (a) cubic, (b) tetragonal, and (c) trigonal crystal symmetry.

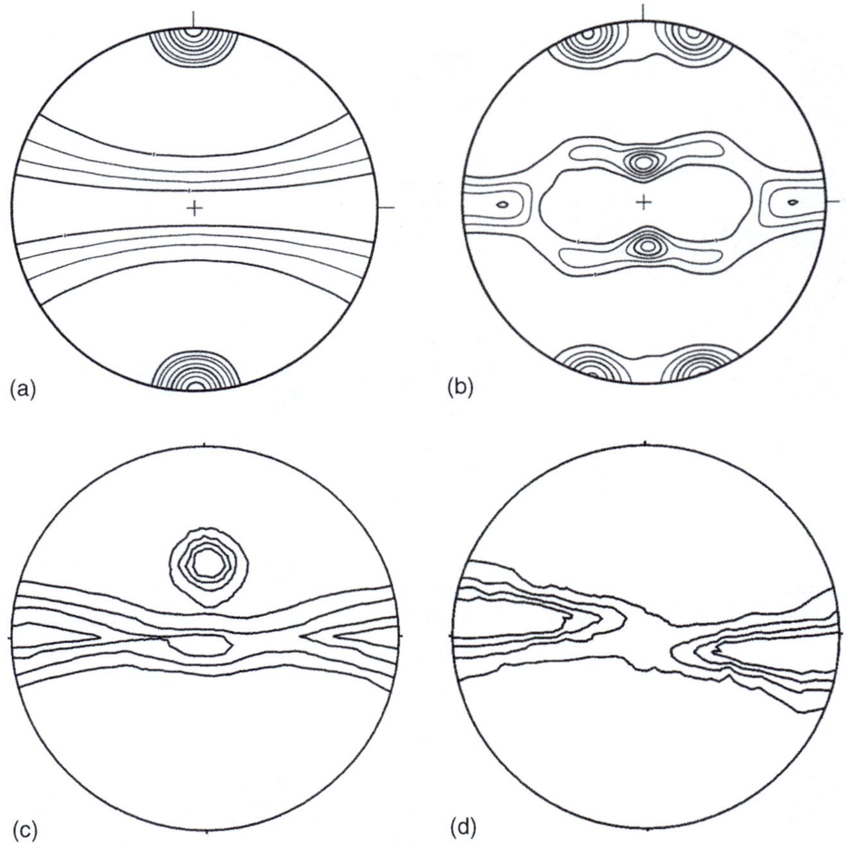

(a) (b) (c) (d)

Figure 3.11 Pole figure symmetry related to the deformation process (a) 96% cold extruded Al, (b) 95% cold rolled Al, (c) 1 pass ECAE-Mg, and (d) profile extrusion Mg.

Firstly, the symmetry of measured pole figures tells the history of the sample deformation. The highest pole figure symmetry is observed for uniaxial deformation such as round extrusion, wire drawing, and uniaxial forging. Pole figure symmetry is rotational with rotation axis parallel to deformation direction (Figure 3.11a). Plain strain deformation results in orthorhombic pole figure symmetry *mmm* with mirror planes ⊥ RD, TD, and ND (Figure 3.11b). Pure plain strain deformation can be found in symmetric rolling techniques, bar extrusion, channel die, and multi-directional forging. In the case of asymmetric rolling, shear strain is present and the pole figure symmetry is reduced to monoclinic (Figure 3.11c). In severe plastic deformation the accumulation of shear is used for grain refinement in the nano-scale, such as equal channel angular extrusion (ECAE), accumulative roll bonding (ARB), high pressure torsion (HPT), or twist extrusion. In many cases where anisotropic flow during processing occurs (tube drawing, profile extrusion, deep drawing), pole figure symmetry is monoclinic or even triclinic (Figure 3.11d).

The interesting information is the deviation from the expected symmetry to control the quality of deformation processes as well as the quality of the texture analysis

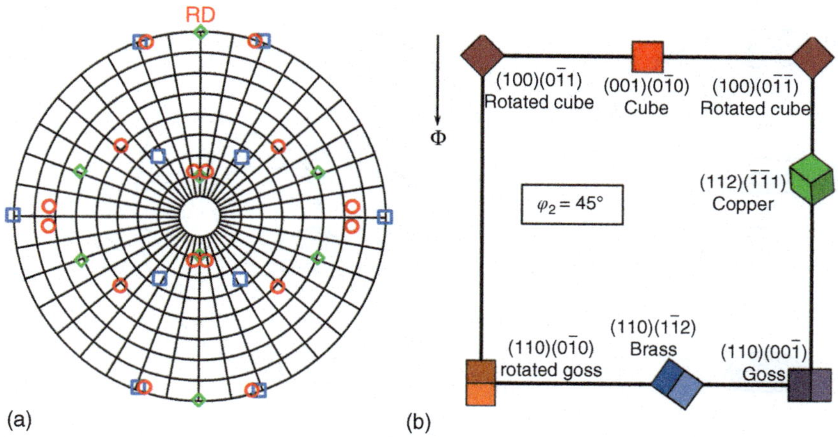

Figure 3.12 Ideal components drawn in the (a) pole figure and (b) the ODF section $\varphi_2 = 45°$.

Table 3.3 Ideal texture components in deformed fcc metals.

Component symbol	{*hkl*}	⟨*uvw*⟩	φ_1	Φ	φ_2
Copper, C	112	111	90	35	45
Dillamore, D	123	634	90	27	45
Brass, B	110	112	55	90	45
Cube	100	100	45	0	45
Goss, G	110	001	90	90	45
S	123	634	59	37	63

itself. It should be pointed out that the crystal symmetry does not affect the pole figure symmetry.

Secondly, ideal components are given by single crystal orientation with {*hkl*}//rolling plane and ⟨*uvw*⟩//RD. Some examples are shown in Figure 3.12a in the (111) pole figure and in the ODF section $\varphi_2 = 45°$ in Figure 3.12b. Ideal components are named by materials where they appear (brass {110} ⟨112⟩ or copper {112} ⟨111⟩), by famous scientists (Goss {100} ⟨110⟩ and Dillamore {4,4,11} ⟨11,11,8⟩) or by the grain shape, for example, cube{001} ⟨100⟩. In Table 3.3, the main texture components of deformed face-centered cubic (fcc) metals are given. It can be seen clearly that texture components overlap inside pole figures. In the ODF, ideal components are separated so that it is easy to calculate the volume fractions of each component. Practical use is, for example, to determine the deformation/recrystallization ratio or the amount of twins.

Another type of texture component is the ODF fiber or ODF band. ODF bands combine orientations *g* ($g = (hkl)[uvw] = (\varphi_1, \Phi, \varphi_2)$) with one (*hkl*) and free [*uvw*] or one fixed [*uvw*] with free (*hkl*). Examples for bcc materials are listed in Table 3.4. Fiber type and intensity distribution along the fiber are very important for the anisotropy of the material and are often used to present texture variations. As one can see in Figure 3.13 the ODF band connects ideal components such as G and B in the α-fiber. A special case

Table 3.4 Orientation band for bcc material inside the ODF.

α fiber ⟨110⟩||RD	From (110) [110] to (001) [110]	From $\varphi_1 = 0°$; $\Phi \rightarrow 0°$ to $90°$, $\varphi_2 = 45°$
γ fiber {111}||ND	From (111) [110] to (111) [112]	From $\varphi_1 \rightarrow 0°$ to $90°$, $\Phi = 55°$, $\varphi_2 = 45°$
ε fiber ⟨110⟩||TD	From (011) [100] to (011] [011]	From $\varphi_1 \rightarrow 0°$ to $90°$, $\Phi = 45°$, $\varphi_2 = 0°$

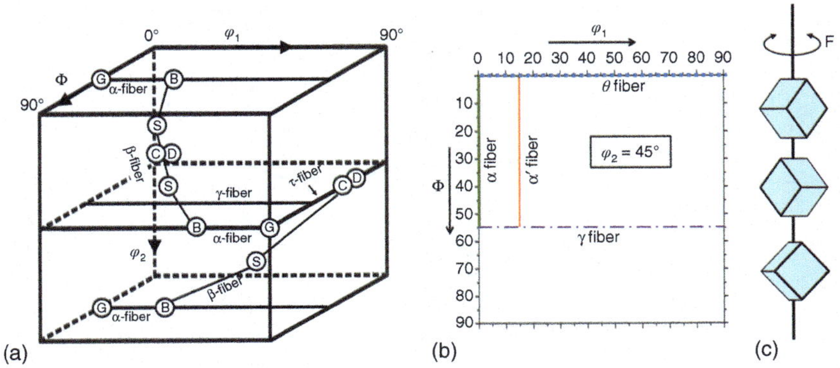

Figure 3.13 ODF bands in the ODF space [24] (a), in ODF section $\varphi_2 = 45°$ (b), and scheme of ideal ⟨110⟩ fiber [1] (c).

of such an ODF band is the ideal fiber as described by Wassermann and Grewen [1]. A fiber texture is defined only by the fiber axis ⟨110⟩ (Figure 3.13c) and homogeneous intensity girdles around this fiber axis. ODF bands or ODF fibers show an intensity variation along the fiber, as already shown in Figure 3.9b.

In general, it can be concluded that mostly a mixture of ideal components, ODF bands, and ideal fibers are needed to describe the quantitative texture, that is, each component with its own volume fraction.

3.5 Errors

As noticed above, the quantitative texture is the statistical information of a well-defined gauge volume at present state. The main errors for all techniques are based on statistical problems and on changing textures during the experiment. Statistical problems are firstly an insufficient number of grains, secondly a wrong step size to collect the texture information, thirdly bad counting statistics, and fourthly an insufficient pole figure coverage.

3.5.1 Grain Statistics

With an estimated number of 10^6 grains, one has a sufficient grain statistics using the volume method. One has to find always a compromise between gauge volume and grain size. Single grain methods need smaller numbers of grains because of the direct measurement of ODF points (e.g., EBSD). Therefore, the analyzing method and gauge volume have to be related to the grain size. Due to high penetration depth and large beam

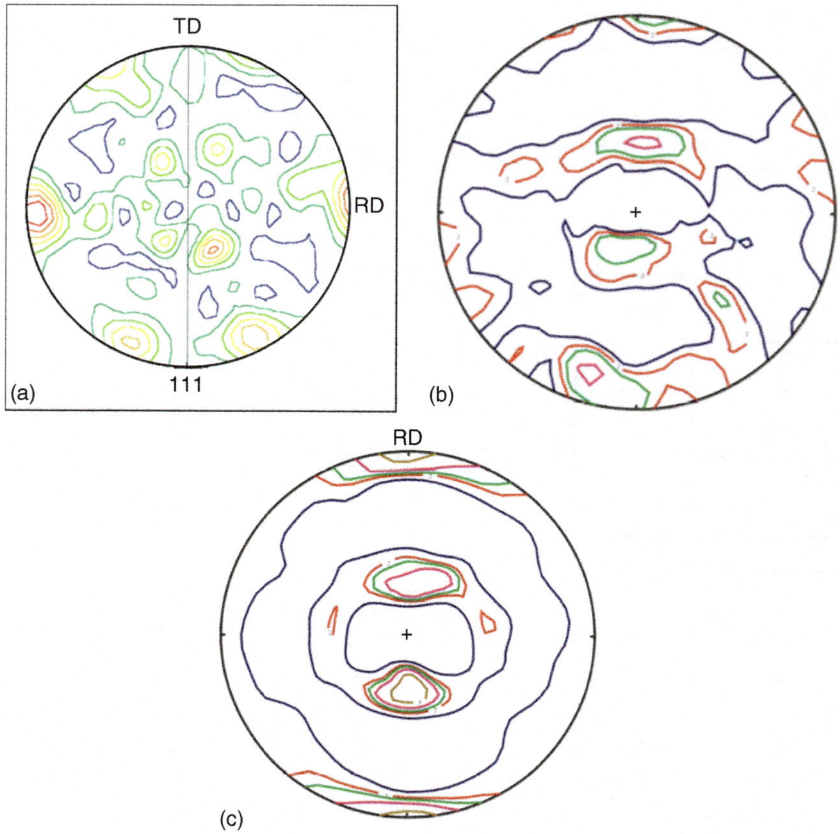

Figure 3.14 Effect of grain statistics (a) EBSD: TiAl alloy, (b) X-ray diffraction: Fe2.6%Si, and (c) neutron diffraction: Fe2.6%Si.

cross-section, neutrons will provide the highest quality on grain statistics (Figure 3.14c) [25], while high brilliant synchrotron beams and EBSD (Figure 3.14a) often have problems to include enough grains. Figure 3.14b shows that also X-rays lead to lower quality pole figures if grain statistics is insufficient.

To obtain the correct texture sharpness is more or less problematic and to get an idea of the texture type needs good knowledge of the investigated material. To overcome the grain statistic problem one has to optimize the experimental conditions to ensure that all grains inside the gauge volume are collected (see pole figure coverage), that the gauge volume must be increased, or that a number of identical samples are added together by a mathematical procedure.

3.5.2 Pole Figure Coverage

For the quality of the quantitative texture the counting grid and the so-called "pole figure window" have to be adjusted to the required resolution in the ODF space. The pole figure window is the integration area on the polar sphere in which all grains were summarized to a pole figure point $P(\alpha, \beta)$. A good comparison of the pole figures for different instruments was given by Moras *et al.* [26]. The pole figure window depends on $\Delta\lambda$, scattering

Table 3.5 Texture sharpness of a AA7020-T6 Al alloy as function of the counting grid.

Counting grid	Pole figure points	Sharpness in ODF (mrd)	Counting time @STRESS-SPEC (h)
$\Delta\alpha = 5°$, $\Delta\beta = 5°$	1 368	20	3
$\Delta\alpha = 2°$, $\Delta\beta = 2°$	8 100	25	6
$\Delta\alpha = 1°$, $\Delta\beta = 1°$	32 400	32	18

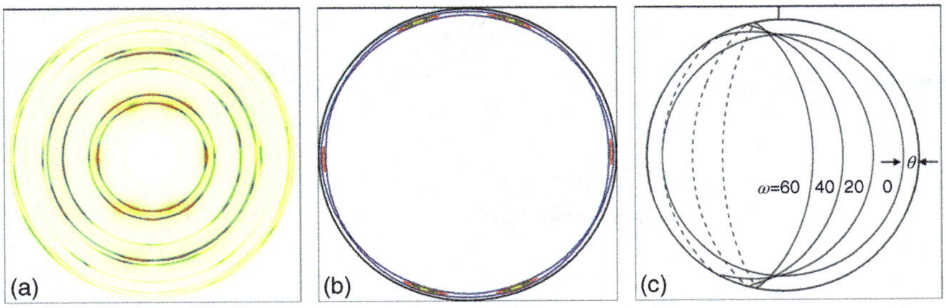

Figure 3.15 Pole figure coverage of a highly brilliant beam (see text).

angle 2θ, and beam divergence. To get a good statistic within a reasonable counting time, many texture diffractometers work with higher beam divergence as compared to that for all other applications such as residual stress analysis. This is one reason why many large-scale facilities run different diffractometers for pole figure and strain measurements. For the instrumental set up of the engineering diffractometer STRESS-SPEC at FRM II (see Chapter 10), the Ge (311) monochromator provides $\Delta\alpha$ from 0.74° to 2.3° while $\Delta\beta$ is only 0.35° [27]. Therefore, even in complete pole figure measurement using a 5×5 counting grid with steps of $\Delta 5°$ in α and $\Delta 5°$ in β, only partial pole figure coverage is achieved. In the case of sharp textures, one needs more pole figure points than in weak textures. An example of strongly oriented material is AA7020-T6, which shows the typical deformation texture of Al alloys (C, B, and S component). However, the texture sharpness depends on the counting grid, see Table 3.5. It has to be noticed that texture sharpness influences directly the degree of materials anisotropy. The development of fast readout and single event based area detectors are a huge step forward to improve quantitative texture analysis.

A special case is the highly brilliant synchrotron beam. As shown by Moras *et al.* [26] the pole figure window is very anisotropic in $\Delta\alpha$ and $\Delta\beta$. Area detector images of an Al sample show a number of Debye–Scherrer rings (Figure 3.15a). Due to the beam brilliance, the area in the pole figure is small (Figure 3.15b) and the higher the brilliance, the fewer the number of grains fulfilling Bragg's condition, and thus the pole figure area decreases. Pole figure measurement by sample rotation (Figure 3.15c) needs finer steps in sample rotation. On the one hand, the high brilliance is excellent for sharp textures and texture gradient studies if the scanning step size is small enough and on the other hand even a small misalignment decreases the quality drastically.

Complete pole figures with equal angular grid from α 0° → 90° and β 0° → 360° are not the usual case. Mostly, one gets incomplete pole figures from measurements due to absorption, defocusing effects, or blind areas from *in situ* devices. To mathematically

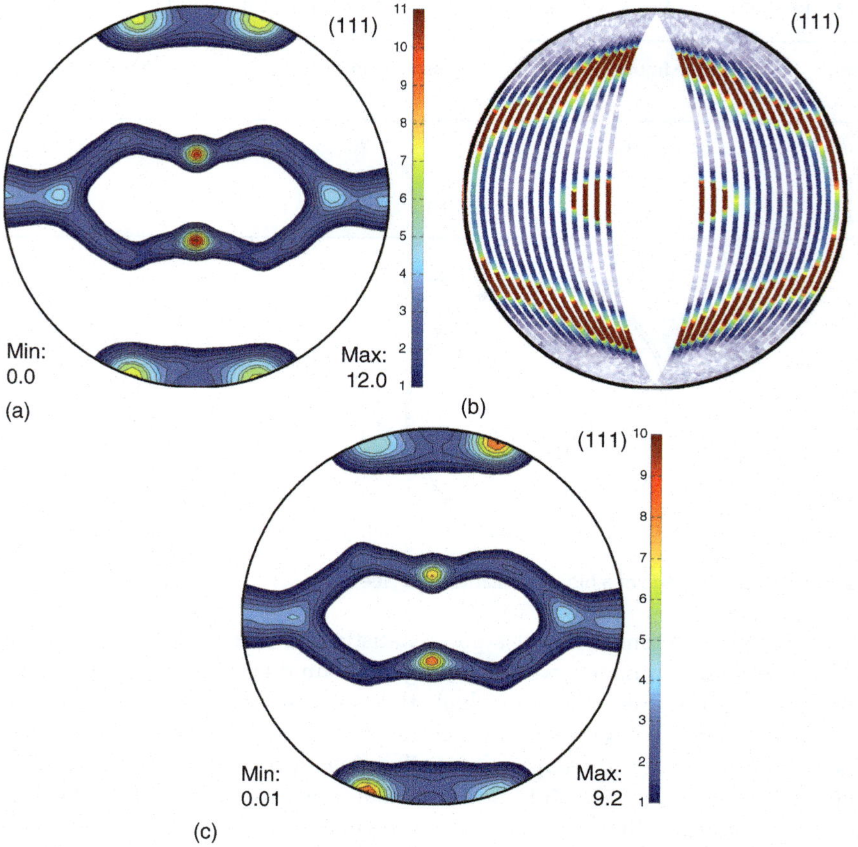

Figure 3.16 (a) Pole figure from complete coverage; (b) counting grid for incomplete pole figures; and (c) pole figure from incomplete coverage.

recalculate complete pole figures from ODF is no problem [2, 12–16]. Due to the quality of experimental pole figure data as well as for complete or incomplete pole figures the calculated ODF can differ from the true ODF. In Figure 3.16 the difference between complete and incomplete pole figures is shown. The pole figure in Figure 3.16a was recalculated from the complete coverage, while the pole figure in Figure 3.16c was recalculated from an incomplete coverage shown in Figure 3.16b. Both pole figures (Figure 3.16a,c) were recalculated after the ODF was calculated from the measured (complete or incomplete) data. The missing part of about 40° in sample rotation (Figure 3.16b) includes strong intensities. Sample symmetry as well as crystal symmetry results in sufficient symmetry equivalent information in the displayed (111) pole figure as well as in other pole figures not shown here, to reproduce all texture components. But the texture sharpness is reduced (complete 12.0 mrd, incomplete 9.2 mrd). Moreover, the pole figure recalculated from the incomplete data set shows less pole figure symmetry than the pole figure recalculated from the complete data set. In general, one can conclude that an increase of the blind area in a measured pole figure will decrease the quality of the quantitative texture determination.

References

1 Wassermann, G. and Grewen, J. (1962) *Texturen metallischer Werkstoffe*, 2. Auflage, Springer, Berlin, Göttingen, Heidelberg.

2 Bunge, H.J. (1993) *Texture Analysis in Materials Science*, Cuvillier Verlag, Göttingen.

3 Kocks, U.F., Tomé, C.N., and Wenk, H.-R. (1998) *Texture and Anisotropy*, Cambridge University Press, Cambridge.

4 Bunge, H.J. and Esling, C. (1986) *Quantitative Texture Analysis*, DGM-Informationsverlag, Oberursel.

5 Engler, O. and Randle, V. (2010) *Introduction to Texture Analysis: Macrotexture, Microtexture, and Orientation Mapping*, CRC Press, Boca Raton, FL.

6 Bunge, H.J. and Schwarzer, R.A. (2001) Orientation stereology – A new branch in texture research. *Adv. Eng. Mater.*, **3**, 25–39.

7 Poulsen, H.F., Garbe, S., Lorentzen, T., Juul Jensen, D., Andersen, N.H., Frello, R., Feidenhans'l, R., and Graafsma, H. (1997) Applications of high-energy synchrotron radiation for structural studies of polycrystalline materials. *J. Synchrotron Radiat.*, **4**, 147–154.

8 Brokmeier, H.-G. (2009) Hard X-rays for in-situ strain and texture measurements. *Part. Part. Syst. Char.*, **26**, 117–124.

9 Merkel, S., Liermann, H.-P., Miyagi, L., and Wenk, H.-R. (2013) In situ radial X-ray diffraction study of texture and stress during phase transformations in bcc-, fcc- and hcp-iron up to 36 GPa and 1000 K. *Acta Mater.*, **61**, 5144–5151.

10 Van Swygenhoven, H. and Van Petegem, S. (2013) In-situ mechanical testing during x-ray powder diffraction. *Mater. Charact.*, **78**, 47–59.

11 Brokmeier, H.-G., Singer, W., and Kaiser, H. (2002) Neutron diffraction – a tool to optimize processing of niobium tubes. *Appl. Phys. A: Mater. Sci. Process.*, **74**, s1704–s1706.

12 Wenk, H.-R., Matthies, S., Donovan, J., and Chateigner, D. (1998) BEARTEX: a Windows-based program system for quantitative texture analysis. *J. Appl. Crystallogr.*, **31**, 262–269.

13 Lutterotti, L. (2016) http://www.ing.unitn.it/∼maud/ (accessed 27 June 2016).

14 Hielscher, R. and Schaeben, H. (2008) A novel pole figure inversion method: specification of the MTEX algorithm. *J. Appl. Crystallogr.*, **41**, 1024–1037. doi: 10.1107/S0021889808030112

15 Dahms, M. and Bunge, H.J. (1989) The iterative series-expansion method for quantitative texture analysis. *J. Appl. Crystallogr.*, **22**, 439–447.

16 Matthies, S., Vinel, G.W., and Helming, K. (1987) *Standard Distributions in Texture Analysis*, Akademie-Verlag, Berlin.

17 Kile, D.E. (2009) The universal stage: the past, present, and future of a mineralogical research instrument. *Geochem. News*, **140**, 1–10.

18 Brokmeier, H.-G., Gan, W.M., Randau, C., Rebelo-Kornmeier, J., and Hofmann, M. (2011) Texture analysis at neutron diffractometer STRESS-SPEC. *Nucl. Instrum. Methods Phys. Res., Sect. A*, **642**, 87–92.

19 Wenhai, Y. (2005) Untersuchungen zum Einfluss thermischer Behandlungen von – 269 °C bis 1000 °C auf die kristallographische Textur und die Eigenspannungen eines Cu-Nb Verbundwerkstoffes. PhD thesis. TU Clausthal, Clausthal-Zellerfeld, Deutschland.

20 Bunge, H.J. (1965) Zur Darstellung Allgemeiner Texturen. *Z. Metallk.*, **56**, 872–874.
21 Roe, R.-J. (1965) Description of crystallite orientation in polycrystalline materials. III. General solution to pole figure inversion. *J. Appl. Phys.*, **36**, 2024–2031.
22 Bunge, H.J. (1969) *Mathematische Methoden der Texturanalyse*, Akademie-Verlag, Berlin.
23 Jayaganthan, R., Brokmeier, H.-G., Schwebke, B., and Panigrahi, S.K. (2010) Microstructure and texture evolution in cryorolled Al 7075 alloy. *J. Alloys Compd.*, **496**, 183–188.
24 Hirsch, J. and Lücke, K. (1988) Mechanism of deformation and development of rolling texture in polycrystalline fcc metals – II. Simulation and interpretation of experiments on the basis of Taylor-type theories. *Acta Metall.*, **16**, 2882–2904.
25 Brokmeier, H.-G. (1991) in *Advances and Application of Quantitative Texture Analysis* (eds H.J. Bunge and C. Esling), DGM-Informationsgesellschaft, Oberursel, pp. 73–86.
26 Moras, K., Fischer, A.H., Klein, H., and Bunge, H.J. (2000) Experimental determination of the instrumental transparency function of texture goniometers. *J. Appl. Crystallogr.*, **33**, 1162–1174.
27 Randau, C. (2012) Entwicklungen am Neutronendiffraktometer STRESS-SPEC für schnelle und lokale Polfigurmessungen zur Bestimmung ortsaufgelöster Texturen. PhD thesis. TU Clausthal, Clausthal-Zellerfeld, Deutschland.

4

Physical Properties of Photons and Neutrons

Andreas Schreyer

4.1 Introduction

Photons in the X-ray wavelength regime and neutrons share a number of physical properties, which make both probes a unique tool for the investigation of materials.

Both probes are available with wavelengths on the order of atomic distances, their strength of interaction with matter is of similar magnitude, and both probes penetrate into matter from micrometer to many centimeter deep. From the interference patterns (e.g., Bragg peaks) generated, when X-rays or neutrons are scattered from matter, essential information on the structure of materials on the atomic level can be obtained in a destruction-free way.

An important distinction from electrons, which are also frequently used for scattering experiments, is the fact that electrons interact more strongly with matter due to the Coulomb interaction and penetrate less deeply into it. Therefore, electrons are only used for the study of surfaces, which is not the focus of the present book. Here we are specifically interested in the capability of X-rays and neutrons to unravel the structure within materials and components in a destruction-free way, that is, to penetrate deep into samples.

Some basic physical properties of X-rays and neutrons are listed in Table 4.1. Both particles are very different in nature: for example, the neutron has a mass, its energy depends on this mass and its velocity, and it is described by the Schrödinger equation. The X-ray photon, on the other hand, has no mass, always moves at the velocity of light, and is described by the Maxwell equations. As a consequence, neutrons of 1 Å wavelength have an energy in the milli electronvolts regime, whereas the energy of X-ray photons is in the kilo electronvolts range. It is not surprising that the physics of the interaction of X-ray photons and neutrons with matter is also very different. However, as will be shown in the following, the formalism for describing this interaction and the strength of the interaction are so similar that the same theories can be used to describe the scattering of photons and neutrons from matter.

Neutrons and Synchrotron Radiation in Engineering Materials Science: From Fundamentals to Applications,
Second Edition. Edited by Peter Staron, Andreas Schreyer, Helmut Clemens, and Svea Mayer.

Table 4.1 Basic physical properties of neutrons and photons.

Neutrons		X-rays (photons)
Three quarks and gluons		Point-shaped
$m_n = 1.675 \times 10^{-27}$ kg ≈ 1 u	Mass	0
0	Charge	0
½	Spin	1
$\mu_n = \gamma \cdot \mu_N$; $\gamma = -1.91$, $\mu_N = 5 \times 10^{-27}\,\mathrm{J\,T^{-1}}$	Magnetic dipole moment	0
Schrödinger equations (particle wave)	*Wave*	*Maxwell equations (electromagnetic wave)*
$\mathbf{p} = m \cdot \mathbf{v} = \hbar \mathbf{k};\ p = \frac{h}{\lambda}$	Momentum	$\mathbf{p} = \hbar \mathbf{k};\ p = \frac{h}{\lambda}$
$E = \frac{1}{2} m \cdot v^2 = \frac{\hbar^2 k^2}{2m} = k_B T$	Energy	$E = h \cdot \nu = \frac{hc}{\lambda}$
$\lambda\,(\text{Å}) = \frac{4000}{v\,(\mathrm{m\,s^{-1}})}$	*Useful data*	$E\,(\mathrm{keV}) = \frac{12.4}{\lambda\,(\text{Å})}$
$E\,(\mathrm{meV}) = \frac{81.8}{\lambda^2\,(\text{Å}^2)} = 0.086T\,(\mathrm{K})$		
"Thermal" neutrons (300 K)	*Example*	$\lambda = 1$ Å $\equiv E = 12.4$ keV
$E = 25$ meV		
$\lambda = 1.8$ Å		
$v = 2200\ \mathrm{m\,s^{-1}}$		

The equations of motions are the Schrödinger equation (neutrons) and the Maxwell equations (photons), respectively, yielding the listed relations for momentum and energy. Some useful formulae and typical values for the energy E, wavelength λ, and velocity v are shown as well. After [1].

4.2 Interaction of X-ray Photons and Neutrons with Individual Atoms

A simplified picture of the process of interaction between X-ray photons or neutrons is visualized in Figure 4.1. A plane wave of wavevector $|\mathbf{k}| = 2\pi/\lambda$ (wavelength λ) along z of amplitude

$$\psi_0 = e^{ikz} \tag{4.1}$$

interacts with an individual scatterer of point dimension, generating a spherical wave

$$\psi_S = b/r\,e^{ikr} \tag{4.2}$$

centered around the scatterer. The amplitude of the scattered spherical wave at the location r is determined by the parameter b which has the dimension of a length; thus b is called the *scattering length*. In general, b is a complex number, where a positive real part describes a repulsive interaction, a negative real part an attractive interaction, and the imaginary part the absorption.

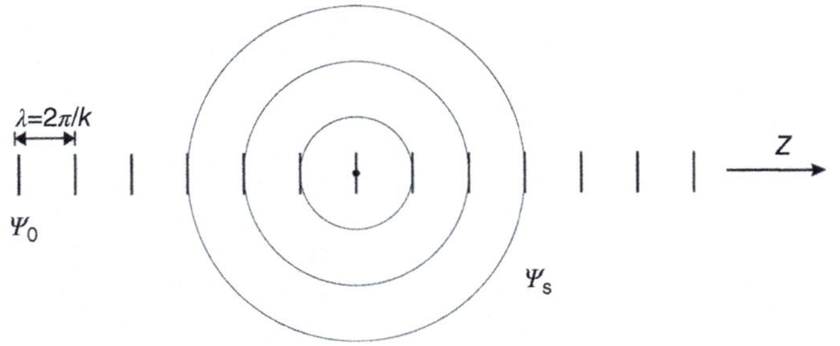

Figure 4.1 Schematic view of the interaction between a neutron or X-ray photon represented by a plane wave ψ_0 of wavelength λ with a point-like particle creating a spherical wave ψ_S.

4.2.1 Neutrons

Let us first consider the scattering length for the scattering of neutrons from an individual atom. Being nuclear particles, neutrons interact via the so-called strong interaction with the particles in the nucleus of the atom, namely with an ensemble of protons and neutrons, the numbers of which are given by the order number Z and the mass number A of the atom. Due to the extremely short range of the strong interaction and the small size of the nucleus (both on the order of fm $= 10^{-15}$ m) compared to the wavelength of the neutron (on the order of Å $= 10^{-10}$ m) the scattering process is actually well described by Figure 4.1, that is, by assuming a scatterer of point dimension.

The values of the neutron scattering length b_n turn out to be on the order of fm, that is, the scattering is weak. b_n depends sensitively on the order number Z and the mass number A of the scattering atom, since Z and A determine the number of protons and neutrons in the nucleus, respectively.

Since b_n varies unsystematically throughout the periodic table, a large scattering contrast between elements with similar Z or between different isotopes (same Z, different A) can be readily obtained with neutrons (see Figure 4.2). This is not possible with X-rays. As opposed to the case of X-rays b_n are negative for a few isotopes.

In the neutron wavelength regime considered here, absorption of neutrons by most isotopes is very weak leading to a very small imaginary part of the scattering length for these cases. For the problems discussed in this book, this property is the most significant advantage of neutrons since this allows the penetration of especially large samples (see, e.g., Chapters 6 and 7).

Since, in general, a sample will contain different isotopes of an element, the disorder in b_n will lead to a so-called incoherent scattering contribution. In addition, in cases where the spin i of a scattering nucleus is not zero, this leads to an additional interaction with the dipole moment of the neutron. Since the nuclear spins are disordered above temperatures in the millikelvin range, that is, for all purposes of this book, this effect will lead to an additional spin-disorder-induced incoherent scattering. Therefore, in the case of neutrons, effective coherent and incoherent scattering lengths have to be distinguished carefully.

However, the interaction of neutrons with atoms is not always limited to the nucleus. In the case of magnetic materials, the magnetic dipole moment of the neutron leads to an additional interaction with any magnetic moment in the shells of the atoms, which is of similar magnitude as the nuclear interaction. For this reason, neutrons are the

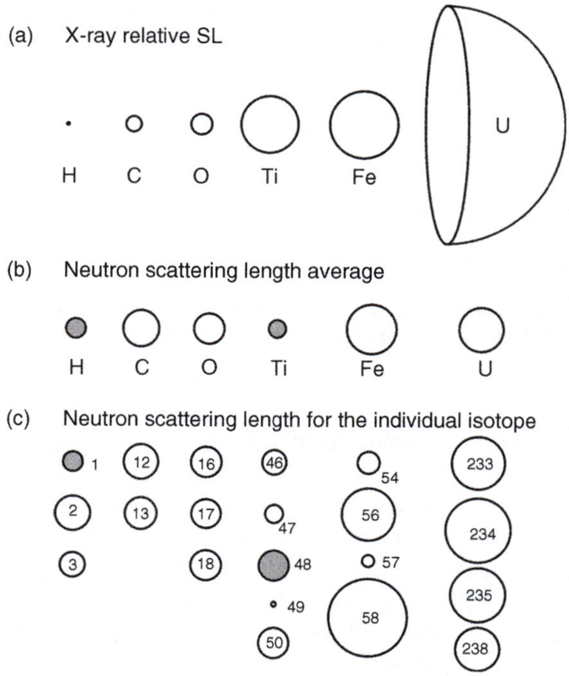

Figure 4.2 Scattering lengths for X-rays (a) and neutrons (b), averaged over the natural isotope distribution and (c) for each isotope of the respective element represented by circles whose radii are proportional to the scattering lengths. The circles representing the neutron scattering lengths have been multiplied by a factor of 2 compared to the X-ray scattering lengths. Shaded circles represent negative values. (After [2].)

classic probe of magnetic structures and magnetism is a main scientific field in neutron scattering. However, since the study of magnetism is not the main scope of this book, we do not provide any details on the magnetic interaction here.

A table of the coherent and incoherent scattering lengths including a comparison with X-ray scattering lengths can be found in Chapter 2 of [2]. An in-depth discussion of the neutron nuclear scattering length is given in the article by Sears in [3]. If not stated otherwise, b_n denotes the coherent nuclear scattering length in the remainder of this book.

4.2.2 X-rays

In the case of X-rays, on the other hand, the interaction takes place with the electronic shells of atoms or the electronic bandstructure, which evolves when atoms form crystals. To begin with, let us consider the interaction of a linearly polarized photon plane wave with an individual electron (Figure 4.1). Being an electromagnetic wave, the oscillating electric-field component of the plane wave, defined to be parallel to x, will couple to the charge of the electron and cause its oscillation along x. This oscillation of the electron in turn generates an electromagnetic dipole field, which is considered to be the scattered wave. Quantitatively, this scattered wave is described by a dipolar field instead of the simpler spherical wave introduced above. In Figure 4.3, this dipolar field in the $x-z$ plane around the oscillating electron is shown schematically.

A very important consequence is that no intensity is emitted along the oscillating direction of the electron, that is, along x. Furthermore, in the $x-z$ plane, the scattered intensity strongly depends on the scattering angle (see Figure 4.3), whereas due to the characteristics of the dipolar field in the $y-z$ plane Figure 4.1 holds, that is, the scattered intensity is independent of the scattering angle.

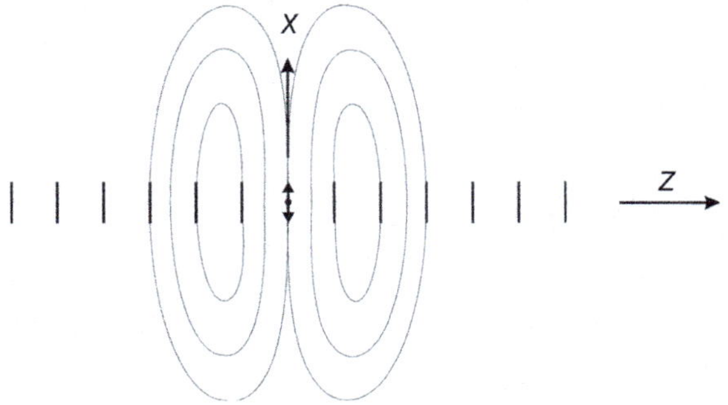

Figure 4.3 Schematic representation of the interaction of an electromagnetic plane wave propagating along z with a free electron. The electric field oscillating along x causes an oscillation of the electron along x which creates a dipolar field. In the orthogonal direction (y–z plane) Figure 4.1 holds.

Since synchrotron sources generate linearly polarized light, it is essential to take care of the principles outlined above. In general, this is achieved by making sure that the scattering experiment is performed in the y–z plane to avoid the anisotropy of the scattering shown in Figure 4.3. Due to their size storage rings are horizontal, producing horizontally polarized light, and, therefore, the scattering angle is usually measured in the vertical direction (y–z plane). In cases, where the scattering angle deviates from the y–z plane (e.g., when large two-dimensional detectors are used) a correction factor has to be applied, which accounts for the form of the dipolar field in the x–z plane [4].

It is actually quite straightforward to calculate the scattering length of the scattering process introduced above [4, 5]. The scattering length of this so-called Thomson scattering from a free electron turns out to be

$$r_e = e^2/m_e c^2 \tag{4.3}$$

where e is the electron charge, m_e is the electron mass, and c is the velocity of light. This Thomson scattering length r_e has the value 2.82 fm, that is, it has the same magnitude as the neutron scattering length b_n.

In the next step we have to take care of the fact that we are interested in the scattering from all electrons in the atomic shells. Since these have a dimension on the order of angstroms, that is, five orders of magnitude larger than r_e, the assumption of a point-like scatterer does not hold for the scattering of X-rays from atoms. To account for this effect, an integral over the scattering contributions from all electrons within the shell accounting for the phase difference due to their different location within the shell has to be performed. This integral over the volume of the shell is the so-called form factor

$$f_e = \int \rho_e(r) e^{iqr} \, dv \tag{4.4}$$

with the electron density ρ_e, and the scattering vector

$$\mathbf{q} = \mathbf{k}_f - \mathbf{k}_i \tag{4.5}$$

where $\mathbf{k}_i$ and $\mathbf{k}_f$ are the incident and final wavevectors before and after the scattering process, respectively (see Figure 4.4) [4, 5]. From simple geometrical considerations it can be shown that

$$|\mathbf{q}| = 4\pi/\lambda \sin\theta \tag{4.6}$$

where 2θ is the scattering angle between $\mathbf{k}_i$ and $\mathbf{k}_f$. The scattering vector $\mathbf{q}$ is the fundamental quantity that is used to describe any scattering process.

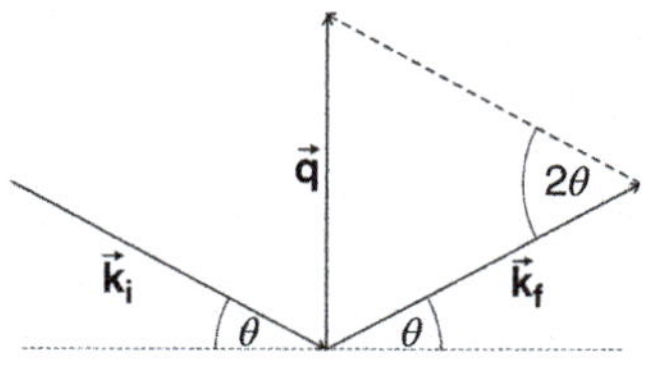

Figure 4.4 Geometrical construction of the scattering triangle defined by $\mathbf{q} = \mathbf{k}_f - \mathbf{k}_i$ with the scattering angle θ.

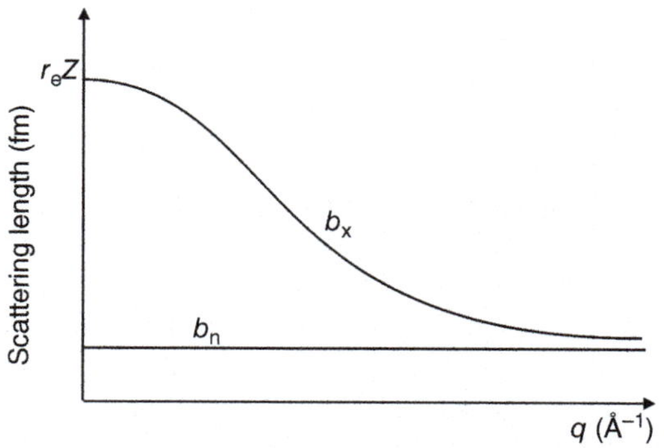

Figure 4.5 Comparison of the q dependence of the scattering lengths for X-rays and neutrons. Whereas b_X strongly depends on q, b_n is constant.

For small $\mathbf{q}$, that is, in the small angle regime $f_e \approx Z$, where Z is the number of electrons (order number) of the atom. With increasing $\mathbf{q}$ a rapid decay of the form factor occurs (see Figure 4.5).

The complete scattering length for X-rays is given by

$$b_X = r_e(f_e + \Delta f_e' + i\Delta f_e'') \tag{4.7}$$

where $\Delta f_e'$ and $\Delta f_e''$ are the real and imaginary parts of the dispersion correction near resonances, where the incoming X-rays can excite core electrons of the atom. The scattering contrast between elements with similar Z can, therefore, be enhanced significantly by tuning the X-ray wavelength to a resonance of one element inducing a major dispersion correction.

The imaginary part of b_X describes the absorption and is related to the more common length absorption coefficient μ_l by

$$\mu_l = 2\lambda n_0 r_e \Delta f_e'' \tag{4.8}$$

with the atomic number density n_0 [6]. Compared to neutrons the absorption of X-rays is significantly higher. According to Eq. 4.8 it can, however, be strongly reduced by reducing λ, that is, by increasing the X-ray energy accordingly. This scheme is commonly used in materials research (see Chapters 8 and 9). In addition, absorption is the most important contrast mechanism for imaging methods, which are of increasing significance in materials research (see Chapters 14–16).

The values of the form factors and the dispersion corrections are tabulated in the International Tables [7]. An excerpt can be found, for example, in [4].

In addition to the Thomson scattering discussed so far, a very much weaker spin-dependent contribution to the scattering of X-rays from magnetic materials exists, which can be enhanced using resonance effects similar to those leading to the dispersion correction [5]. As for the case of neutrons, we do not elaborate on this contribution to the scattering here.

In Figure 4.5, a qualitative comparison of the q-dependence of b_X and b_n is shown for the case of a negligible dispersion correction. While b_n is constant b_X increases significantly for small q and can be larger than b_n for large Z.

Another useful comparison is made in Figure 4.2, where b_n and b_X are plotted for increasing order number Z from left to right. While b_X increases continuously with Z, b_n varies irregularly and also depends on the mass number A. The latter property actually allows a variation of b_n by isotope substitution. This technique is very important for the investigation of hydrogenous materials, where the exchange of hydrogen (negative scattering length, see Figure 4.2) against deuterium (large positive scattering length) allows an isotope labeling.

4.3 Scattering of X-ray Photons and Neutrons from Ensembles of Atoms

We have seen that b_n and b_X are of similar magnitude and quite small. This has the important consequence that the probability that one photon or neutron is scattered more than once by an ensemble of atoms is small as well, that is, multiple scattering effects can often be neglected. Based on this assumption the scattering from an ensemble of atoms can be calculated similar to the case of the form factor introduced above. Instead of an integral, a discrete sum over the scattering contributions from all atoms accounting for the phase difference due to their different location within the ensemble is performed. The amplitude of the scattered wave then is

$$A(\mathbf{q}) = \sum b\,\mathrm{e}^{iqr} \tag{4.9}$$

where the sum runs over all atoms with the scattering length b_n or b_X, respectively. The scattered intensity $I(\mathbf{q})$ is then obtained by multiplying with its complex conjugate

$$I(\mathbf{q}) = A(\mathbf{q})A^*(\mathbf{q}) = \left|\sum b\,\mathrm{e}^{iqr}\right|^2 \tag{4.10}$$

For this book the specific case of scattering from a crystal, that is, from a regular periodic three-dimensional array of atoms is most relevant. It can be shown [2, 4, 5, 8] that for an infinite rigid crystal

$$I(\mathbf{q}) \sim \sum \delta(\mathbf{q} - \mathbf{G})|F(\mathbf{G})|^2 \tag{4.11}$$

with the structure factor

$$|F(\mathbf{G})| = \sum b_m\,\mathrm{e}^{iG\rho} \tag{4.12}$$

where ρ are the positions of the atoms within a unit cell describing the crystal. The sum runs over all potentially different atoms within this unit cell. The so-called reciprocal lattice vectors $\mathbf{G}$ are related to the lattice via the lattice spacings d_{hkl}:

$$|\mathbf{G}_{hkl}| = 2\pi/d_{hkl} \tag{4.13}$$

According to Eq. 4.11 the scattered intensity $I(\mathbf{q})$ has strong interference maxima (δ-functions), the so-called Bragg peaks, whenever

$$\mathbf{q} = \mathbf{G} \tag{4.14}$$

where **G** is directly related to the lattice spacings d_{hkl} via Eq. 4.13. For details, on the definition of the unit cell and the Miller indices *hkl*, which describe the lattice planes in a crystal, see, for example, [2, 4, 5, 8, 9]. The structure factor (Eq. 4.12) contains information on the symmetry of the unit cell and determines the relative intensity of the Bragg peaks. A very important consequence is that for certain reflections the structure factor can become zero, implying that certain reflections do not occur for a given unit cell. In [4] the structure factors are calculated for the most common crystal structures.

From Eqs. 4.6, 4.13, and 4.14 the well-known Bragg equation is obtained as

$$n\lambda = 2d_{hkl} \sin\theta \tag{4.15}$$

From the measured $I(\mathbf{q})$ the crystal structure can be obtained including the position of all atoms (phase analysis), variations in d_{hkl} as a function of location within the sample can be measured (strain analysis, see Chapters 6–9), or the orientation distribution of crystallites in a sample can be determined (texture analysis, see Chapters 3 and 10).

For a real nonrigid crystal at finite temperatures Debye Waller factors must be included to account for the temperature-induced deviation of the atoms from their lattice positions [2].

In practice, crystals are neither infinitely large nor perfect. Instead of infinitely sharp Bragg peaks (δ-functions, Eq. 4.11) the finite number of coherently scattering planes leads to peaks of, for example, Gaussian or Lorenzian shape, depending on the kind of disorder. The width of the peak is a measure for the number of coherently scattering planes. A common approximation for the determination of the crystal size is the Debye Scherrer formula given, for example, in [4].

The formalism outlined above can also be modified to describe scattering from non-crystalline matter, as detailed, for example, in Chapter 11 for small angle scattering.

The scattering theory introduced here is called the *kinematical scattering theory*. As mentioned above it neglects multiple scattering. In large perfect crystals, however, multiple scattering can occur even for small scattering lengths since the photon or neutron has a non-negligible probability to be scattered more than once. The reason is that a scattered neutron/photon still fulfills the Bragg condition, and thus its probability to be scattered more than once increases with the quality and size of the crystal. Consequently, a second scattering process can occur, leading to a reduction of the scattered beam intensity resulting from the first scattering process. This effect is called *extinction*. Therefore, extinction corrections have to be applied in these cases within the kinematical theory [6, 9].

Alternatively, the dynamical scattering theory can be used, which takes multiple scattering into account [4, 5]. It is based on the solution of the Schrödinger/Maxwell equation(s) for the given sample for neutrons/X-rays, respectively (see Table 4.1). Unlike the kinematical theory, the dynamical theory also takes into account optical effects like refraction and total reflection, which can become dominant for small scattering vectors **q**, when surfaces or interfaces are involved [5, 6].

For the purpose of this book, which deals with the investigation of polycrystalline materials, the simpler kinematical scattering theory is sufficient in general. As outlined above, the same formalisms apply for neutrons and X-rays, making a complementary use of both probes easy for the user.

Acknowledgment

The drawing of the figures by Danica Solina is gratefully acknowledged.

References

1 *Magnetische Schichtsysteme in Forschung und Anwendung*, Hrsg. Institut für Festkörperforschung des Forschungszentrums Jülich GmbH, ISBN 3-89336-235-5.
2 Bacon, G.E. (1975) *Neutron Diffraction*, Clarendon Press, Oxford.
3 Sköld, K. and Price, D.L. (eds) (1986) *Neutron Scattering in Methods of Experimental Physics, Part A*, vol. **23**, Academic Press, Orlando, FL.
4 Warren, B.E. (1990) *X-Ray Diffraction*, Dover, New York.
5 Als-Nielsen, J. and McMorrow, D. (2011) *Elements of Modern X-Ray Physics*, John Wiley & Sons, Inc., Hoboken, NJ.
6 Varley, F. (1989) *Sears, Neutron Optics*, Oxford University Press, Oxford.
7 International Union of Crystallography (2006) International Tables for X-Ray Crystallography, vol. **C**, http://it.iucr.org (accessed 27 June 2016).
8 Stephen, W. (1987) *Lovesey, Theory of Neutron Scattering from Condensed Matter*, vol. **1 and 2**, Clarendon Press, Oxford.
9 Willis, B.T.M. and Carlile, C.J. (2009) *Experimental Neutron Scattering*, Oxford University Press.

5

Radiation Sources

5.1 Generation and Properties of Neutrons

Ina Lommatzsch, Wolfgang Knop, Philipp K. Pranzas, and Peter Schreiner

5.1.1 Introduction

Neutron scattering is one of the most effective ways to study microstructure and dynamics of condensed matter. A wide scope of problems, ranging from different fields like physics, chemistry, biology, medicine, environmental research, geology to materials science can be investigated with neutrons. For this reason, a high neutron flux $\Phi_{th} > 10^{13}\ n\,s^{-1}\,cm^{-2}$ with different spectra including cold, thermal, and hot neutrons is required of the neutron supplier/operator. Aside from the important neutron scattering techniques, nondiffractive methods like imaging techniques as well as isotope production can also be applied. Their relevance for industrial applications has increased over the past decades.

5.1.2 Generation of Neutrons

Two production methods for a reliable supply with high neutron fluxes $>10^{13}\ n\,s^{-1}\,cm^{-2}$ are at disposal: research reactors and spallation sources [1].

5.1.2.1 Research Reactors

Figure 5.1.1 shows how neutrons are produced by the nuclear fission process [3] that is used in research reactors.

The Uranium-235 nucleus has a high fission cross-section for thermal neutrons, thus it breaks into lighter elements and liberates two to three neutrons for every fissioned nucleus (average value 2.5).

The fast neutrons generated during this nuclear fission have an energy within the mega electronvolts range. In order to decelerate these fast neutrons, aiming to reach the thermal energy region (million electronvolts), impacts with moderator atoms are necessary (Figure 5.1.2). Only thermal neutrons can initiate new fission reactions, maintaining the process. The fission neutrons can be exploited for experiments.

A good moderator medium should have small neutron absorption as well as a high probability of scattering (deceleration by impact). Table 5.1.1 lists typical moderator material characteristics.

Neutrons and Synchrotron Radiation in Engineering Materials Science: From Fundamentals to Applications, Second Edition. Edited by Peter Staron, Andreas Schreyer, Helmut Clemens, and Svea Mayer.

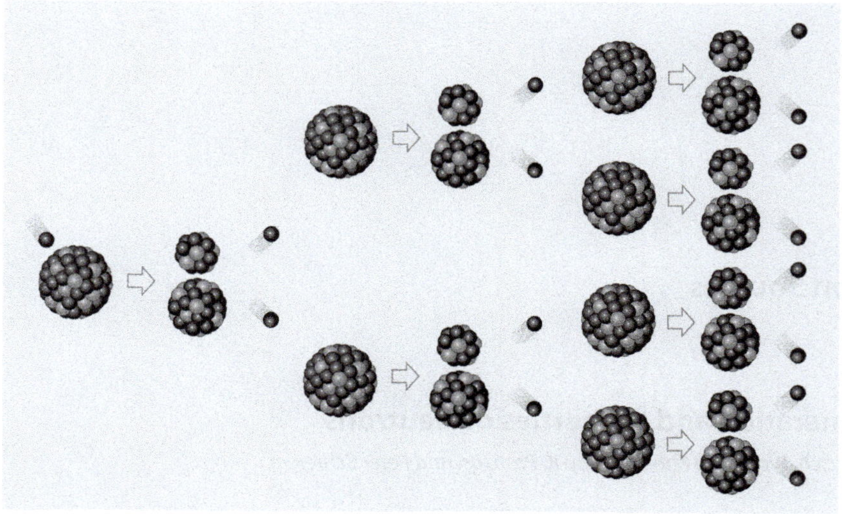

Figure 5.1.1 The fission process in a nuclear reactor (following [2]). A thermal neutron hits a U-235 nucleus, which decomposes into two lighter fission products, two to three neutrons, and gamma radiation (not shown in the figure).

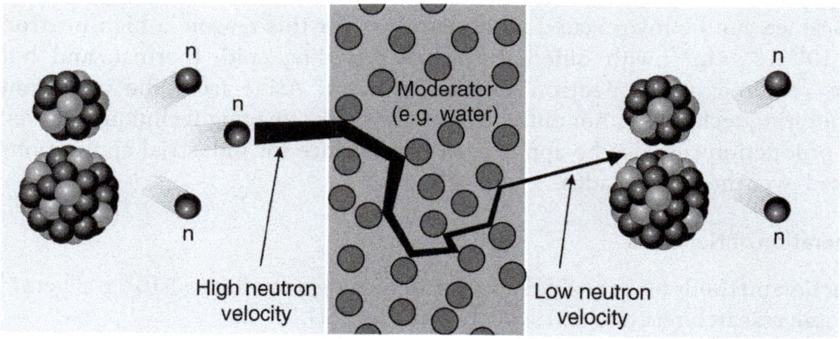

Figure 5.1.2 Moderation of neutrons (following [2]). Fast neutrons transfer energy to moderator molecules by collisions.

Table 5.1.1 Characteristics of moderator materials.

Element	Absorption (cm^{-24})	Scattering (cm^{-24})
H	0.33	82
D	0.00052	7.64
Be	0.0077	7.63
C	0.0035	5.55

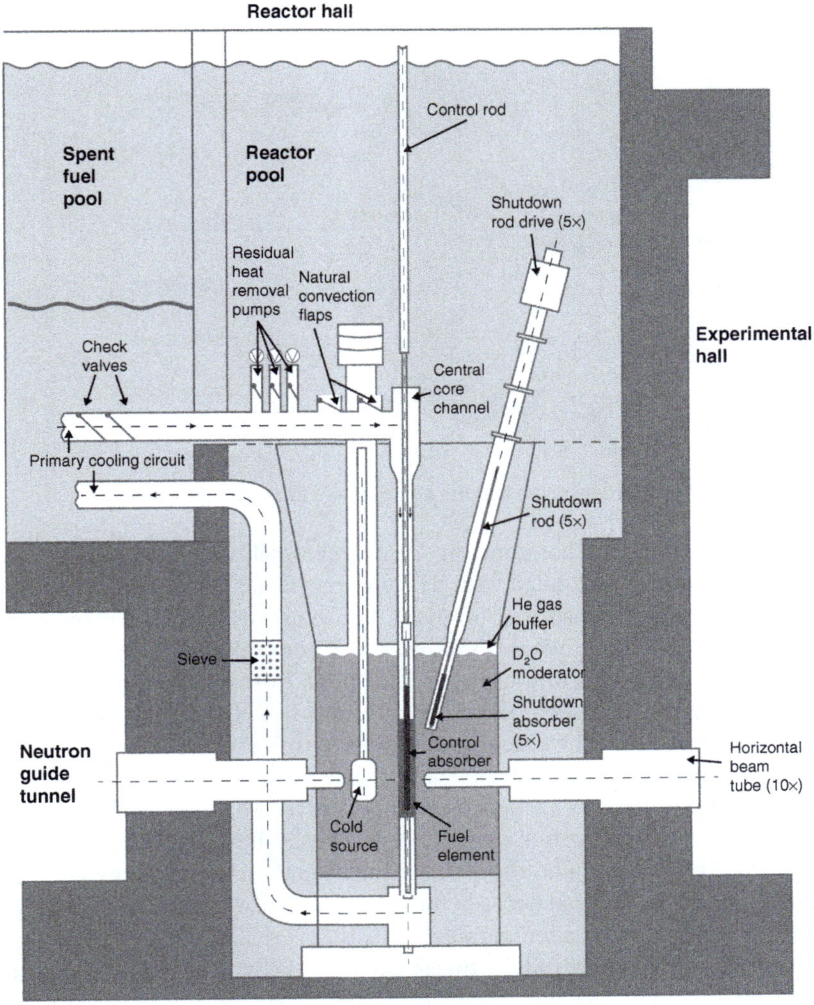

Figure 5.1.3 Sketch of the high flux reactor FRM II in Garching near Munich, Germany [4].

Table 5.1.1 reveals that deuterium and beryllium are very good moderator materials (small absorption coefficient and strong scattering effect). Hydrogen (or actually water) may possess a high scattering length but also – and this is more important – a higher neutron absorption coefficient. Therefore, heavy water reactors or beryllium-encased cores are mainly used for moderation.

An example for the principal layout of a research reactor is depicted in Figure 5.1.3. The chart shows the high flux neutron research source Heinz Maier-Leibnitz (FRM II) in Garching near Munich, Germany [4]. It is the most powerful source in Germany and reaches the highest neutron flux ($8 \cdot 10^{14}$ n s^{-1} cm^{-2}) compared to its thermal power (20 MW). The FRM II uses the concept of a compact core and thus, a single cylindrical fuel element is sufficient for 60 days of reactor operation. It is located in a vertical tube ("central channel"), cooled with light H_2O. According to Table 5.1.1, hydrogen (H)

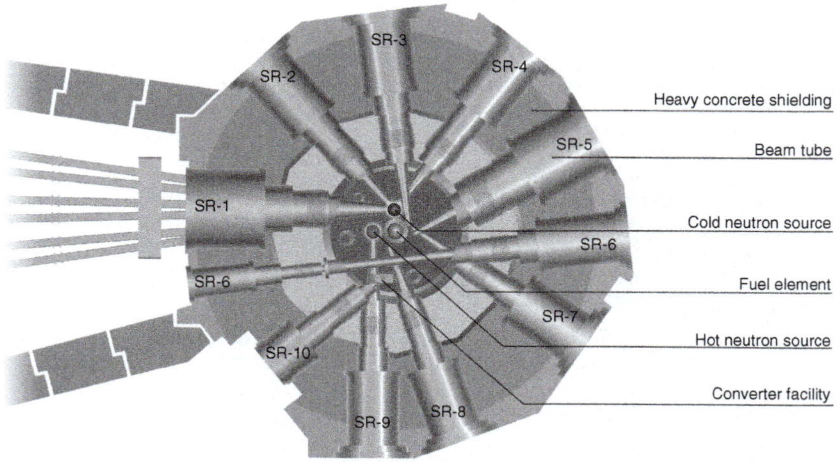

Figure 5.1.4 Sketch of the FRM II compact core and the arrangement of beam tubes (SR).

has a high absorption coefficient. Therefore, the "central channel" with the fuel element is surrounded by a moderator tank filled with heavy water (D_2O). For D_2O, the loss of neutrons is significantly smaller than that for H_2O. Consequently, more neutrons can be caught by the beam tubes and guided from the moderator tank to the experiments. Figure 5.1.4 shows a horizontal section of the reactor pool. In the center, the fuel element in the moderator tank can be seen. The moderator tank has a height and a diameter of 2.5 m. The tank is enclosed by the reactor pool, filled with 700 m^3 of highly purified water. In order to enable a large variety of applications (see Table 5.1.2), the instruments hosted at the FRM II are supplied by various secondary sources slowing down or speeding up the neutrons after leaving the fuel element. As the chart shows, the cold source is located directly above the fuel element, the distance between them is only 8 cm and this is very close to the maximum thermal neutron flux. It consists of a volume of 16 l of liquid deuterium (heavy hydrogen) at a temperature of 25 K (−250 °C). Through collisions with the deuterium atoms, the energy of the thermal neutrons is shifted to lower values, that is, they are "cooled." The beam tubes 1, 2, and 4 point at this source and make the cold neutrons available for half of the instruments. Beam tube 1 delivers neutrons to all instruments in the Neutron Guide Hall West. Within its in-pile unit, the beam is divided into six neutron guides that are mostly split aiming to distribute the neutrons to as many instruments as possible.

Table 5.1.2 Neutrons needed for experimental facilities.

Cold, hot, and thermal neutrons

- The fast neutrons generated during nucleus fission exhibit an energy of about 1 meV – too high for practical use
- Thus, they are forced to collide with a special material (moderator) and lose energy down to useable energy levels
- For scientific experiments we need
  - hot neutrons (0.1–0.5 eV) moderated at 2000 °C
  - thermal neutrons (0.01–0.1 eV) moderated at 40 °C
  - cold neutrons (0–0.01 eV) moderated at −250 °C

Next to the fuel element, the hot source can be found. It uses a graphite block as moderator, being well insulated and heated up to about 2300 K (1700 °C) due to the gamma radiation in the fuel element arising from the uranium fission. The hot source also heats the thermal neutrons to speeds up to several thousand kilometers per second. These hot neutrons are passed through beam tube 9. In front of the tip of beam tube 10, an arrangement of uranium plates is inserted as a converter. By this, a high-intensity neutron beam with an unmoderated fission spectrum is obtained. The through-going beam tube 6 will be used by the ultra-cold neutron source, whereas the remaining beam tubes are placed into the highest neutron flux taking up the thermal neutrons.

5.1.2.2 Spallation Sources

The principle process that takes place in a spallation source [5] is schematically shown in Figure 5.1.5: Protons, generated in a large accelerator facility, dislodge one or two nucleons with high energy from a nucleus through direct impact and transfer a large amount of energy into the entire volume of the nucleus. This energy is then released from the nucleus by evaporating up to 10 spallated neutrons, which has to be compared with the two to three neutrons for each fission process.

Independently of how the neutrons are produced (nuclear fission, spallation), the remaining facilities for both ways are similar in principle. The emerging fast neutrons must be moderated (Figure 5.1.2) in order to use them for different experiments.

Information about the new European Spallation Source that is currently built in Lund, Sweden, can be found at [7].

5.1.3 Instrumentation

Neutrons are used as radiation source for structural investigations in a large number of different scattering instruments. As an example, the Neutron Guide Halls (West and East) as well as the Experimental Hall of the FRM II at Garching are shown in Figure 5.1.6.

The Heinz Maier-Leibnitz Zentrum (MLZ) represents the cooperation of the Technische Universität München (TUM) and three research centers of the Helmholtz Association, namely Forschungszentrum Jülich (FZJ), Helmholtz-Zentrum Geesthacht (HZG), and Helmholtz-Zentrum Berlin (HZB, inactive partner), to exploit the scientific use of the research reactor FRM II in Garching. In addition, MLZ relies on a network of strong

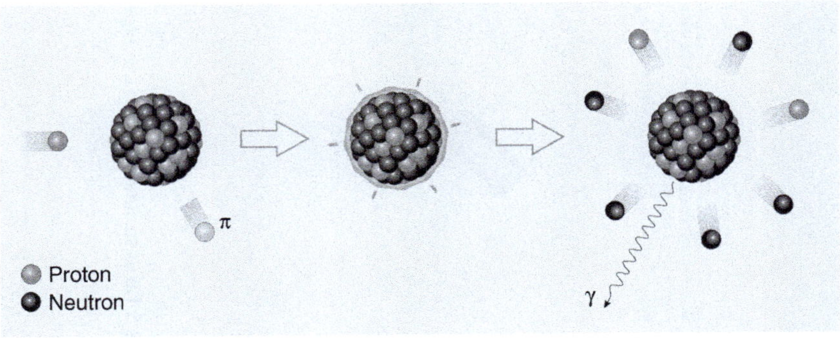

Figure 5.1.5 The spallation process (following [6]).

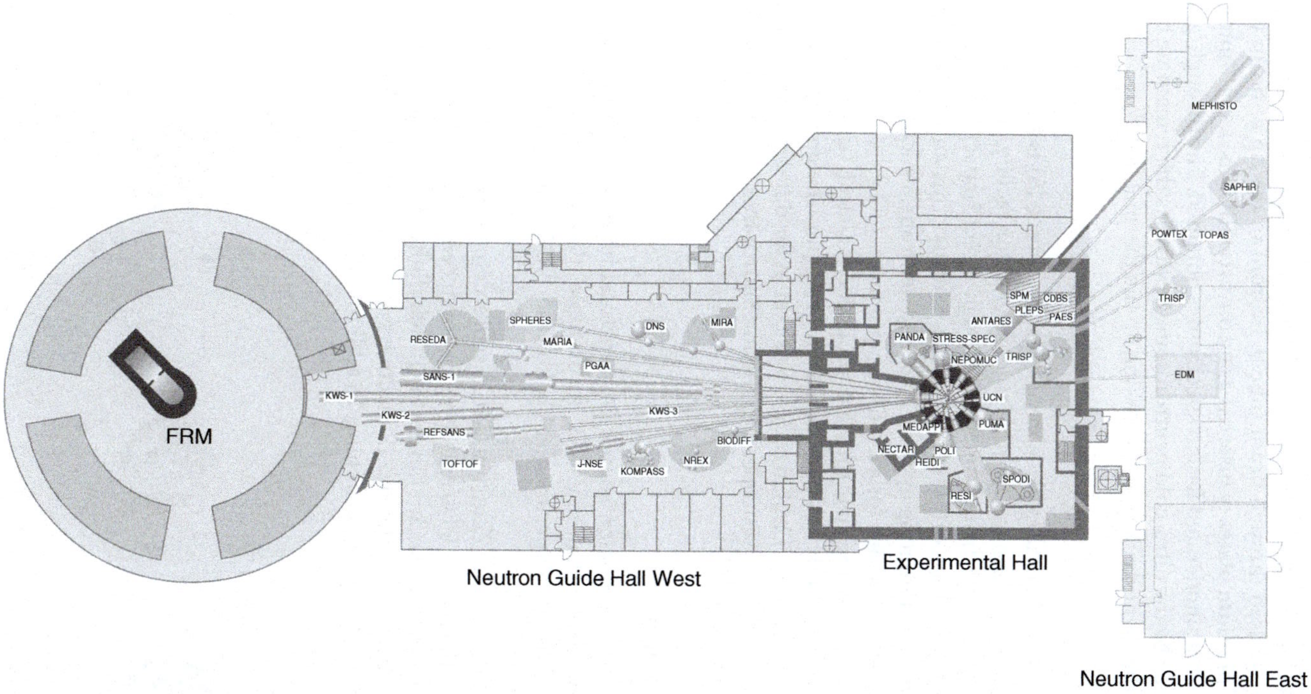

Figure 5.1.6 Instrumentation in the experimental and neutron guide halls of the FRM II [4].

partners among the German universities and institutes of the Max-Planck-Society. The result is a suite of world-class instruments available for the benefit of scientific users around the world. In the following text, the instruments grouped by their main application are described in brief.

5.1.3.1 Structure Research

Instruments to investigate atomic positions, lattice spacings, and magnetic structures are defined as single crystal diffractometers (including protein crystallography) and powder diffractometers (including engineering diffractometers). They mainly use a fixed, monochromatic incoming wavelength. Common to these instruments is the analysis of scattered neutrons, which does not include an energy separation. An optional analyzer might separate elastic from otherwise diffuse inelastic scattering; the spectroscopic measurement of atomic movements is, however, not the focus of these instruments.

5.1.3.2 Large-Scale Structures

All the small-angle neutron scattering (SANS) instruments and reflectometers share a similar forward focusing scattering geometry, and accordingly the structural sizes they investigate are also similar. When performing grazing incidence experiments (GISANS) on a SANS machine, the distinction between these types of instruments vanishes.

5.1.3.3 Spectroscopy

Investigating the movements of atoms on a time scale from pico- to microseconds is the domain of neutron spectroscopy. The bandwidth is wide enough to cover the measurement of excitation energies on three-axes instruments and time-of-flight spectrometers through to the measuring of slow, sometimes diffusive motion using high-resolution methods such as backscattering or spin-echo. Combining these methods reveals unprecedented insights into the dynamics of condensed matter. Using polarized neutrons, a detailed investigation of magnetic excitations becomes feasible.

5.1.3.4 Imaging and Analysis

Neutrons have a high penetration depth for many materials that are practically opaque for X-rays; furthermore, excellent and even isotope-specific contrast is achieved for many light materials. This fact renders neutron radiography and tomography powerful tools complementary to X-ray imaging techniques, which have encountered a tremendous increase in popularity in nondestructive testing applications. In addition, the magnetic moment of a neutron gives access to new methods such as the spatially resolved investigation of magnetic structures and magnetic fields, an area which up to now has been barely explored. Neutron imaging combined with elemental analysis using prompt-gamma activation analysis opens up a large variety of applications from engineering and archeometry through to solid-state physics.

5.1.3.5 Particle Physics

As a hadron, the neutron is involved with all four elementary forces of the Standard Model. These forces range from the largest scales in astronomy to the smallest dimensions in nuclear and particle physics. To distinguish between the different competing theories describing the origin of our universe, the underlying fundamental constants of these models must be determined as accurately as possible. The most prominent examples in these fields of experiments with cold and ultra-cold neutrons are the decay and lifetime of the free neutron and the question of the electric dipole moment of the neutron itself.

An up-to-date overview of neutron facilities worldwide is available in [8].

References

1 Dianoux, A.J. and Lander, G. (eds) *Neutron Data Booklet*, Institute Laue Langevin, http://www.ill.fr. (accessed 29 June 2016).
2 Volkmer, M. (2013) *Basiswissen Kernenergie*, Deutsches Atomforum e.V., Berlin, http://www.kernenergie.de (accessed 29 June 2016).
3 Glasstone, S. and Edlund, M.C. (1952) *Elements of Nuclear Reactor Theory*, Van Nostrad, Princeton, NJ.
4 Heinz Maier-Leibnitz Zentrum. http:// www.mlz-garching.de (accessed 29 June 2016).
5 Carpenter, J.M. (1977) Pulsed spallation neutron sources for slow neutron scattering. *Nucl. Instrum. Methods*, **145**, 91–113.
6 http://neutron.neutron-eu.net.
7 European Spallation Source. http://www.esss.se (accessed 29 June 2016).
8 Neutron Scattering Web. http://www.neutron.anl.gov/ (accessed 29 June 2016).

5.2 Production and Properties of Synchrotron Radiation

Rolf Treusch

5.2.1 Introduction

It is well known that charged particles emit electromagnetic radiation when accelerated. One example from classical electrodynamics is the accelerated (and decelerated) motion of electrons in an antenna yielding the well-known $\cos^2\Theta$ - distribution of the Hertz dipole (see, e.g., [1]). The same effect also forms the basis for the currently widely used Synchrotron radiation (SR). Synchrotron radiation (in the following mostly "SR" for brevity) is generated in particle accelerators when charged particles are accelerated to relativistic velocities (with a Lorentz factor γ of typically a few hundreds to several thousands) and deflected in magnetic fields, for example, in order to keep them on a circular orbit. Due to their high ratio of charge to mass (q/m), SR is particularly pronounced for electrons and positrons. SR got its name because it was first observed at the General Electric 70 MeV Synchrotron in Schenectady, New York in 1947 [2].

More than 100 years ago, at the end of the nineteenth century, Larmor already worked out the theoretical description of the radiation of electrons on a circular orbit [3]. In contrast to the Hertz dipole, motion and acceleration vectors are not parallel, but perpendicular here. The centripetal force which keeps the electron on its orbit by accelerating it toward the center is typically the Lorentz force of a dipole magnet. Larmor's work dealt with slow, nonrelativistic electrons, and yielded a doughnut-like pattern for the spatial distribution of the radiation (Figure 5.2.1a). For the transition to the case of relativistic

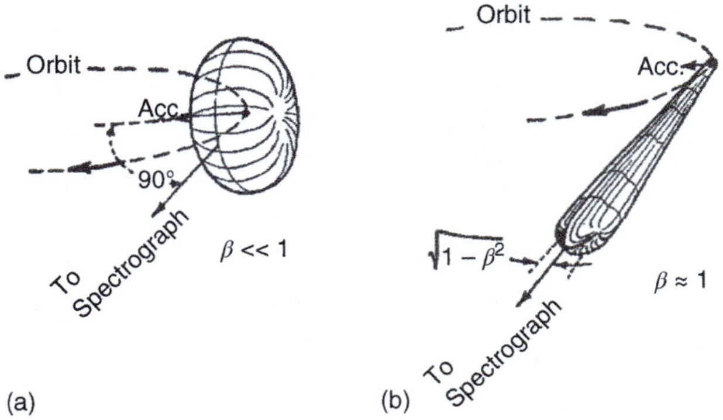

Figure 5.2.1 Angular distribution of radiation emitted from charged particles on a circular orbit: (a) nonrelativistic velocities and (b) relativistic case. (Tomboulian 1956 [4]. Reproduced with permission of American Physical Society.)

motion, basically a Lorentz transformation has to be applied to Larmor's classical treatment. As worked out by Ivanenko and Pomeranchuk [5] and Schwinger [6], this Lorentz transformation results in a strongly forward directed radiation cone with a typical opening angle of $1/\gamma$ in the range of a few milliradian down to some tenths of a milliradian (Figure 5.2.1b).

In the first particle accelerators, which were mainly used for nuclear and high-energy physics, SR was a rather undesirable byproduct, since it significantly affected the machine performance. The radiation power per particle is

$$P \simeq \frac{2e^2c}{3\rho^2} \cdot \beta^4 \cdot \left(\frac{E}{mc^2}\right)^4 = \frac{2e^2c}{3\rho^2} \cdot \beta^4 \cdot \gamma^4 \tag{5.2.1}$$

where E is the particle energy, e its charge, mc^2 is the particle rest mass, $\beta = v/c$ is the speed of the particles with respect to the speed of light, and ρ is the orbit radius,

$$\rho(\text{m}) = 3.34\frac{E(\text{GeV})}{B(\text{T})} \tag{5.2.2}$$

It is evident that for proton accelerators this radiation loss is in most cases negligible, since – due to the much higher mass – it is about 10^{13} times smaller than for electrons or positrons. For the latter two, the radiative energy loss per turn is (in practical units)

$$I_{\text{tot}}(\text{W}) \simeq 88.5\frac{E^4(\text{GeV})}{\rho(\text{m})}j(\text{mA}) \tag{5.2.3}$$

where j is the circulating current. For today's machine energies of a few gigaelectronvolts and currents of some 100 mA it typically is in the range of several 100 kW to few megawatts and hence, apart from the more technical problems in accelerator design, a substantial part of the electricity bill. Scaling with E^4/ρ it also sets a "natural" limit to the upscaling in energy for particle physics accelerators which have today, for example, with the 27 km circumference of LHC at CERN, already reached enormous dimensions. For much higher energies, it is obviously advisable to scale up the radius ρ to infinity, that is, use a linear accelerator in order to avoid the production of SR. Whereas in the 1960s–1970s, SR has been mostly used in a "parasitic" mode at the particle physics accelerators, the growing demand during the past 40 years resulted in a continuously growing number of facilities ("storage rings," see Section 5.2.3) which are exclusively dedicated to the generation of SR. Presently, about 50 facilities around the world [7] serve thousands of users in a wide variety of fields in fundamental and applied research, ranging from physics, biology, chemistry across crystallography, materials, and geological sciences to medical applications. SR is well suited for all these fields since it possesses the following outstanding properties:

1) continuous spectrum from the infrared (IR) to X-rays
2) high intensities
3) small source size
4) collimation in forward direction
5) high degree of linear polarization in the orbit plane
6) elliptical or circular polarization above and below the orbit plane (for bending magnets)
7) well-defined, pulsed time structure (short electron/positron bunches in machine)
8) quantitatively known characteristics
9) clean (ultra high vacuum (UHV)) environment.

These properties will now be detailed (additional background information, also on beamline optics and experimental techniques can be found in [8–12]).

5.2.2 Properties of Synchrotron Radiation

(1) Spectrum

Due to the strong forward collimation of the radiation and the circular motion of the electron/positron bunches in the storage ring (see Section 5.2.3), it very much resembles a sweeping search light. An external, stationary observer who looks tangentially onto the circular orbit sees only a short flash of radiation every time the bunches pass by. Due to the shortness of the pulses, one observes besides the rotation frequency ω_o also its higher harmonics up to a typical frequency of

$$\omega_{typ} \simeq \frac{3\pi c\gamma^3}{2\rho} = \frac{3\pi}{2}\omega_o\gamma^3 \tag{5.2.4}$$

Since the rotation frequency ω_o is not exactly constant due to so-called betatron and synchrotron oscillations around the nominal orbit (see, e.g., [13]), the IR harmonics are already smeared out to a continuum. For bending magnet radiation, the spectrum as a function of photon energy can be displayed in "universal curves" (Figure 5.2.2) with one characteristic parameter, the critical energy

$$\varepsilon_c\,(\mathrm{keV}) = 0.665 \cdot E^2\,(\mathrm{GeV}) \cdot B_0\,(\mathrm{T}) \tag{5.2.5}$$

which is defined such that 50% of the total radiation energy is emitted above and below ε_c, respectively. B_o is the magnetic field of the bending magnet which is typically around 1 T.

The continuous spectrum of SR allows to tune the photon energy to the experimental demands using an appropriate monochromator, for example, a grating or crystals. Figure 5.2.3 illustrates the length scales corresponding to typical wavelength ranges of today's SR sources. Storage rings with energies below 1 GeV are best suited for the IR to ultraviolet (UV) part of the spectrum, whereas multi-gigaelectronvolts storage rings are

Figure 5.2.2 "Universal curves" for the spectral distribution of photon flux as a function of machine energy.

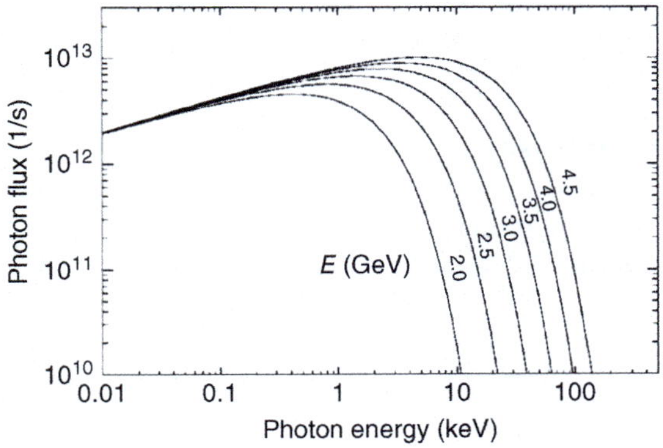

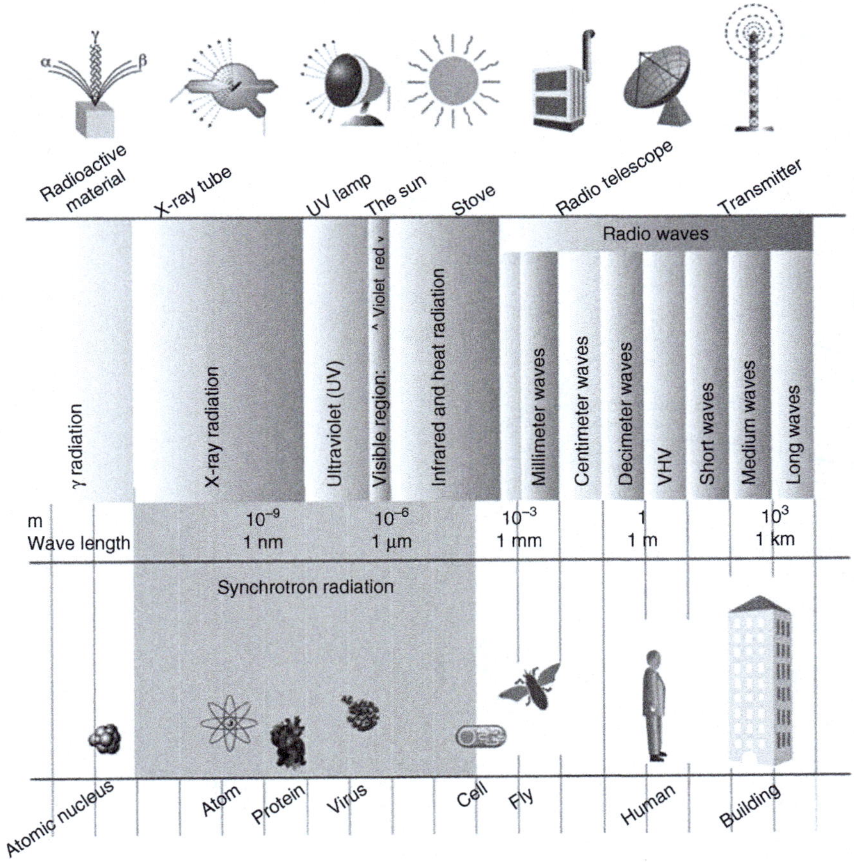

Figure 5.2.3 Length and wavelength scales from macroscopic across microscopic toward nuclear dimensions.

required for the X-ray range with wavelengths around 1 Å and access to corresponding atomic length scales for scientific investigations.

(2) Intensities

Usually, intensities of SR sources are compared in terms of brilliance, which is the number of photons per time interval (s), per source size (mm^2), per opening angle ($mrad^2$), and per 0.1% spectral bandwidth. Figure 5.2.4 compares a few typical examples of SR

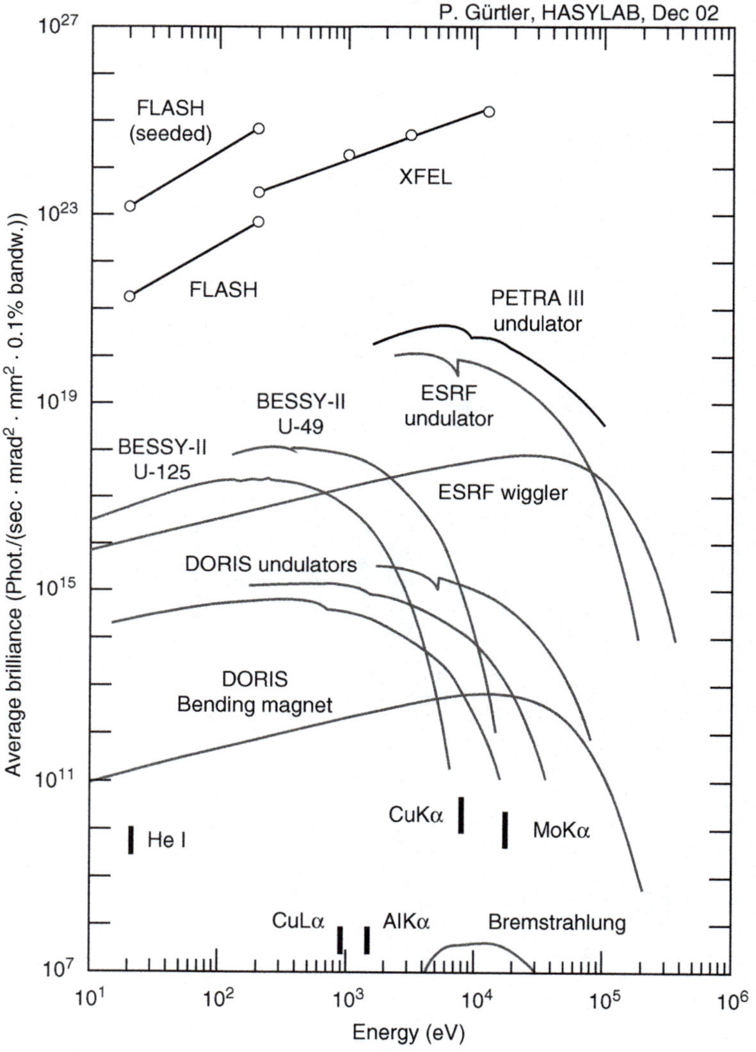

Figure 5.2.4 Comparison of different radiation sources from X-ray tubes across synchrotron radiation storage rings toward the free electron lasers (FELs) at DESY (FLASH and the future European XFEL) in terms of average brilliance (see units at left) as a function of photon energy. Note that the practical conversion from energy to wavelength is $\lambda(\text{Å}) = 12.398/E(\text{keV})$.

sources (from DORIS[1] and PETRA III at DESY, the ESRF in Grenoble and BESSY II in Berlin) to other sources, such as X-ray tubes, He lamps and above all of that, the newest technological developments – free electron lasers (FELs). Compared to X-ray tubes with rotating anodes, SR is superior by 5–10 orders of magnitude which, besides the general advantage of continuous tunability, allows to get better signals in shorter time with less sample volume required and/or better energy resolution and spatial resolution. A short outlook on FELs which yield another five orders of magnitude enhancement of average brilliance will be given at the end of this chapter.

(3) Source Size and Collimation

Typical electron beam diameters and, hence, SR source sizes of modern storage rings are in the range of 100 μm. This small size along with the high degree of collimation which has already been illustrated in the introduction of this chapter (Figure 5.2.1) makes the SR beam well suited for focusing onto small samples or for space resolved techniques. Beam sizes of a few 10 μm at the sample can easily be achieved and even submicron focusing is feasible with the appropriate optics.

(4) Polarization

A typical example for the polarization properties of SR is depicted in Figure 5.2.5. Shown are the horizontal (σ), vertical (π), and total polarization component of bending magnet radiation for different wavelengths as a function of the elevation angle ψ relative to the orbital plane of the storage ring. In the orbit plane, the beam is only linearly (horizontally) polarized and with growing elevation angle it becomes elliptically polarized

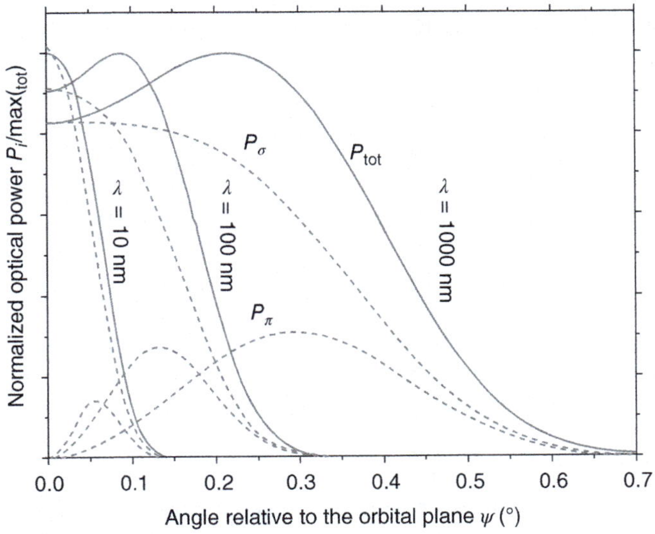

Figure 5.2.5 Example of angular distribution and polarization of synchrotron radiation from a bending magnet as function of emitted wavelength (http://physics.nist.gov/MajResFac/SURF/SURF/sr.html).

1 Update note for the second edition of this book: after 38 years of operation, DORIS was shut down in January 2013. The operation for photon users ended already on October 22, 2012.

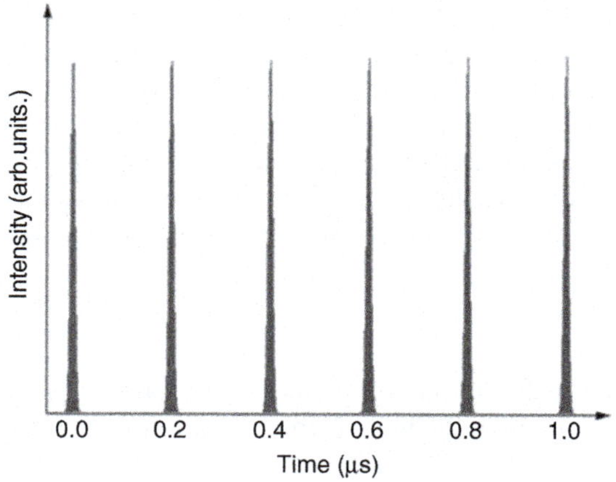

Figure 5.2.6 Illustration of the pulsed time structure of synchrotron radiation. For details see the text.

and finally obtains full circular polarization. The longer the wavelength (the lower the photon energy) the wider is the angular distribution.

Note that the polarization properties of radiation from (symmetric) wigglers and undulators are different, as will be discussed later.

(5) Time Structure

In SR storage rings, from one up to several hundred electron bunches circulate at almost the speed of light. A ring circumference of, for example, 300 m implies that each individual bunch runs around at 1 MHz repetition rate and all bunches together yield SR pulse repetition rates of one up to several 100 MHz. This pulsed structure of SR is illustrated in Figure 5.2.6. The typical spacing between the pulses (depending on the storage ring and its specific operation mode/filling pattern) ranges from 2 ns to 1 μs and the typical pulse length (electron/positron bunch length divided by *c*) from 10 to 100 ps.

(6) Quantitative Description of SR

Some examples with formulas are given in the text, for additional information the reader is referred to [8–12].

(7) Clean Environment

Since the storage rings are operated under UHV conditions (typically 10^{-9} mbar) and since no debris is produced – as, for example, from X-ray laser targets – one can directly connect UHV chambers, for example, for surface science experiments.

5.2.3 Sources of Synchrotron Radiation

Today's machines solely dedicated to the production of SR are mostly so-called storage rings where the particles are kept on a closed orbit at constant energy. A few of the basic ingredients of a SR storage ring are sketched in Figure 5.2.7. Electrons or positrons, respectively, are running around on a closed orbit at about 99.999% the speed of light and emit SR when they pass bending magnets or special magnetic insertion devices. Since a substantial amount of kinetic energy of the particles is lost via SR, the bunches have to be reaccelerated every turn running through accelerating radio frequency cavities.

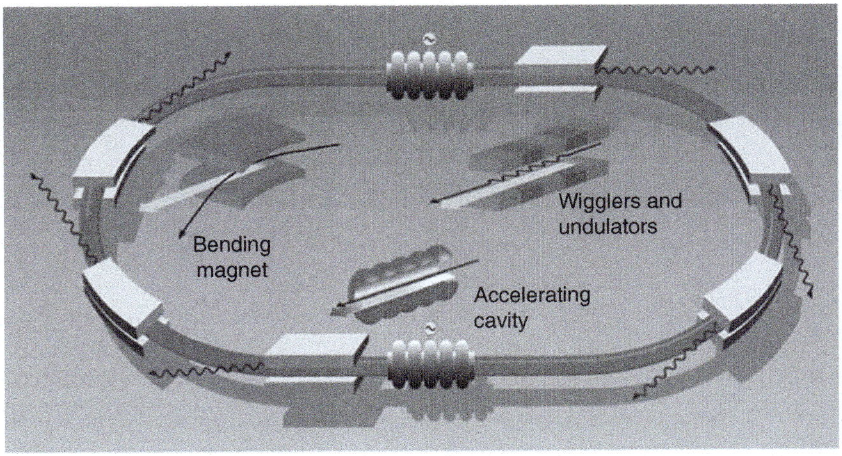

Figure 5.2.7 Sketch of a synchrotron radiation storage ring where either electron or positron bunches are traveling through a closed system of evacuated tubes with dipole magnets to force the particles onto a closed orbit. Energy losses due to synchrotron radiation are compensated in radio-frequency cavities which reaccelerate the particles. Note that this is a very simplified picture, in particular omitting all other magnets, such as quadrupoles and sextupoles and all the beam diagnostics around the ring.

During the last 30 years, advanced magnetic insertion devices for SR, called *wigglers* and undulators, became more and more the key components of SR storage rings. Wigglers and undulators are periodic magnetic multipole structures of typically a few meters length which are put into straight sections of the storage rings. The repetitive left and right turns of the electron/positron bunches result in no net deviation from the original direction but in a significant enhancement of the emitted SR as will be outlined below. A simplified overview of the difference between the sources of SR is given in Figure 5.2.8 and will be discussed in detail in the following section.

5.2.3.1 Bending Magnets

For a *bending magnet*, the horizontal fan of radiation corresponds roughly to the net deflection angle which is typically in the order of a few up to several ten degrees. This means that even several beamlines can collect radiation (typically a few milliradian horizontally each) from different points of the bent trajectory. The characteristics of SR from a bending magnet can be accurately calculated via the Schwinger equation [6]

$$I(\lambda,\psi) = \frac{27e^2c}{32\pi^3\rho^3} \cdot \left(\frac{\lambda_c}{\lambda}\right)^4 \cdot \gamma^8 \cdot (1+(\gamma\psi)^2)^2 \cdot \left(K_{2/3}^2(\xi) + \frac{(\gamma\psi)^2}{1+(\gamma\psi)^2} \cdot K_{1/3}^2(\xi)\right) \tag{5.2.6}$$

with ρ, the bending radius of the trajectory according to Eq. (5.2.2), and

$$K_X(\xi) \tag{5.2.7}$$

modified Bessel functions, where the first term with $K_{2/3}$ corresponds to radiation polarized parallel to the orbit plane and the second term with $K_{1/3}$ corresponds to the radiation component polarized perpendicular to the orbit plane with

$$\xi = \frac{\lambda_c}{2\lambda} \cdot (1+(\gamma\psi)^2)^{3/2} \tag{5.2.8}$$

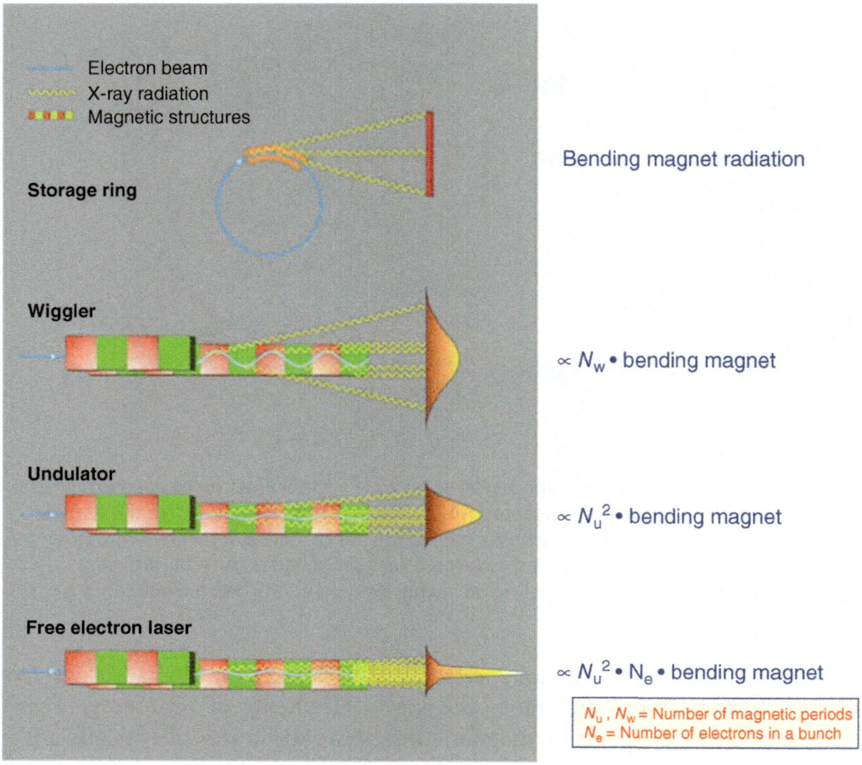

Figure 5.2.8 Sources of synchrotron radiation.

where λ_c is the critical wavelength, which can be derived from the critical energy described before as

$$\lambda_c\,(\text{Å}) = \frac{12.398}{\varepsilon_c\,(\text{keV})} = 5.59\frac{\rho\,(\text{m})}{E^3\,(\text{GeV})} \tag{5.2.9}$$

where γ is the usual Lorentz parameter and ψ again the elevation angle from the orbit plane.

5.2.3.2 Wigglers and Undulators

As already mentioned above, wigglers and undulators are special magnetic multi-pole devices with alternating magnetic field, where the electrons/positrons are forced on a usually sinusoidal trajectory (Figure 5.2.9) in order to enhance the radiation output.

A wiggler with its N magnetic periods is basically a superposition of a certain number $2N$ of bending magnet sources with a bending radius given by the strength of the alternating magnetic field. Wigglers and undulators are described by the wiggler parameter

$$K := \alpha \cdot \gamma = \frac{eB_0\lambda_0}{2\pi m_e c} \tag{5.2.10}$$

which gives the ratio of deflection angle α of the particles in the B-field with respect to opening cone $1/\gamma$ of the emitted radiation. λ_0 is the period length of the magnetic field

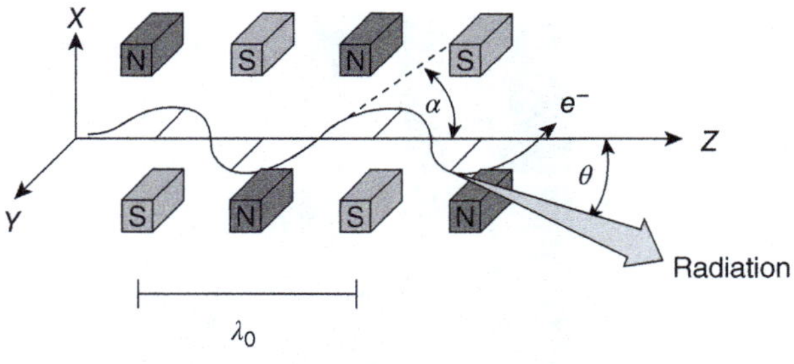

Figure 5.2.9 Principle of the periodic motion of electrons in the magnetic field of a wiggler/undulator along with the characteristic parameters as used in the text.

and B_0 its peak value. In practical units

$$K = 0.934 \cdot \lambda_0\,(\text{cm}) \cdot B_0\,(\text{T}) \tag{5.2.11}$$

If $K \gg 1$, one speaks of a *wiggler.* Here, the intensity enhancement is about $2N$-fold compared to a single dipole of same strength and the radiation is emitted in forward direction with typical horizontal opening angles in the milliradian range.

If $K \lesssim 1$, the device is called an *undulator*. In this case, it is not just a pure superposition of $2N$ independently radiating bending magnet sources. Since the net deflection is similar or smaller than the typical SR opening cone, radiation emitted from different poles can interfere constructively. An undulator can be treated in a simple geometrical picture where one just calculates the path difference between the photons emitted from consecutive poles. This setup very much resembles the emission of radiation from a $2N$-fold slit and it is hence not surprising that the intensity enhancement of an undulator is not just $2N$-fold but scales with N^2 and results in certain resonance wavelengths of undulator emission given by

$$\lambda_i = \frac{\lambda_0}{2i\gamma^2} \cdot \left(1 + \frac{K^2}{2} + \gamma^2\Theta^2\right) \tag{5.2.12}$$

where λ_i is the wavelength of the i-th harmonic, λ_0 the period length of the magnetic field, and Θ the observation angle with respect to the forward direction.

To obtain the energy and angular distribution of wiggler/undulator radiation, one has to put the (sinusoidal) motion of electrons in the periodical magnetic field of the wiggler/undulator in the far field approximation of the radiation equation

$$\frac{\mathrm{d}I(\omega)}{\mathrm{d}\Omega} = \frac{e^2\omega^2}{4\pi^2 c}\left|\int_{-\infty}^{\infty} \vec{n} \times (\vec{n} \times \vec{\beta})\ \exp(i\omega(t - \vec{n} \cdot \vec{r}(t)/c))\ \mathrm{d}t\right|^2 \tag{5.2.13}$$

(Eq. (14.67) from [1]). A detailed treatment of this procedure and the results can, for example, be found in [14]. Figure 5.2.10 shows examples of a wiggler spectrum and an undulator spectrum with the characteristic "undulator lines." At the right, the fluorescence caused by a white (=full spectrum) wiggler beam traveling through air is depicted.

This also illustrates that SR can be really intense, with up to several kilowatts of radiation power in a "white," non-monochromatized wiggler beam and power densities up to the order of kW mm^{-2}. This heatload is quite a challenge, advanced beamline optics, beamstops and filters, and so on, have to deal with, for example, via special cooling designs or adaptive (distortion compensating) optics.

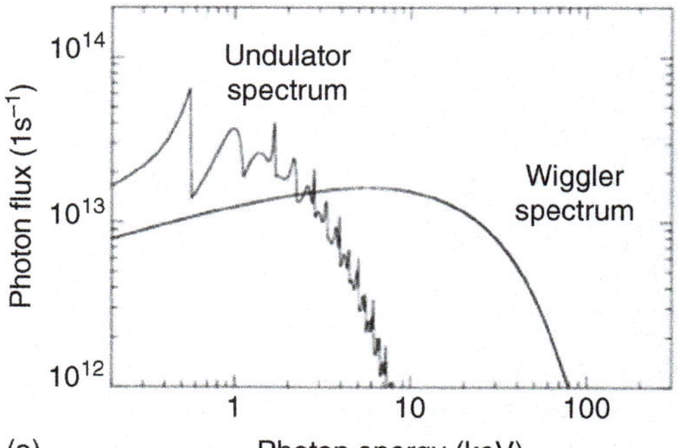

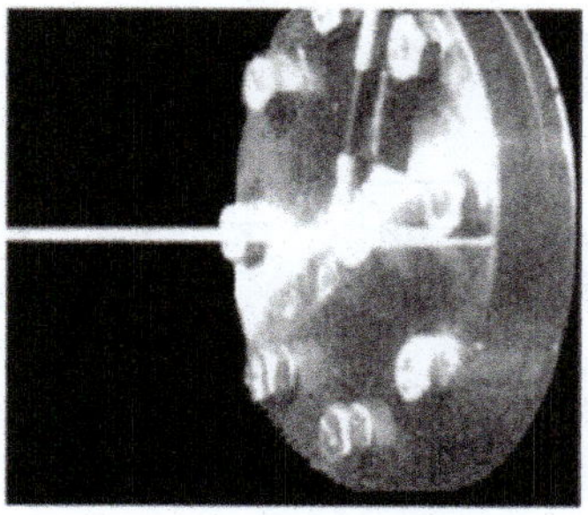

Figure 5.2.10 Typical (angle-integrated) wiggler and undulator spectra (a) along with the trace of a "white" (= non-monochromatized) wiggler beam producing intense visible fluorescence while traveling through air. (From ID11 at the ESRF, Grenoble) (b).

Note that for wigglers and undulators, since the trajectory has an equal number of left and right turns, left- and right-handed circular polarization above and below the orbit plane, respectively, cancel out. On the other hand, insertion devices with full polarization control, such as helical undulators or asymmetric wigglers are also available today [15].

5.2.4 Outlook: Free Electron Lasers

What will be presented here is mainly a short excerpt from [16]. More background information for the interested reader can, for example, be found in [17].

The basic principle of the FEL can be described within the standard picture for the generation of SR: while traveling with relativistic velocity ($v \simeq c,\ \gamma \simeq 10^2 - 10^5$) through the undulator, the electrons are accelerated in the direction transverse to their propagation due to the Lorentz force introduced by the magnetic field. They propagate along a sinusoidal path and emit SR in a narrow cone in the forward direction. In the undulator, the deflection of the electrons from the forward direction is comparable to the opening angle of the SR cone. Thus the radiation generated by the electrons while traveling along the individual magnetic periods overlaps (as described before). This interference effect is reflected in the formula for the wavelength λ of the first harmonic of the spontaneous, on-axis ($\Theta = 0$) undulator emission (cf. also Eq. (5.2.12))

$$\lambda = \frac{\lambda_0}{2\gamma^2}\left(1 + \frac{K^2}{2}\right) \tag{5.2.14}$$

The interference condition basically means that, while traveling along one period of the undulator, the electrons slip by one radiation wavelength with respect to the (faster) electromagnetic field. This is one of the prerequisites for the SASE (self-amplified spontaneous emission) process of the FEL. To obtain an exponential amplification of the spontaneous SR emission present in any undulator, some additional criteria have to be met: One has to guarantee a good electron beam quality and a sufficient overlap between radiation pulse and electron bunch, which generates it, along the undulator. To achieve that, one needs a low emittance, low-energy spread electron beam with an extremely high charge density in conjunction with a very precise magnetic field and accurate beam

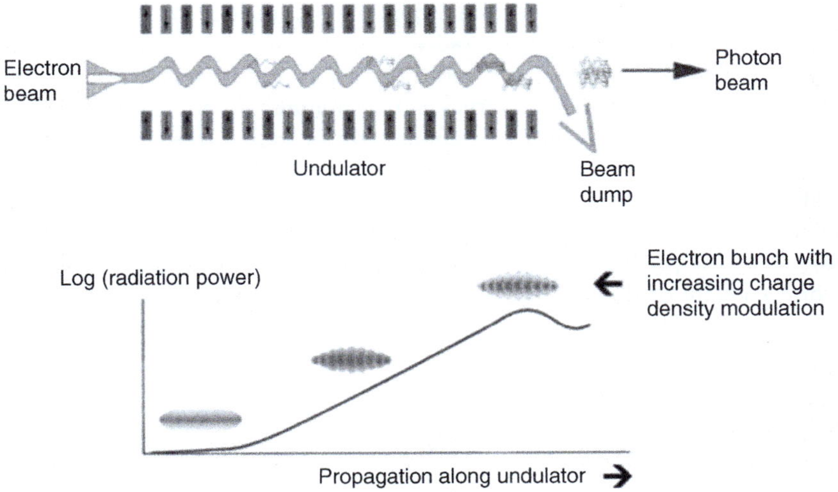

Figure 5.2.11 Sketch of the self-amplification of spontaneous emission (SASE) in an undulator resulting from the interaction of the electrons with the synchrotron radiation they emit. In the lower part of the figure the longitudinal density modulation (microbunching) of the electron bunch is shown together with the resulting exponential growth of the radiation power along the undulator.

steering through a long undulator. Note that electron bunches of the high quality that is necessary to get lasing in the vacuum-ultraviolet (VUV) or X-ray range cannot be produced and transported in a storage ring. It needs a state-of-the-art linear accelerator, for example, the one built in TESLA technology that is used for FLASH (Free Electron Laser in Hamburg) at DESY [18].[2]

Oscillating through the undulator, the electron bunch then interacts with its own electromagnetic field created via spontaneous undulator emission. Depending on the relative phase between radiation and electron oscillation, electrons experience either a deceleration or acceleration: electrons that are in phase with the electromagnetic wave are retarded while those with opposite phase gain energy. Through this interaction a longitudinal fine structure, the so-called micro-bunching, is established which in turn amplifies the electromagnetic field (Figure 5.2.11).

The longitudinal distribution of electrons in the bunch is "cut" into equidistant slices with a separation corresponding to the wavelength λ of the emitted radiation which causes the modulation. More and more electrons begin to radiate in phase, which results in an increasingly coherent superposition of the radiation emitted from the microbunched electrons. The more intense the electromagnetic field gets, the more pronounced the longitudinal density modulation of the electron bunch and vice versa. In the beginning – without microbunching – all N_e electrons in a bunch ($N_e \geq 10^9$) can be treated as individually radiating charges with the power of the spontaneous emission $\propto N_e$ (this is the usual SR from an undulator!). With complete microbunching, all electrons radiate almost in phase. They basically behave like a single point charge $N_e \cdot e$. This leads to a radiation power $\propto N_e^2$ (see also Figure 5.2.8) and thus an amplification of

2 At the time of writing, lasing down to a shortest wavelength of 4.1 nm in the EUV/soft X-ray range has been obtained at FLASH with pulse energies up to 500 μJ and pulse lengths from 200 fs down to about 10 fs.

many orders of magnitude with respect to the spontaneous emission of the undulator. Due to the progressing microbunching, the radiation power of such a SASE FEL grows exponentially with the distance z along the undulator (Figure 5.2.11, bottom).

The peak brilliance of such SASE FELs surpasses the spontaneous undulator radiation from today's state-of-the-art SR facilities by about eight or more orders of magnitude while the average brilliance is about five orders of magnitude higher (see Figure 5.2.4). The peak brilliance is the brilliance scaled to the length of a single pulse while average brilliance is normalized to seconds at the highest possible repetition rate.

In summary, with the present FLASH facility and future XFELs around the world, in comparison to a state-of-the-art SR source (spontaneous undulator radiation) one obtains

- $10^8 \times$ more peak brilliance
- short pulses ($\approx$ 100 fs vs 100 ps)
- full transverse coherence
- partial/almost full temporal coherence.

This opened the door to a completely new class of experiments, ranging from non-linear X-ray physics, across femto (second)chemistry to plasma physics and biological (protein) dynamics. FELs will give scientists, for example, insight into hitherto unknown properties of materials.

5.2.5 Summary

Today's SR sources as well as the upcoming FELs provide unique capabilities for a wide range of applications from physics and material sciences across geology, environmental sciences, chemistry, and biology toward medical and industrial applications. With their unique combination of properties such as the wide wavelength spectrum (respectively tunability) from the IR to the X-ray range, a tightly collimated beam, high intensities, and short pulses, these radiation sources allow us to study the subtleties of the structure of matter down to atomic length scales and with the FELs even at femtosecond time scales.

References

1 Jackson, J.D. (1962) *Classical Electrodynamics*, John Wiley & Sons, Inc., New York.
2 Elder, F.R., Gurewitsch, A.M., Langmuir, R.V., and Pollock, H.C. (1947) Radiation from electrons in a synchrotron. *Phys. Rev.*, **71**, 829–830.
3 Larmor, J. (1897) On the theory of the magnetic influence on spectra and on the radiation from moving ions. *Philos. Mag.*, **44**, 503–512.
4 Tomboulian, D.H. and Hartman, P.L. (1956) Spectral and angular distribution of ultraviolet radiation from the 300 Mev Cornell synchrotron. *Phys. Rev.*, **102**, 1423–1447.
5 Ivanenko, D. and Pomeranchuk, I. (1944) On the maximal energy attainable in a betatron. *Phys. Rev.*, **65**, 343.
6 (a) Schwinger, J. (1946) Electron radiation in high energy accelerators. *Phys. Rev.*, **70**, 798–799; (b) Schwinger, J. (1949) On the classical radiation of accelerated electrons. *Phys. Rev.*, **75**, 1912–1925.

7 See, e.g., DESY http://photon-science.desy.de/research/studentsteaching/sr_and_fel_labs/index_eng.html (accessed 29 June 2016).
8 Koch, E.E. (ed.) (1983) *Handbook on Synchrotron Radiation*, vol. **1**, North-Holland, Amsterdam.
9 Marr, G.V. (ed.) (1987) *Handbook on Synchrotron Radiation*, vol. **2**, North-Holland, Amsterdam.
10 Brown, G.S. and Moncton, D.E. (eds) (1991) *Handbook on Synchrotron Radiation*, vol. **3**, North-Holland, Amsterdam.
11 Ebashi, S., Koch, M., and Rubinstein, E. (eds) (1991) *Handbook on Synchrotron Radiation*, vol. **1**, North-Holland, Amsterdam.
12 Kunz, C. (ed.) (1979) *Synchrotron Radiation – Techniques and Applications*, Springer, Berlin.
13 Krinski, S. *et al* (1983) in *Handbook on Synchrotron Radiation*, Chapter 3, vol. **1a** (ed. E.E. Koch), North-Holland, Amsterdam.
14 Krinski, S. *et al* (1983) in *Handbook on Synchrotron Radiation*, Chapter 2, vol. **1a** (ed. E.E. Koch), North-Holland, Amsterdam.
15 Clarke, J.A. (2004) *The Science and Technology of Undulators and Wigglers*, Oxford Series on Synchrotron Radiation, vol. **4**, Oxford University Press, Oxford.
16 Materlik, G. and Th. Tschentscher (eds) (2001) TESLA Technical Design Report, Part V, DESY Report 2001-011/TESLA-FEL 2001-05, Hamburg. The introduction to the SASE FELs is also available at http://photon-science.desy.de/research/studentsteaching/sr_and_fel_basics/fel_basics/index_eng.html (accessed 29 June 2016)
17 A list of references is given on the FLASH user information pages at http://photon-science.desy.de/facilities/flash/the_free_electron_laser/how_it_works/basic_papers_and_books/index_eng.html (accessed 29 June 2016)
18 (a)See, e.g., DESY http://photon-science.desy.de/facilities/flash/index_eng.html (accessed 29 June 2016);(b) Ackermann, W. *et al* (2007) Operation of a free electron laser in the wavelength range from the extreme ultraviolet to the water window. *Nat. Photonics*, **1**, 336–342. doi: 10.1038/nphoton.2007.76 for the European XFEL project see http://www.xfel.eu.

Part II

Methods

6

Stress Analysis by Angle-Dispersive Neutron Diffraction

Peter Staron

6.1 Introduction

Diffraction-based techniques for the analysis of residual stress (RS) states in crystalline solids have been used for a long time [1]. Angle-dispersive X-ray diffraction using laboratory sources are widely employed for determining surface stresses [2]. The great advantage that comes along with neutrons is their large penetration depth, which is about 1000 times larger than that of X-rays from a Cu-tube in most materials. Therefore, with neutrons, information can be obtained nondestructively from the interior of a specimen. Residual stress analysis (RSA) with neutrons is comprehensively described in [3]. When high-energy X-rays from synchrotron sources became available, they were also used for RSA. The main advantage of synchrotron sources is the high intensity, which facilitates high spatial resolutions up to single-grain measurements or fast *in situ* measurements of time-dependent processes. These three techniques – using X-ray tubes, neutrons, or medium and high-energy X-rays from synchrotron sources – should, in general, be regarded as complementary techniques, each offering its special advantages.

The specialty of neutrons for RSA is the *combination* of large penetration depth and wavelengths that enable a scattering angle of 90°. For example, measuring internal strains throughout the interior of a 3 cm-thick steel plate is possible with neutrons within reasonable data acquisition times. In addition to absorption (σ_a), also scattering power (σ_{coh}) and incoherent background (σ_{inc}) determine data acquisition times. Consequently, RSA can be performed in thick aluminum alloys, while it is much more difficult in titanium alloys because of strong absorption and high incoherent background combined with low scattering power (Table 6.1). The maximum penetration thickness also depends on the strength of the source and the spatial resolution that is required. As concerns the diffraction geometry, the 90° scattering angle not only leads to a near-cubic gauge volume, but also makes strain measurements possible even in large parts with a complicated geometry. This can be more difficult with high-energy X-rays because of the much smaller scattering angles (cf. Chapter 8).

Angle-dispersive diffractometers for RSA are usually limited to a single lattice reflection at a time. In contrast, with energy-dispersive neutron diffraction (Chapter 7) a complete spectrum is recorded with one measurement, giving information about several reflections. The latter can be especially useful for multiphase materials or when microstresses are to be analyzed. However, time-of-flight spectrometers generally work most efficient at pulsed sources, while angle-dispersive spectrometers are well suited

Neutrons and Synchrotron Radiation in Engineering Materials Science: From Fundamentals to Applications, Second Edition. Edited by Peter Staron, Andreas Schreyer, Helmut Clemens, and Svea Mayer.

Table 6.1 Coherent (σ_{coh}), incoherent (σ_{inc}), and absorption (σ_{abs}) cross-sections [1] as well as 50% attenuation length ($l_{1/2}$) for neutrons with a wavelength of 1.6 Å in several engineering materials.

Material	Al	Ti	Fe	Ni	Cu
σ_{coh} (barn)	1.495	1.485	11.22	13.3	7.485
σ_{inc} (barn)	0.008	2.87	0.4	5.2	0.55
σ_{abs} (barn)	0.205	5.41	2.28	3.99	3.36
$l_{1/2}$ (cm)	6.74	1.24	0.59	0.34	0.72

for continuous sources. In addition, for many engineering applications the amount of information gained with a single-peak measurement is sufficient.

Fields of application for angle-dispersive neutron strain scanning are, generally speaking, the determination of RS in the interior of components. To be more specific, one broad field of engineering application is welding, because RS and distortion are always connected with welding; they can decrease load carrying capacity and the lifetime of a component and, thus, information about stress states in welds is often crucial. Another field of application is the verification of models such as finite element models (FEMs). Models are used for stress prediction, for example, in welds, in thermally treated, or in deformed material. In Section 6.4, two examples will be given for RS in a weld and FEM predictions of RS in water-quenched components.

6.2 Diffractometer for Residual Stress Analysis

6.2.1 Setup of a Diffractometer for Strain Scanning

Today, dedicated instruments are used for RSA, or strain scanning. They are optimized for using a small gauge volume, precisely defined by slits, and for achieving a resolution of lattice parameters better than 10^{-4}, corresponding to a stress resolution of about 20 MPa for steel. For this, thermal neutrons are used with wavelengths between about 1 and 3 Å for obtaining the 90°-diffraction geometry for most relevant engineering materials. Strain scanners are often found close to the neutron source where they can make use of the large vertical divergence of the neutron beam for increasing the intensity. Another possibility is a place at a thermal neutron guide with the advantage of lower background.

The basis of diffraction strain measurements is Bragg's equation, $n\lambda = 2d\sin\theta$, where λ is the neutron wavelength, d is the lattice spacing, 2θ is the scattering angle, and n is a positive integer. A change Δd in the lattice spacing due to internal strains will result in a shift of a Bragg peak position $\Delta\theta$ when a single wavelength is used. The shift of the peak positions is of the order of 0.01–0.1°; thus, diffraction strain measurements are precision measurements and all components of the diffractometer are required to be sufficiently precise and stable.

The main components of a strain scanner are monochromator, collimating system, slit system, sample stages, and detector. Figure 6.1 shows the setup of the strain scanner

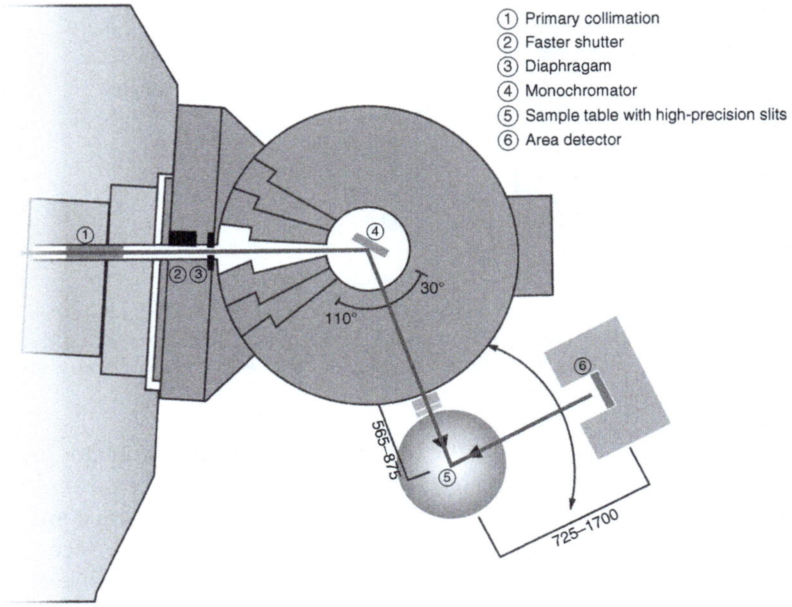

Figure 6.1 Schematic drawing of the setup of the constant-wavelength strain scanner STRESS-SPEC at MLZ (research reactor FRM II).

STRESS-SPEC at the Maier-Leibnitz Zentrum (MLZ) in Garching, Germany [4, 5]. The collimator defines the divergence of the primary beam. For an instrument position at a thermal neutron guide, the divergence is defined by the transmission of the guide. Since RS distributions are always inhomogeneous, a precise and reproducible positioning of a sample in the beam is required. For this, the precision of the positioning devices has to be sufficient and appropriate sample positioning procedures have to be used. In the following text, the most important components will be introduced.

6.2.2 Monochromator

Neutron sources offer a broad spectrum of wavelengths; however, neutrons of a single wavelength are needed for an angle-dispersive diffractometer. These can be delivered from a monochromator using either perfect single crystals or mosaic crystals. A good choice of a monochromator for a strain scanner is the use of elastically bent perfect Si single crystals [6, 7]. As an example, the monochromator currently used at the strain scanner STRESS-SPEC at the FRM II at MLZ is shown (Figure 6.2a). It essentially consists of a bending device for motor-controlled variable elastic bending of the silicon crystals, which is used for horizontal focusing. Several stacked long crystals are used for vertical focusing.

The advantage of perfect single crystals for strain scanning is that no collimators are required [6]. The diffracted beam emerging from a polycrystalline sample is nearly parallel when the radius of curvature is adapted to the specific diffraction angle because of the angle–wavelength correlation established by the perfect crystals (Figure 6.2b). When the sample does not produce significant broadening of diffraction peaks, the divergence

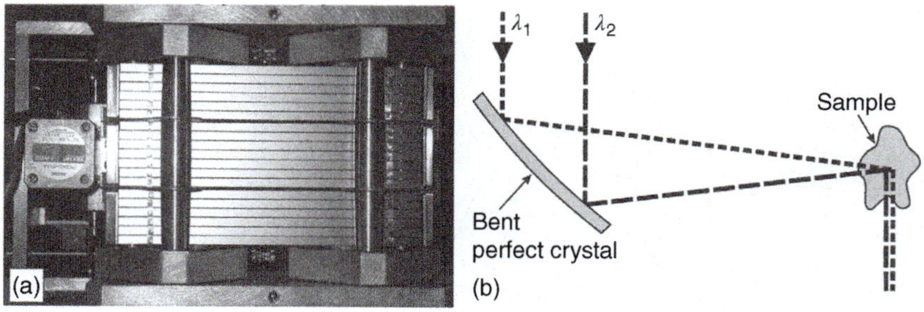

Figure 6.2 (a) Photograph of a horizontally and vertically focusing monochromator at the strain scanner STRESS-SPEC at MLZ with elastically bent perfect Si single crystals. (By courtesy of Michael Hofmann, Technical University of Munich.) (b) Quasiparallel diffracted beam emerging from a sample when perfect crystals are used as monochromator.

of the diffracted beam is mainly given by the thickness of the monochromator crystals [6]. Thus, a well-defined gauge volume can be achieved.

Conventional monochromators are used with mosaic crystals that contain domains of crystallites with slightly varying orientations, increasing the reflected intensity. Often, pyrolytic graphite (PG) mosaic crystals are used for monochromators because of their good reflectivity. Stacked Ge wafers also have good reflectivity. However, usually collimators have to be used with mosaic crystals to obtain a good spatial resolution.

6.2.3 Slit System

The slit system is used to define the size and shape of the gauge volume within a sample. It consists of two slits: the first one in the incoming beam and the second one in a diffracted beam. Typical slit widths are between 0.5 and 5 mm; slit heights can be from 1 mm up to about 30 mm (Figure 6.3). The true gauge volume is larger than the nominal gauge volume given by the size of the slits due to the divergence of the incoming and diffracted beam. The scattering angle should be around 90° to produce a gauge volume with a rectangular cross-section. Both slits have to be mounted on linear devices for adapting

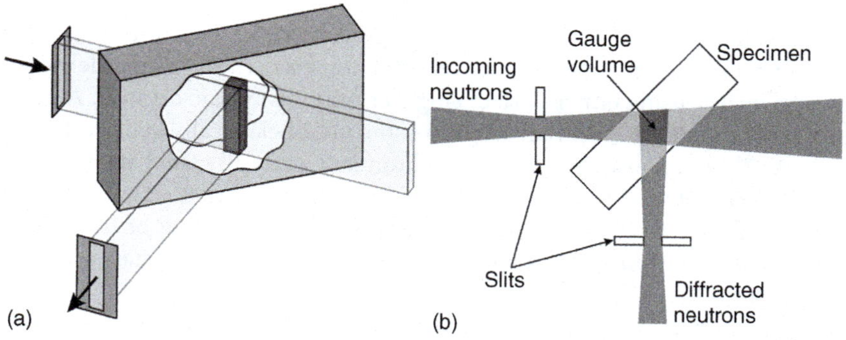

Figure 6.3 (a) Schematic drawing of a matchstick-like gauge volume within a sample, formed by two slits. (b) Cross-section of the gauge volume for divergent beams.

their distance from the sample. The second slit can be replaced by a radial collimator, improving the gauge volume definition, for example, when peaks are broadened by the sample or for measurements close to surfaces.

6.2.4 Sample Positioning

"Strain scanning" is actually performed by moving a specimen through the beam and recording the position of a diffraction peak for different specimen positions. A specimen has to be moved in three perpendicular directions x, y, and z for bringing a specific location within the sample into the beam. Thus, three linear movements are needed. When two linear tables are mounted on top of a goniometer, they can be rotated to align the specimen movement with the scattering vector $\mathbf{q}$. A useful accessory is a Eulerian cradle for orienting and positioning of samples with masses up to about 5 kg.

6.2.5 Detector

Area detectors with a height of 200–300 mm cover a significant part of a diffraction cone for collecting as much intensity as possible without reducing the resolution of the strain direction too much (Figure 6.4). The readout of the detector electronics gives an intensity matrix (Figure 6.5); therefore, a calibration is needed for the calculation of angles.

Figure 6.4 Schematic drawing of the first seven diffraction cones of Al at a wavelength of 1.64 Å. The box is at a detector position of 70° for recording the (220) reflection.

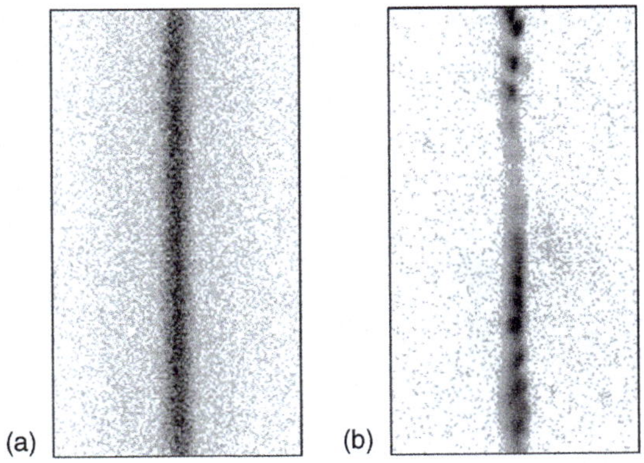

Figure 6.5 Detector images of the (a) (311) reflection of fine-grained IN718 (cf. Section 6.4.2) and a nominal gauge volume of 27 mm^3 ($2\theta \approx 100°$) and (b) (220) reflection of a coarse-grained Cu alloy and a nominal gauge volume of 90 mm^3 ($2\theta \approx 80°$). The width of the images covers a scattering angle interval of approximately 10°.

6.3 Measurement and Data Analysis

6.3.1 Gage Volume and Sample Positioning

A matchstick-like gauge volume can be used for scanning strains in directions where the symmetry of the stress state allows integration over one direction (cf. Figure 6.3a) [8]. For example, in welded sheets this would be the longitudinal direction parallel to the weld. This is done to increase the intensity for saving beam time as well as to increase the number of diffracting grains. The size of the gauge volume can range from 1 mm^3 up to more than 100 mm^3, depending on the material, specimen thickness, grain size, and available intensity.

The cross-section of a gauge volume can be measured by moving a wire through it and recording the diffracted intensity. With a monochromator using elastically bent perfect crystals, the divergence of the diffracted beam is small, resulting in a very small increase of the gauge volume in that direction (Figure 6.6).

In the application of diffraction-based RSA it is often tacitly assumed that the assumptions the method is based on are fulfilled. One assumption is that the diffracting grains are representative of the material. However, crystallites are not isotropic, nor are their elastic properties; this fact is accounted for by the concept of *hkl*-dependent diffraction elastic constants (DECs; cf. Chapters 7 and 9). But this concept only works when the diffracting subset of grains is truly representative of the material, which is only the case when the number of grains in the gauge volume is large enough. The reason is that diffraction is not sensitive to a grain rotation around an axis parallel to q, although the elastic properties of the grain may be. So, only when a sufficient number of these grain orientations are included, the experimental strain value will be a representative one.

With a fine-grained specimen like the Ni-alloy discussed in Section 6.4.2 (mean grain size 5 μm) the number of diffracting grains is large and the intensity distribution on the diffraction cone is homogeneous (Figure 6.5a). In contrast, when the grains are large, reflections from single grains will become visible on the area detector and the intensity distribution will become spotty (Figure 6.5b). In such a case, measures have to be taken to increase the number of diffracting grains. Otherwise, one should be aware of the fact that the error bar for stress results can be much larger than just that resulting from the statistical error of the measurement. Possible ways to increase the number of diffracting grains are (i) increasing the gauge volume, (ii) moving the sample in an appropriate direction during the measurement, or (iii) rotating the sample for small angles during the measurement, which also brings new grains into diffraction condition.

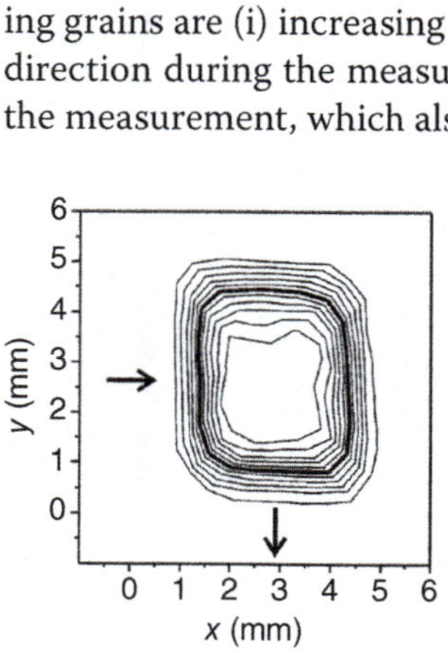

Figure 6.6 Measured cross-section of a gauge volume formed by 3 mm wide slits at a distance of 90 mm from the center. The thick isointensity line marks half of the maximum intensity giving a size of 3.8 mm × 3.0 mm. The arrows mark the directions of incoming and diffracted beams.

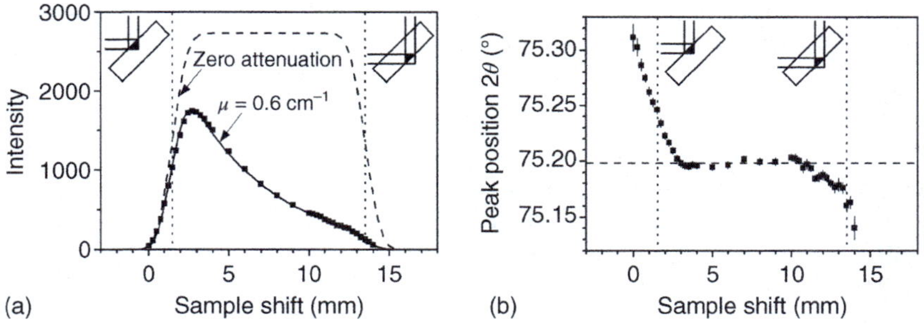

Figure 6.7 Recorded peak intensity (a) and position (b) for a 12 mm thick container filled with Cu powder that is shifted through the gauge volume. The solid and dashed lines in (a) are calculated functions.

Utmost care has to be taken while positioning a sample in the beam. Especially when large stress gradients are present in a sample, positioning errors can change results drastically. Different accessories like a plumb-line, lasers, telescopes, or video cameras with cross-hairs can be used for supporting sample positioning. However, the best results are usually obtained by using the neutron beam for positioning.

For this, a sample is shifted through the gauge volume and the integral intensity of the desired reflection is recorded. Several scans can be required, depending on the geometry of the part. The result of such a scan could be a curve like that shown in Figure 6.7a, although usually measured with less intensity for positioning purposes. The whole curve can be fitted or the position of just one sample surface can be determined from the left part of the curve. Note that for the configuration shown in Figure 6.7a absorption leads to a shift of the position where the intensity is half of the maximum intensity relative to the true sample surface. This shift depends on the material and the size of the gauge volume and has to be corrected for precise sample positioning.

Partial immersion of the gauge volume in the sample leads to shifts in the peak position (Figure 6.7b). The reason is that the centroid of the immersed volume shifts when the gauge volume is only partially filled. Therefore, care has to be taken that the gauge volume is always completely within the sample during normal measurements. Depending on the configuration of the diffractometer, the use of a radial collimator can reduce or eliminate such shifts [9].

6.3.2 Data Reduction and Analysis

Only a few important equations will be reported in the following text; comprehensive descriptions are found in the textbooks [1–3]. In the following discussions, it will be assumed that the investigated material is an isotropic single phase material with a close-to-random texture and microstresses can be neglected.

6.3.2.1 Data Reduction and Peak Fitting

The raw data delivered from an area detector is an intensity matrix. At scattering angles different from 90° the curvature of the cone section is visible on the area detector and has to be taken into account for integrating the intensity vertically. Detector channels have to be converted to angles. Note that the wavelength has to be known with a high precision only when absolute lattice parameters need to be determined, which is normally not required for strains measurements. For the latter, only the difference between a peak position and the position of the stress-free reference has to be determined with high precision.

The recorded diffraction peak is a convolution of the diffractometer resolution function and the reflectivity of the sample. In many cases, simple Gaussian peak profiles are observed. When there is significant peak broadening due to the microstructure of the sample, other peak profile functions can be involved. Peak profile analysis is a topic itself and shall not further be discussed in this context; a comprehensive treatment can be found, for example, in [10, 11].

It is important to fit the background under a peak as well. In many cases it is sufficient to account only for a linear background; however, in some cases a parabolic contribution should be taken into account as well. In these cases, using a linear background can lead to systematic deviations of the calculated peak position from the true position.

When the background is small, a few hundred counts in the peak maximum are sufficient to achieve small enough errors in the peak position. Increasing the data acquisition time only slowly reduces this error; therefore, the required precision should be determined before setting the data acquisition time in order to save beam time. In general, the required counting time will depend on the signal-to-background ratio.

6.3.2.2 Calculation of Stresses

The strain ε in the direction normal to the reflecting lattice plane (direction of the scattering vector) is obtained from measured peak positions 2θ by

$$\varepsilon = \frac{d - d_0}{d_0} = -\frac{\theta - \theta_0}{\tan \theta_0} \tag{6.1}$$

where d_0 is the stress-free lattice parameter. The error in the strain can be calculated from the errors in the lattice parameters by

$$(\Delta\varepsilon)^2 = \left(\frac{\Delta d}{d}\right)^2 + \left(\frac{\Delta d_0}{d_0}\right)^2 \tag{6.2}$$

Since the stress state is described by a symmetric third rank tensor, in general six unknowns have to be determined. Therefore, strain measurements in at least six directions would be required; over-determining the problem by measuring more than six directions reduces the error. However, when the principal stress axes are known, only strain measurements in these three perpendicular directions are required to fully determine the stress tensor, which is diagonal in this case. For specimen geometries such as sheets, plates, tubes, cylinders, or disks the principal stress axes often are assumed to coincide with the principal geometrical axes of a specimen. However, one has to bear in mind that this must not always be true, especially in inhomogeneous material regions such as, for example, welds. But even if the axes chosen for strain measurements are not the true principal stress axes within the specimen, the stresses in the chosen orthogonal directions will be determined.

Triaxial residual stresses in an isotropic polycrystal are calculated from the strains measured in three perpendicular directions by Allen *et al.* [12]

$$\sigma_i = \frac{E_{hkl}}{1 + \nu_{hkl}} \left[\varepsilon_i + \frac{\nu_{hkl}}{1 - 2\nu_{hkl}}(\varepsilon_1 + \varepsilon_2 + \varepsilon_3)\right] \tag{6.3}$$

Table 6.2 Diffraction angle of lattice plane families for residual stress analysis for a wavelength of 1.6 Å and corresponding DEC [3] for some engineering materials.

Material	*hkl*	*d* (Å)	2θ (1.6 Å) (°)	E_{hkl} (GPa)	ν_{hkl}
α-Fe	211	1.170	86.3	225.5	0.28
γ-Fe	311	1.083	95.2	183.5	0.31
Al	311	1.221	81.9	70.2	0.35
Ni	311	1.063	97.6	185.0	0.33
Cu	311	1.090	94.4	122.0	0.35

where $i = 1, 2, 3$. DECs have to be used for the modulus of elasticity E_{hkl} and Poisson number ν_{hkl} (Table 6.2). As found in the literature, this can often be calculated to a good approximation from single-crystal elastic constants, or can be determined for a specific material in uniaxial tensile tests.

When the error in the stresses σ_i is calculated from the errors in the strains $\varepsilon_{\lambda i}$, the correlations via d_0 have to be taken into account and the full covariance matrix has to be calculated. Therefore, it is much easier to calculate the error in the stresses from the errors in the uncorrelated lattice parameters (or peak positions) according to

$$(\Delta\sigma_i)^2 = \left(\frac{A}{d_0}\right)^2 \left[B(\Delta d_i)^2 + C^2(\Delta d_0)^2 + D^2((\Delta d_1)^2 + (\Delta d_2)^2 + (\Delta d_3)^2)\right]$$
$$A = \frac{E}{1+\nu} \quad B = \frac{1}{1-2\nu} \quad C = \frac{1+\nu}{1-2\nu} \quad D = \frac{\nu}{1-2\nu} \tag{6.4}$$

Stresses can be separated into hydrostatic and deviatory parts; the former is the mean of the diagonal elements of the stress tensor, the latter is the deviation from the mean value in each direction. Note that d_0 is not required for calculating deviatoric stresses. Moreover, using Eq. 6.3 it is easy to see that also the difference of any two of the three stress components as well as the effective stress after Von Mises can be calculated without a precise knowledge of d_0.

Strong textures can have a significant influence on the measured strains [2, 13]. The recommendation is to measure the texture, if possible, and add this information to the documentation of the RS results [14]. Methods have been developed for the treatment of RSA in textured materials; some of them are introduced in the textbook by Hauk [2].

6.3.2.3 Macro and Microstresses

The stresses determined by diffraction methods are the sum of macro and microstresses that lead to peak shifts [15]. Microstresses can result, for example, from plastic deformation of a material. Microstresses can lead to peak broadening as well as peak shifts. Since the engineer often is interested only in macrostresses, which influence load carrying capacity, or distortion, macrostresses have to be separated from microstresses.

Intergranular microstresses have different influences on different lattice planes due to the elastic anisotropy of the crystallites [16]. Therefore, besides the fact that this effect can be used to study intergranular stresses connected with the plastic history of a material, it is recommended to avoid certain reflections for the determination of macroscopic stresses but to use instead such reflections for which the influence of intergranular strains is small [14]. With these reflections, the best approximation to the macrostresses is obtained. In face-centered cubic (fcc) and body-centered cubic (bcc) materials the use of the (200) reflection is strongly discouraged; instead the use of, for example, the (111) or (311) reflection (fcc) or the (110) or (211) reflection (bcc) is recommended. In hexagonal materials such as, for example, Ti the use of the pyramidal planes (10−12) or (10−13) is recommended while the use of the basal (0002) and prism planes (10−10) or (10−13) can be problematic.

6.3.2.4 Stress-Free Reference

Determination of strains requires the knowledge of the unstrained (stress-free) lattice parameter d_0. This is sometimes problematic because it has to be made sure that the specimen that is chosen as stress-free reference actually has the same chemical composition as the specimen under investigation. In welds, the heat treatment that is introduced by welding can significantly change the chemical composition in the weld and the heat affected zone (HAZ), depending on the material. One way to obtain reliable reference samples is to cut small pieces out of the specimen under investigation at those positions where stresses are to be determined. The macrostresses are known to relax largely in such small pieces [17, 18].

Other methods like making use of conditions for the balance of forces inside a specimen can also be used to determine d_0. In thin specimens with a plane stress state, d_0 can be calculated by setting $\sigma_z = 0$ in Eq. (6.3) [12, 19].

Errors in d_0 lead to hydrostatic stress components in the results, deviatoric stresses remain unchanged.

6.4 Examples

6.4.1 Residual Stresses in Friction Stir Welded Aluminum Sheets

A broad field of application of RSA comes from welding. During welding the material is heated with a point-like heat source and cools again while the neighboring material is heated. This inhomogeneous heating and cooling induces RS and distortion. RS can reduce the life or load carrying capacity of a weld or influence the service performance of the welded components with respect to fatigue and corrosion properties; therefore, the stress state of a weld has to be assessed before usage. In addition, the effectiveness of stress-reducing measures like post-weld heat treatments can be checked.

Friction stir welding (FSW) is a solid state joining process that takes place below the melting point of the materials to be joined [20]. FSW uses a rotating cylindrical tool with a pin to heat the material by friction. The tool pin stirs the plasticized material and joins two pieces together when it is moved along the welding line. Advantages of this technique include, for example, joining of materials that are difficult to fusion weld, low distortion, and excellent mechanical properties. A potential field of application is mainly Al alloys used in modern aerospace structures where cost and weight are to be reduced by using new joining techniques [21, 22].

Aluminum sheets (AA2024-T351) of size 125 mm × 380 mm and 6.3 mm thickness were butt-joined together by FSW at the BAE SYSTEMS, UK. The welding and rotation speeds of the tool were 95 mm min^{-1} and 350 rpm. Residual strain measurements

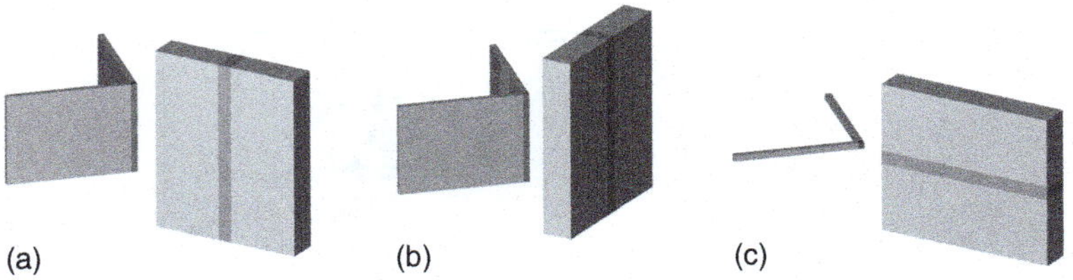

Figure 6.8 Orientations of a welded sheet relative to the incoming and diffracted beams for strain measurements in the (a) transverse, (b) normal, and (c) longitudinal directions.

had been performed with the neutron diffractometer ARES at HZG [23]. Neutrons of 0.164 nm wavelength from an elastically bent perfect silicon (311) monochromator were used. Strain measurements in the three perpendicular directions x (longitudinal direction, parallel to the weld), y (transverse), z (through-thickness) were performed with the orientations of a sheet as shown in Figure 6.8. A matchstick-like gauge volume with a length of 30 mm was used for scanning ε_y and ε_z for improving the intensity [8], assuming that the stresses are constant along x over the length of the volume element. A cubic gauge volume was used for scanning ε_x since larger stress gradients are present along the transverse (y) direction. The nominal cross section of the gauge volume was $3 \times 3\,\text{mm}^2$. The Al (311) diffraction peak was recorded with an area detector at an angle of about 84°. The scan line was 150 mm away from the edge near the weld finishing point. The crystallographic anisotropy of the elastic constants is small for Al and, therefore, the macroscopic elastic constants were used ($E = 73$ GPa and $\nu = 0.33$).

AA2024 is a heat treatable alloy in which the solute concentration (mainly Cu) in the matrix depends on the heat treatment. But as the lattice parameter can change considerably with the solute content, the d_0-value also depends on the heat input the material has experienced during welding [24]. This problem can be solved, for example, by cutting a "comb" with thin teeth out of the weld region by spark erosion in which most of the macroscopic stresses are known to relax [17]. Such a comb can, therefore, serve as unstrained reference material. However, this procedure requires quite an effort and unfortunately is a destructive method. Therefore, it was assumed that there is a plane stress state in the relatively thin sheets [12, 19]. From the condition $\sigma_z = 0$, d_0 was calculated for each point. The relative change of d_0 in the weld determined in this way is up to -4.6×10^{-4} (Figure 6.9a). This is a significant magnitude in terms of strain and thus has to be accounted for.

There are tensile stresses in the longitudinal direction with peak values of about 130 MPa (Figure 6.9b). The double-peak feature of the stress distribution with maximum stress in the HAZ and a minimum in the weld nugget has already been observed earlier [26] and seems to be typical for FSW joints in Al sheets of this thickness. As expected, no significant stresses are present in the transverse direction.

Welding a sheet under an external longitudinal tensile load and releasing the load after welding leads to large compressive longitudinal stresses in the weld (Figure 6.9b) [25, 27].

6.4.2 Residual Stresses in Water-Quenched Turbine Disks

Another field of application of experimental RSA is the verification and optimization of numerical models. Finite element (FE) models are often used for the simulation of technical processes. These can be processes involving thermal treatments (e.g., welding), or

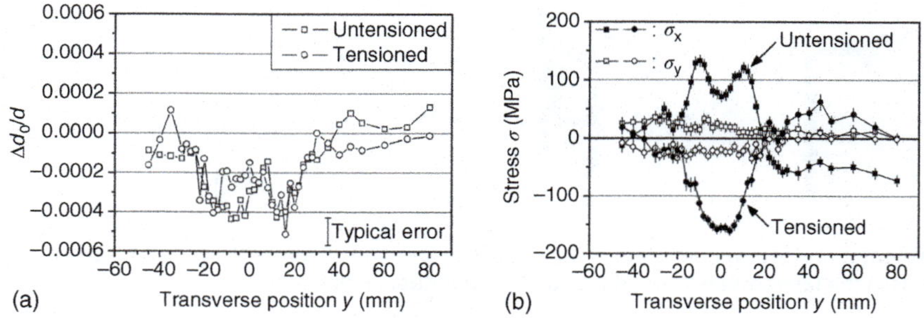

Figure 6.9 (a) Variations in d_0 as calculated from the plane stress assumption for the data of the AA2024 sheets welded with and without tensioning. (b) Residual stresses for sheets welded without and with an external tensile longitudinal load applied during welding. (Staron 2004 [25]. Reproduced with permission of Elsevier.)

mechanical treatments (e.g., deforming), or a combination of both (e.g., forging). Models are often very good in describing the development of RS states; nevertheless, they have to be checked against measurements in a real component before their predictions can be fully trusted.

Nickel-based superalloys are being employed for aeroengine components due to their unique combination of high-temperature strength and considerably high fatigue strength [28, 29]. A special thermomechanical processing route has been developed for the Inconel alloy 718 (IN718) in order to obtain enhanced high-temperature strength in turbine disks used for aeroengines. The so-called direct age process requires water quenching directly after forging [30, 31], in contrast to the conventional heat treatment where forging is followed by slow air cooling [32]. A high number of dislocations are frozen in by quenching, acting as potential nucleation sites for the strengthening precipitates that are formed in the subsequent annealing treatment. These precipitates are fine semicoherent γ''-Ni_3(Nb, Ti) and coherent γ'-Ni_3(Al, Ti) particles.

However, the high cooling rate connected with water quenching leads to much higher RS than the standard slow air cooling. Therefore, problems can arise during machining of a disk to its final shape when RS are not completely relieved during the subsequent annealing treatment, which often seems to be the case. If the RS state within the disk is known prior to turning, the process can be adjusted in order to minimize distortion considerably.

Therefore, an FE model was used for the simulation of the water quenching step. In order to verify the RS predictions of the FE simulation, RS in a turbine disk that was taken out of the commercial production (Böhler Schmiedetechnik GmbH & Co KG) after forging and water quenching were studied by neutron diffraction. Due to the relatively large dimensions of the part (∅ 320 mm and thickness up to 25 mm), neutron diffraction is the only experimental technique to study bulk residual stresses within the disks. Forging leads to a fine-grained microstructure with a mean grain size of 5 μm [29]. Therefore, the number of grains within the gauge volumes used for neutron measurements is high enough for representing the bulk.

Strain measurements had been performed with the neutron diffractometer ARES at HZG, using a constant wavelength of $\lambda = 0.164$ nm [23]. For the measurement of water-quenched IN 718, which is dominated by the fcc Ni-matrix, the (311) reflection

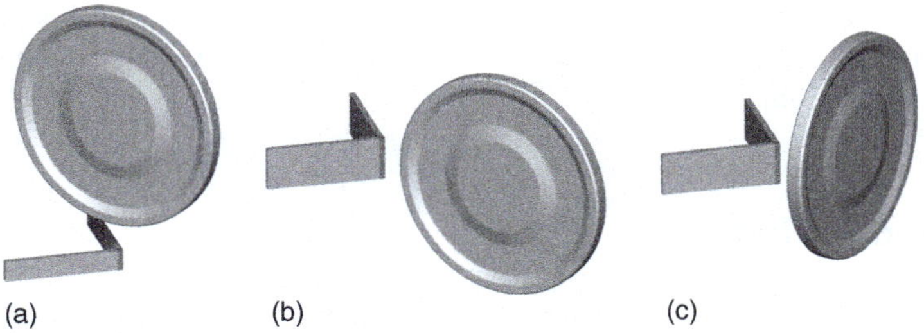

Figure 6.10 Orientations of a disk relative to the incoming and diffracted beams for strain measurements in the (a) tangential, (b) radial, and (c) axial directions.

was used, because it is generally recommended for the evaluation of macrostrains in fcc lattices [15, 16].

In the case of the disk the symmetry axes (radial, tangential, axial) were used as a coordinate system and the measurements were performed accordingly (Figure 6.10). A matchstick-like gauge volume with a cross-section of $2 \times 2\,mm^2$ was chosen in order to increase intensity; its length was up to 30 mm for measurements of axial and radial strains and 10–15 mm for tangential strains. It is thus assumed that the stresses are constant in the hoop direction over the length of the gauge volume [12]. The DECs $E_{hkl} = 195$ GPa and $\nu_{hkl} = 0.31$ for the (311) reflection were calculated using the Kröner model [33, 34] with the following single crystals values: $C_{11} = 248$ GPa, $C_{22} = 152$ GPa, $C_{44} = 125$ GPa [35].

The unstrained lattice parameter d_0 was determined by measuring small cubes with 4–5 mm edge length. The cubes were cut out of the disks by electro discharge machining with their faces parallel to the axial, radial, and hoop direction. Due to their small size they were considered to be free of macrostresses [18].

The FE model predicts tensile radial and tangential stresses up to about 500 MPa in the interior of the water-quenched disk; these tensile stresses are balanced by compressive stresses near the surfaces of the disk. The distributions of radial and tangential stresses are very similar except for the outer rim of the disk. Axial (through-thickness) stresses are zero except for a region near the circumference of the disk. Repeated measurements along three different radii, 120° apart from each other, showed a rotational symmetry of the stress state (Figure 6.11a) [37].

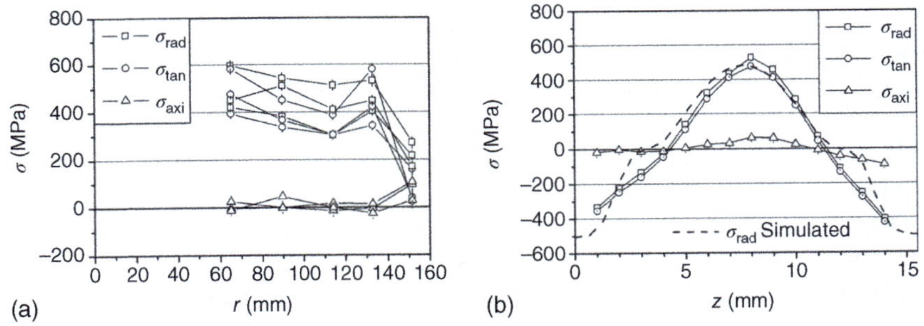

Figure 6.11 (a) Stress distributions in three different radial directions of a disk, separated by 120° from each other. (b) Stresses through the thickness of a disk (symbols: as measured; dashed line: FE result). (Cihak 2006 [36].)

A through-thickness scan was conducted at the thin section of a disk at a radius of 110 mm. A comparison shows that the FE model describes the measured stresses quite accurately (Figure 6.11b) [36, 38].

6.5 Summary and Outlook

Angle-dispersive neutron RSA is a powerful method for the nondestructive determination of RS states in the interior of large specimens. The technique has been developed for about 25 years and is now used routinely. This makes the technique more and more interesting for industrial users.

The need for strain measurements in "real-life" components leads to sophisticated designs of new dedicated strain scanners with enough space at the sample position and load carrying capacity for precise positioning of components of several hundred kilograms.

The availability of new neutron sources, like the European Spallation Source (ESS) [39], with more intensity at the sample position will be another step toward measurements in "real-life" components producing not only one-dimensional stress scans, but also two- or three-dimensional stress maps that will contribute to a more comprehensive understanding of stress distributions inside components.

References

1 Noyan, I.C. and Cohen, J.B. (1987) *Residual Stress, Measurement by Diffraction and Interpretation*, Springer, New York.

2 Hauk, V. (1997) *Structural and Residual Stress Analysis by Non-Destructive Methods*, Elsevier, Amsterdam.

3 Hutchings, M.T., Withers, P.J., Holden, T.M., and Lorentzen, T. (2005) *Characterization of Residual Stress by Neutron Diffraction*, Taylor & Francis, Boca Raton, FL.

4 Lommatzsch, I. and Hesse, C. (eds.) (2013) *Experimental Facilities*, 2nd rev. edn, Heinz Maier-Leibnitz Zentrum, Garching, http://www.mlz-garching.de/instrumente.html (accessed 29 June 2016).

5 Hofmann, M., Seidl, G.A., Rebelo-Kornmeier, J., Garbe, U., Schneider, R., Wimpory, R.C., Wasmuth, U., and Noster, U. (2006) *Mater. Sci. Forum*, **524–525**, 211–216.

6 Wagner, V., Mikula, P., and Lukas, P. (1993) *Nucl. Instrum. Methods Phys. Res., Sect. A*, **338**, 53–59.

7 Mikula, P. and Wagner, V. (2000) *Mater. Sci. Forum*, **347–349**, 113–118.

8 Pintschovius, L., Jung, V., Macherauch, E., and Vöhringer, O. (1983) *Mater. Sci. Eng.*, **61**, 43–50.

9 Pirling, T. (2000) *Mater. Sci. Forum*, **347**, 107–112.

10 Snyder, R.L., Fiala, F., and Bunge, H.J. (1999) *Defect and Microstructure Analysis by Diffraction*, Oxford University Press, New York.

11 Mittemeijer, E.J. and Scardi, P. (eds) (2004) *Diffraction Analysis of the Microstructure of Materials*, Springer-Verlag, Berlin, Heidelberg.

12 Allen, A.J., Hutchings, M.T., Windsor, C.G., and Andreani, C. (1985) *Adv. Phys.*, **34**, 445–473.

13 van Houtte, P. (1993) *Mater. Sci. Forum*, **133–136**, 97–110.

14 ISO (2001) *Polycrystalline Materials–Determination of Residual Stresses by Neutron Diffraction*, ISO/TTA 3:2001, International Organization for Standardization (ISO), Geneva.

15 Holden, T.M., Tome, C.N., and Holt, R.A. (1998) *Metall. Mater. Trans. A*, **29**, 2967–2973.

16 Clausen, B., Lorentzen, T., and Leffers, T. (1998) *Acta Mater.*, **46**, 3087–3098.

17 O'Brien, E. (2000) *Proceedings of the Sixth International Conference on Residual Stresses ICRS-6*, IOM Communications Ltd., London, p. 13.

18 Krawitz, A.D. and Winholtz, R.A. (1994) *Mater. Sci. Eng., A*, **185**, 123–130.

19 Albertini, G., Bruno, G., Dunn, B.D., Fiori, F., Reimers, W., and Wright, J.S. (1997) *Mater. Sci. Eng., A*, **224**, 157–165.

20 Dawes, C.J. and Thomas, W.M. (1996) *Weld. J.*, **75**, 41.

21 Williams, S.W. (2001) *Air Space Eur.*, **3** (**3/4**), 64–66.

22 Wang, X.L., Feng, Z., David, S.A., Spooner, S., and Hubbard, C.R. (2000) *Proceedings of the Sixth International Conference on Residual Stresses ICRS-6*, IOM Communications Ltd., London, p. 1408.

23 Staron, P., Ruhnau, H.-U., Mikula, P., and Kampmann, R. (2000) *Physica B*, **276–278**, 158–159.

24 Priesmeyer, H.G. (1992) in *Measurement of Residual and Applied Stress Using Neutron Diffraction*, NATO ASI Series E, vol. **216** (eds M.T. Hutchings and A.D. Krawitz), Springer, Netherlands, pp. 277–284.

25 Staron, P., Kocak, M., Williams, S., and Wescott, A. (2004) *Physica B*, **350**, e491–e493.

26 Staron, P., Kocak, M., and Williams, S. (2003) *Proceedings of the Sixth International Conference on Trends in Welding Research, Pine Mountain, GA*, ASM International, pp. 253–256.

27 Price, D.A., Williams, S.W., Wescott, A., Harrison, C.J.C., Rezai, A., Steuwer, A., Peel, M., Staron, P., and Koçak, M. (2007) *Sci. Technol. Weld. Joining*, **12**, 620–633.

28 Loria, E.A. (ed.) (2005) *Superalloys 718 2005*, The Minerals, Metals & Materials Society (TMS), Warrendale, PA.

29 Betteridge, W. (1992) *Materials Science and Technology*, vol. **7**. (ed. F.B. Pickering), Wiley-VCH Verlag GmbH, Weinheim, p. 641.

30 Horvath, W., Zechner, W., Tockner, J., Berchthaler, M., Weber, G., and Werner, E.A. (2001) *Superalloys 718, 625, 706 and Derivates*, The Minerals, Metals & Materials Society (TMS) (ed. E.A. Loria), Warrendale, PA, pp. 223–228.

31 Rao, G.A., Kumar, M., Srinivas, M., and Sarma, D.S. (2003) *Mater. Sci. Eng., A*, **355**, 114.

32 Hong, S.J., Chen, W.P., and Wang, T.W. (2001) *Metall. Mater. Trans. A*, **32**, 1887–1901.

33 Kröner, E. (1958) *Z. Phys.*, **151**, 504–518.

34 Bollenrath, F., Hauk, V., and Müller, E.H. (1967) *Z. Metallk.*, **58**, 76–82.

35 Dye, D., Roberts, S.M., Withers, P.J., and Reed, R.C. (2000) *J. Strain Anal. Eng. Des.*, **35**, 247–259.

36 Cihak, U., Staron, P., Clemens, H., Homeyer, J., Stockinger, M., and Tockner, J. (2006) *Mater. Sci. Eng., A*, **437**, 75–82.

37 Cihak, U., Staron, P., Marketz, W., Leitner, H., Tockner, J., and Clemens, H. (2004) *Z. Metallk.*, **95**, 663–667.

38 Staron, P., Cihak, U., Clemens, H., Stockinger, M., and Schreyer, A. (2011) *Adv. Mater. Res.*, **278**, 102–107.

39 ESS (2014) http://www.esss.se (accessed 29 June 2016).

7

Stress Analysis by Energy-Dispersive Neutron Diffraction

Javier Santisteban

7.1 Introduction

An energy-dispersive diffraction experiment is conceptually very simple: a polychromatic neutron beam hits a sample, and the neutrons scattered at a fixed scattering angle are discriminated in energy by a detector. For a polycrystalline sample, the spectrum recorded in the detector consists of several diffraction peaks at the incident wavelengths that fulfill the Bragg condition. The study of internal and applied stresses by the execution and analysis of such experiments is the topic of this chapter.

7.2 Time-of-Flight Neutron Diffraction

Pulsed sources are especially suited for energy-dispersive diffraction experiments because the neutrons are created over a wide energy range, and because their wavelength is easily defined by the time-of-flight (TOF) technique. The speed v of a neutron is related to its wavelength λ by the de Broglie relation $h\lambda = mv$, with h the Planck constant and m the neutron mass, so that v (m s^{-1}) = 3956/λ (Å). As all neutrons leave the source at the same time, the wavelength of a neutron traveling a distance L_1 from source to sample and L_2 from sample to detector, is readily defined from its TOF t

$$\lambda = \frac{h}{mv} = \frac{h}{m}\frac{t}{(L_1 + L_2)} \tag{7.1}$$

Neutrons with a wavelength of 1 Å take around 2.5 ms to travel 10 m, a time easily resolved by modern detection systems. The TOF spectrum from a stainless steel specimen for a detector 50 m away from the source is shown in Figure 7.1. The spectrum is typical of a polycrystalline sample, showing several diffraction peaks corresponding to different (hkl) families of lattice planes, as given by Bragg's law $\lambda_{hkl} = 2d_{hkl} \sin\theta_B$. So, the d-spacing for each family is directly obtained from the position t_{hkl} of the peak in the TOF spectrum,

$$d_{hkl} = \frac{h}{2\sin\theta_B m(L_1 + L_2)} t_{hkl} \tag{7.2}$$

where θ_B is given by the detector position, mostly $2\theta_B = \pm 90°$.

Neutrons and Synchrotron Radiation in Engineering Materials Science: From Fundamentals to Applications, Second Edition. Edited by Peter Staron, Andreas Schreyer, Helmut Clemens, and Svea Mayer.

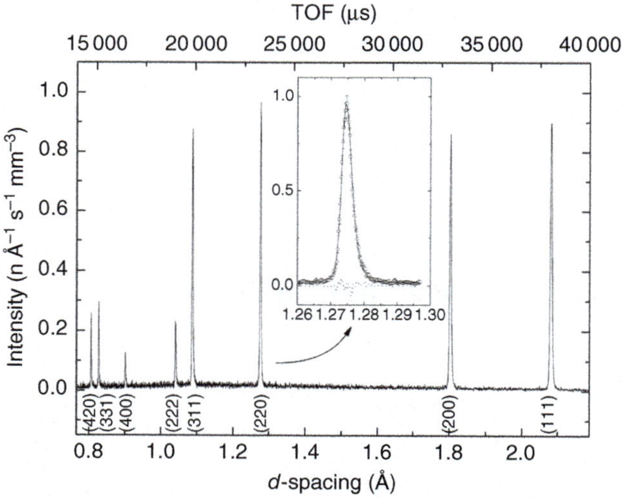

Figure 7.1 A typical time-of-flight (TOF) diffraction spectrum, in this case for a stainless steel specimen measured on ENGIN-X. Elastic strains are measured from the shift observed in the position of the diffraction peaks, defined by least-squares refinement as that shown in the inset. Alternatively, average or macroscopic strain can be defined through the change in lattice parameter obtained from a full-pattern refinement. (Santisteban 2006 [14]. Reproduced with permission of International Union of Crystallography.)

Peak positions are very precisely defined by least-squares refinement of the peaks, with a typical sensitivity of $\delta\varepsilon = \Delta t_{hkl}/t_{hkl} = \Delta d_{hkl}/d_{hkl} \sim 50\ \mu\varepsilon$ ($1\ \mu\varepsilon = 10^{-6}$). A typical fit is displayed in the inset of Figure 7.1. Basically, the atomic planes are used as microscopic strain gauges and the elastic strain is determined from the variation in the interplanar distances d_{hkl}, as compared to the value d^0_{hkl} measured in the unstressed condition, $\varepsilon_{hkl} = (d_{hkl} - d^0_{hkl})/d^0_{hkl}$.

Each diffraction peak provides information from only a small fraction of all the crystallites within the sampled volume, that is, only those favorably oriented to fulfill the Bragg condition. By averaging the strains ε_{hkl} recorded for several crystal families, TOF neutron diffraction provides a very good approximation to the macroscopic or type I elastic strain. This is because, ultimately, the macroscopic strain corresponds to the average response of *all* crystallites within the sampled volume [1]. Besides this, as several crystal families are measured simultaneously, TOF neutron diffraction is especially suited for the study of intergranular or type II stresses, which develop as a result of the elastic or plastic anisotropy of the individual crystallites. The presence of intergranular stresses manifest as different strains ε_{hkl} in the different crystal families, which can be simultaneously measured in TOF neutron diffractometers. Finally, peak profile analyses using TOF neutron strain scanners have been increasingly used to determine the density and type of the dislocations present within grains, that is, the main source of type III stresses.

7.2.1 TOF Peak Shape and Data Analysis Packages

TOF diffraction spectra display asymmetric peak profiles, as a result of experimental uncertainties preventing the precise definition of the travel time of the neutrons. Pulsed sources create high-energy neutrons (~MeV), which need to be slowed down to energies suitable for Bragg diffraction (~meV) before bringing them into the sample. This is achieved within a moderator, a hydrogenous material where neutrons lose their energy through inelastic collisions with protons, broadening the initially sharp pulse. Hence, the total TOF of a neutron is actually divided into two parts

$$t = t_{\mathrm{m}} + t_{\mathrm{L}} = t_{\mathrm{m}} + \frac{m}{h} L\lambda \tag{7.3}$$

where t_{m} is the time spent in the moderator, t_{L} is the flight time from moderator to detector, and $L = L_1 + L_2$ is the total flight path. The sharpness and the energy distribution of the neutron pulse depend on the choice of moderator material and temperature (e.g., water, methane, liquid hydrogen). The emission time from the moderator (t_{m}) has associated a very asymmetric probability distribution, usually described as a truncated exponential decay profile (Figure 7.2a), with a decay constant τ that depends on the neutron wavelength. On the other hand, the uncertainties in flight path and neutron wavelength produce a rather symmetric probability distribution for t_{L}. Several profiles have been proposed for the actual experimental peak shape, resulting from the convolution of the two distributions. The simplest model is that from Kropff and collaborators [2], consisting of the convolution of a truncated exponential with a Gaussian profile describing the symmetric broadening. A more popular profile is that proposed by Jorgensen *et al.*, consisting of separate rising and falling exponentials for the moderator term, convolved with a Gaussian representing the symmetric broadening [3]. More complex profiles are needed to describe the peaks from high-resolution instruments at small d-spacings [56].

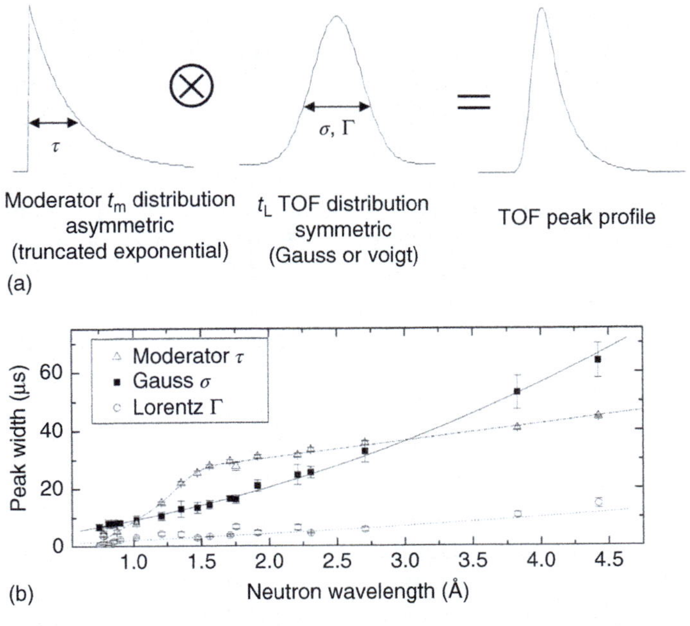

Figure 7.2 (a) Contributions to the TOF peak shape. In this case, the profile has been represented as the convolution of a truncated exponential (described by the decay time τ) and a Voigt function (described by the widths σ and Γ). (b) Variation of the peak-width parameters for the ENGIN-X diffractometer, as measured on a CeO_2 reference.

Whatever the model, the instrumental peak asymmetry must be properly incorporated in the analysis of the peak positions, as large shifts in peak positions or *pseudo strains* can arise in least-squares fits that use wrongly defined peak shapes. The shape and width of the diffraction peaks vary with neutron wavelength. Figure 7.2b shows such a variation for the ENGIN-X strain scanner, described in the following section. In this case, the peak profile has been defined as the convolution of a truncated exponential, characterized by a decay constant τ; with a Voigt function, which itself is the convolution of a Gauss (of width σ) and a Lorentzian function (of width Γ). The figure shows the variation of these parameters obtained from a reference CeO_2 sample, considered to be a good approximation to the instrument resolution. As sample-induced broadenings are typically symmetric, strain analysis of real specimens are performed by least-squares fits keeping the asymmetry parameter τ fixed to its calibrated value.

The main appeal of TOF diffraction lies in exploiting all the information contained in the complete diffraction spectrum. This is achieved through a least-squares refinement of the full diffraction pattern, as originally proposed by Rietveld for angle-dispersive neutron diffraction [4]. The main advantage of the Rietveld method is that it incorporates information from overlapping or idle-defined peaks, which for strain analysis translates into a reduction of counting times over a single-peak based analysis. There are several computer packages that can deal with TOF diffraction spectra. Maybe the most popular is the General Structure Analysis System (GSAS) code developed by Larson and Von Dreele [5], which is very powerful but not terribly user-friendly, so a graphical user interface (EXPGUI) has been recently developed for it [6]. Other popular programs include Fullprof [7], that is, very strong for magnetic scattering; and MAUD (Materials Analysis Using Diffraction) oriented to materials science studies and capable of determining orientation distribution functions from TOF diffraction data [8].

Regarding the use of full pattern refinements for strain analysis, Daymond and collaborators [9] have shown that engineering strains can be well approximated from the change in the average lattice parameters obtained from Rietveld refinements of TOF diffraction spectrum, even neglecting the elastic anisotropy of the material. This method is very popular because Type I or *engineering* stresses can be reasonably calculated using the macroscopic elastic constants of the material. Phenomenological demonstrations of this approach for face centered and hexagonal materials have been given in [9, 10], respectively, while a theoretical discussion has been presented in [11]. On the other hand, a method to obtain the full strain tensor from simultaneous Rietveld refinement of spectra collected at different specimen orientations has been implemented in the GSAS code [12].

7.3 TOF Strain Scanners

A TOF neutron strain scanner is essentially a TOF diffractometer, but with some special features to perform measurements at precise locations within the bulk of a specimen. As seen in Chapter 6, this is achieved by defining a small *instrumental gauge volume* (IGV) with the help of slits and collimators. The strain distribution within a sample is easily explored by moving the sample across the IGV using a translation stage. The neutron

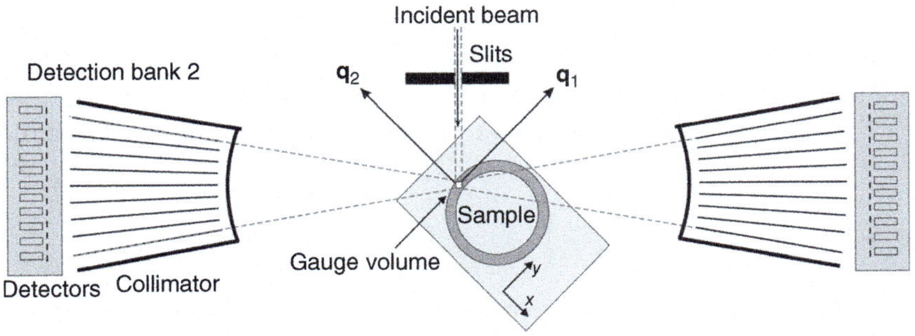

Figure 7.3 The ENGIN-X strain scanner, located at the ISIS Facility, Rutherford Laboratory, United Kingdom. The instrument gauge volume (IGV) is defined by slits and collimators. The use of two detection banks allows two perpendicular strain directions (identified by $\mathbf{q}_1$ and $\mathbf{q}_2$ in the figure) to be measured simultaneously. The sample is scanned through the IGV using a positioning table.

strain scanning technique is conceptually very simple but its practical application can be time-consuming. Very long counting times are required when small gauge volumes and large penetration depths are involved, which effectively dictates the range of problems that can be studied with this technique. Withers [13] has defined a *maximum acceptable acquisition time* for neutron and synchrotron strain scanning when measuring strain deep within materials. This is an important concept because neutron strain scanning experiments are performed on large-scale facilities, where the experimenter is granted only a limited amount of beam time, under competitive review. Hence, a minimization of the experimental time required for an accurate determination of peak positions (and hence strain) can be considered the first goal of a neutron strain scanner.

Figure 7.3 shows a schematic diagram of ENGIN-X, an optimized TOF strain scanner installed at the Isis Facility, United Kingdom [14]. The basic layout is equivalent to other instruments such as SMARTS [15] at LANSCE, Vulcan at SNS (Spallation Neutron Source) [16] (both in the United States), and Takumi [17] at J-Parc (Japan). The instrument is composed of two diffraction banks, plus slits and collimators used to define the IGV as the intersection of the incident and diffracted beams. The measured strain gives the component of the strain tensor along the direction of the scattering vector $\mathbf{q}$, which bisects the incident and diffracted beams. So, in a TOF strain scanner two strain directions can be measured simultaneously. In the following text it will be explained how the different instrumental parameters (diffraction angle, incident and diffracted beam paths and collimations, detector size) have been chosen, as the criteria are also applicable to other TOF strain scanners.

The choice of diffraction angle is driven by the geometric shape of the IGV. For most problems, the principal strain axes can be inferred from the geometry of a sample. So in practice, macroscopic strain measurements are usually performed only along three mutually perpendicular directions. The detectors are placed at approximately $2\theta_B = \pm 90°$ from the incident beam because this results in a cubic IGV, ensuring that the same volume of the specimen is measured at each sample orientation by both detectors. On the other hand, in an optimized instrument, the choice of primary and secondary flight paths and the detectors size are not independent, as it is learnt from a minimization of the experimental counting times.

7.3.1 Counting Times and Resolution

Ideally, the uncertainty in the measured strain is directly proportional to the uncertainty in the determination of the position of the Bragg peaks. Withers *et al.* [18] have shown that the time T required to measure the position of a zero-background peak with a strain precision $\delta\varepsilon$ is

$$T = \frac{1}{I_{hkl}}\left(\frac{\delta t_{hkl}}{t_{hkl}\delta\varepsilon}\right)^2 = \frac{1}{I_{hkl}}\left(\frac{\delta_\varepsilon t_{hkl}}{\delta\varepsilon}\right)^2 \tag{7.4}$$

where I_{hkl} is the integrated peak intensity per unit time, t_{hkl} is the peak position, and δt_{hkl} the peak width. On the right hand side of Eq. (7.4), T is expressed in terms of the relative peak width $\delta_\varepsilon t_{hkl} = (\delta t_{hkl}/t_{hkl})$, a dimensionless parameter that allows direct comparison between peaks appearing at different TOF, as well as with the required strain precision $\delta\varepsilon$. Through this chapter, the notation $\delta_\varepsilon X$ will be used to represent the dimensionless relative width of the probability distribution of the variable X. The relative peak width, or *resolution*, of a TOF diffractometer is [19]

$$(\delta_\varepsilon t_{hkl})^2 = (\delta_\varepsilon t)^2_{\text{mod}} + \frac{\cot^2\theta_{\text{B}}}{4}\left[\left(\alpha^x_{\text{in}}\right)^2 + \left(\alpha^z_{\text{det}}\right)^2\right] + (\delta_\varepsilon d_{hkl})^2 \tag{7.5}$$

where $(\delta_\varepsilon t)_{\text{mod}}$ is the moderator contribution to the peak width, α^x_{in} and α^z_{det} are the divergences in the diffraction plane for the incident and detected beams respectively, and $(\delta_\varepsilon d_{hkl})$ is the sample-induced broadening that appears from microstresses or a local variation in composition. On the other hand, the integrated intensity of a TOF diffraction peak [20] depends on the scattering power and absorption of the sample as well as on instrumental parameters such as the incident neutron flux Φ_0 (expressed in neutrons $\text{s}^{-1}\,\text{cm}^{-2}\,\text{Å}^{-1}\,\text{sr}^{-1}$), the total divergence in the diffraction plane, the incident beam divergence perpendicular to the diffraction plane α^y_{in}, the size of the illuminated gauge volume δV, and the detector characteristics

$$I_{hkl} = \Phi_0\alpha^y_{\text{in}}\left(\frac{h_{\text{det}}\%_{\text{det}}}{\pi L_2}\right)\sin^2\theta_{\text{B}}\left[\left(\alpha^x_{\text{in}}\right)^2 + \left(\alpha^z_{\text{det}}\right)^2\right]^{1/2} P_{hkl}\,\text{e}^{-\mu_{hkl}l}\,\delta V \tag{7.6}$$

where h_{det} and $\%_{\text{det}}$ are the detector height and efficiency, respectively. The beam attenuation is given by the absorption coefficient μ_{hkl}, and l is the neutron flight path inside the sample. The factor P_{hkl} depends only on the material

$$P_{hkl} = \frac{m_{hkl}\left|F^2_{hkl}\right| d^4_{hkl}}{v_0^2} \tag{7.7}$$

with v_0 the volume per atom, F^2_{hkl} the structure factor (including the Debye-Waller factor), and m_{hkl} the multiplicity of the reflection. Equation (7.6) can be used to normalize integrated TOF peak intensities in order to assess the effect of texture as exemplified in [21] for a Zircaloy-4 weld. By replacing Eqs. (7.5) and (7.6) in Eq. (7.4), it can be shown that the minimum in counting time is achieved when

$$\frac{\cot^2\theta_{\text{B}}}{4}\left(\left(\alpha^x_{\text{in}}\right)^2 + \left(\alpha^z_{\text{det}}\right)^2\right) = (\delta_\varepsilon t)^2_{\text{mod}} + (\delta_\varepsilon d_{hkl})^2 \tag{7.8}$$

So in an optimized TOF strain scanner, the total divergence in the diffraction plane must match the combined peak broadenings introduced by the neutron moderation

Table 7.1 Estimated counting times (in minutes) for strain measurements on ENGIN-X with an uncertainty of 50 $\mu\varepsilon$. For each material, the estimated times are those from a Rietveld refinement of the complete diffraction pattern.

Gauge volume	$2 \times 2 \times 2\,mm^3$				$4 \times 4 \times 4\,mm^3$			
Depth (mm)	2	10	20	30	4	20	40	50
Al	5.9	6.4	7.1	7.8	0.8	0.9	1.1	1.2
γ-Fe	0.9	2.2	6.5	19.7	0.1	0.8	7.4	22.5
α-Fe	0.9	2.1	6.1	18.3	0.1	0.8	6.8	20.4
Ni	0.9	4.2	31.5	233.4	0.2	3.9	216.2	1602.9
Cu	1.3	2.9	7.9	21.1	0.2	1	7.1	19.3
Zr	1.5	1.6	1.9	2.2	0.2	0.2	0.3	0.4
Ti	6.5	11.4	23.1	46.7	0.9	2.9	11.8	24

process, and by the sample. The moderator contribution to the peak width is inversely proportional to the total neutron flight path [19],

$$(\delta_\varepsilon t)^2_{\text{mod}} = \left(\frac{\hbar \Delta t_0}{(L_1 + L_2) m \lambda}\right)^2 \tag{7.9}$$

where Δt_0 is the intrinsic width of the neutron pulse, a measure of the time spent by the neutrons in the moderator. In practical terms, the moderator contribution to the resolution is defined by a careful choice of the moderator and the incident flight path L_1. Increasing the secondary flight path L_2 is not effective because there is an associated decay in the intensity of the form $(1/L_2)^2$. On the other hand, the primary flight path L_1 can be increased without a significant loss of intensity by bringing the neutrons to the sample position using a neutron guide. The actual choice of L_1 will depend on a variety of parameters including building costs and the materials likely to be studied with the diffractometer. Besides this, Eq. (7.8) calls for an instrument having a variable divergence, which could be adjusted to match the peak broadening introduced by the sample – since typically we are willing to increase the divergence if the incident flux can also be increased.

The ISIS methane moderator was chosen for ENGIN-X due to its combination of narrow pulse width and high flux over the 1–3 Å wavelength range relevant to most engineering materials. With this choice, Johnson and Daymond showed that the counting times are minimized when the primary flight path of the instrument is ~50 m and the horizontal angular divergence of the diffracted beam is 0.002 rad [22]. For detectors located at a typical secondary flight path of 1.5 m, this gives a detector width $\delta z \sim 3$ mm. The height of the detector δy can be increased considerably (~200 mm) as the divergence perpendicular to the diffraction plane does not affect the resolution. In ENGIN-X, the incident beam divergence can be adjusted with two sets of slits located at 1.5 and 4 m from the IGV. Typical counting times in ENGIN-X for a 50 $\mu\varepsilon$ resolution are presented in Table 7.1 [14]. A typical 20× reduction in counting times on ENGIN-X over its predecessor ENGIN [23] was achieved through this optimized design.

7.3.2 Neutron Optics and Time Focusing

The measured strain corresponds to an average over the volume δV and solid angle $\delta\Omega$ sampled by the experiment. Figure 7.4a schematically shows the main parameters affecting the shape and size of the IGV. The beam cross-section is defined by slits of height S_y and width S_x located at a distance S_z from the geometric center of the diffractometer. The incident beam has a divergence α^x_{in} in the diffraction plane and α^y_{in} perpendicular to it. Due to this divergence, the beam cross-section is larger at the IGV center than at the slits position and the edges get blurred. The edge profile is described by an error function, with a width defined by the beam divergence and slit distance.

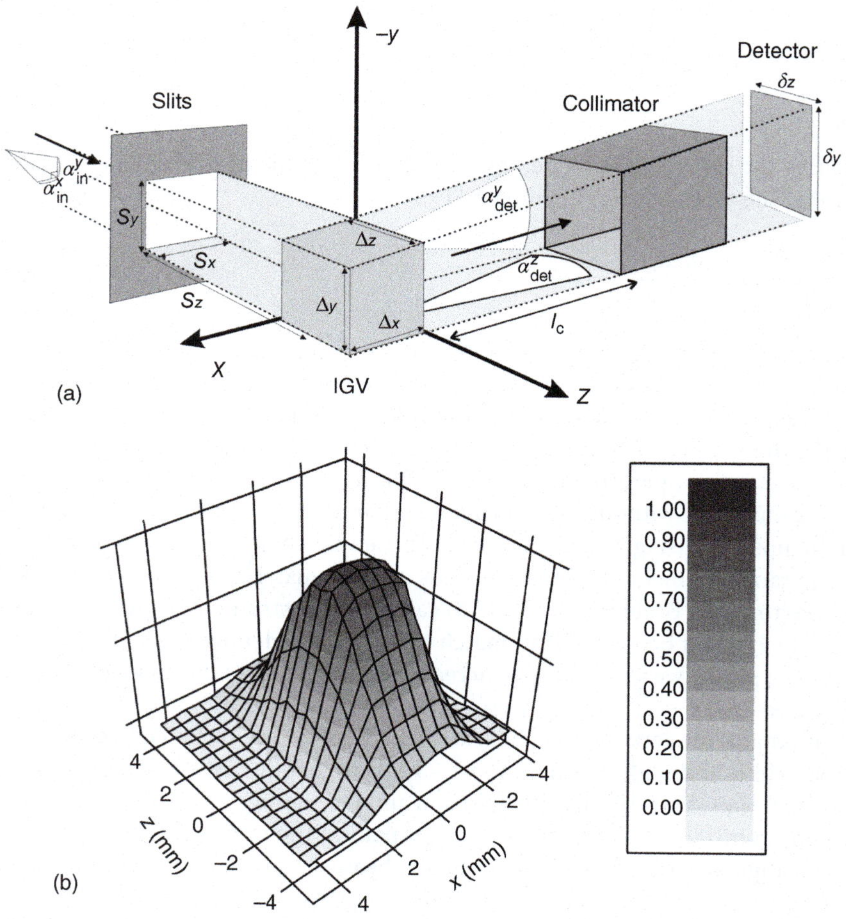

Figure 7.4 (a) Schematic representation of the instrumental gauge volume (IGV) in a TOF neutron strain scanner. The figure identifies the experimental parameters responsible for the shape and dimensions of the IGV. (b) Experimental intensity map on the horizontal plane of a $2 \times 2 \times 2\,\text{mm}^3$ IGV, measured on ENGIN-X. (Santisteban 2006 [14]. Reproduced with permission of International Union of Crystallography.)

The FWHM (Full-Width at Half-Maximum) of the neutron beam profile at the IGV center is [24]

$$\Delta x, y = S_{x,y}\left[1 + 2.35\left(\frac{S_z}{S_{x,y}}\tan\alpha_{x,y}\right)^3\right]^{1/3} \tag{7.10}$$

Hence, the effective dimensions of the beam – as defined by the FWHM – increase rather slowly with the slit distance S_z, even when the actual cross-section of the beam increases markedly. The size of the IGV along the beam direction Δz is usually defined by a collimator of the specified divergence. But as TOF neutron count rates are relatively low (compared to angle-dispersive diffraction) usually large arrays of detectors covering horizontal and vertical ranges of about 30° need to be used. Radial collimators made of thin absorbing blades are employed for this purpose (Figure 7.3), to collimate the diffracted beam and ensure that all detectors look at exactly the same volume of the sample [24]. Each ENGIN-X detection bank is composed of 1200 TOF detectors, arranged in 240 columns across $2\theta_B$ angles of 75° and 105°. The intensity profile along the beam direction seen from a radial collimator is triangular for the line passing through the focal point, and gradually becomes a Gaussian profile at increasing distances from this point. A horizontal map of the experimental intensity distribution for the ENGIN-X IGV is shown in Figure 7.4b.

From the present discussion, it follows that a precise definition of the IGV is a rather complex matter. For operational purposes, we can define the IGV as a cuboid of sides Δx, Δy, and Δz, so the illuminated gauge volume becomes $\delta V = \Delta x \Delta y \Delta z$. The dimensions of Δx and Δy are the FWHM's given by Eq. (7.10), and Δz is the width of the intensity profile seen from the collimators. For measurements performed near surfaces, the IGV may be only partially immersed within the specimen, and the effective centroid of the IGV differs from its geometrical center. This may introduce an instrumental peak shift that needs to be discounted from the total peak shift.

The TOF spectra recorded by each individual detector are transformed into a common d-spacing scale and added together in a single diffraction spectrum, in a process commonly known as *electronic time focusing* [3]. Larger angular arrays are not feasible due to space and access requirements, and due to the increased uncertainty in the direction of the measured strain [25].

7.4 A Virtual Laboratory for Strain Scanning

In parallel to the optimization of the neutron optics, fast and reliable positioning can be achieved by using detailed 3D models of the samples, and a pair of theodolites for precise alignment on the instrument. In ENGIN-X the processes of planning, alignment, and data analysis have been simplified by writing SSCANSS [26], a computing program providing a "virtual laboratory," that is, a 3D representation of ENGIN-X that can be easily manipulated through a graphical user interface. This dedicated software: (i) helps in setting up the initial measuring strategy, (ii) automates the sample alignment and all the routine aspects of the measurement process, and (iii) provides data analysis in almost real-time, allowing decisions on changes to the measurement strategy.

The capabilities of SSCANSS are better exemplified by a case study: the residual stresses introduced by weld repairs in nuclear power plants, a work in collaboration with British Energy and described in [27]. Repair welds are introduced into structures either to remedy initial fabrication defects found by routine inspection, or to rectify in-service degradation of components. A typical example is the repair-welded component shown in position on ENGIN-X in Figure 7.5a. The pipe was fabricated by manual metal arc girth welding of two ex-service stainless steel power station steam headers. In order to simulate a typical repair process, a section of the original weld was removed, and subsequently filled with new material, as schematically shown by Figure 7.6a. Scans of the full strain tensor were performed along several lines in the heat-affected zone and in the original weld. A map of the axial strain in the heat affected zone (HAZ) of the repair weld is shown in Figure 7.6c, which displays two main concentrations of tensile strain beneath the outer surface of the pipe, at the stop-end, and about midway. These short-range stress concentrations revealed the importance of stop-end effects for a correct description of the repair process.

Critical to this work has been the use of SSCANSS for the planning and execution of the experiment. The experimental configuration for the measurement of the axial strain is shown in Figure 7.5a, together with the virtual laboratory model of Figure 7.5b. The 3D model of the laboratory includes the collimators, the slits, and the positioning table. The 3D model of the sample can be produced from basic primitive objects (cuboids, cylinders, pipes, etc.), exemplified in this case by the simple model used to describe the pipe. The program allows exploring alternative experimental arrangements for measuring the different strain components, and alerts in case of collisions. It also provides an estimation of the required counting time using the expressions described in the previous section. For more complex shapes or regions requiring improved spatial resolution, a very precise model can be created using a coordinate measuring machine (CMM). Figure 7.6b shows such a model for the region of interest of the pipe, that is, in the vicinity of the weld repair. The shape of the IGV (dark cube) and the measurement positions (stars) are also displayed within the 3D model of the sample. The points to be measured are defined interactively in a graphical visualization of the sample, or alternatively their (x, y, z) coordinates in the sample system can be imported from a file.

The positioner movements required to bring the sample to the correct position for measurement may be generated automatically. This facility requires the determination of the transformation relating the sample and laboratory coordinate systems. In order to do this, the positions of a number of fiducial points on the surface of the sample are measured using a pair of precision theodolites [26]. A least-squares procedure is then used to find the transformation matrix that most closely maps the fiducial points on the sample to their measured positions in the laboratory. Since the sample is a rigid body, a minimum of three non-collinear fiducial points are sufficient for this purpose, though using a larger number may improve accuracy. A positional accuracy of better than 0.1 mm is usual for this alignment procedure.

The spectra recorded at each measurement position by the detectors belonging to the same bank are time-focused into a single diffraction spectrum, as the one shown in Figure 7.1. The program provides automatic single peak and full pattern refinement of the diffraction spectra specially devised for strain determination, through

Figure 7.5 The ENGIN-X virtual laboratory. (a) Measurement of axial strain for a 190 kg pipe containing a repair weld and (b) the corresponding experiment simulation within SSCANSS. (c) Simulation for the radial and hoop strain directions. (Santisteban 2006 [14]. Reproduced with permission of International Union of Crystallography.)

(a)

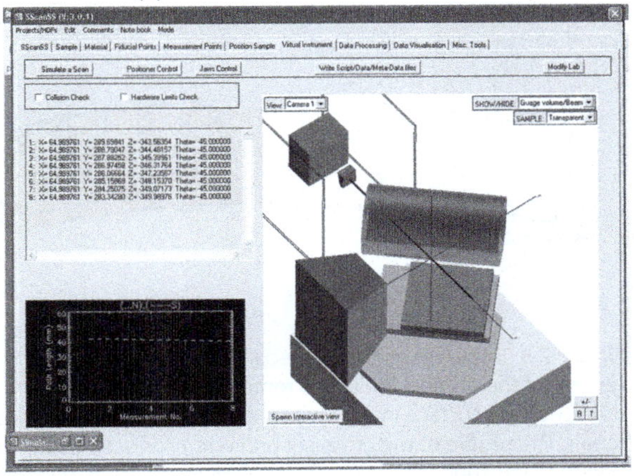

(b)

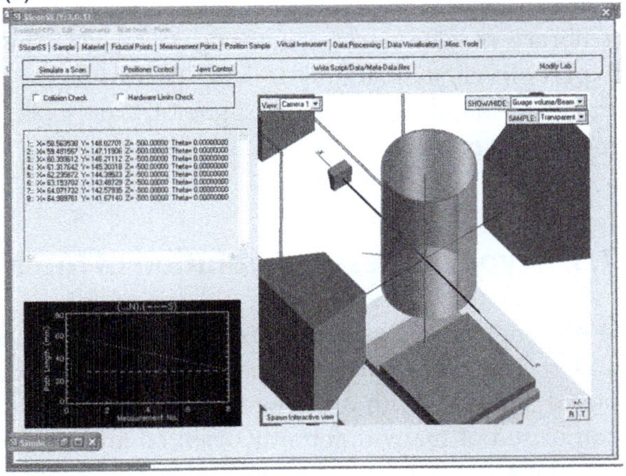

(c)

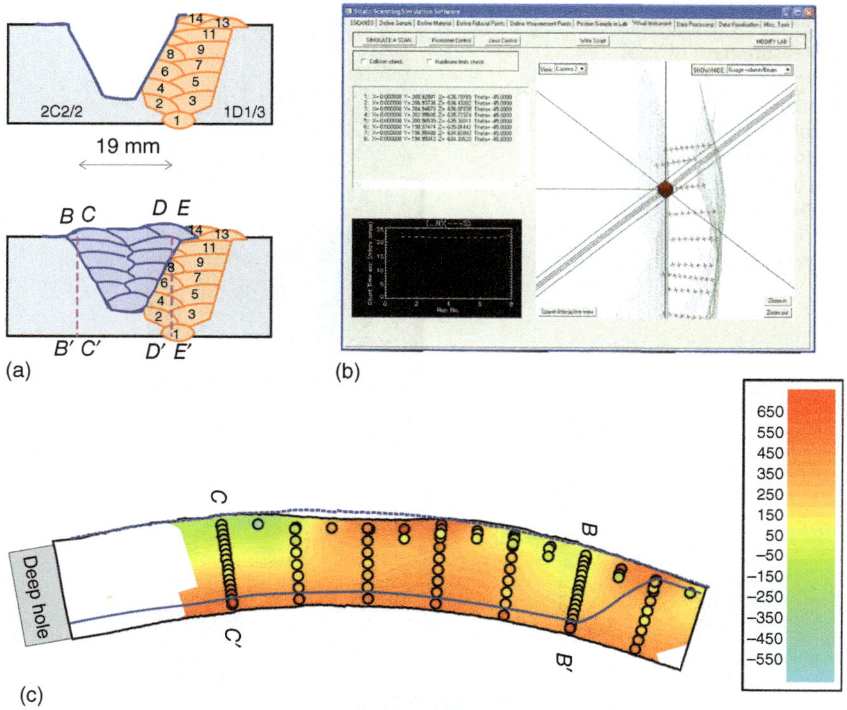

Figure 7.6 Strain mapping in a repaired weld. (a) Hoop view of repairs introduced in the weld. A section of the original weld was removed and then filled with new material. (b) Detailed 3D model of the specimen surface in the vicinity of the repair. (c) Axial strain map in the heat affected zone of the repair. The profile of the cross-section was produced using a coordinate measuring machine. (Santisteban 2006 [14]. Reproduced with permission of International Union of Crystallography.)

a library of common engineering materials. This enables researchers who are not experts in crystallography to keep pace with the experiment, and be able to modify the experimental plan in the light of experience gained during earlier parts of the measuring process.

7.5 Type II Stresses: Evolution of Intergranular Stresses

While engineering materials are commonly assumed to be mechanically homogeneous and isotropic, in reality this is never the case. For example, composite materials are composed of phases with differing mechanical properties, and even single phase alloys are comprised of many differently oriented anisotropic crystallites with different properties along a particular sample direction. Due to incompatibility between the constituent parts, interphase and intergranular stresses develop as functions of applied stress or plastic strain. The collection of neutron diffraction profiles *in situ* during mechanical testing provides an effective means of tracking the evolution of these internal stresses. Energy dispersive diffraction is especially suited in this respect, because a single TOF spectrum is both phase- and orientation-selective, as different phases and different grain orientations contribute to different diffraction peaks. This is because only those grains oriented with an (*hkl*) plane normal parallel to the scattering vector **q** contribute to the *hkl* diffraction peak.

A typical *in situ* testing experiment is exemplified by Oliver *et al.* [28]. In that work, intergranular and interphase strains were compared in a single (ferrite) phase low carbon steel and a two phase (ferrite + spheroidized cementite) high carbon steel. A hydraulic stress rig was mounted on the beamline with its loading axis oriented horizontally at an angle of 45° to the incident beam, allowing simultaneous collection of diffraction profiles for lattice planes oriented parallel and perpendicular to the loading axis. Tensile tests were performed in a stepwise fashion – increasing the applied load and interrupting for neutron data acquisition. In more modern instruments, such as Vulcan [16], neutrons are recorded in so-called event-based data acquisition mode, where each neutron carries a time stamp. So, instead of stepwise scans, measurements are done continuously, and data synchronization is performed later according to neutron arrival time at the sample position. In [28] phase-specific elastic strains were determined by Rietveld full profile refinement of the lattice parameters, using the lattice parameters determined prior to loading as the stress-free references. Grain family-specific elastic strains were determined by fitting of individual peaks. The elastic phase strains observed along the axial direction are shown as functions of applied stress in Figure 7.7a. As seen, the ferrite elastic strain continues to increase almost linearly with applied stress even after yielding in the single phase low carbon steel. However, in the two phase steel, while both phases develop similar elastic strains before yield (indicating that they have similar elastic constants), at the yield point there is a sudden re-distribution of strain from the majority ferrite phase to the minority cementite (Fe_3C) phase. This can be understood on the basis that the ferrite yields while the cementite continues to deform elastically. Thus, a strain misfit develops between the phases, introducing compressive residual stress (−170 MPa) into the ferrite and tensile residual stress (+640 MPa) into the cementite. Similar effects can be observed in other composite or multiphase materials such as Al/SiC metal matrix composites or γ/γ' Ni superalloys.

Figure 7.7b shows the development of grain family-specific elastic strains with applied stress in the ferrite phase of the low carbon steel. It is noticeable that differently oriented families of grains behave in significantly different ways. Firstly, the different slopes prior to yield reflect the significant elastic anisotropy of ferrite single crystals, which are stiffest along the ⟨111⟩ direction and most compliant along ⟨100⟩. Secondly, the strains diverge during yielding. This reflects a similar effect at the grain level to that causing the interphase stress between ferrite and cementite in the high carbon steel. The experimental intergranular strains can be used to validate models of intergranular stress evolution. The continuous lines in Figure 7.7b represent calculations of an elastoplastic self-consistent model, which is in good agreement with the experimental results.

7.6 Type III Stresses: Dislocation Densities

The microscopic deformation of the lattice due to type III stresses manifests through broadening of the diffraction peaks. Many types of point, linear, planar, or volumetric defects can be responsible of such deformation. A thorough analysis of the effects of defects on diffraction can be found in [29]. Among these defects, dislocations have

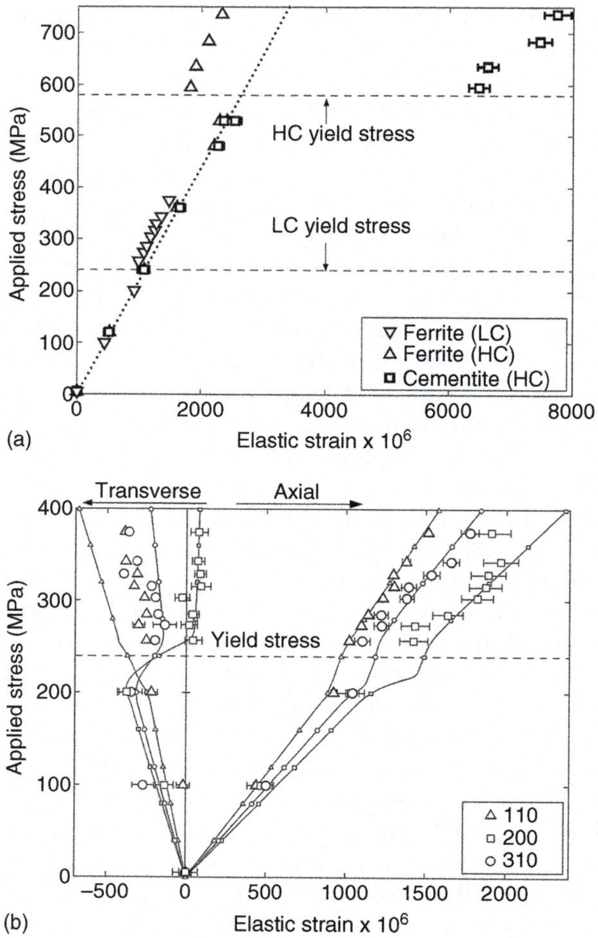

Figure 7.7 (a) Phase-specific axial strains versus applied uniaxial tensile stress. The dotted line is the linear elastic response of the ferrite phase in high carbon steel; the dashed lines mark the yield stresses of high carbon (HC) and low carbon (LC) steels. (b) Modeled and experimental evolution of average grain family elastic strains.

received special attention, due to their ubiquitous presence in metals; and because they play a major role in the response of materials to thermo-mechanical treatments. The dislocation density is a measure of the number of dislocations in a unit volume of a crystalline material, which, under some assumptions, can be effectively determined from the width of diffraction peaks. The most simple analysis of peak broadening is the Williamson–Hall (WH) plot [30], in which the *physical* broadening is analyzed in the reciprocal space using the variable $K = 1/d_{hkl}$. Figure 7.8a shows such a WH plot for Zr-2.5%Nb pressure tubes used in CANDU nuclear power plants, before installation and after 7 years of service [31], from experiments performed at the SMARTS beamline [15]. The type and densities of dislocations of as-fabricated pressure tubes are carefully controlled due to their role in the anisotropic irradiation growth of the tube observed during service. TOF neutron strain scanners are especially suited for these studies

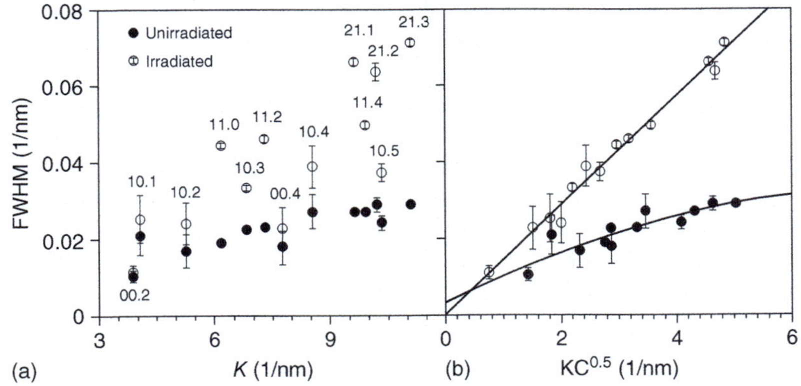

Figure 7.8 (a) Traditional and (b) modified Williamson–Hall plot of the irradiated and unirradiated Zr2.5%Nb samples. The traditional plot shows that diffraction vectors corresponding to the more broadened reflections are perpendicular or nearly perpendicular to the *c*-axis of the hexagon. This suggests that most irradiation-induced dislocations have Burgers vectors within the basal plane. (Balogh 2012 [31]. Reproduced with permission of Elsevier.)

because many reflections can be accessed simultaneously, with rather high and nearly constant instrumental resolution. The physical broadening in the plot, FWHM, is the peak broadening due only to microstructural features of the sample ($\delta_\varepsilon d_{\mathrm{hkl}}$ in Eq. (7.5)), and is obtained from the measured experimental broadening ($\delta_\varepsilon t_{\mathrm{hkl}}$) after proper subtraction of instrumental contributions. Such contribution is determined from measurements on a reference specimen, usually a ceramic powder, obtained under exactly the same experimental conditions. Several methods are available to subtract the instrumental contribution [29]. The Warren–Averbach method, which uses a Fourier series to describe the peak, is maybe the most versatile as it can deal with an arbitrary peak profile [32]. Assuming an isotropic distribution of the deformation, the slope (m) obtained by a least-squares fit gives a measure of the mean-square strain in the material, $\langle\varepsilon^2\rangle^{1/2} = m/\sqrt{2\pi}$. An estimation of the density of dislocations within the material can be obtained by assuming that the observed $\langle\varepsilon^2\rangle$ is solely due to dislocations. If only one type of dislocation is present, their density can be calculated as, $\rho = K\langle\varepsilon^2\rangle^{1/2}/b^2$ with b the Burgers vector and K a constant that includes details of the dislocation type [33]. However, the FWHM values shown in Figure 7.8a display a consistent orientation dependence across the pattern. Following irradiation, the {11.0}, {11.2}, {21.1}, and {21.3} reflections have broadened significantly more than {00.2}, {00.4}, {10.3}, {10.4}, and {10.5} reflections. The less broadened reflections all have diffraction vectors along or near the *c*-axis (00.1) of the reflecting crystallites, while those that broaden more have diffraction vectors within or close to the basal planes of the hexagonal cell. Thus, from inspection of the WH plot, we can conclude qualitatively from the raw data that most of the dislocations generated during irradiation have Burgers vectors within the basal plane, the so-called $\langle a\rangle$-type dislocations in hexagonal materials. Further analysis requires a more elaborated calculation of the effect of dislocations on the measured peak width, that is, their actual contribution to the mean-square strain $\langle\varepsilon^2\rangle$. This contribution can be quantified in terms of a contrast factor C for each specific dislocation type, which depends on the relative orientation between the Burgers vector (**b**) and line vector (**l**) of the dislocation, and the diffraction vector (**G**), in a similar way as in transmission electron microscopy (TEM) [29]. A very popular method that implements such contrast factors is the extended Convolutional Multiple Whole Profile (eCMWP), developed by Ungár and co-workers [34–36]. The method uses the Warren–Averbach series expansion [32] and can provide information about the type and density of the dislocations present in the material. Specific details are beyond the scope of this work, but can be obtained from the references above. After calculation of such a contrast factor, the FWHM can be analyzed in terms of a "modified" WH plot [34], shown in Figure 7.8b. In both the pre- and post-irradiated data, the contrast factor correction collapses the FWHM onto a line, providing strong validation of the assumption that the broadening is indeed due to dislocations. Such analysis showed that fast neutron irradiation resulted in a profound change in the dislocation arrangement, due to a fourfold increase in type $\langle a \rangle$-dislocation densities. Studies of dislocation densities are being increasingly performed on TOF neutron strain scanners [31, 37, 38].

7.7 Strain Imaging by Energy-Dispersive Neutron Transmission

In a TOF diffractometer, the neutrons that pass through the sample also contain a wealth of crystallographic information about the specimen. For instance, a neutron beam transmitted by a polycrystalline sample displays sudden, well-defined increases in intensity as a function of neutron wavelength (Figure 7.9c). These *Bragg edges* occur because for a given (*hkl*) reflection, the Bragg angle increases as the wavelength increases until $2\theta_B$ is equal to 180°. At wavelengths greater than this critical value, no scattering by this particular (*hkl*) lattice spacing can occur, and there is a sharp increase in the transmitted intensity. From Bragg's law, the wavelength at which this occurs is $\lambda = 2d_{hkl}$, giving a measure of the (*hkl*) d-spacing in the direction of the incoming beam.

By placing a detector just behind the sample, the position and magnitude of these edges can be determined by the TOF technique. As the position, height, and width of a Bragg-edge depend on essentially the same laws as the position, intensity, and width of conventional diffraction peaks, information about the stress state [39–46, 57], texture [47, 48] or the phases present in the sample [49–51] is readily available from an analysis of the transmission pattern. The transmission Tr is simply the ratio between the beam intensities recorded with and without the sample in front of the detector. It is related to the microscopic properties of the sample by $\mathrm{Tr}(\lambda) = \exp(-nw\sigma_{\mathrm{tot}}(\lambda))$, where n is the number of scattering centers per unit volume, w is the sample thickness, and σ_{tot} is the microscopic total cross-section of the material. In practice, the Bragg edges are broadened by the finite resolution of the instrument and by imperfections within the material, so the transmission near a Bragg edge is described by a characteristic edge profile [40]

$$Tr(t) = \exp(-nw\sigma_0)\left\{ \exp(-nw\sigma_{hkl}) + (1 - \exp(-nw\sigma_{hkl})) \times \frac{1}{2} \right.$$
$$\left. \times \left[\mathrm{erfc}\left(-\frac{t - t_{hkl}}{\sqrt{2}\sigma} \right) - \exp\left(-\frac{t - t_{hkl}}{\tau} + \frac{\sigma^2}{2\tau^2} \right) \mathrm{erfc}\left(-\frac{t - t_{hkl}}{\sqrt{2}\sigma} + \frac{\sigma}{\tau} \right) \right] \right\} \tag{7.11}$$

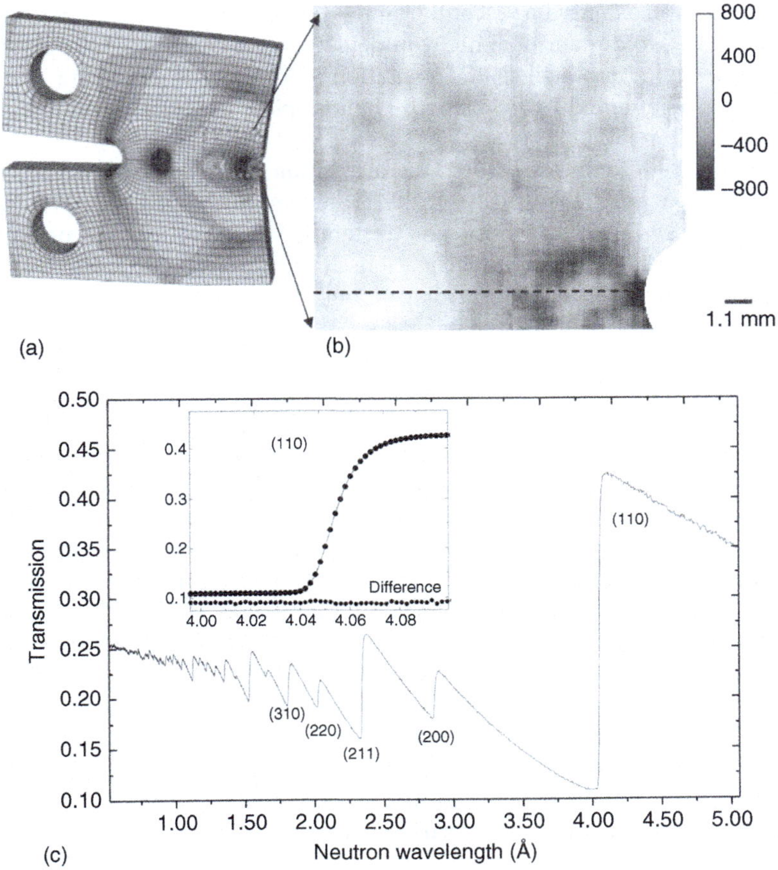

Figure 7.9 Out-of-plane strain in a compact-tension crack sample (a) from a finite element model and (b) measured by Bragg-edge transmission within the area shown by the dashed rectangle. The legend bar indicates strain values in $\mu\varepsilon$ and applies to both the transmission measurement and the finite element model. The transmission values result from the shift measured in the position of the Bragg edges that appear in the transmitted spectra, as illustrated in (c). (Panels (a,b): Tremsin 2012 [39]. Reproduced with permission of Wiley. Panel (c): Santisteban 2001 [40]. Reproduced with permission of the International Union of Crystallography.)

with n and w defined above. The edge position is t_{hkl}, and the edge width is defined by τ and σ, the asymmetric and symmetric broadenings discussed in Section 7.2.1. A fit for the (110) edge of iron using this profile is shown in the inset of Figure 7.9c. The edge position t_{hkl} can be precisely defined by a least-squares fit using the edge profile of Eq. (7.11), and converted into d-spacing after a simple linear calibration (with $\Delta d_{hkl}/d_{hkl} \sim 50\ \mu\varepsilon$). This opens up the possibility of using an array of detectors to produce a radiographic "image" of the strain variation in the sample [43], as shown in Figure 7.9b for a compact-tension crack sample [39]. However, it also means that the measurement is an average over the complete transmission path through the sample, making the method useful for examining plates and other essentially two-dimensional objects. With such applications in mind, ENGIN-X was provided with a pixelated transmission detector consisting of a 10×10 array of $2 \times 2\ \text{mm}^2$ scintillator detectors on a 2.5 mm pitch

[43]. Such strain imaging capability has recently gained renewed interest [39, 46], due to technical advances in two dimensional neutron detectors and more intense pulsed neutron sources. At present, both high spatial resolution ($\sim 55 \times 55\,\mu m^2$) and high TOF resolution ($\sim 1\,\mu s$) can be attained using a micro-channel plate detector, doped with boron for neutron detection [52].

The $\sin^2\psi$ technique, traditionally used in X-ray diffraction, can also be successfully implemented in TOF neutron transmission. In [44], Steuwer and collaborators performed uniaxial tensile tests on both ferritic and austenitic steel specimen, recording the transmitted spectra at different sample inclinations. Their results showed that TOF transmission can be used to determine anisotropic strains, diffraction elastic constants, the residual and applied stress state, and the unstrained lattice parameter [45].

The energy-dispersive neutron transmission of single crystals is rather different to the polycrystalline case. Instead of Bragg edges, the transmitted beam presents a series of dips in intensity as a result of reflection in the crystal planes [53, 58]. The position of these dips can be exploited for the definition of the crystal orientation with a resolution of 1 min of arc and to quantify internal or applied stress, and the width gives a measure of the crystal mosaicity. In [53], a Cu monochromator was plastically deformed by uniaxial tension along the [110] direction *in situ*, and the evolution of crystal orientation and mosaicity was tracked by analyzing the transmitted TOF spectra. Several crystal reflections at different locations of the sample were simultaneously studied during the experiments, revealing the anisotropic increase in mosaicity. On the other hand, the transmission of thermal neutrons through textured materials is strongly dependent on the neutron wavelength as well as on the specimen orientation [47]. So, large variations exist in the neutron total cross section of naturally occurring and manmade objects, as opposed to the simple Bragg edges expected for randomly oriented crystallites. Such differences have been exploited for the identification of the manufacturing process of ancient bronze artifacts [54] and for imaging variations in texture in Zr-based components [48].

A new instrument under construction at ISIS, IMAT [55], will combine imaging options for mapping microstructural properties (provided by transmission detectors), with phase analysis, residual strain analysis, and texture analysis possibilities provided by neutron diffraction, as in a TOF neutron strain scanner. A key point of the design is the "tomography-driven" diffraction, with software that will enable user-friendly operation of the instrument to study geometrically complex samples.

7.8 Conclusions

Energy-dispersive neutron diffraction presents some real advantages for the study of residual and applied stresses. This is because the elastic strain along different sample directions and for different phases or crystal families can be measured simultaneously, which represents an ideal scenario for the study of type-II stresses. Thus, *in situ* loading experiments have provided unique experimental information about deformation mechanisms in polycrystalline materials. Besides this, the technique can also provide a very good approximation to the macroscopic or engineering elastic strain. As several crystal families (and hence several crystal orientations) are measured on a single diffraction spectrum, the average of the experimental single-peak strains is usually very close to the macroscopic response of the material. This allows the use of macroscopic elastic constants for the calculation of stresses, thereby providing a rather reliable and simple technique for non-destructive stress analysis.

Pulsed neutron sources are especially suited for energy dispersive experiments because the TOF technique offers a simple energy analysis of the diffracted beam. Important reductions in counting times in TOF neutron strain scanning can be achieved by making the right choice of neutron flight paths and beam collimation. Besides conventional neutron diffraction, energy-dispersive neutron transmission at pulsed neutron sources can also provide information about the spatial variation of strain across plate specimens. It must be remembered however that in this case the information corresponds to the through-thickness average of the out-of-plane elastic strain.

Acknowledgments

I am indebted to Ed Oliver, Lyndon Edwards, Mark Daymond, Axel Steuwer, John Bouchard, Sven Vogel, Phil Withers, Jon James, Mike Fitzpatrick, and Hans Priesmeyer, for sharing with me most of the knowledge contained within this chapter.

References

1 Kamminga, J.D., De Keijser, T.H., Mittemeijer, E.J., and Delhez, R. (2000) *J. Appl. Crystallogr.*, **33**, 1059–1066.

2 Kropff, F., Granada, J.R., and Mayer, R.E. (1982) *Nucl. Instrum. Methods Phys. Res.*, **198**, 515–521.

3 Jorgensen, J.D., Faber, J. Jr., Carpenter, J.M., Crawford, R.K., Haumann, J.R., Hitterman, R.L., Kleb, R., Ostrowski, G.E., Rotella, F.J., and Worlton, T.G. (1989) *J. Appl. Crystallogr.*, **22**, 321–333.

4 Rietveld, H.M. (1969) *J. Appl. Crystallogr.*, **2**, 65–71.

5 Larson, A.C. and Von Dreele, R.B. (1994) General Structure Analysis System (GSAS). Los Alamos National Laboratory Report LAUR 86–748.

6 Toby, B.H. (2001) *J. Appl. Crystallogr.*, **34**, 210–213, EXPGUI and GSAS at http://www.ncnr.nist.gov/programs/crystallography/software/expgui/expgui.html (accessed 29 June 2016).

7 Rodriguez-Carvajal, J. (1993) *Physica B*, **192**, 55–69, FullProf http://www-llb.cea.fr/fullweb/fp2k/fp2k.htm (accessed 29 June 2016).

8 Matthies, S., Pehl, J., Wenk, H.-R., Lutterotti, L., and Vogel, S.C. (2005) *J. Appl. Crystallogr.*, **38**, 462–465, MAUD http://www.ing.unitn.it/~maud (accessed 29 June 2016).

9 Daymond, M.R., Bourke, M.A.M., Von Dreele, R.B., Clausen, B., and Lorentzen, T. (1997) *J. Appl. Phys.*, **82**, 1554–1562.

10 Daymond, M.R. and Bourke, M.A.M. (1999) *J. Appl. Phys.*, **85**, 739–747.

11 Daymond, M.R. (2004) *J. Appl. Phys.*, **96**, 4263–4272.

12 Balzar, D., Von Dreele, R.B., Bennet, K., and Ledbetter, H. (1998) *J. Appl. Phys.*, **84**, 4822–4833.

13 Withers, P.J. (2004) *J. Appl. Crystallogr.*, **37**, 596–606.
14 Santisteban, J.R., Daymond, M.R., James, J.A., and Edwards, L. (2006) *J. Appl. Crystallogr.*, **39**, 812–825.
15 Bourke, M.A.M., Dunand, D.C., and Ustundag, E. (2002) *Appl. Phys. A*, **74**, s1707–s1709.
16 Wang, X.L., Holden, T.M., Stoica, A.D., An, K., Skorpenske, H.D., Jones, A.B., Rennich, G.Q., and Iverson, E.B. (2010) *Mater. Sci. Forum*, **652**, 105–110.
17 Harjo, S., Ito, T., Aizawa, K., Arima, H., Abe, T., and Kamiyama, T. (2011) *Mater. Sci. Forum*, **681**, 443–448.
18 Withers, P.J., Daymond, M.R., and Johnson, M.W. (2001) *J. Appl. Crystallogr.*, **34**, 737–743.
19 Windsor, C.G. (1981) *Pulsed Neutron Scattering*, Taylor & Francis, London.
20 Buras, B. and Gerward, L. (1975) *Acta Crystallogr., Sect. A*, **31**, 372–374.
21 Carr, D.G., Ripley, M.I., Holden, T.M., Brown, D.W., and Vogel, S.C. (2004) *Acta Mater.*, **52**, 4083–4091.
22 Johnson, M.W. and Daymond, M.R. (2002) *J. Appl. Crystallogr.*, **35**, 49–57.
23 Johnson, M.W., Edwards, L., and Withers, P.J. (1997) *Physica B*, **234**, 1141–1143.
24 Withers, P.J., Johnson, M.W., and Wright, J.S. (2000) *Physica B*, **292**, 273–285.
25 Daymond, M.R. (2001) *Physica B*, **301** (**3-4**), 221–226.
26 James, J., Santisteban, J. Edwards, L., and Daymond, M. (2004) *Physica B: Condens. Matter*, **350** (**1–3**), Supplement, E743–E746.
27 Bouchard, P.J., George, D., Santisteban, J.R., Bruno, G., Dutta, M., Edwards, L., and Smith, D.J. (2005) *Int. J. Press. Vessels Pip.*, **82**, 299–310.
28 Oliver, E.C., Mori, T., Daymond, M.R., and Withers, P.J. (2003) *Acta Mater.*, **51**, 6453–6464.
29 Snyder, R.L., Fiala, J., Bunge, H.J., and International Union of Crystallography (1999) *Defect and Microstructure Analysis by Diffraction*, Oxford University Press, Oxford.
30 Williamson, G.K. and Hall, W.H. (1953) *Acta Metall.*, **1**, 22–31.
31 Balogh, L., Brown, D.W., Mosbrucker, P., Long, F., and Daymond, M.R. (2012) *Acta Mater.*, **60**, 5567–5577.
32 Warren, B.E. and Averbach, B.L. (1950) *J. Appl. Phys.*, **21**, 595–599.
33 Williamson, G.K. and Smallman, R.E. (1956) *Philos. Mag.*, **1**, 34.
34 Ungár, T., Ott, S., Sanders, P.G., Borbély, A., and Weertman, J.R. (1998) *Acta Mater.*, **46**, 3693–3699.
35 Ungár, T., Dragomir, I., Révész, A., and Borbély, A. (1999) *J. Appl. Crystallogr.*, **32**, 992–1002.
36 Dragomir, I.C. and Ungár, T. (2002) *J. Appl. Crystallogr.*, **35**, 556–564.
37 Ungár, T., Stoica, A.D., Tichy, G., and Wang, X.-L. (2014) *Acta Mater.*, **66**, 251–261.
38 Use of Neutron Beams for Materials Research Relevant to the Nuclear Energy Sector-Companion CD (2015) IAEA TecDoc-1773, pp 77–84, International Atomic Energy Agency.
39 Tremsin, A.S., McPhate, J.B., Steuwer, A., Kockelmann, W., Paradowska, A.M., Kelleher, J.F., and Feller, W.B. (2012) *Strain*, **48**, 296–305.
40 Santisteban, J.R., Edwards, L., Steuwer, A., and Withers, P.J. (2001) *J. Appl. Crystallogr.*, **34**, 289–297.

41 Priesmeyer, H.G. (1992) in *Measurement of Residual and Applied Stress Using Neutron Diffraction* (Proceedings NATO ARW Oxford) (eds M.T. Hutchings and A.D. Krawitz), Kluwer Academic Publishers, Dordrecht, pp. 389–394.
42 Steuwer, A., Withers, P.J., Santisteban, J.R., Edwards, L., Bruno, G., Fitzpatrick, M.E., Daymond, M.R., Johnson, M.E., and Wang, D. (2001) *Phys. Status Solidi A*, **185**, 221–230.
43 Santisteban, J.R., Edwards, L., Fitzpatrick, M.E., Steuwer, A., Withers, P.J., Daymond, M.R., Johnson, M.W., Rhodes, N., and Schooneveld, E.M. (2002) *Nucl. Instrum. Methods Phys. Res., Sect. A*, **481**, 255–258.
44 Steuwer, A., Santisteban, J.R., Withers, P.J., Edwards, L., and Fitzpatrick, M.E. (2003) *J. Appl. Crystallogr.*, **36**, 1159–1168.
45 Santisteban, J.R., Steuwer, A., Edwards, L., Withers, P.J., and Fitzpatrick, M.E. (2002) *J. Appl. Crystallogr.*, **35**, 497–504.
46 Strobl, M., Woracek, R., Kardjilov, N., Hilger, A., Wimpory, R., Tremsin, A., Wilpert, T., Schulz, C., Manke, I., and Penumadu, D. (2012) *Nucl. Instrum. Methods Phys. Res., Sect. A*, **680**, 27–34.
47 Santisteban, J.R., Edwards, L., and Stelmukh, V. (2006) *Physica B*, **385–386**, 636–638.
48 Santisteban, J.R., Vicente-Alvarez, M.A., Vizcaino, P., Banchik, A.D., Vogel, S.C., Tremsin, A.S., Vallerga, J.V., McPhate, J.B., Lehmann, E., and Kockelmann, W. (2012) *J. Nucl. Mater.*, **425**, 218–227.
49 Steuwer, A., Santisteban, J.R., Withers, P.J., and Edwards, L. (2005) *J. Appl. Phys.*, **97**, 074903.
50 Meggers, K., Priesmeyer, H.G., Trela, J., and Dahms, M. (1994) *Mater. Sci. Eng., A*, **188**, 301–304.
51 Vogel, S., Ustundag, E., Hanan, J.C., Yuan, V.W., and Bourke, M.A.M. (2002) *Mater. Sci. Eng., A*, **333**, 1–9.
52 Tremsin, A.S., McPhate, J.B., Vallerga, J.V., Siegmund, O.H.W., Hull, J.S., Feller, W.B., and Lehmann, E. (2009) *Nucl. Instrum. Methods Phys. Res., Sect. A*, **604**, 140–143.
53 Santisteban, J.R. (2005) *J. Appl. Crystallogr.*, **38**, 934–944.
54 Siano, S., Bartoli, L., Santisteban, J.R., Kockelmann, W., Daymond, M.R., Miccio, M., and De Marinis, G. (2006) *Archaeometry*, **48**, 77–96.
55 Kockelman, W., Zhang, S.Y., Kelleher, J.F., Nightingale, J.B., Burca, G., and James, J.A. (2013) *Physics Procedia*, **43**, 100.
56 Ikeda, S. and Carpenter, J.M. (1985) *Nucl. Instrum. Methods Phys. Res., Sect. A*, **239**, 536–544.
57 Santisteban, J.R., Edwards, L., Steuwer, A., Withers, P.J., and Fitzpatrick, M.E. (2002) *Appl. Phys. A*, **75**, 1433–1436.
58 Malamud, F. and Santisteban, J.R. (2016) *J. Appl. Crystallogr.*, **49**, 348–365.

8

Residual Stress Analysis by Monochromatic High-Energy X-rays

René V. Martins

In Chapters 6 and 7, the use of the angle dispersive and time-of-flight method for residual stress analysis using neutron diffraction was explained. This chapter and Chapter 9 will deal similarly with residual stress analysis using monochromatic high-energy synchrotron radiation (angle dispersive setup) and white high-energy synchrotron radiation (energy dispersive setup), respectively.

8.1 Basic Setups

The aim of all the diffraction experiments described here is to determine the (residual) lattice strain from diffraction peak positions with an accuracy of at least 100×10^{-6} (100 μ strain). Figure 8.1 shows schematically four basic experimental configurations for residual strain measurements using high-energy synchrotron radiation. The setups aim at different length scales, ranging from the macroscopic to the size of individual grains. They are designed for use in transmission geometry, meaning that the incoming beam penetrates the sample at the front side and exits it at the backside delivering diffraction cones. The incoming beam is usually blocked by a beam stop behind the sample (not always shown in Figure 8.1) while the diffracted rays are detected with a detector system such as a point, line, or area detector. The area detector usually speeds up data acquisition considerably and allows, depending on its size and the experimental technique employed, the monitoring of complete diffraction rings.

The simplest setup (Figure 8.1a) consists of a sample mounted on a positioning device, of a pair of slits or a pinhole to confine the incoming beam, and of a detector. The detector can either be a point, line, or area detector. The gauge volume size can only be horizontally and vertically defined by the beam size. Therefore, only a two-dimensional (2D) spatial resolution can be achieved. The data obtained represents information integrated along the beam path and no depth-resolved information is contained. Because of the small Bragg angle of high-energy synchrotron radiation, the scattering vectors for the diffraction cones lie almost in a plane (e.g., for the Al {3 1 1} reflection the diffraction angle 2θ is about 7.3° at a photon energy of 80 keV; the scattering vector is then only by the Bragg angle of 3.65° out of plane). Therefore, one data image with at least a quarter of a diffraction ring containing the principal axes allows already the calculation of an in-plane biaxial stress if it can be assumed that the stress component parallel

Neutrons and Synchrotron Radiation in Engineering Materials Science: From Fundamentals to Applications, Second Edition. Edited by Peter Staron, Andreas Schreyer, Helmut Clemens, and Svea Mayer.

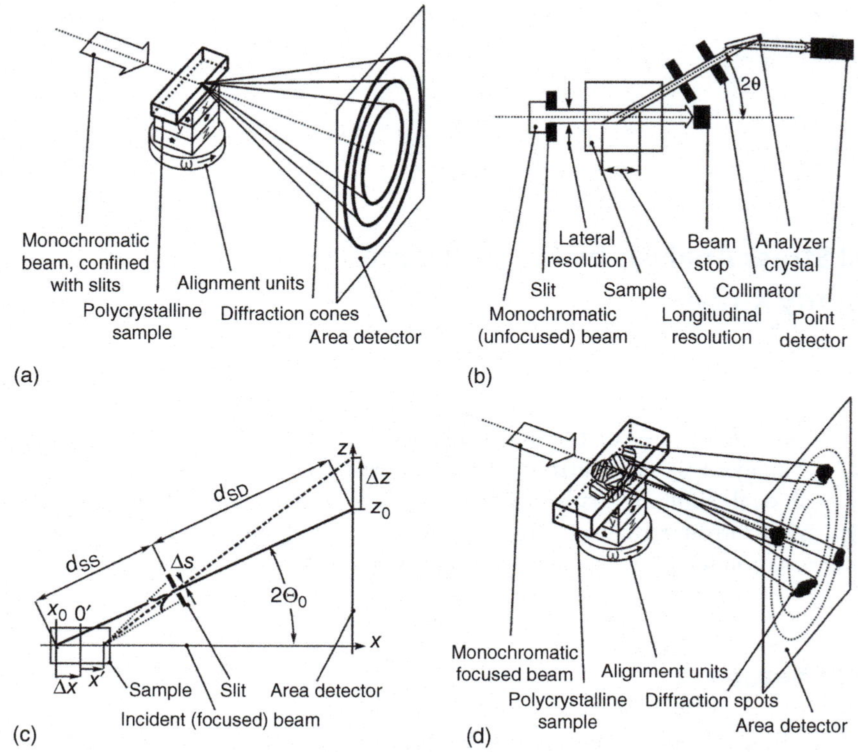

Figure 8.1 Basic experimental setups in transmission mode for strain measurements: (a) integrating over the sample thickness; (b) crossed beam technique with collimator; (c) crossed beam principle with slit imaging showing ray diagram and coordinate system describing the relations between sample position, gauge volume position, and beam positions on the area detector placed perpendicular to the *x*-axis at the position of the *z*-axis; and (d) bulk grain studies applying 3DXRD technique.

to the incoming beam is zero, which is, for example, often valid for thin samples. If two additional images are recorded with the sample rotated around one of the principal axes by 45° and 90°, the complete strain tensor information is available. To avoid systematic errors due to changes of the sample-to-detector distance it is very useful to attach directly to the sample a strain-free reference sample made from a different material than the one investigated to avoid peak overlap between the different materials (e.g., Al/Cu is a useful combination). The measurement of line shifts relative to the reference peaks reduces the influence of errors originating from wavelength instabilities and uncontrolled sample movements. For the determination of the strain-free lattice parameter d_0 please consult Chapter 9.

The same data can be used for texture analyses provided an image collection for a sufficient number of sample orientations is available (cf. Chapters 3 and 10).

Setups of this kind are typically found at second-generation synchrotrons or bending magnet beamlines. The beams available there can have sizes in the centimeter range. This can be an advantage when a coarse grain structure makes macroscopic measurements difficult (see Section 8.6). Narrowing down such a beam in the direction where steep stress or texture gradients are expected in the sample, but leaving it large in the other direction allows to increase the number of illuminated grains without sacrificing spatial resolution in the direction of interest.

The setups in Figures 8.1b,c can be grouped under the term *crossed-beam techniques*. They are characterized by an adjustable 3D spatial resolution. The lateral resolution (i.e., perpendicular to the incoming beam) is given by the beam size and the longitudinal one (i.e., parallel to the beam) is defined by an optical element placed behind the sample. Figure 8.1b shows a setup with a collimator behind the sample, thus defining an elongated diamond-shaped gauge volume whose geometrical size depends on the dimensions of the incoming beam, the diffraction angle, and the collimator opening behind the sample. A collimator can be made, for example, from two distant slits that meet the condition that the opening angle is smaller than the divergence of the radiation to be collimated. The first slit should be as close to the sample as possible. The geometrical elongation of the gauge volume can be calculated according to Eq. (8.1):

$$l = \frac{b + a \cdot \cos(2\theta)}{\sin(2\theta)} \tag{8.1}$$

l : gauge volume length
a : width of the incoming beam
b : width of the collimator
2θ : diffraction angle

The obviously lower spatial resolution in longitudinal direction as compared to the horizontal and vertical one can sometimes be compensated by a clever sample alignment. In the case of strain investigations, one would try to align the sample such that the longitudinal gauge volume axis lies in the direction of a shallow strain gradient. Equation (8.1) is also useful for a rough estimation in the case of the other experimental configurations (Figure 8.1c). However, it has to be noted that the effective resolution can be worse by a factor of 3 or more, depending on the beam divergence and the energy-dependent spatial resolution of the detector used (cf. Section 8.3.2).

In the case of the collimator setup, the diffraction peak position for the gauge volume defined is measured by scanning the collimator and the point detector across the peak. It is clear that a simultaneous monitoring of a reference signal is not possible and that the system is very sensitive to geometrical errors in the collimator orientation. Furthermore, the technique is rather time consuming since only a small fraction of on diffraction ring can be measured at a time. The MAXIM setup reduced this shortcoming with a capillary array matched to a CCD chip, with each CCD pixel capturing the information of only one capillary [1]. After the shutdown of the DORIS III ring at Hamburger Synchrotronstrahlungslabor/Deutsches Elektronen-Synchrotron (HASYLAB/DESY), this setup was discontinued.

In [2] a setup with two rotating slits was reported more recently, resulting in a collimator that accepts diffracted photons from a specific gauge volume as a function of their azimuth and diffraction angle. The depth resolution is in the millimeter range.

To circumvent the influence of geometrical changes due to sample surface movements or collimator movements, an analyzer crystal can be inserted between the collimator and the detector. This is shown in Figure 8.1b. Again, each peak has to be scanned individually. This experimental setup is one of the standard configurations available, for example, at beamline ID22 of the European Synchrotron Radiation Facility (ESRF). The data quality obtained there allows also peak profile analyses for the investigation of microstrains and microstructural phenomena. An example for depth-resolved studies using this technique can be found in [3].

Figure 8.1c shows a so-called slit imaging setup. In contrast to the before-mentioned setups with a collimator, the slit as the gauge volume-defining element stays fixed and the sample is scanned along the beam. At each sample position, an image of the diffracted signal is recorded. The implications for the data analysis are addressed in Section 8.2.

Figure 8.1d shows a setup for the simultaneous investigation of several individual bulk grains, applying the three-dimensional X-ray diffraction technique (3DXRD) (cf. Chapter 17). In principle, the setup resembles the one in Figure 8.1a to a large extent. However, investigations on this length scale require special arrangements and approaches in the setup and the data analysis. Examples are presented in Section 8.5.

8.2 Principle of Slit Imaging and Data Reconstruction

Depth-resolved strain measurements by slit imaging aim at determining, as a function of the sample position, the variation of diffraction peak positions emanating from an illuminated local volume with respect to the peak positions of an ideally unstrained reference sample. In Figure 8.1c, this reference beam position is represented by the solid line at an angle of $2\Theta_0$ to the incoming beam. The origin of this line coincides here with the origin of the laboratory coordinate system x_0 and the point where it impinges on the area detector, placed at the z-axis, is the detector reference point z_0. The beam passes the slit that defines the local gauge volume within the strained sample. The slit opening is Δs. The respective distances between sample, slit, and detector are called d_{SS} and d_{SD}. The strained sample is positioned with its center $0'$ at the position Δx. Due to the finite width of the reflection, only part of the rays emanating at the position x' are recorded on the detector at the distance Δz, relative to the reference position z_0. It is clear that beams impinging on the detector can have their origin at different positions x'. Therefore, it is not possible to obtain the peak position for a specific x' from a single recorded intensity distribution. A series of images have to be taken in a sufficient step width and range along x in order to assure that the signal originating from x' can be completely detected, although it may be split up in several images. The diffraction peaks for each position x' can now be reconstructed using Eq. (8.2) [4]. The equation permits the calculation of the source point x' of each beam detected at a position Δz on the detector.

$$x' = b \cdot \Delta z \left[1 + \frac{\Delta z}{d_{SD} \cdot \sin(2\theta_0)}\right]^{-1} - \Delta x \quad \text{with} \quad b = \frac{d_{SS}}{d_{SD} \cdot \tan(2\theta_0)} \tag{8.2}$$

d_{SS} : sample-to-slit distance
d_{SD} : sample-to-detector distance
$2\theta_0$: diffraction angle of unstrained sample
Δx : sample displacement along incoming beam
Δz : displacement on detector
Δs : slit opening

For the practical application of the formula, an approximate value can be used for $2\theta_0$ instead of its exact value. Furthermore, the data reconstruction is only necessary, when

d_{SD} is less than 10 times d_{SS}. At larger d_{SD} values, the coupling between sample position and peak position becomes negligible and the data correction can be omitted.

8.3 The Conical Slit

8.3.1 Working Principle

The conical slit setup (Figure 8.2) is a special case of the slit imaging setup (cf. Sections 8.1 and 8.2). The idea behind the conical slit is to use a set of concentric conical apertures with a common apex instead of a slit made from a pair of straight blades, aligned for a specific reflection at a specific azimuthal angle on the diffraction ring. The advantage of the conical slit is obvious: instead of observing just a small fraction of a diffraction cone emanating from a gauge volume it becomes possible to monitor simultaneously different complete diffraction rings emanating from the same gauge volume. The conical apertures have to match the diffraction cones of the material to be investigated. The distance of their common apex to the slit surface can be considered as the focal length of the system. Note that usually a series of images have to be taken to achieve the highest spatial resolution, after data reconstruction (cf. Section 8.2).

8.3.2 Capabilities

Presently, the conical slit devices have evolved to their third-generation type. The first generation, made for fcc materials with a focal distance of 10 mm, is not in use anymore. A description, which is, in principle, also valid for the other conical slits, can be found in [5]. The second-generation slit has a focal length of 100 mm, thus enabling the easy accommodation of a bulky sample environment. It was designed for fcc and bcc materials. An overview of the reflections accepted, including the experimentally determined gauge lengths and geometrical parameters of the slit system, are given in Table 8.1.

Each of the seven apertures has a gap of 25 μm. Due to the fixed accepted diffraction angles, the energy of the incoming beam has to be adjusted so that the diffraction

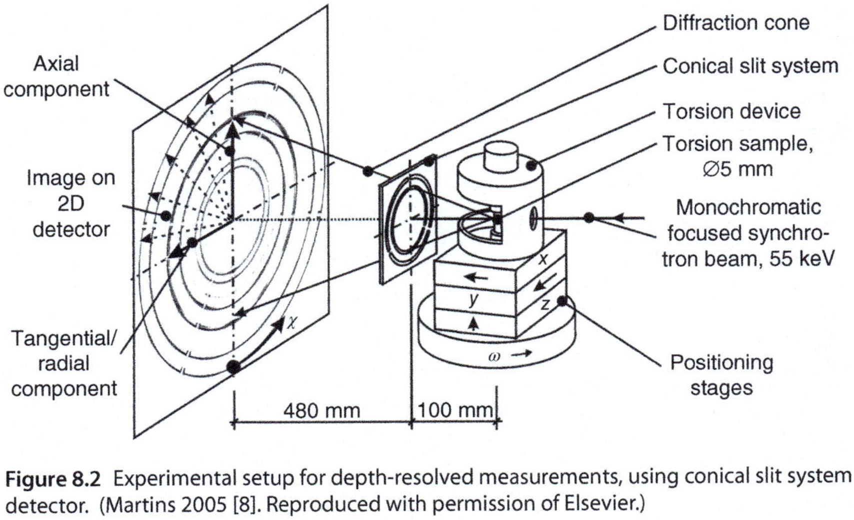

Figure 8.2 Experimental setup for depth-resolved measurements, using conical slit system and 2D detector. (Martins 2005 [8]. Reproduced with permission of Elsevier.)

cones match the aperture pattern. For Al this is 55 keV and for α-Fe it is 77.7 keV. In case of other crystal symmetries, it is still possible to adjust the energy in a way so that at least one-diffraction ring is passing the slit. This second-generation slit is available at the beamline ID11 at the ESRF.

Third-generation slits are mechanically improved, and have a slit gap of 20 μm. They are optimized either for cubic or for hexagonal structures, and are available with a focal length of 50 mm at beamline 1-ID of the Advanced Photon Source (APS), or with a focal length of 100 mm at the High-Energy Materials Science (HEMS) beamline at PETRA III, DESY, operated by Helmholtz-Zentrum Geesthacht.

The commissioning of the second-generation slit showed that the dimensions of the incoming beam should be around 10 μm × 10 μm, to achieve a high-spatial resolution parallel to the incoming beam. A major requirement is, therefore, the focusing of the incoming beam. Just cutting down the beam size with slits would result in a very low photon flux and long exposure times. Looking at Eq. (8.1) and the gauge lengths given in Table 8.1 it can be seen that the gauge lengths, determined experimentally for the second-generation slit, are a factor of about three bigger than the geometrical ones. The main reasons for that is the 2θ and optics-dependent beam divergence behind the sample and the energy-dependent spatial resolution of the detector. In the present case, the highest divergences behind the sample ranged between 0.696 mrad for the {1 1 1} and 1.41 mrad for the {4 2 2} reflection at the beamline ID11 at the ESRF with a bent Laue crystal and an elliptically bent multilayer [6] for vertical and horizontal focusing respectively. The spatial resolution for a MAR345 image plate detector (with a nominal resolution set to 100 μm) was found to be 335 μm at 55 keV. Improving the energy-dependent detector resolution by a factor of 2 will lead to an improvement of the gauge length by the same factor.

Recent results from measurements with a third-generation slit at PETRA III show a spatial resolution of 700 and 640 μm for the Fe (1 1 0) and (2 0 0) reflections, respectively [7]. The incoming beam with a photon energy of 63.4 keV was shaped with slits to a size

Table 8.1 Overview of the reflections selected by the second-generation conical slit, geometrical slit parameters, and the corresponding measured longitudinal gauge lengths.

hkl	Material	2θ (°)	Radius (mm)	Gauge length (μm)
1 1 1	Al[a)]	5.529	9.6755	750
2 0 0	Al/α-Fe	6.383	11.1854	640
2 1 1	α-Fe	7.821	13.7314	560[b)]
2 2 0	Al/α-Fe	9.029	15.8931	500
2 2 2	Al/α-Fe	11.064	19.5573	420
3 3 1	Al	13.934	24.8159	370
4 2 2	Al	15.676	28.0596	320

a) For Al, a photon energy of 55 keV has to be used and for α-Fe 77.7 keV to match the conical slits.
b) Interpolated value.

of 50 μm. Smaller beam sizes did not yield significant improvements in spatial resolution, which was imputed to the energy resolution given by the monochromator.

The great advantage of the conical slit is the possibility to acquire almost complete depth-resolved orientation space data (texture), which is, in addition, suitable for strain analyses. The data acquisition is orders of magnitude faster than with conventional crossed beam techniques.

A drawback is that each setup has to be adjusted via the wavelength to the unit cell of the material investigated. Therefore, in most cases, phase transformations cannot be observed.

8.3.3 Example

A typical setup with a conical slit is shown in Figure 8.2. It consists of sample translation and rotation devices, the conical slit, and a large area detector. The conical slit itself is mounted on a mechanically stable high-precision positioning unit to allow the alignment within 1 μm.

The incoming beam is focused below a spot size of 10 μm × 10 μm. The available detector for high-energy X-rays allowed a maximum sample-to-detector distance of about 600 mm for the monitoring of all available complete Debye–Scherrer rings. In the present example, torsion samples were investigated in an *in situ* sample deformation device. The samples were made from aluminum alloy AlMg3. Aspects such as radial-dependent texture, macrostrain, and microstrain evolution were investigated [8]. The setup also allowed investigations of the dynamic texture and strain tensor evolution during deformation [9]. Figure 8.3 shows the radial dependence of the six macrostrain tensor components as a function of the radius. Apart from the expected maximum compressive shear strain close to the sample center a tensile axial strain component close to the center could be observed. The axial strains can be explained by the texture-related, radially dependent development of axial forces.

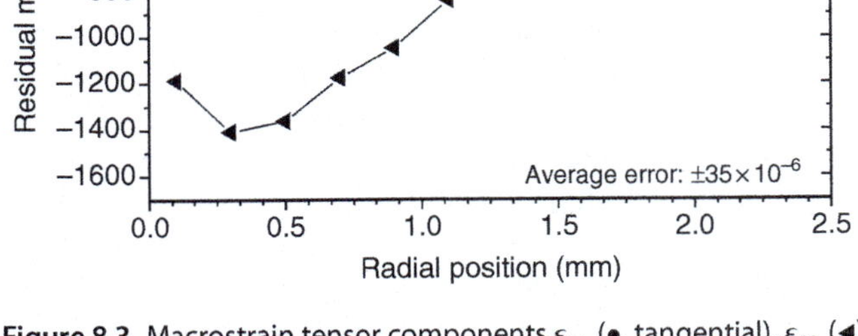

Figure 8.3 Macrostrain tensor components ε_{11} (•, tangential), ε_{12} (◄), ε_{13} (▼), ε_{22} (■, axial), ε_{23} (▲), and ε_{33} (♦, radial) as a function of the radius within a solid torsion sample made from Al alloy AlMg3 with a diameter of 5 mm and deformed to $\gamma = 1.5$.

8.4 The Spiral Slit

The spiral slit [10] was developed at the beamline ID15 at the ESRF and is available there for user experiments. As the conical slit, the spiral slit is a special case of the slit imaging technique as well (cf. Sections 8.1 and 8.2).

8.4.1 Functional Principle

Each of the presently available spiral slits consist of 12 equidistant concentric spiral apertures cut into a Tungsten plate. All paths through the 2-mm-thick tungsten plate lie on cones, which have one common apex (i.e., the focal point of the slit system; slits for distances of 50 and 100 mm are available). Depending on the slit model, the gap is 50 μm or smaller.

Instead of having just one slit selecting a gauge volume within the sample for a given (small) azimuthal range and one 2θ angle, rays are accepted between $2\theta = 1.5°$ and $35°$, depending on the slit model. However, in contrast to the conical slit system discussed in Section 8.3, the spiral slit does not allow the observation of complete rings. Each spiral cuts out a small fraction of the diffraction cone. The great advantage of the spiral slit is the fact that one single image of the diffraction pattern behind the slit contains all the depth- and phase-resolved information for the sample volume illuminated by the X-ray beam. This can be explained in the following way: Considering Figure 8.1a, no slit at all would give rise to information integrated over the sample thickness. By inserting a single slit in the beam path behind the sample as shown in Figure 8.1c, depth-resolved information can be obtained. But, to collect the complete depth-resolved information either the sample needs to be translated along the incoming beam or the slit has to be moved to different positions and the diffraction data has to be recorded at each position of either the sample or the slit. Now, the spiral slit makes this movement unnecessary, because each point on the spiral accepts a different scattering angle and azimuth (angle around the incoming beam, cf. Figure 8.2). In a manner of speaking, the spiral form of the aperture itself leads to a movement. In addition, the depth acceptance of several millimeters to centimeters of the spiral slit (grossly comparable to the depth of focus of an optical lens system) allows the rays from different depths along the beam path in the sample to pass the slit (see inset in Figure 8.4). As a result, the depth-resolved signal is azimuthally spread over the image.

8.4.2 Capabilities

The depth- and phase-resolved diffraction information contained in one single image makes the spiral slit very interesting for the fast depth-resolved investigation of phase transformations [11] and residual strain. The time resolution for the observation of dynamic processes can be in the range of a second, depending on the detector used.

An inherent drawback of the spiral slit technique is that the diffraction rings recorded are not complete. Therefore, in many cases, depth-resolved texture investigations and strain tensor measurements for individual grains (cf. Section 8.6) should preferably be made by other means, for example, with a conical slit. However, if time is not an issue a

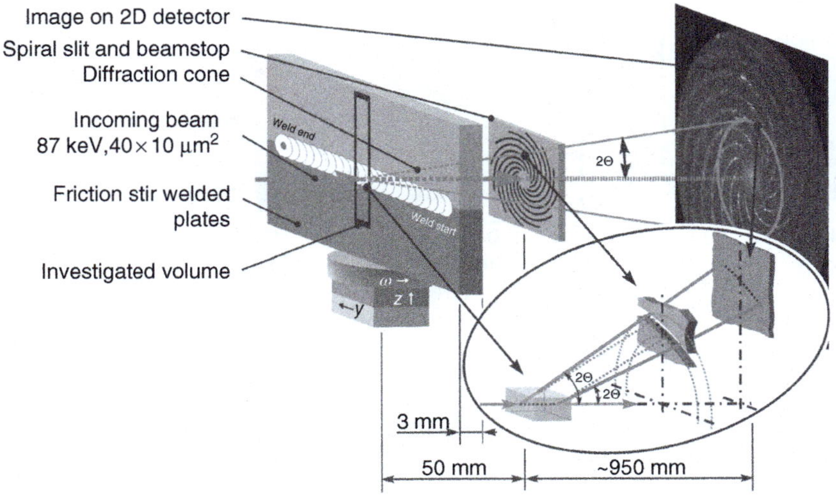

Figure 8.4 Schematic of the experimental set-up with spiral slit and area detector for depth-resolved investigations. The inset illustrates the functional principle of the spiral slit technique. (Martins 2003 [10]. Reproduced with permission of Taylor and Francis.)

solution to obtain quasi-continuous rings is either by the rotation of the sample around the beam in several steps or by a scanning of the sample within the depth acceptance of the slit. Then, for each sample position an image has to be recorded.

8.4.3 Example

Figure 8.5 shows experimental results from the investigation of a dissimilar friction stir welded sample [12] made from AA2024 in T3 condition and AA6082 in T6 condition (single-phase alloys) [10]. A sketch of the experimental setup is shown in Figure 8.4 (cf. Section 8.4.1).

Due to the depth acceptance parallel to the X-ray beam, a spinning capillary (not shown here) containing standard reference Si powder and Al powder could be placed before the sample to provide the reference information for each recorded diffraction pattern. The patterns were monitored with a large area detector system. Each recorded image contains the depth-resolved information for a 2D strain distribution and the material composition. The strain distribution was measured along a line crossing the weld at 80 mm from the weld start. In order to improve the grain averaging, the sample was oscillated during the data collection over a range of 10 mm parallel to the weld.

Residual macrostrain tensor components were determined by fitting the stress-induced distortion of the recorded fractioned Debye–Scherrer rings. The depth-resolved phase distribution was derived from the differences in the recorded diffraction patterns of the two materials. A part of the results is shown in Figure 8.5. For better comprehension, the tool is schematically drawn in Figure 8.5a. The depth-resolved strain map for the direction parallel to the weld is presented in Figure 8.5a. It is particularly inhomogeneous on the retreating side with high tensile strains up to $+2000 \times 10^{-6}$

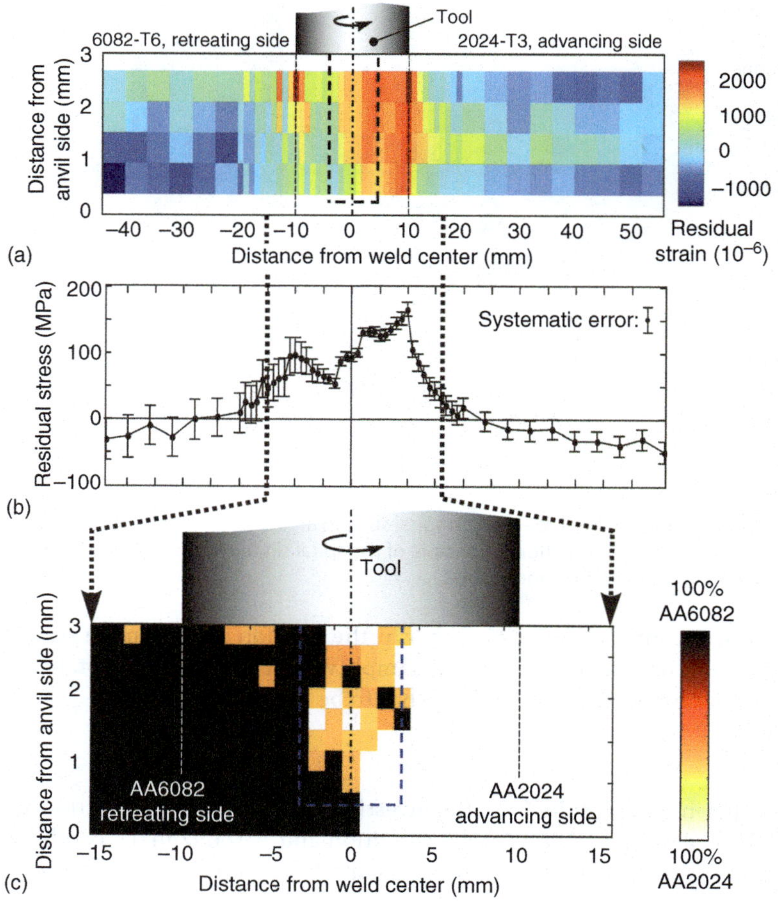

Figure 8.5 (a) Depth-resolved residual macrostrains in the longitudinal direction (parallel to weld); (b) through thickness longitudinal stress, calculated from strain measurements in three dimensions; and (c) enlarged view of the plate material distribution (Al alloys: AA6082 and AA2024) in the stirring zone. (Martins 2003 [10]. Reproduced with permission of Taylor and Francis.)

under the tool shoulder (about −10 mm from the center) and values around zero close to the anvil side. The distribution of the longitudinal residual stress across the weld and integrated over the sample thickness can be seen in Figure 8.5b. The stress rises up to +160 MPa under the tool shoulder at +10 mm. The strong asymmetries of the strain and stress distributions arise from the mixing of the two materials and the different response of the material to the tool force. The material distribution is shown in Figure 8.5c and a close correlation with the depth-resolved strain distribution can be observed.

A more recent experiment demonstrated that also coarse grained specimens can be investigated (cf. Section 8.7). The residual strain was mapped with 3D spatial resolution in an entire 316L austenitic stainless steel weld specimen [13]. The full data set, comprising over 2200 images, with 4 s exposure time each, was collected in a total time of 30 h.

8.5 Simultaneous Strain Measurements in Individual Bulk Grains

An introduction to the principles of 3DXRD applied here as well as its capabilities can be found in Chapter 17. Here, the first demonstration is presented for the simultaneous strain measurement in individual bulk grains (i.e., strain of kind II) of a tensile test sample [14]. The setup resembles the one in Figure 8.1a. It has much higher requirements concerning the accuracy of sample positioning than the other setups featured in Figure 8.1. Furthermore, the difference to the experiments where the information is integrated over the thickness is that the number of illuminated grains is sufficiently small so that distinct diffraction spots arise rather than continuous diffraction rings. In fact, the technique only works well, as long as the spots can be separated from each other, and, as a consequence, can be analyzed individually. To reduce the number of grains observed, one can also use a conical slit described in Section 8.3.

The experiment was carried out using the 3DXRD microscope [6] at beamline ID11 at the ESRF. The synchrotron X-ray beam with a photon energy of 54.9 keV was focused vertically to 100 μm with a bent Laue monochromator and confined by slits to 150 μm in horizontal direction.

The sample was mounted in an *in situ* stress rig, which in turn was placed on lateral and angular alignment units. The tensile axis of the sample was aligned perpendicularly to the X-ray beam at an angular position of $\omega = 0°$ (cf. Figure 8.1d). The volume illuminated by the X-ray beam is given by the horizontal and vertical dimensions of the beam and by the path length of the beam through the sample. The diffraction spots originating from grains fulfilling the diffraction condition are recorded with a detector system, 180 mm in diameter, with an effective pixel size of 87.5 μm and placed 442 mm from the sample center. Space constraints limited the rotation range to $-44° \leq \omega \leq 46°$.

The data presented in Figure 8.6 is a subset of the complete set collected at five lateral sample positions for improved grain statistics and reduced spot overlap. It shows the strain evolution in a deeply embedded grain as a function of load. The total plastic strain was 2.12%.

Upon unloading, the measured strain components are close to zero, indicating no or very little plastic deformation for this grain. In total, the data for 10 individual grains

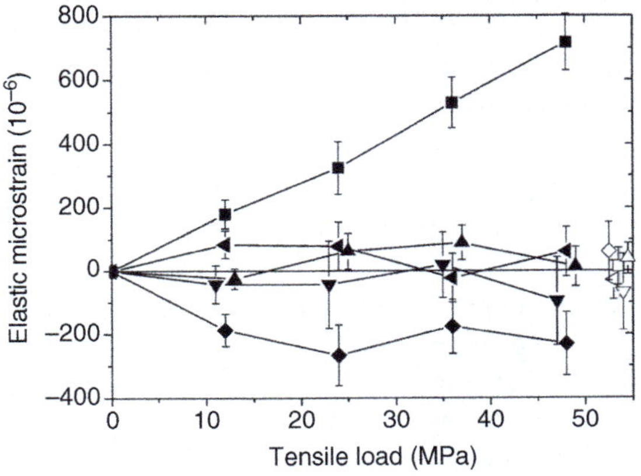

Figure 8.6 Strain evolution measured for one of the investigated grains as a function of load for the components ε_{12} (◀), ε_{22} (■, axial), ε_{13} (▼), ε_{23} (▲), and ε_{33} (♦, transverse). Unfilled symbols relate to the strain state after the final unloading. A slight scatter along the abscissa is imposed in the plot to enhance the visibility of the error bars.

were evaluated. The results were grouped according to slip patterns as expected for the single crystals after the Taylor model. The different groups exhibited significant differences in the strain evolution (data not shown here).

The more recent development of new algorithms, bundled in the FitAllB suite for operation in the FABLE software package [15], has made the robust measurement and analysis of strain and orientation data of ensembles of ~2000 grains possible [16].

An example for an investigation of intragranular strains, that is, strain of kind III, can be found in [17].

8.6 Coarse Grain Effects

One of the general problems when performing strain or texture measurements with X-rays using a small gauge volume is the average grain size of the material investigated. Conclusions on the macroscopic level can be erroneous due to the fact that too few grains have contributed to the measured interference line. Estimations of the influence of the grain size and the gauge volume have been proposed [18, 19]. Here, an approach is presented which is optimized for small gauge volumes. To diminish the influence of the coarse grain effects, the sample can be oscillated, either in rotation or in translation but it has to be noted that a combination of several oscillations requires an increased speed for each additional movement in order to irradiate the resulting gauge volume homogeneously. More than one oscillation direction becomes impractical when detector exposure times are of the order of a few seconds. Furthermore, the oscillation range should not significantly deteriorate the spatial resolution. In the experiments discussed here, tests have shown that the linear oscillation was by far the most effective as compared to angular oscillation.

To assess how many grains within the irradiated effective volume are contributing to the signal detected at a specific position, first the number of illuminated grains has to be calculated. In fact, it should be considered how many grains are intersected by the effective gauge volume. In the experiments discussed here, the horizontal size of the beam is much smaller than the length of the illuminated effective volume and its height, given by the linear oscillation range. The division of the gauge volume by the average grain volume would give misleading results. This is easily understandable when considering a gauge volume of say 10 μm × 10 μm × 500 μm and a grain size of 60 μm. The result would be that less than one grain contributes to the diffraction pattern. Instead, based on the intersections, the number of (partially) irradiated grains can be estimated according to:

$$N_g = \text{ceil}\left(\frac{l}{d_{gl}}\right) \cdot \text{ceil}\left(\frac{v}{d_{gv}}\right) \cdot \text{ceil}\left(\frac{h}{d_{gh}}\right) \tag{8.3}$$

N_g : number of (partially) illuminated grains
l : gauge volume length
v : effective vertical gauge volume size, for example, the linear oscillation range
h : effective horizontal gauge volume size
d_{gx} : grain size in longitudinal, vertical, and horizontal direction

The expression *ceil* means that the result of the division is rounded toward the greater integer; therefore, it will be always at least 1.

In the next step, it has to be estimated how many of these grains are actually diffracting, when considering only one reflection. This depends on the mosaic spread of the material, the multiplicity of the reflection of interest, and, if applied, the range of the angular oscillation. The fraction of diffracting grains can then be estimated based on Eq. (8.4) [18], assuming a uniform texture.

$$N_{dg} = N_g \cdot \frac{J(\Delta\omega_g + \Delta\omega_o)}{2} \tag{8.4}$$

N_{dg} : number of diffracting grains, assuming uniform texture
N_g : number of (partially) illuminated grains
J : multiplicity of the reflection
$\Delta\omega_g$: mosaic spread of the grains in radians
$\Delta\omega_o$: angular oscillation range in radians

This number relates to the complete diffraction ring. Therefore, in the last step it has to be estimated how many grains are contributing to the observed azimuthal section of the diffraction ring. This can be, for example, the section of the diffraction ring accepted by a point detector or in the case of an area detector the azimuthal range over which the diffraction rings are averaged for the later evaluation [19].

$$N_{odg} = N_{dg} \cdot \frac{\chi_a}{2\pi} \tag{8.5}$$

N_{odg} : number of observed diffracting grains
N_{dg} : number of diffracting grains
χ_a : azimuthal averaging range in radians

Values for the mosaic spread of Al grains of up to 0.12 rad were reported by Margulies *et al.* [20, 21].

According to a common rule of thumb, 1000 illuminated grains are considered to be sufficient for macroscopic investigations when using a reflection with high index and low symmetry (e.g., the Al {3 1 1} reflection). However, modeling efforts by Clausen *et al.* [22] seem to indicate that the number should be considerably higher and that in some cases a low-index and high-symmetry reflection might yield better macroscopic results. Their investigations related to single-phase materials with no initial texture. For that case, 10 000 sampled grains started to give reliable results.

8.7 Analysis of Diffraction Data from Area Detectors

Each image contains usually several complete Debye–Scherrer rings. After background subtraction and determination of the beam center, that is, the common center of all the rings (note that this must be done only once for one data set), the images are transformed to polar coordinates using, for example, the software Fit2D [23]. A fine azimuthal slicing (in the order of 1000 slices), as proposed by Wanner and Dunand [24], allows the analysis of diffraction rings that have no smooth azimuthal intensity distribution. Before the least square fitting of a linear background and Pseudo Voigt profiles to the peak pattern of single pixel rows, a reconstruction procedure might need to be applied to the data [4] (cf. Section 8.2). Following this, the results of the fit are rebinned to azimuthal sections of 5°.

Applying the appropriate coordinate transformations and absorption corrections to the integrated intensities, obtained from the least square fitting, the data can be used to calculate the orientation distribution function (ODF) (cf. Chapter 3).

The strain tensor can be calculated for each measured reflection from the distortion of the diffraction rings. This requires the knowledge of the unstrained lattice parameter d_0 (cf. Chapter 9), and the correction of the data for systematic distortions, caused by, for example, the detector system itself and the geometry of the experimental setup. The distortions are corrected by calculating the strains relative to the supposedly strain-free diffraction rings of a powder reference. The strain tensor components are then determined, for example, by fitting the parameters of the general Eq. (8.6) [25] to the corrected peak positions.

$$f_{11}\varepsilon_{11} + f_{12}\varepsilon_{12} + f_{22}\varepsilon_{22} + f_{13}\varepsilon_{13} + f_{23}\varepsilon_{23} + f_{33}\varepsilon_{33} = -{}^{\Delta\theta}/_{\tan\theta} \tag{8.6}$$

ε_{ij} are the components of the strain tensor, f_{ij} are scattering vector and sample orientation dependent factors, and θ is the Bragg angle of the observed $h\ k\ l$.

When, for example, with a spiral slit setup, no powder-like diffraction rings but only distinct spots from individual grains are detected, another approach has proven to yield acceptable results for the measurement of strains of kind I [13]: Based on the geometry of the setup and photon energy, the lattice spacing of individual grains is derived from the center-of-mass positions of distinct diffraction spots and their origin within the sample. All data points obtained for one desired volume are then used to determine the strain tensor components by fitting the parameters of Eq. (8.6).

8.8 Matrix for Comparison and Decision Taking Which Technique to Use for a Specific Problem

When planning an experiment one has to decide which length scale the experiment should aim at and on which time scale. Macroscopic investigations with no time resolution will require a different setup than single grain investigations with a time resolution of minutes or seconds. Furthermore, the application of some techniques described before also depends on the light source. Focusing of high-energy X-rays, for example, is usually not available at second-generation synchrotron sources, though exceptions exist [26]. Special techniques involving, for example, a spiral or conical slit are still only available at specific beamlines. Table 8.2 is meant to provide an informative basis concerning the capabilities of the techniques and their performance in terms of time and spatial resolution. The numbers given in the table can vary considerably, for example, because of the material investigated or the detector system used. In general, one can say that the numbers change for the better because of continuous improvements in setup and detector systems.

Table 8.2 Overview of different experimental techniques and their capabilities.

Technique	Application	Spatial resolution	Acquisition time for one data set	Remarks
Transmission with integrated signal	Strain, texture, and phase mapping	Lateral: beam size	Seconds to min/image/sample position	2D information, no depth resolution
Crossed beam with collimator	Strain scanning	Longitudinal: ≥50 μm	≥60 s/peak/depth position/sample orientation	Focused beam, 1D-strain info., scanning of diffraction peak
Slit imaging	Strain scanning	Longitudinal: ≥50 μm	≥1 s/image/peak/depth position/sample orientation	Focused beam, 1D-strain info., imaging of peak
Conical slit system	Texture and strain mapping, single grain studies	Longitudinal: ≥260 μm	≥1 s/image/depth position/sample orientation	Focused beam, 2D-strain info./image, recording of complete rings
Spiral slit system	Strain and phase mapping	Longitudinal: ≥80 μm	≥1 s/sample orientation	Focused beam, 2D-strain info. with complete depth info. in one image
3DXRD	Simultaneous strain measurements in individual grains	Grain size	≤2.5 s/image/ω-pos. ≤600 s for data set with 240 ω positions and a $\Delta\omega$ of 0.5°	Focused beam, grain indexing provides orientation of each grain, grain position can be calculated

Note that exceptions are always possible and that the table is just meant as a rough guideline.

References

1 Wroblewski, T. and Bjeomikhov, A. (2005) *Nucl. Instrum. Methods Phys. Res., Sect. A*, **538**, 771–777.

2 Suzuki, K., Shobu, T., Shiro, A., and Toyokawa, H. (2014) *Mater. Sci. Forum*, **772**, 15–19.

3 Webster, P.J., Djapic Osterkamp, L., Browne, P.A., Hughes, D.J., Kang, W.P., Withers, P.J., and Vaughan, G.B.M. (2001) *J. Strain Anal. Eng. Des.*, **36** (**1**), 61–70.

4 Lienert, U., Martins, R., Grigull, S., Pinkerton, M., Poulsen, H.F., and Kvick, Å. (2000) *Mater. Res. Soc. Symp. Proc.*, **590**, 241–246.

5 Nielsen, S.F., Wolf, A., Poulsen, H.F., Ohler, M., Lienert, U., and Owen, R.A. (2000) *J. Synchrotron Radiat.*, **7**, 103–109.

6 Lienert, U., Poulsen, H.F., and Kvick, Å. (1999) Proceedings of the 40th Conference of the American Institute of Aeronautics and Astronautics (AIAA) on Structures, Structural Dynamics and Materials, St. Louis, 1999, pp. 2067–2075.

7 Staron, P., Fischer, T., Eims, E.-H., Frömbgen, S., Schell, N., Daneshpour, S., Martins, R.V., Müller, M., and Schreyer, A. (2014) *Mater. Sci. Forum*, **772**, 3–7.
8 Martins, R.V., Lienert, U., Margulies, L., and Pyzalla, A. (2005) *Mater. Sci. Eng., A*, **402**, 278–287.
9 Martins, R.V., Lienert, U., Margulies, L., and Pyzalla, A. (2002) *Mater. Sci. Forum*, **404–407**, 115–120.
10 Martins, R.V. and Honkimäki, V. (2003) *Texture Microstruct.*, **35** (**3/4**), 145–152.
11 Hagen, A., Poulsen, H.F., Klemmsø, T., Martins, R.V., Honkimäki, V., and Buslaps, T. (2006) *Fuel Cells*, **5**, 361–366.
12 Thomas, W.M., Nicholas, E.D., Needham, J.C., Murch, M.G., Temple-Smith, P., and Dawes, C.J., *Friction-stir butt welding* (1991) International Patent PCT/GB92/02203, GB Patent 9125978.8 (1991), US Patent 5,460,317 (1995).
13 Martins, R.V., Ohms, C., and Decroos, K. (2010) *Mater. Sci. Eng., A*, **527**, 4779–4787.
14 Martins, R.V., Margulies, L., Schmidt, S., Poulsen, H.F., and Leffers, T. (2004) *Mater. Sci. Eng., A*, **387–389**, 84–88.
15 Sourceforge http://sourceforge.net/projects/fable/ (accessed 03 Oct. 2016).
16 Oddershede, J., Schmidt, S., Poulsen, H.F., Sørensen, H.O., Wright, J., and Reimers, W. (2010) *J. Appl. Crystallogr.*, **43**, 539–549.
17 Pantleon, W., Poulsen, H.F., Almer, J., and Lienert, U. (2004) *Mater. Sci. Eng., A*, **387–389**, 339–342.
18 Lienert, U., Grigull, S., Kvick, Å., Martins, R.V., and Poulsen, H.F. (2000) Proceedings of the International Conference on Residual Stress (ICRS)-6, Oxford, United Kingdom, vol. **2**, pp. 1050–1057.
19 Pyzalla, A. (2000) *J. Nondestr. Eval.*, **19** (**1**), 21–31.
20 Margulies, L., Winther, G., and Poulsen, H.F. (2001) *Science*, **291**, 2392–2394.
21 Poulsen, H.F., Juul Jensen, D., Tschentscher, T., and Wcislak, E.M. (2001) *Texture Microstruct.*, **35** (**1**), 39–54.
22 Clausen, B., Leffers, T., and Lorentzen, T. (2003) *Acta Mater.*, **51**, 6181–6188.
23 Hammersley, A.P., Svensson, S.O., Hanfland, M., Fitch, A.N., and Häusermann, D. (1996) *High Pressure Res.*, **14**, 235–248.
24 Wanner, A. and Dunand, D.C. (2000) *Mater. Res. Soc. Proc.*, **590**, 157–162.
25 He, B.B. and Smith, K.L. (1998) Proceedings of the Society for Experimental Mechanics (SEM) Spring Conference on Experimental and Applied Mechanics, Houston, TX, June 1–3, 1998, pp. 217–220.
26 Zhong, Z., Kao, C.C., Siddons, D.P., and Hastings, J.B. (2001) *J. Appl. Crystallogr.*, **34**, 646–653.

9

Residual Stress Analysis by Energy-Dispersive Synchrotron X-ray Diffraction

Christoph Genzel and Manuela Klaus

9.1 Introduction

Perhaps the most striking advantage of white beam, that is, energy-dispersive (ED) X-ray diffraction on polycrystalline materials, which was introduced independently in 1968 by Giessen and Gordon [1] and Buras [2], arises from the fact that it provides complete diffraction patterns (including the fluorescence lines of the contributing elements) under fixed but arbitrary scattering directions, 2θ. Thus, it is not surprising that the early ED diffraction experiments were focused on crystal structure analysis, which is based on the evaluation of a large number of diffraction lines. While most of the progress made in the first decade after the introduction of the ED method was achieved by employing conventional laboratory X-ray sources [3], the modern third-generation synchrotron radiation facilities available today provide much better possibilities for advanced ED X-ray diffraction studies.

The first ED X-ray stress analysis (XSA) experiments were reported in the early 1980s [4, 5]. Since that time various methodological developments have been made, which make use of the special features of the ED diffraction technique. The vast majority of these investigations were based on the application of white high-energy synchrotron radiation in transmission geometry and were aimed at analyzing the residual stress state in the bulk of the material, for example, see [6, 7]. The respective measuring and data evaluation techniques were borrowed from neutron diffraction. A combination of white beam diffraction and tomography for the simultaneous *in situ* characterization of creep-damaged materials was reported in [8]. Pioneering work concerning the use of the white beam technique for residual stress analysis in reflection geometry was carried out by Ruppersberg, who applied his Universal plot method [9] in the early 1990s to the ED case of diffraction [10, 11].

Since the publication of the first edition of this book in 2008, significant progress has been made in the field of ED-XSA (energy-dispersive-X-ray stress analysis), which applies equally to the measurement techniques and the data evaluation methods. Starting with some fundamental relations of ED X-ray (residual) stress analysis (which of course have not changed since 2008), this chapter aims at updating the examples that were given in the first edition. The corresponding experiments were performed at the materials science beamline EDDI@BESSY II [12]. With the selection of examples presented in the following sections we try to give an overview of the most important ED-XSA methods that have been developed in the recent past. It will be shown in

Neutrons and Synchrotron Radiation in Engineering Materials Science: From Fundamentals to Applications, Second Edition. Edited by Peter Staron, Andreas Schreyer, Helmut Clemens, and Svea Mayer.

particular that – depending on the quality of the investigated material and the data collected during an XSA measurement – the ED diffraction method allows data evaluation and residual stress analysis at different levels of approximation. Therefore, it is hoped that this chapter helps to convey an idea about the powerful features of the ED diffraction method to the interested reader and/or potential user to solve his/her scientific questions concerning residual stress analysis.

9.2 Fundamentals of Energy-Dispersive X-ray Diffraction Stress Analysis

9.2.1 The Basic Equation of Energy-Dispersive X-ray Diffraction

Bragg's equation which reads in its angle-dispersive (AD) form $n\lambda = 2d^{hkl} \sin \theta^{hkl}$, relates for a fixed wavelength λ a set of lattice planes hkl with spacing d^{hkl} to a certain angle θ^{hkl} (the Bragg angle). Recording diffraction patterns in the $\theta-\theta$- or $\omega-2\theta$-mode, interference lines are observed for different scattering angles $2\theta^{hkl}$. On the other hand, while using a white beam with a continuous photon energy spectrum in ED diffraction, the diffraction angle θ and the scattering angle 2θ under which the diffracted energy spectrum is observed by means of an energy-dispersive solid state germanium or silicon detector, can be chosen freely and remain fixed during the measurement. The correlation between the lattice spacing d^{hkl} and the corresponding diffraction line E^{hkl} on the energy scale follows immediately by inserting the Planck equation $E = h\nu = hc/\lambda$ into the Bragg equation

$$d^{hkl} = hc/(2 \sin \theta E^{hkl}) = \text{const.}/E^{hkl} \tag{9.1}$$

(h – Planck's constant, c – velocity of light). An important conclusion to be drawn from Eq. (9.1) is that an increase of θ leads to a "compression" of the diffracted energy spectrum, that is, a shift of the diffraction lines E^{hkl} toward smaller energies and vice versa (Figure 9.1).

9.2.2 Near-Surface Depth Profiling in the Energy-Dispersive Diffraction Mode

For the depth-resolved analysis of any structural property such as residual strain and stress, chemical composition or texture, ED diffraction offers different opportunities.

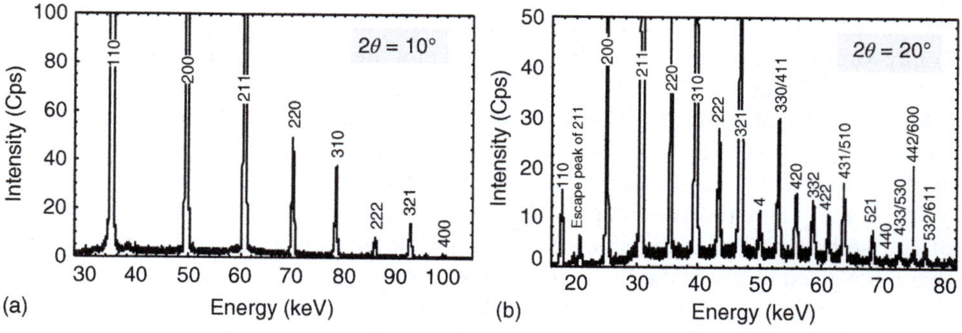

Figure 9.1 ED diffraction patterns of ferritic steel (bcc crystal structure) recorded for scattering angles 2θ of 10° (a) and 20° (b), respectively. The total counting time for each ED spectrum was 60 s.

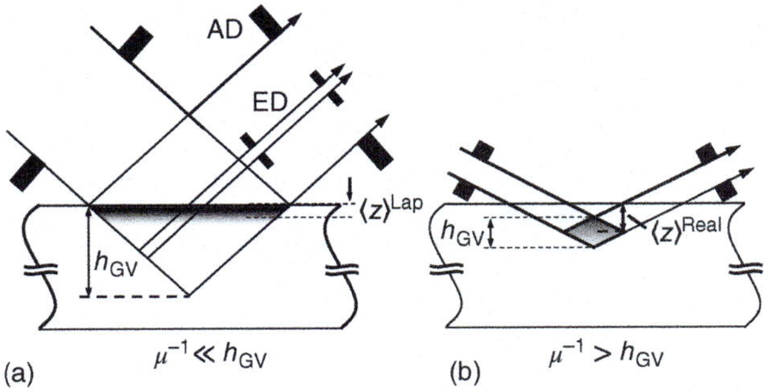

Figure 9.2 Illustration of the definition of the information depth. (a) Laplace space depth z^{Lap} and (b) real space depth z^{Real}. h_{GV} is the height of the diffracting part of the gauge volume. AD and ED should indicate the different slit apertures which are usually used in angle and energy dispersive diffraction, respectively, to confine the equatorial divergence of the diffracted beam.

The specific method that can be applied for X-ray depth profiling is defined by the aperture of the slits in the primary and diffracted beam, which form the gauge volume (GV) that is used to probe the sample. In general, the GV has a stretched rhombic shape with an aspect ratio that depends on the Bragg (diffraction) angle θ. Only that part of this volume, which immerses in the sample, contributes with its height h_{GV} to the diffraction signal. Figure 9.2 shows the two main scenarios that may occur depending on the ratio of the inverse of the linear absorption coefficient, μ^{-1}, and the height h_{GV}.

If at least one of the slit systems that is used to confine the beam cross-section in the primary and/or the diffracted beam, respectively, has an aperture that is much larger than μ^{-1}, the condition $\mu^{-1} \ll h_{\mathrm{GV}}$ is fulfilled and the information contained in the individual diffraction line profiles E^{hkl} predominantly originates from a surface layer with the depth

$$\langle z \rangle^{\mathrm{Lap}} = \tau = \int_0^{\infty} z\,\mathrm{e}^{-z/\tau}\,\mathrm{d}z \Big/ \int_0^{\infty} \mathrm{e}^{-z/\tau}\,\mathrm{d}z \tag{9.2}$$

The superscript "Lap" should indicate that the depth z^{Lap} to which the diffraction signal in the configuration shown in Figure 9.2a has to be assigned is the (normalized) Laplace transform of the actual depth z below the surface. This depth which is also called $1/e$ information depth τ is defined by the condition, that the intensity of the X-rays passing through the material is $1/e$ of the primary intensity I_0. With respect to the ED diffraction mode it is important to note that each reflection E^{hkl} in ED diffraction patterns originates from another information depth τ because the absorption coefficient μ depends on the photon energy E. A general formulation for τ which is valid for any diffraction geometry is given by

$$\tau = (\sin^2\theta - \sin^2\psi + \cos^2\theta \sin^2\psi \sin^2\eta) / [2\mu(E) \sin\theta \cos\psi] \tag{9.3}$$

Genzel [13]. In this equation, ψ is the inclination between the diffraction vector $\mathbf{g}^{hkl}$ (i.e., the measuring direction) and the sample surface normal $\mathbf{S}_3$, whereas η describes the rotation of the sample around $\mathbf{g}^{hkl}$, see Figure 9.3.

Figure 9.4 demonstrates that the ED diffraction mode allows a very flexible variation of the information depth τ in two respects: (i) According to Eq. (9.1), it is possible to adapt the information depth to some region of interest below the sample surface by

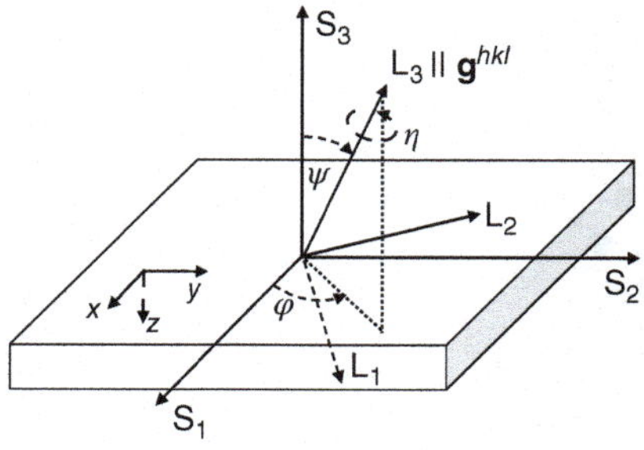

Figure 9.3 Diffraction geometry in X-ray stress analysis. S and L denote the sample reference and the laboratory system, respectively. Note, that the positive z-axis of the sample system is directed toward the interior of the material. The angle set (φ, ψ) defines the orientation of the diffraction vector $\mathbf{g}^{hkl}$ with respect to S, whereas η describes the rotation of the sample around $\mathbf{g}^{hkl}$. (Genzel 2011 [14]. Reproduced with permission of Sage Publications.)

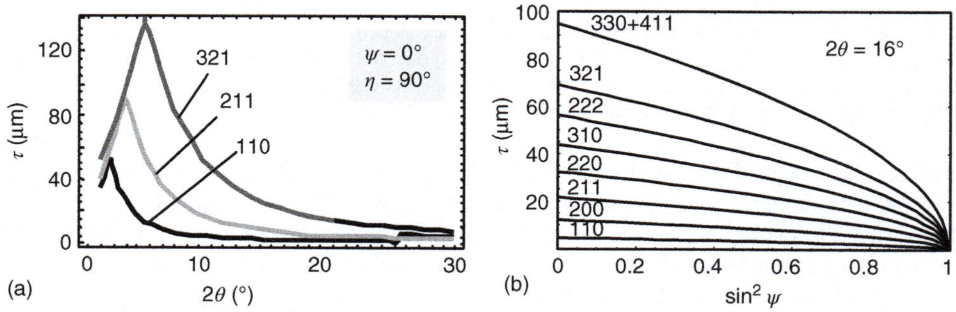

Figure 9.4 The dependence of the information depth τ on (a) the scattering angle 2θ for $\psi = 0°$ and (b) the inclination angle ψ, respectively. The calculations were done for selected reflections hkl of a ferritic steel according to Eq. (9.2) for $\eta = 90°$, which corresponds to the symmetrical Ψ-mode of XSA. (Genzel 2004 [15]. Reproduced with permission of Elsevier.)

choosing an appropriate diffraction angle θ, see Figure 9.4a. (ii) For a fixed diffraction angle θ, (quasi) continuous depth profiling according to Eq. (9.3) can be performed by means of the angles ψ (Ψ- or Ω-mode of XSA [16]) and η (scattering vector mode [13]), respectively, see Figure 9.4b.

On the other hand, if the GV is confined by narrow slits to very small values with a height h_{GV} of only a few microns, see Figure 9.2b, absorption within the GV becomes less significant and the recorded diffraction signal can be assigned to the distance between the surface and the geometrical center of gravity of the GV:

$$\langle z\rangle^{\mathrm{Real}} = \iiint_{V_{\mathrm{GV}}} z g(x,y,z)\, \mathrm{d}x\, \mathrm{d}y\, \mathrm{d}z \Big/ \iiint_{V_{\mathrm{GV}}} g(x,y,z)\, \mathrm{d}x\, \mathrm{d}y\, \mathrm{d}z \tag{9.4}$$

In the above equation, the superscript "Real" means that depth profiling with such a small GV is performed in the real- or z-space.

9.2.3 Principles of Depth-Resolved X-ray Stress Analysis and Application to the Energy-Dispersive Case of Diffraction

XSA in quasi-isotropic polycrystalline materials is based on the measurement of diffraction line profiles and the evaluation of strains

$$\varepsilon_{\varphi\psi}^{hkl} = (d_{\varphi\psi}^{hkl} - d_0^{hkl})/d_0^{hkl} = (E_0^{hkl} - E_{\varphi\psi}^{hkl})/E_{\varphi\psi}^{hkl} = \Delta E_{\varphi\psi}^{hkl}/E_{\varphi\psi}^{hkl} \tag{9.5}$$

(d_0^{hkl}, E_0^{hkl} strain-free lattice spacing and corresponding diffraction line position on the energy scale according to Eq. (9.1), respectively) for different azimuth angle sets (φ, ψ)

of the diffraction vector $\mathbf{g}^{hkl}$ with respect to the sample reference system **S** (Figure 9.3). The relation between the elastic lattice strain $\varepsilon_{\varphi\psi}^{hkl} = \{\varepsilon_{33}^{\mathrm{L}}\}_{\varphi\psi}^{hkl}$ (the superscript L indicates tensor components related to the laboratory system L) obtained for the *diffracting* crystallites within the probed sample volume and the average mechanical stress related to the sample reference system S, $\langle\sigma_{ij}^{\mathrm{S}}\rangle$, of *all* crystallites in this volume fraction is given by the fundamental equation of XSA. Considering that both strain and stress can vary more or less significantly with the depth z below the surface, this equation reads

$$\begin{aligned}
\varepsilon_{\varphi\psi}^{hkl}(z) &= \{\varepsilon_{33}^{\mathrm{L}}\}_{\varphi\psi}^{hkl}(z) \\
&= \frac{1}{2}S_2^{hkl}\sin^2\psi\,[\langle\sigma_{11}^{\mathrm{S}}\rangle(z)\cos^2\varphi + \langle\sigma_{22}^{\mathrm{S}}\rangle(z)\sin^2\varphi + \langle\sigma_{12}^{\mathrm{S}}\rangle(z)\sin 2\varphi] \\
&\quad + \frac{1}{2}S_2^{hkl}[\langle\sigma_{13}^{\mathrm{S}}\rangle(z)\cos\varphi + \langle\sigma_{23}^{\mathrm{S}}\rangle(z)\sin\varphi + \langle\sigma_{33}^{\mathrm{S}}\rangle(z)\cos^2\psi] \\
&\quad + S_1^{hkl}[\langle\sigma_{11}^{\mathrm{S}}\rangle(z) + \langle\sigma_{22}^{\mathrm{S}}\rangle(z) + \langle\sigma_{33}^{\mathrm{S}}\rangle(z)] \\
&= F_{ij}^{hkl}(\varphi,\psi)\langle\sigma_{ij}^{\mathrm{S}}\rangle(z)
\end{aligned} \tag{9.6}$$

where S_1^{hkl} and $\frac{1}{2}S_2^{hkl}$ are the diffraction elastic constants (DECs). To facilitate matters the superscript S as well as the different types of brackets $\{\ldots\}$ and $\langle\ldots\rangle$ will be omitted in the following equations. The DEC and the stress factors $F_{ij}^{hkl}(\varphi,\psi)$ used in the concise form in the last line of Eq. (9.6) can be determined either experimentally or by theoretical calculation on the basis of different grain interaction models like Voigt, Reuss, Hill–Neerfeld, Eshelby/Kroener, Vook/Witt, and its inversion as well as – especially in the case of thin films – by appropriate combinations of these models, see, for example, [17].

Taking into account Beer's exponential attenuation law, one has to distinguish between the actual (real space) z-depth profiles on the one hand and the exponentially weighted Laplace space- or τ-depth profiles, which are obtained experimentally, on the other hand. They are correlated with each other by the same relation that exists between real space depth z and the Laplace space depth τ (cf. Eq. (9.2)):

$$\begin{aligned}
\varepsilon_{\varphi\psi}^{hkl}(\tau) &= \int_0^\infty \varepsilon_{\varphi\psi}^{hkl}(z)\,\mathrm{e}^{-z/\tau}\,\mathrm{d}z \Big/ \int_0^\infty \mathrm{e}^{-z/\tau}\,\mathrm{d}z \\
\sigma_{ij}(\tau) &= \int_0^\infty \sigma_{ij}(z)\,\mathrm{e}^{-z/\tau}\,\mathrm{d}z \Big/ \int_0^\infty \mathrm{e}^{-z/\tau}\,\mathrm{d}z
\end{aligned} \tag{9.7}$$

which have the form of a Laplace transform with respect to $1/\tau$ [16]. Within the framework of this chapter, we will focus on the materials near surface zone and assume a biaxial residual stress state with $\sigma_{i3} = 0 (i = 1, 2, 3)$ to be present within the information depth of the synchrotron X-rays. In this case, taking into account Eq. (9.7), the fundamental equation of XSA (Eq. (9.6)) reduces for measurements that are carried out in the azimuths $\varphi = 0°$ and $90°$, respectively, to

$$\begin{aligned}
\varepsilon_{0\psi}^{hkl}(\tau) &= \frac{1}{2}S_2^{hkl}\sigma_{11}(\tau)\sin^2\psi + S_1^{hkl}[\sigma_{11}(\tau) + \sigma_{22}(\tau)] \\
\varepsilon_{90\psi}^{hkl}(\tau) &= \frac{1}{2}S_2^{hkl}\sigma_{22}(\tau)\sin^2\psi + S_1^{hkl}[\sigma_{11}(\tau) + \sigma_{22}(\tau)]
\end{aligned} \tag{9.8}$$

Employing the ED diffraction method in the wide-slit-configuration shown in Figure 9.2a, a determination of the residual stress depth distribution of the in-plane components σ_{11} and σ_{22} in the real (z)- and/or the Laplace (τ)-space is possible on different levels of approximation, if the well-known and widely used $\sin^2\psi$-technique [18] is applied. Thus, recording ED diffraction patterns in the azimuths $\varphi = 0°$ and $90°$ for various tilt angles ψ, $\varepsilon_{\varphi\psi}^{hkl} - \sin^2\psi$-distributions are obtained for each diffraction line E^{hkl} in the patterns. These strain data allow for the evaluation of the residual stress depth profiles in different ways [19]:

1) *Modified multi-wavelength method* [12, 15]: In this first approximation the residual stress values obtained from the slopes of the linear regression lines that were fitted to the individual $\varepsilon_{\varphi\psi}^{hkl}$ versus $\sin^2\psi$-distributions are plotted versus an average information depth $\overline{\tau}^{hkl}$ given by

$$\overline{\tau}^{hkl} = [\tau(\psi_{\min}, E^{hkl}) + \tau(\psi_{\max}, E^{hkl})]/2 \tag{9.9}$$

with $\tau(\psi, E^{hkl})$ from Eq. (9.3).

2) *Universal plot method* [9–12, 15]: Solving the Eq. (9.8) for the unknown residual stress depth profiles yields

$$\sigma_{11}(\tau) = \frac{1}{2}[f^+(\tau) + f^-(\tau)]$$
$$\sigma_{22}(\tau) = \frac{1}{2}[f^+(\tau) - f^-(\tau)] \tag{9.10}$$

with

$$f^+(\tau) = [\varepsilon_{0\psi}^{hkl}(\tau) + \varepsilon_{90\psi}^{hkl}(\tau)] / \left[\frac{1}{2}S_2^{hkl}\sin^2\psi + 2S_1^{hkl}\right]$$
$$f^-(\tau) = [\varepsilon_{0\psi}^{hkl}(\tau) - \varepsilon_{90\psi}^{hkl}(\tau)] / \left[\frac{1}{2}S_2^{hkl}\sin^2\psi\right] \tag{9.11}$$

The right hand side of Eq. (9.10) contains the pure experimental information, whereas the unknown in-plane stresses stand alone on the left hand side. Therefore, these equations are of universal nature, and all experimental data, independently of the radiation and/or reflections E^{hkl} used, can be plotted versus the corresponding information depth τ into one "universal"- or "master"-plot [9]. For a more detailed discussion of the universal-plot method and its application to the ED-case of diffraction, the reader is referred to [15, 19].

3) *Rietveld method*: The principle of this method [20, 21] is to minimize the weighted sum of squared residuals $WSS = \sum_i w_i(y_i^{\text{obs}} - y_i^{\text{calc}})^2$ of the observed intensities y_i^{obs} in a diffraction pattern and the calculated intensities y_i^{calc} of a model that describes the materials crystal structure and its microstructural properties. For ED diffraction patterns the following approach is used [22]:

$$y_i^{\text{calc}} = S\sum_{hkl}\left[W(E^{hkl})A(E^{hkl})P(E^{hkl},\theta)M^{hkl}|F^{hkl}|^2(E^{hkl})^{-3}G(E_i - E^{hkl})\right] + y_i^{bkg} \tag{9.12}$$

(S – scale factor, W, A, P, M – wiggler spectrum, absorption-, polarization-, and multiplicity factor, respectively, F^{hkl} – structure factor, G – line profile function, y_i^{bkg} – background intensity). Taking into account the depth dependency of the strains and stresses, and making use of the Eqs. (9.6) and (9.7), the line profile function G becomes:

$$G(E_i - E^{hkl}_{\varphi\psi}) = G\{E_i - E^{hkl}_0[1 - \varepsilon^{hkl}_{\varphi\psi}(\tau)]\}$$
$$= G\left\{E_i - E^{hkl}_0\left[1 - F^{hkl}_{ij}(\varphi,\psi)\left(\int \sigma_{ij}(z)e^{-z/\tau}\,dz \Big/ \int e^{-z/\tau}\,dz\right)\right]\right\} \tag{9.13}$$

Apel *et al.* [23]. This equation correlates the peak position $E^{hkl}_{\varphi\psi}$ of any diffraction line *hkl* in the ED diffraction pattern obtained for a measuring direction (φ, ψ) directly with the actual depth profiles $\sigma_{ij}(z)$ of the residual stress tensor in the real space. Based on the approach (Eq. (9.13)) *all* diffraction patterns measured in a $\sin^2\psi$-based ED-XSA experiment can be refined *simultaneously* by a nonlinear least squares fit algorithm to obtain the set of parameters used to describe the individual stress depth profiles $\sigma_{ij}(z)$ [23].

In Section 9.4.1 we demonstrate the application of the methods introduced above for near-surface residual stress depth profiling with practical examples.

9.3 Experimental Setup

Figure 9.5 shows the setup of a white beam synchrotron experiment for ED diffraction. The high-energy X-ray beam is provided by a superconducting multipole wiggler, which emits photons in a horizontal fan of some degree but highly collimated in the vertical plane. The emitted energy spectrum mainly depends on the electron storage ring parameters as well as on the magnetic field of the wiggler. The optical elements of the beamline consist of an absorber mask to predefine the primary beam cross-section, a filter system to attenuate low-energy photons giving rise to sample heating due to absorption and a cross-slit system to define the final beam shape. The primary beam meets the sample, which can be mounted on different sample positioning units that enable to realize any diffraction geometry applied in stress and texture analysis.

Because synchrotron radiation is linearly polarized in the horizontal plane (i.e., in the storage ring plane), synchrotron diffraction experiments in most cases are performed in the vertical plane. In the ED mode using high-energy photons rather small scattering angles 2θ between about 4° and 20° are used in most experiments, since higher angles 2θ would shift and compress the diffracted energy spectrum toward smaller energies (see Eq. (9.1) and Figure 9.1). This would lead to a reduction of the information depth τ in case of reflection mode experiments (cf. Figure 9.4) as well as to diffraction line overlapping, which prevents a separate evaluation of individual reflections E^{hkl}.

The experimental resolution in ED diffraction experiments is given by [24]

$$\Delta E/E = \sqrt{\Gamma_0^2 E^{-2} + 5.55F\epsilon E^{-1} + (\cot\theta\Delta\theta)^2} \tag{9.14}$$

(Γ_0 – noise (about 100 eV), F – Fano factor (0.1–0.13), and ϵ – energy for creating an n^-/n^+ pair (3 eV for germanium). For germanium solid state detectors the intrinsic resolution is about 160 eV for $E = 10$ keV and about 420 eV for $E = 100$ keV, respectively. Consequently, the line broadening $\Delta E/E$ resulting from the detector electronics (i.e., the first two terms in Eq. (9.14)) lies between about 5×10^{-3} and 1×10^{-2} which is rather

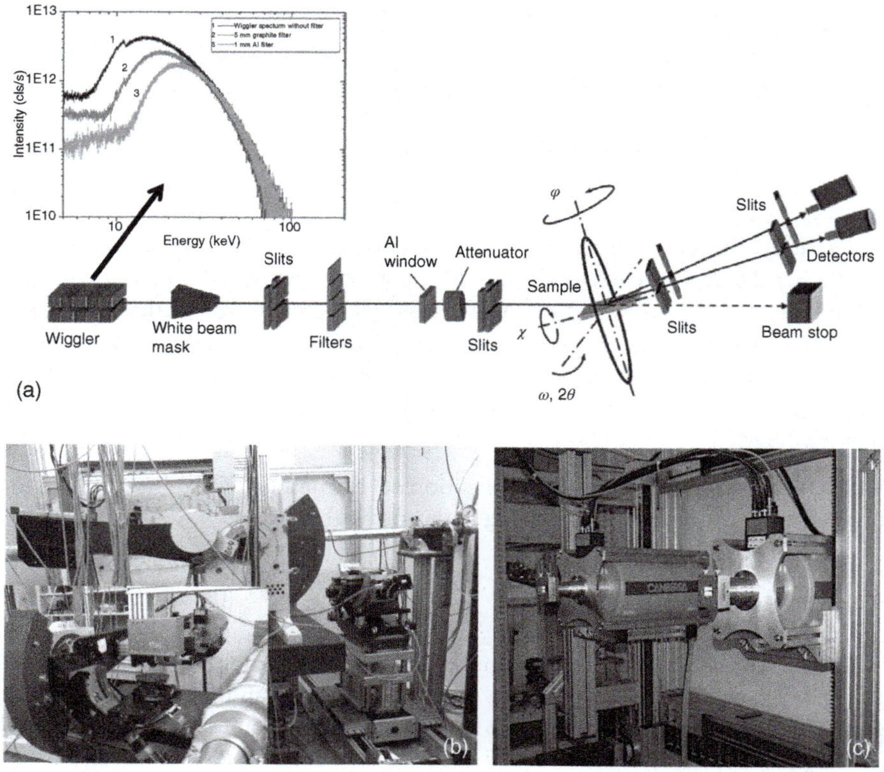

Figure 9.5 Setup of the white beam synchrotron beamline EDDI@BESSYII for ED diffraction. (a) Schematic view of the beamline components and spectrum emitted by a superconducting 7T multipole wiggler for different absorber conditions. (b) Experimental hutch with four- and five-axes sample positioning units. (c) Two-detector setup for simultaneous data acquisition in different measuring directions.

large on the absolute scale. Relative shifts ΔE^{hkl} of the diffraction lines which are the basis of strain analysis (cf. Eq. (9.5)), however, do not depend on this absolute intrinsic energy broadening. They can be measured with high accuracy, if the equatorial divergence $\Delta\theta$ of the diffracted beam(s) is confined by means of double slit systems (cf. Figure 9.5) to very small values of $< 0.005°$. Furthermore, it is important to note that the position of the diffraction lines on the energy scale strongly depends on the detector parameter settings and especially on the dead time (DT) of the detector [25]. In Ref. [23] it was shown that the systematical line shift ΔE toward smaller energies with increasing DT has to be taken into account even in the detector calibration, which allocates the channels of the detector's multichannel analyzer to the energy.

9.4 Examples for Energy-Dispersive Stress Analysis

9.4.1 Near Surface Residual Stress Depth Profiling

With the example of a shot-peened roller bearing steel 100Cr6 (for details of the sample treatment see [26]) we will demonstrate the suitability of the $\sin^2\psi$-based ED methods

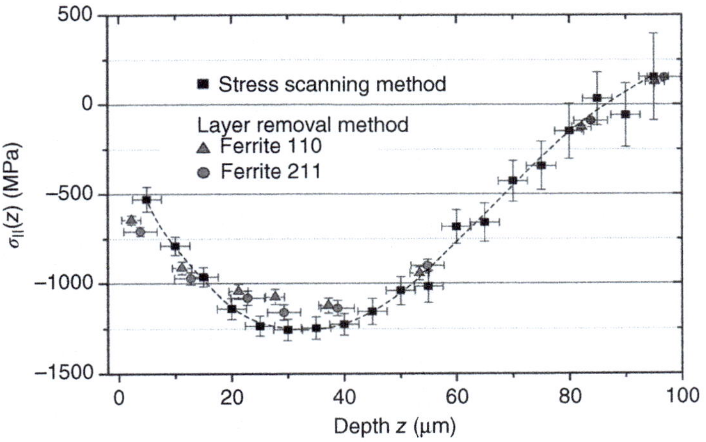

Figure 9.6 Real space in-plane residual stress depth distribution in the near surface region of a shot-peened steel 100Cr6. See text for further details. (After Refs. [26, 27].)

for detecting nonuniform residual stress fields below the sample surface. As a reference for the actual (i.e., the real space) residual stress state $\sigma(z)$ we take the results achieved by means of two different methods, see Figure 9.6. The first method is the stress scanning method [27] which is based on through surface strain scanning in the ED diffraction mode using a small GV of about 10 μm in height. Performing depth scans for both, vertical and inclined diffraction geometries (the latter realized either by shifting the detector along a horizontal line, or by employing the two-detector setup shown in Figure 9.5c), lattice strain depth profiles are obtained for different angles ψ without changing the orientation between the GV and the sample, that is, without losing spatial resolution [27, 28]. Applying the $\sin^2\psi$-method to the data sets $\varepsilon_{\varphi\psi}^{hkl}(z)$ in each depth, z yields a discrete profile of the in-plane residual stress component. The second method, applied after all nondestructive investigations had been carried out, was the layer removal method using conventional AD diffraction and standard $\sin^2\psi$-analysis [26].

To compare the different "multiple line methods" for residual stress gradient analysis introduced in Section 9.2.3 (i.e., methods that make use of *all* reflection lines in the diffraction patterns), series of ED diffraction patterns were measured in the symmetrical Ψ-mode of XSA for various azimuths φ up to large inclination angles ψ, for details see [19]. Based on these data, the ED diffraction method allows for the evaluation of $d_{\varphi\psi}^{hkl} - \sin^2\psi$ plots for each diffraction line $E_{\varphi\psi}^{hkl}$ observed in the patterns. From the examples shown in Figure 9.7a,b it can be seen that the respective $\sin^2\psi$-distributions differ from each other for low ($E^{110} = 22$ keV) and high ($E^{321} = 58$ keV) energy lines. The reason for this result is to be found in the residual stress gradient itself and in the different information depths τ, which are covered by the individual reflection E^{hkl} during the $\sin^2\psi$-measurement, see Figure 9.4b.

Application of the $\sin^2\psi$-method [18] to each of the $d_{\varphi\psi}^{hkl} - \sin^2\psi$ data sets and plotting the corresponding residual stress values versus the average information depth $\overline{\tau}^{hkl}$ according to Eq. (9.9) yields the discrete depth profile $\sigma(\overline{\tau}^{hkl})$ shown in Figure 9.7c. This modified multi-wavelength-plot (MWP) has to be considered as a "Laplace-like" residual stress depth profile, which increasingly differs from the real space profile $\sigma(z)$ with

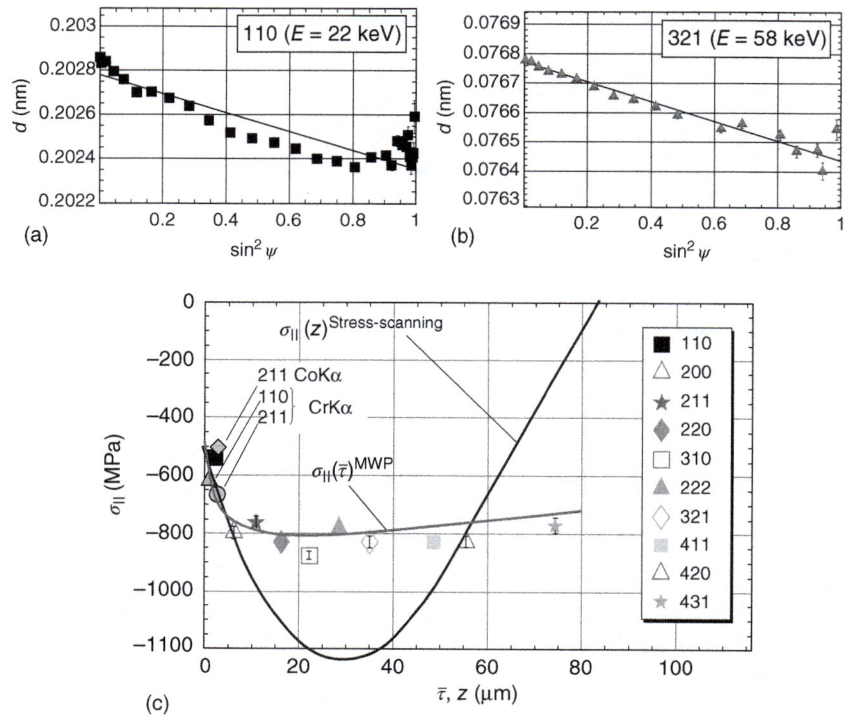

Figure 9.7 ED-XSA on shot-peened steel 100Cr6: (a,b) examples of $d_{\psi}^{hkl} - \sin^2\psi$ plots obtained for different reflections E^{hkl}. The straight lines are the fit curves obtained by linear regression. (c) Modified multi-wavelength plot (MWP). The empirical solid curve laid through the discrete stress data is provided to illustrate the profile. The stress values obtained from angle-dispersive XSA measurements performed in the lab with CrKα and CoKα radiation as well as the real space stress depth profile from Figure 9.6 are displayed for comparison. (After Ref. [19].)

increasing steepness in the actual residual stress gradient. Therefore, the modified MWP plots should be taken as a usable first approximation in order to compare, for example, large sample series with respect to the near surface residual stress state.

An enhanced second approximation for the residual stress depth distribution may be obtained by means of the universal plot formalism outlined in Eqs. (9.10) and (9.11). Figure 9.8 shows the application of the universal plot approach to the $d_{\psi}^{hkl}(\tau)$-data sets evaluated from the ED diffraction patterns as well as from AD measurements performed in the laboratory using CoKα radiation. The lattice parameters d_0^{hkl} required for the evaluation of the strains $\varepsilon_{\varphi\psi}^{hkl} = (d_{\varphi\psi}^{hkl} - d_0^{hkl})/d_0^{hkl}$ were taken from the strain-free directions of the $d_{\varphi\psi}^{hkl} - \sin^2\psi$ plots, which is given for a biaxial residual stress state by $\sin^2\psi^{*,hkl} = -2S_1^{hkl}/\frac{1}{2}S_2^{hkl}$. The results confirm that the universal plot method is able to detect even strongly nonuniform residual stress depth distributions, at least within a depth range that contains most of the experimental data (in the present case about 75 μm).

Strain and stress analysis in the framework of the modified multi-wavelength method and the universal plot method is based on the analysis of the diffraction line shift $\Delta E_{\varphi\psi}^{hkl}$ of *individual* diffraction lines and, therefore, for the evaluation of stress values in *discrete* (average) information depths, a more comprehensive approach is used in

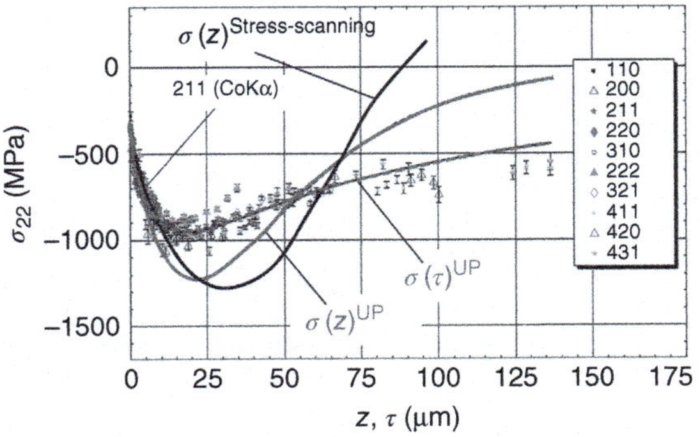

Figure 9.8 ED-XSA on shot-peened steel 100Cr6: Universal plot (UP) evaluation of the $\sin^2\psi$-data. The real space in-plane residual stress depth profile $\sigma(z)$ was described by $\sigma_{\|}(z) = (a_0 + a_1 z + a_2 z^2)e^{-a_3 z}$. Its Laplace transform which is $\sigma_{\|}(\tau) = a_0/(a_3\tau + 1) + a_1\tau/(a_3\tau + 1)^2 + 2a_2\tau^2/(a_3\tau + 1)^3$ according to Eq. (9.7) was fitted to the discrete stress data to refine the unknown coefficients $a_0, \ldots, a_3$. For comparison the real space stress depth profile obtained by the stress scanning and layer removal method is also shown. ((After Ref. [19]), see also Ref. [27].)

Rietveld-based ED residual stress gradient analysis. Figure 9.9 summarizes the results achieved by *simultaneous* refinement of *all* ED diffraction patterns that were recorded during the $\sin^2\psi$-analysis performed on the shot-peened steel sample. It is clearly seen from Figure 9.9b that the recalculated $d^{hkl}_{\varphi\psi} - \sin^2\psi$-curves fit much better to the corresponding experimental distributions than the linear regression lines obtained by applying the conventional $\sin^2\psi$-method (cf. Figure 9.7a,b). A further advantage of the Rietveld-based XSA which can be realized from Figure 9.9c, is that the achieved results are very stable and, in contrast to the universal plot method, almost independent of the model function (here: exponentially damped polynomials of different degree) used in the refinement procedure.

9.4.2 Fast *In situ* Stress Analysis by Means of Energy-Dispersive Diffraction

Another advantage of ED diffraction consists in the possibility of conducting diffraction experiments with a complex sample environment, where it is difficult to "thread" both the incoming and the diffracted beam(s) through the apparatus. Such experiments benefit from two features of high-energy ED diffraction: (i) the measuring direction(s) can be chosen arbitrarily and in such a way that the beams can pass special entrance and exit windows of the devices and (ii) the high-energy photons can penetrate even thick chamber walls and windows made, for example, of aluminum, which may simplify the device construction. Figure 9.10 shows the experimental setup used at the EDDI beamline for real-time studies on the rapid thermal sulfurization of Cu-In thin film systems [29]. ED-XSA investigations were performed on a 500 nm thick Mo-film that serves as a back contact for $Cu(In,Ga)S_2$ and $Cu(In,Ga)Se_2$ solar cells. For experimental details the reader is referred to [14].

For *in situ* stress analysis the sample was subjected in the vacuum chamber to the temperature profile displayed in the upper diagram of Figure 9.11b. During the temperature cycle, diffraction patterns were recorded with an integration time of 20 s employing the two-detector setup shown in Figure 9.10a (see also Figure 9.5c). From the diffraction

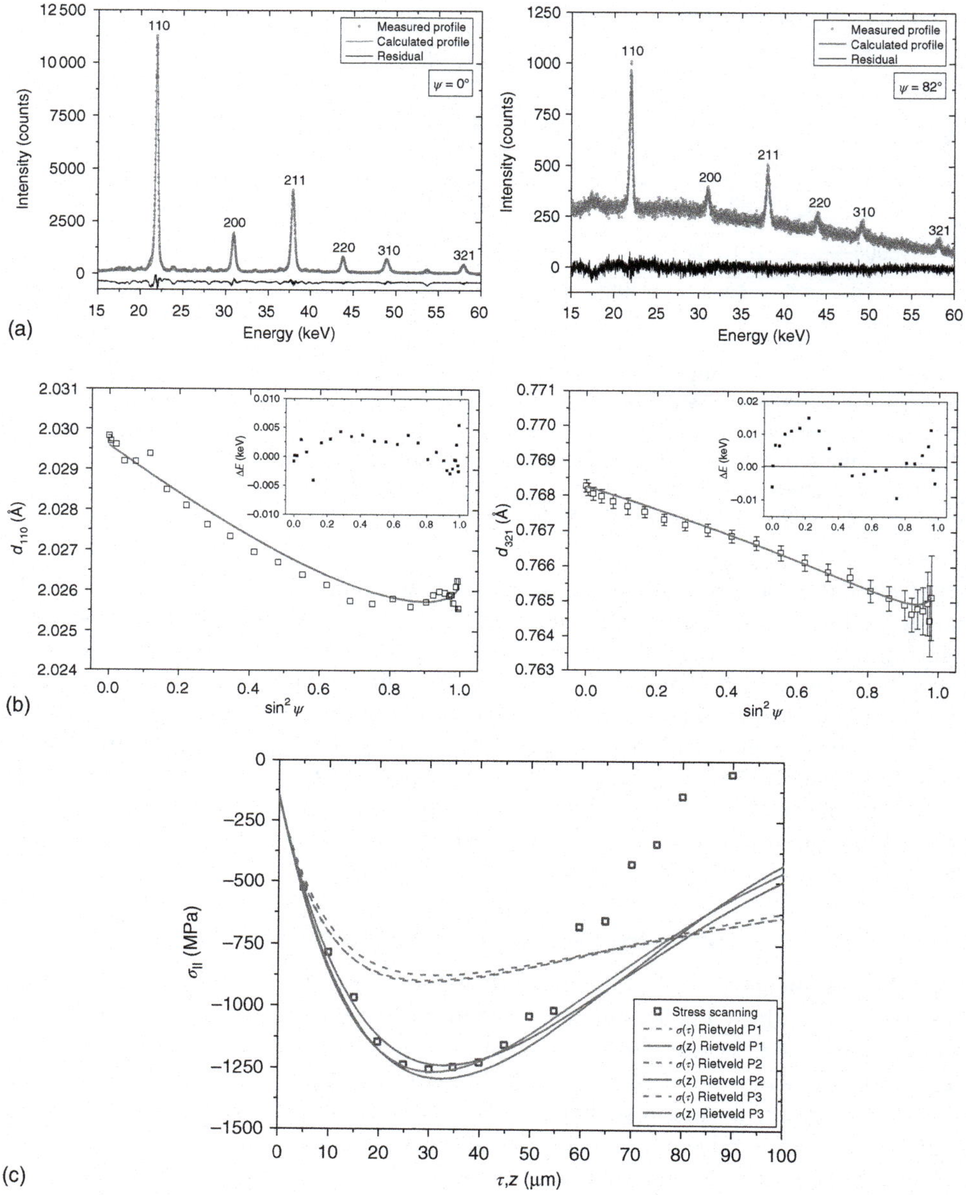

Figure 9.9 ED-XSA on shot-peened steel 100Cr6: (a) Simultaneously Rietveld-refined ED diffraction patterns (selected ψ-positions). The real space in-plane residual stress state was described by an exponentially damped first order polynomial function $\sigma_{||}(z) = (a_0 + a_1 z)e^{-a_2 z}$. (b) Comparison of the $d_{\psi}^{hkl,\mathrm{exp}} - \sin^2\psi$-distributions obtained by conventional whole pattern fitting and the $d_{\psi}^{hkl}[\tau(\sin^2\psi)]$ curves that were recalculated according to $d_{\psi}^{hkl}(\tau) = \left[\left(\frac{1}{2}S_2^{hkl}\sin^2\psi + 2S_1^{hkl}\right)\sigma_{||}(\tau) + 1\right] d_0^{hkl}$ with $\sigma_{||}(\tau)$ being the Laplace transform of the refined model for $\sigma_{||}(z)$. The insets show the corresponding residuals on the energy scale. (c) Results obtained for the refinement of models using exponentially damped polynomial functions of first (P1), second (P2), and third (P3) order. For comparison, the residual stress profile from the stress scanning method (cf. Figure 9.6) is shown as well. (After Ref. [23].)

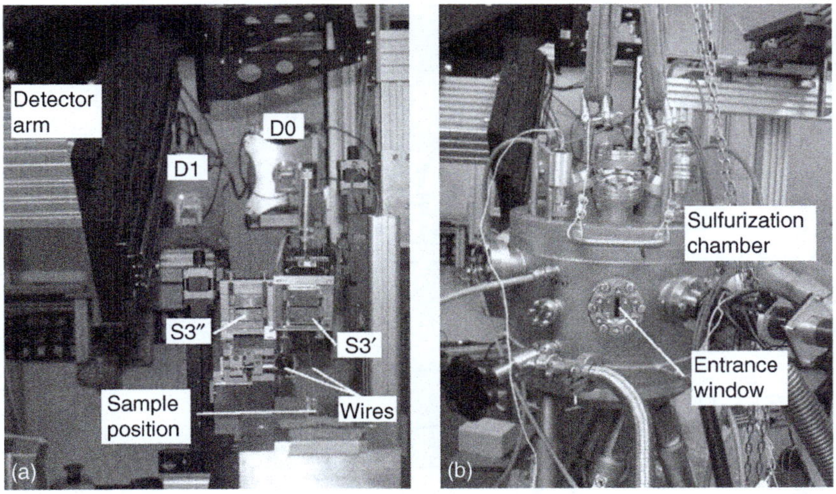

Figure 9.10 (a) The "two-entrance-slit-setup" at the EDDI beamline (cf. Figure 9.5a) for *in situ* stress analysis seen in propagation direction of the synchrotron beam. (b) Reaction chamber for rapid thermal processing of thin films for solar cell fabrication mounted on a hexapod stage in front of the secondary slit systems shown in (a). (After Ref. [14].)

patterns it can be seen that the larger diffraction angle used for data acquisition in detector D1 leads to a compression and shift of the diffraction lines toward smaller energies, whereas the position of the fluorescence lines remains unchanged. Stress evaluation on the basis of the $\sin^2\psi$ – method was performed for the Mo-110 reflection. The resulting $\sigma_{\parallel}^{Mo}(T)$ plot shown in Figure 9.11b is due to the different thermal expansion coefficients of the Mo-film ($\alpha^{Mo} = 5 \times 10^{-6}\,K^{-1}$) and the glass substrate ($\alpha^{glass} = 9.5 \cdot 10^{-6}\,K^{-1}$). Starting at room temperature from a high intrinsic compressive residual stress level caused by the sputtering process, an onset of stress release is observed immediately after the beginning of the heating process. The $\sigma_{\parallel}^{Mo}(T)$ curve reaches the maximum value of stress release at the temperature plateau of 500 °C, before it drops down at the final temperature in the process of 120 °C. Assuming a temperature dependency of $\Delta\alpha^{Mo,glass}E^{Mo} \approx 1.5\,MPa\,K^{-1}$ for the thermal film stress, extrapolation to room temperature would yield about −1300 MPa, which is close to the result obtained afterward by AD-XSA (angle-dispersive-X-ray stress analysis) performed in the laboratory.

9.5 Final Remarks

Compared with the long tradition of AD methods in XSA, the application of ED diffraction methods is strongly associated with the availability of powerful synchrotron sources of the third generation and, therefore, came up only some 10 years ago. ED diffraction has a versatile field of applications, since it combines high flexibility regarding the multitude of possible experimental setups and configurations with a high information content contained in the diffraction patterns. The most important difference between AD and ED diffraction consists in the mode of data acquisition. In the latter case, the positions of the sample and the detector remain fixed during the measurement and complete

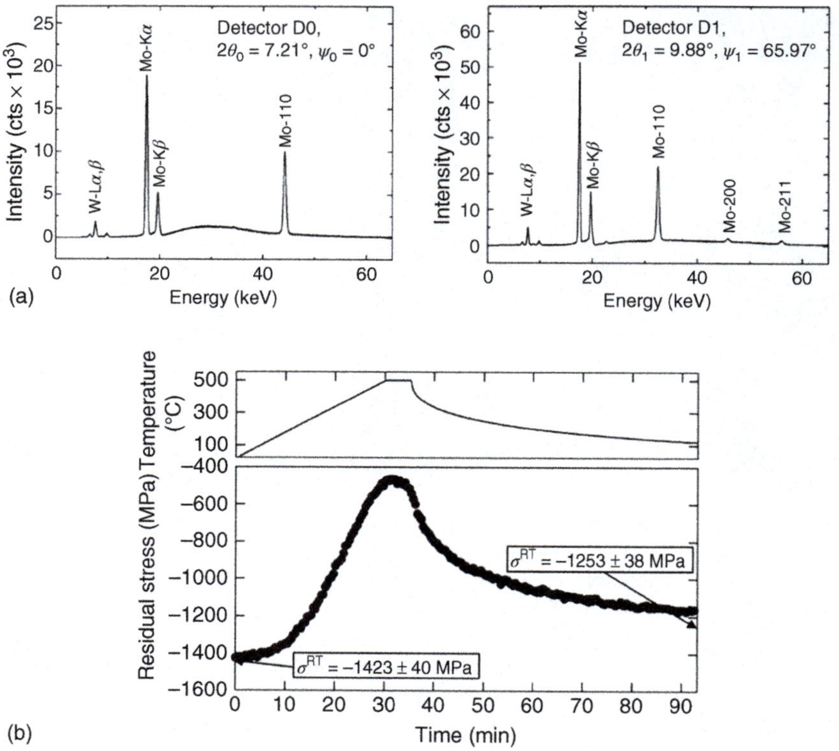

Figure 9.11 (a) ED diffraction patterns recorded simultaneously with two detectors for a 500 nm thick molybdenum film on a glass substrate. (b) $\sin^2\psi$-based *in situ* film stress analysis performed using the lattice spacings $d^{110}_{\psi=0°}(T)$ and $d^{110}_{\psi=65.97°}(T)$ which were evaluated from the diffraction patterns shown in (a). Evaluation of the exact angular positions $2\theta_0$ (detector D0) and $2\theta_1$ (detector D1) was done by means of stress-free LaB_6 powder. σ^{RT} denotes the film residual stress measured at room temperature before and after the heating cycle in the AD diffraction mode in the laboratory. (Genzel 2011 [14]. Reproduced with permission of Sage Publications.)

diffraction patterns with a multitude of diffraction lines are recorded simultaneously in a well-defined scattering direction within very short times of some seconds up to a few minutes.

Apart from this difference, the same experimental methods used in AD-XSA can be applied to the ED diffraction mode. However, the higher information content in the form of complete diffraction patterns instead of single diffraction lines allows for decisive advancements of the stress evaluation procedures. With respect to XSA performed in the reflexion mode, this especially concerns new possibilities for strain/stress depth profiling. In this chapter, with the example of a conventional $\sin^2\psi$-measurement, we have shown how the extensive information included in the ED diffraction patterns can be used to evaluate near surface residual stress depth profiles at different levels of approximation.

However, there are still other possible methods that can be applied for strain and stress depth profiling in ED diffraction with white synchrotron radiation. From Figure 9.4, for example, it follows immediately, that a variation of the (free) parameter θ can be used to adjust the information depth τ thereby keeping it constant even for different inclination angles ψ. That way it becomes possible to perform residual stress analysis in predefined depths below the sample surface [15]. But, as shown recently in Ref. [30], it is also possible to realize the "constant information depth mode" in ED diffraction without changing the Bragg angle, if the strain depth profiling is done in the "scattering vector mode," that is, by stepwise sample rotation η around the diffraction vector for various inclination angles ψ (cf. Figure 9.3). In this way, composition gradients giving rise to a depth variation of the strain-free lattice parameter, $d_0^{hkl}(z)$, can be separately evaluated in addition to gradients of the in-plane stress, $\sigma_{\|}(z)$. Moreover, in Ref. [23] it has been demonstrated very recently that ED diffraction in combination with the Rietveld method also allows for a full triaxial analysis of the residual stress tensor, that is, including the normal component, $\sigma_{33}(z)$, the detection of which is known to be challenging by other XSA methods (see also [15]).

The feature of ED diffraction to provide complete diffraction patterns in fixed but arbitrary directions can be used in different ways for residual stress analysis. Employing the two-detector setup shown in Figure 9.5 in connection with slit systems of small aperture, residual stress depth profiles $\sigma(z)$ in the real space can be determined directly [14, 27] and serve as reference for the assessment of the results obtained by other (rather indirect) methods as demonstrated in Section 9.4.1. On the other hand, by applying the two-detector setup with a complex sample environment as shown in Figure 9.10, *in situ* stress monitoring can be carried out. Thus, together with further information on crystallographic texture and microstructure which is also available from the ED diffraction patterns, a comprehensive *in situ* study of growth mechanisms, for example, in thin films becomes possible.

References

1 Giessen, B.C. and Gordon, G.E. (1968) New high-speed technique based on X-ray spectrography. *Science*, **159**, 973–975.

2 Buras, B., Chwaszczewska, J., Szarras, S., and Szmid, Z. (1968) Fixed Angle Scattering/FAS/Method for X-ray Crystal Structure Determination. Report No.894/II/PS, Institute of Nuclear Research, Warsaw, 10 pp.

3 Laine, E. and Lähteenmäki, I. (1980) The energy dispersive X-ray diffraction method: annotated bibliography 1968–1978. *J. Mater. Sci.*, **15**, 269–278.

4 Bechtoldt, C.J., Placious, R.C., Boettinger, W.J., and Kuriyama, M. (1982) X-Ray residual stresses mapping in industrial materials by energy dispersive diffractometry. *Adv. X-Ray Anal.*, **25**, 329–338.

5 Black, D.R., Bechtoldt, C.J., Placious, R.C., and Kuriyama, M. (1985) Three dimensional strain measurement with X-ray energy spectrometry. *J. Nondestr. Eval.*, **5**, 21–25.

6 Reimers, W., Broda, M., Brusch, G., Dantz, D., Liss, K.-D., Pyzalla, A., Schmackers, T., and Tschentscher, T. (1998) Evaluation of residual stresses in the bulk of materials by high energy synchrotron diffraction. *J. Nondestr. Eval.*, **17**, 129–140.

7 Korsunski, A.M., Collins, S.P., Owen, R.A., Daymond, M.R., Achtioui, S., and James, E. (2002) Fast residual stress mapping using energy-dispersive synchrotron X-ray diffraction on station 16.3 at the SRS. *J. Synchrotron Radiat.*, **9**, 77–81.

8 Pyzalla, A., Camin, B., Buslaps, T., Di Michiel, M., Kaminski, H., Kottar, A., Pernack, A., and Reimers, W. (2005) Simultaneous tomography and diffraction analysis by creep damage. *Science*, **305**, 92–95.

9 Ruppersberg, H., Detemple, I., and Krier, J. (1989) Evaluation of strongly non-linear surface-stress fields $\sigma_{xx}(z)$ and $\sigma_{yy}(z)$ from diffraction experiments. *Phys. Status Solidi A*, **116**, 681–687.

10 Ruppersberg, H. and Detemple, I. (1993) Evaluation of the complex stress field in a ground steel plate from energy dispersive X-ray diffraction experiments. *Mater. Sci. Eng. A*, **161**, 41–44.

11 Ruppersberg, H. (1994) Complicated average stress-fields and attempts at their evaluation with X-ray diffraction methods. *Adv. X-Ray Anal.*, **37**, 235–244.

12 Genzel, C., Denks, I.A., Gibmeier, J., Klaus, M., and Wagener, G. (2007) The materials science synchrotron beamline EDDI for energy-dispersive diffraction analysis. *Nucl. Instrum. Methods Phys. Res. A*, **578**, 23–33.

13 Genzel, C. (1994) Formalism for the evaluation of strongly non-linear surface stress fields by X-ray diffraction in the scattering vector mode. *Phys. Status Solidi A*, **146**, 629–637.

14 Genzel, C., Denks, I.A., Coelho, R., Thomas, D., Mainz, R., Apel, D., and Klaus, M. (2011) Exploiting the features of energy dispersive synchrotron diffraction for advanced residual stress and texture analysis. *J. Strain Anal.*, **46**, 615–625.

15 Genzel, C., Stock, C., and Reimers, W. (2004) Application of energy-dispersive diffraction to the analysis of multiaxial residual stress fields in the intermediate zone between surface and volume. *Mater. Sci. Eng., A*, **372**, 28–43.

16 Hauk, V. (1997) *Structural and Residual Stress Analysis by Nondestructive Methods*, Elsevier, Amsterdam.

17 Welzel, U. and Mittemeijer, E.J. (2003) Diffraction stress analysis of macroscopically elastically anisotropic specimens: on the concepts of diffraction elastic constants and stress factors. *J. Appl. Phys.*, **93**, 9001–9011.

18 Macherauch, E. and Müller, P. (1961) Das $\sin^2\psi$ – Verfahren der röntgenographischen Spannungsmessung. *Z. Angew. Phys.*, **13**, 305–312.

19 Genzel, C., Apel, D., Klaus, M., Genzel, M., and Balzar, D. (2014) Residual stress gradient analysis by multiple diffraction line methods. *Mater. Sci. Forum*, **768–769**, 3–18.

20 Rietveld, H.M. (1967) Line profiles of neutron powder-diffraction peaks for structure refinement. *Acta Crystallogr.*, **22**, 151–152.

21 Rietveld, H.M. (1969) A profile refinement method for nuclear and magnetic structures. *J. Appl. Crystallogr.*, **2**, 65–71.

22 Apel, D., Klaus, M., Genzel, C., and Balzar, D. (2011) Rietveld refinement of energy-dispersive synchrotron measurements. *Z. Kristallogr.*, **226**, 934–943.

23 Apel, D., Klaus, M., Genzel, M., and Genzel, C. (2014) Rietveld based energy-dispersive residual stress evaluation: analysis of complex stress fields $\sigma_{ij}(z)$. *J. Appl. Crystallogr.*, **47**, 511–526.

24 Cox, D.E. (1991) in *Handbook on Synchrotron Radiation*, vol. 3, Chapter 5 (eds G. Brown and D.E. Moncton), Elsevier Science Publishers B.V.

25 Denks, I.A. and Genzel, C. (2007) Enhancement of energy-dispersive residual stress analysis by consideration of detector electronic effects. *Nucl. Instrum. Methods Phys. Res. B*, **262**, 87–94.

26 Manns, Th. (2010) Analyse oberflächennaher Eigenspannungszustände mittels komplementärer Beugungsverfahren. Doctoral thesis, Universität Kassel.

27 Denks, I. A. (2008) Entwicklung einer Methodik zur Erfassung randschichtnaher Eigenspannungsverteilungen $\sigma(z)$ in polykristallinen Werkstoffen mittels energiedispersiver Diffraktion. Doctoral thesis, Universität Kassel.

28 Genzel, C., Denks, I.A., and Klaus, M. (2012) in *Modern Diffraction Methods* (eds E.J. Mittemeijer and U. Welzel), Wiley-VCH Verlag GmbH, pp. 127–154.

29 Rodriguez-Alvarez, H., Kötschau, I.M., and Schock, H.-W. (2008) Pressure-dependent real-time investigations on the rapid thermal sulfurization of Cu–In thin films. *J. Cryst. Growth*, **310**, 3638–3644.

30 Jegou, S., Christiansen, T.L., Klaus, M., Genzel, C., and Somers, M.A.J. (2013) Determination of composition, residual stress and stacking fault depth profiles in expanded austenite with energy-dispersive diffraction. *Thin Solid Films*, **530**, 71–76.

10

Texture Analyses by Synchrotron X-rays and Neutrons

Sangbong Yi, Weimin Gan, and Heinz-G. Brokmeier

For an effective use of texture analysis as a sensitive indicator of the various metal-physical mechanisms, it is essential to describe accurately the texture evolution under various materials conditions (see Chapter 3). In addition, a numerical prediction of anisotropic behavior and crystal plasticity simulation require also a quantitative texture analysis with high accuracy. A quantitative texture analysis needs a mathematical interpretation of the experimental inputs to achieve the three-dimensional orientation distribution function (ODF) through pole figure inversion [1]. Currently available texture analysis softwares [2–6] that have been developed mostly based on the series expansion [1, 6], for example, the Williams-Imhof-Matthies-Vinel (WIMV) [5] algorithm or component method [7], allow the mathematical processing of the experimental data sets from different diffraction techniques. The data sets with arbitrary format, from neutron, X-rays, or electron diffraction experiments can be processed in modern texture analysis softwares. More detailed descriptions regarding the mathematical algorithms and available software are found elsewhere [1, 5, 6, 8]. In this chapter, various texture measuring techniques are introduced, ranging from laboratory machines to large facilities. By comparing the characteristics of each method, a guideline for selecting adequate equipment for users' demands is provided. The use of large-scale facilities, for example, neutron and synchrotron sources, opens a wide range of possibilities of *in situ* texture analysis, due to the relatively high freedom to install an *in situ* device in the beam and the high penetration power of neutrons and high-energy X-rays into solid matter. In the last part of this chapter, important features of texture measurements at large facilities are demonstrated by giving selected examples, especially from high-energy X-ray diffraction.

10.1 Texture Measurements on Laboratory Scale

10.1.1 X-ray Diffraction

X-ray diffraction, as standard laboratory equipment, is most commonly used for texture measurement. The essential setup of an X-ray texture diffractometer is constructed with an X-ray tube, a detector, and a four-circle goniometer that includes an Eulerian cradle. The pole figure measurement is conducted either in reflection or transmission

Neutrons and Synchrotron Radiation in Engineering Materials Science: From Fundamentals to Applications,
Second Edition. Edited by Peter Staron, Andreas Schreyer, Helmut Clemens, and Svea Mayer.

geometry, see Figure 10.1. Because X-rays are strongly absorbed by matter, the transmission geometry is applicable to thin foils or materials with low absorption. The sample is rotated around two axes, tilting (χ in reflection and ω in transmission) and rotation (φ in reflection and χ in transmission), at a fixed detector position (2θ), corresponding to a crystallographic plane $\{hkl\}$. The diffraction intensity varies depending on the density of the $\{hkl\}$ planes that align to satisfy the diffraction condition at the corresponding sample position, $\{\chi, \varphi\}$ or $\{\omega, \chi\}$. Figure 10.2 represents the pole figure grid of $5° \times 5°$ in tilting and rotation, and the pole figure coverage at one sample position by using different type of detectors. More than one pole figure can be measured simultaneously by using one-dimensional position sensitive detector (line detector) covering more than one diffraction peak. In case of a two-dimensional position sensitive detector (area detector), several measuring points in each pole figure can be covered simultaneously, where the degree of coverage depends on the setup, for example, detector size, distance, and X-ray wavelength as well as sample materials.

Because of the low penetration power and small beam size in laboratory X-ray diffraction, when compared to those for neutrons, poor grain statistics are achieved in coarse-grained materials. To increase the grain statistics, sample oscillation over some millimeters or more is used during the measurement under the assumption of homogeneous texture in the oscillated area. The addition of several pole figures measured at more than one sample position (or several samples) is also considered for increasing the grain statistics of the texture analysis. Due to the strong defocusing effect and diffraction geometry, a full coverage of the experimental pole figure cannot be obtained by laboratory X-ray diffraction. In reflection geometry, the pole figure

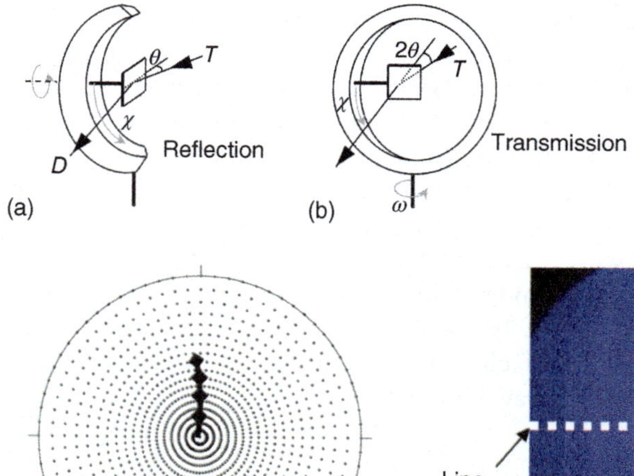

Figure 10.1 Diffraction geometry in pole figure measurement; (a) in reflection and (b) in transmission geometry (T: X-ray tube, D: detector, χ: sample tilting, ϕ: sample rotation, ω: goniometer rotation, and θ: diffraction angle).

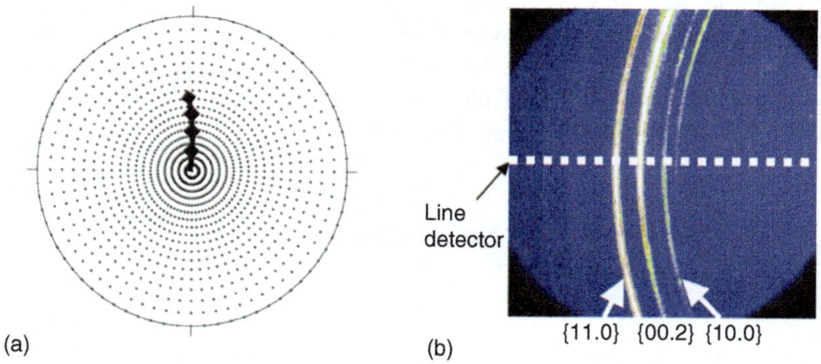

Figure 10.2 (a) Pole figure measuring grid with $5° \times 5°$ in tilting and rotation. Each grid point is measured individually by point or line detector. (b) Diffraction image of an extruded Mg on an area detector. The pole figure coverage by one diffraction line on this image, for example, (00.2), is marked in (a) with a black line.

is measured in general up to $\chi = 70°$ or at most up to $\chi = 85°$. On the contrary, the peripheral part of the pole figure is covered in transmission geometry. In transmission, pole figures are measured up to $\omega = 55°$ or at most up to $\omega = 60°$. A complete pole figure can be obtained by diverse methods, for example, recalculation of the pole figure from the ODF, combination of transmission with reflection method, or the use of oblique samples [9].

However, the last two methods are barely used due to the difficulties in measurement and sample preparation, while the ODF calculation has become a standard procedure in X-ray texture analysis. It is often found in practical measurements that some diffraction peaks are located very close to each other, or totally overlap, especially in multi-phase samples and/or materials with low crystal symmetry. In such cases, the overlapped peaks are hardly to be used for pole figure measurements, unless mathematical peak fitting can be employed to separate the peaks. In spite of the above limitations, laboratory X-ray diffraction is most widely used for texture analysis, because of the easiness in machine usage, low cost, and the relatively easy sample preparation.

10.1.2 Electron Diffraction

From the bulk diffraction using (hard) X-rays or neutrons, the orientation information is measured in average over a large sample volume and a large number of grains. For the investigation of local texture or the orientation of individual grains, the orientation measurement on the grain scale is necessary and can be performed by the electron diffraction technique in scanning or transmission electron microscopes. Orientation determination by electron diffraction is accomplished by indexing of the so-called Kikuchi patterns or the spot diffraction patterns, for example, selected area diffraction (SAD) or small-angle convergent beam electron diffraction (SCBED). Figure 10.3 illustrates the formation of Kikuchi line and SCBED spot pattern. The Kikuchi patterns in SEM and TEM are formed in a similar way. The only difference is that the Kikuchi lines in SEM are formed from the diffraction of the backscatter electrons leaving the sample surface opposite to the direction of the incident beam, while in TEM

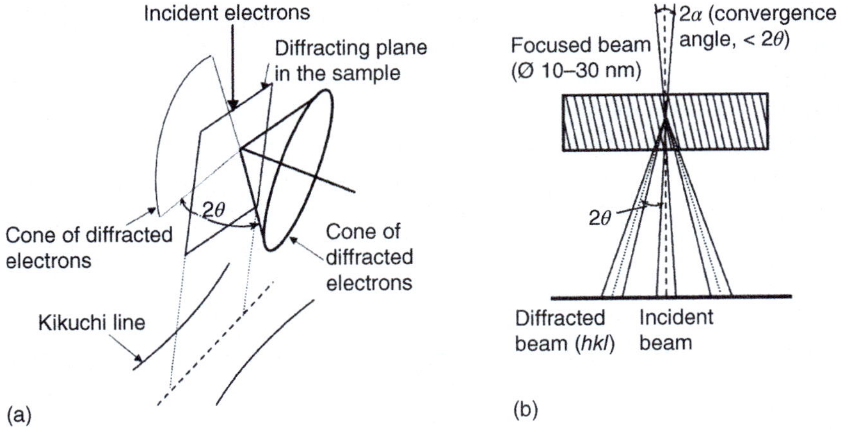

Figure 10.3 (a) Kikuchi line formation in TEM. (b) Spot pattern in TEM-SCBED, Kossel–Möllenstedt pattern, in case of $2\alpha < 2\theta$ [10].

transmitted electrons form the Kikuchi pattern. When an electron beam is directed onto a crystalline sample, electrons are scattered in all directions. There must be some electrons satisfying the diffraction conditions, according to the Bragg's law, at every set of lattice plane. These inelastically scattered electrons emerge along the diffraction cones having their opening angle of $180° - 2\theta_B$, where θ_B is the Bragg angle. When the cones of diffracted electrons intersect a detecting screen, they form Kikuchi lines on the screen. The use of SCBED in TEM results in a significant improvement in spatial resolution under 10 nm, while SAD delivers the average information from about 0.5 μm [11, 12]. Once the electron beam is focused such that the convergence angle (2α) is smaller than the diffraction angle (2θ), the nonoverlapped spot pattern by elastic diffraction of the electrons is observed on the TEM screen, called Kossel–Möllenstedt pattern. More details on electron diffraction can be found in Ref. [10, 13].

With the aid of the current developments in electron microscopy, computer techniques, and automated pattern indexing, for example, Kikuchi pattern indexing based on the Hough transformation [13] or SAD spot pattern indexing by template matching [14], electron diffraction has become an important tool for microstructure and texture analysis. The real power of the electron diffraction in materials analysis lies in the simultaneous determination of both orientation and microstructure with a high angular and lateral resolution at the defined sample area. Also, energy dispersive X-ray chemical analysis (EDX) can be carried out simultaneously with orientation mapping. Based on the above possibilities, the texture components can be evaluated separately with respect to the microstructural elements, such as chemical composition, grain size, phase, position, and so on. Another important feature of electron diffraction is the possibility of direct texture measurement, that is, the three-dimensional orientation at the measuring point is determined from indexing the diffraction patterns. Therefore, the results from electron diffraction can be presented directly in the orientation space without additional mathematical calculations, which are necessary in X-ray and neutron measurements. Based on this feature, possible errors occurring during mathematical treatments can be avoided in electron diffraction. However, long-time and large-scale mapping is necessary to reach satisfactory statistics for determining the global texture, which is hardly to obtain by means of TEM. The diffraction quality is very sensitive to sample conditions, for example, sample surface flatness, grain size, dislocation density, electric conductivity, and so on. In electron backscatter diffraction (EBSD) in SEM, the sample is tilted by angles of typically 60–70° that allows more electrons to be diffracted and escape toward the detector. Due to the steep tilt, the backscatter electrons emerge from very small depth, in order of nanometers, such that the pattern quality is even more sensitive than in TEM. It is generally recommended to use high beam currents for getting a high signal to noise ratio in SEM-EBSD and also to apply electro-polishing for preventing sample damages during preparation.

10.2 Texture Measurements at Large Scale Facilities

10.2.1 Neutron Diffraction

The advantages of neutron diffraction are the combination of large penetration depth, on centimeter level for almost all materials, and relatively large beam cross-sections

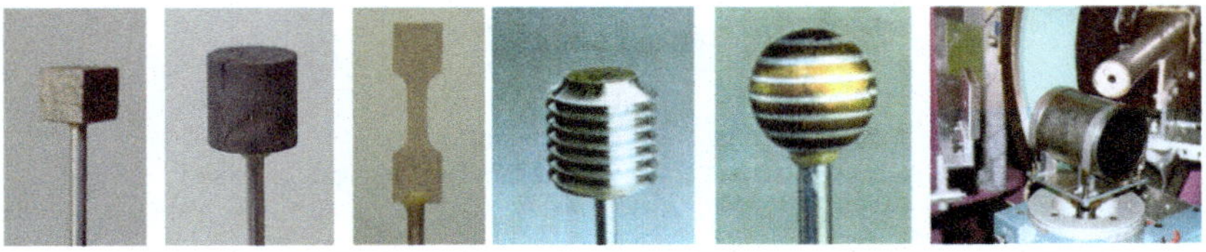

Figure 10.4 Various sample geometries for neutron texture measurements. The sample volume varies from some cubic millimeter to cubic meter, depending on the sample materials.

[15]. Thus, texture measurements are performed without a special sample preparation and sample geometries are typically different to those for laboratory X-ray and EBSD measurements, Figure 10.4. Moreover, a complete pole figure is obtained without further experimental or mathematical treatments, and measured pole figures can be used for ODF calculation after correction of the background intensity. These characteristics make neutron diffraction a powerful tool for texture analysis in geological materials as well as in metallic samples having a complex shape or a large volume. Due to the low absorption of neutrons, a quantitative texture analysis can be performed with satisfactory grain statistics even for coarse-grained material. For the same reason, non-destructive texture analysis of a semi-finished product as well as *in situ* pole figure measurements under various sample environments are also possible. Even though the basic principle of pole figure measurement in neutron diffraction is analogous to that of the laboratory X-ray reflection, a longer counting time is often necessary with neutrons because of the low neutron flux from the reactor.

Because of long counting time and possible flux variations depending on reactor conditions, a flux-controlled measuring scheme is generally employed, that is, constant neutron flux at a measuring point, not constant measuring time. As an effective way to reduce the measuring time in neutron diffraction, pole figure measurement is typically carried out on an equal area grid, by which a full coverage of the pole figure is obtained with a smaller number of measuring points similar to that obtained on an equal-angular grid [9].

Figure 10.5 shows the setup of the materials science neutron diffractometer STRESS-SPEC located at the FRM II (Garching, Germany) [16]. For pole figure measurements, a monochromatic neutron beam with relatively high flux ($\sim 8 \times 10^7$ n cm^{-2} s^{-1}) is used [17]. Together with three different monochromators (Ge, bent Si, pyrolytic graphite (PG)) and the continuous take-off angle variation between 30° and 110°, it allows a wavelength range of 1–2.4 Å. The gauge volume can be defined via slits and/or a radial collimator for better background reduction. A ^{3}He area detector (30 × 30 cm^2) was installed for recording more than one measuring point simultaneously, see Figure 10.2, and it covers a total 2Θ range between 30° and 120°. An Eulerian cradle and a newly developed robot-based system are both standardized equipment for pole figure measurements using step scan or continuous scan mode at STRESS-SPEC [16].

Robotic-based sample handling and positioning system at modern large-scale facilities offer more flexible options such as to run an automatic sample changer (Figure 10.6a), to combine residual strain and texture analyses on complex or large industry components (Figure 10.6b) [18], and to install a sample environment (e.g., electric field, furnace, etc.) for *in situ* pole figure measurement (Figure 10.6c). With

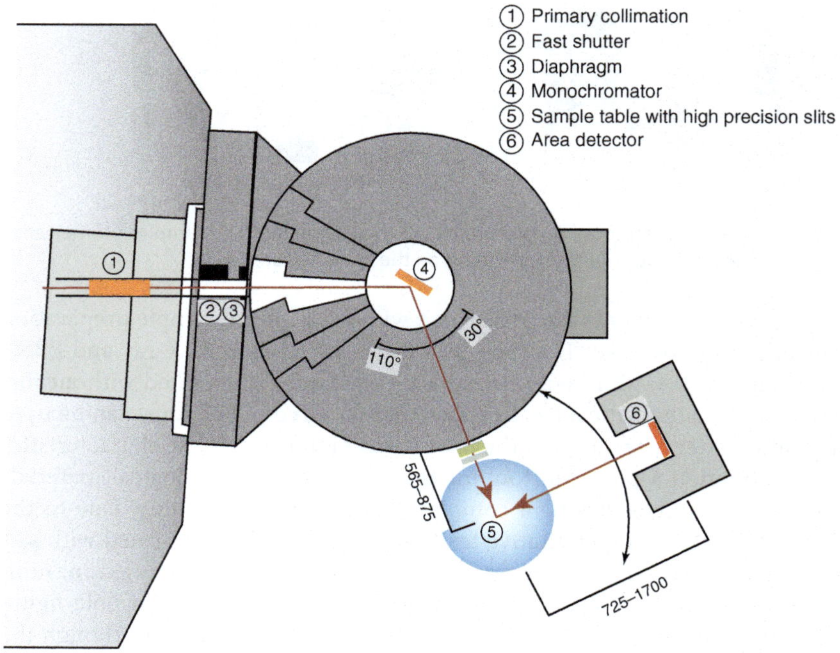

Figure 10.5 The materials science neutron diffractometer STRESS-SPEC at FRM II, Garching, Germany. (http://www.mlz-garching.de/stress-spec. Used under CC-BY 4.0 license https://creativecommons.org/licenses/by/4.0/legalcode.)

implementation of a laser or a video camera control system, the target positioning accuracy of the robotic system can reach about 50 μm. Application of the robotic technique is also attractive for modern spallation sources that have much higher flux and can realize fast texture analysis. Moreover, it is easy to combine texture and strain analyses with a robot.

Another way to cover several reflections simultaneously is the use of time of flight (TOF) neutron diffraction [19, 20]. The wavelength of a neutron is inversely proportional to its velocity so that a polychromatic neutron beam can be analyzed with regard to the flight time of the neutrons. In TOF measurements the diffraction angle (θ in Bragg's law) remains constant, hence a lattice spacing (d in Bragg's law) profile is achieved based on the wavelength pattern. A long flight path of the neutrons is necessary for high resolution in flight time analysis. In a typical setup with flight lengths of approximately 10–50 m time differences of the order of milliseconds are to be resolved. Since the whole diffraction profile is measured at once, for example, 0.5–4 Å of wavelength range corresponds to the d-spacing of 0.35–2.83 Å at fixed $\theta = 45°$, the TOF technique is well suited for low crystal symmetric as well as for multi-phase samples. Diffraction profile covering a wide range of the diffraction peaks can be directly evaluated by a whole spectrum analysis, such as Rietveld analysis of texture and structure [2, 21]. Combining TOF technique with position-sensitive detectors, the time for pole figure measurements can be substantially reduced. In this combination complete pole figure measurement is conducted by sample revolution around only one axis. Based on a relatively large freedom in sample environment and the fast measuring time, *in situ* texture analysis

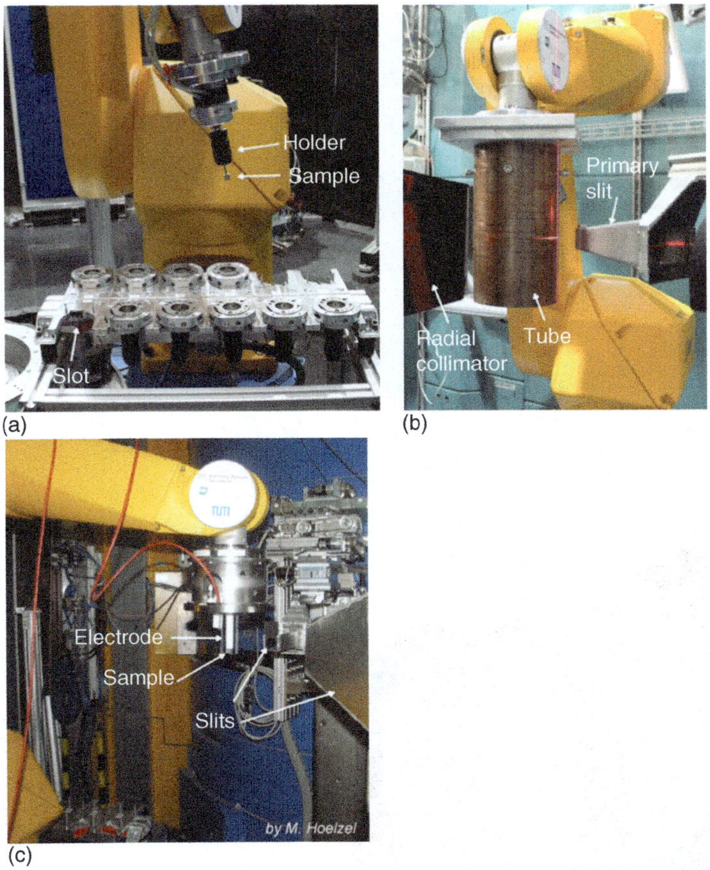

Figure 10.6 Applications of the robotic system at STRESS-SPEC: (a) as an automatic sample changer with 12 storage slots; (b) combined texture and strain analyses on a commercial 15 kg Cu tube; and (c) pole figure measurement under electric field. (http://www.mlz-garching.de/stress-spec. Used under CC-BY 4.0 license https://creativecommons.org/licenses/by/4.0/legalcode.)

has been accomplished at neutron TOF facilities. Figure 10.7 shows two different TOF diffractometers installed with position-sensitive detectors, the SKAT (from Russian: Spektrometer Kolitshestvennovo Analiza Tekstury) diffractometer with 19 detectors at the IBR2 Dubna Research Centre in Russia and HIPPO (High Pressure Preferred Orientation) diffractometer with the total of 1240 detectors arranged on 53 detector panels at Los Alamos Laboratory in the United States.

10.2.2 Texture Analysis Using Synchrotron X-rays

High-energy X-rays from a synchrotron are highly attractive for texture analysis because of high spatial resolution with excellent brilliance and high penetration depth through materials in the same order as that for thermal neutrons. The high spatial resolution results from the fine-focused high-intensity X-ray beam with a large range of photon energies and beam sizes down to a few micrometers. In addition, the divergence of synchrotron X-rays is extremely small so that an orientation analysis with high

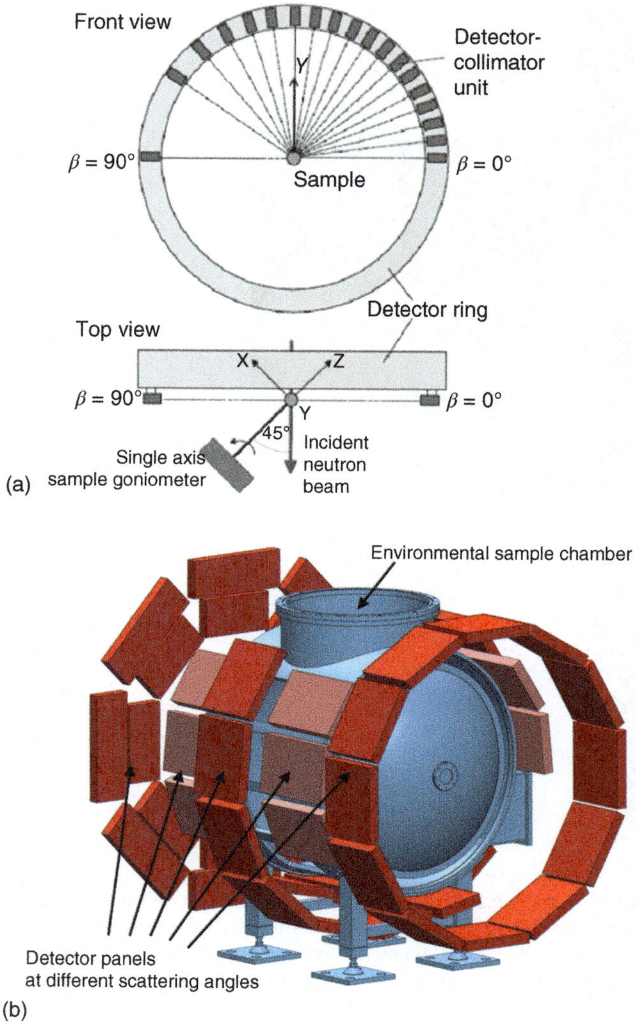

Figure 10.7 Examples of TOF diffractometers in combination with position sensitive detectors. (a) SKAT in Dubna, Russia. (b) HIPPO equipped with 53 detector panels on 5 rings with nominal diffraction angles between 40° to 144° at Lujan Center in Los Alamos laboratory, USA. (Panel (a): Ullemeyer 1998 [19]. Reproduced with permission of Elsevier. Panel (b): Vogel 2004 [20]. Courtesy of Dr. S. Vogel at Los Alamos Neutron Science Center)

angular resolution ($<0.01°$) is available. On the basis of the very high photon flux of the incident beam a minimum percentage of transmission is already enough for a satisfactory intensity at the detector. For this reason, texture analysis of heavy elements in large volume is possible also in transmission geometry. Texture analysis using high-energy X-rays at synchrotron actually began with the advent of area detectors in the mid-1990s. The use of two-dimensional area detectors in transmission geometry, in which full Debye–Scherrer rings are recorded simultaneously, leads to a significant reduction of the measuring time and high angular resolution. The combination of synchrotron X-rays with area detectors has opened a wide range of possibilities in texture analysis.

Texture analysis at localized volumes in ranges of micrometers to centimeters is possible in a nondestructive way. Although neutron diffraction allows nondestructive texture analysis, its spatial resolution is limited to the order of millimeters. With regard to its advantages of fast measuring time, high penetration power, and a narrow beam, high-energy X-ray diffraction at synchrotron offers a wide possibility for *in situ* texture measurements under various sample environments. Pole figure measurement is carried out by sample revolution around one axis, which offers high freedom for sample environment devices.

Depending on research interests and machine availabilities, for example, stationary and moving detector modes [22, 23], conical slit system using a monochromatic beam [24], and energy dispersive method using a polychromatic beam various measuring methods have been developed by different research groups [25]. The texture measurement using a monochromatic beam and area detector will be introduced in the next section. This basic measuring technique, which is called *stationary detector mode*, is available at almost all synchrotron X-ray beam lines without employing special slit systems. Figure 10.8 illustrates the basic beam line setup for texture measurement at the high-energy materials science (HEMS) beam line P07 at PETRA III DESY in Hamburg, Germany [26].

The primary white beam from the undulator enters the optic hutch, where the beam energy is tuned to a desired wavelength using a monochromator located in a vacuum chamber. The monochromatic beam is then guided to the experimental station. The intensity variation of the incident beam, which would occur by a decay of the synchrotron beam, is monitored at a diode (gas counter) for the calibration of recorded intensity at the detector. The incident beam size from the optic hutch is adjustable by motorized horizontal and vertical slits installed at the experimental station. The monochromatic beam with well-controlled size is irradiated at the sample that is positioned on the ω-rotation table mounted on a XYZ-translation stage. This setup allows a fine alignment of the sample position as well as a serial measurement at different sample positions.

The principle of pole figure measurement using synchrotron X-rays is similar to that of laboratory X-ray measurement that was introduced already in the beginning

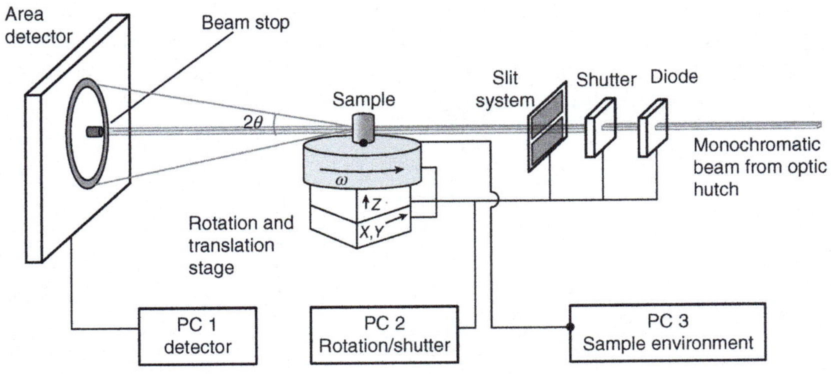

Figure 10.8 Experimental setup for texture measurement at a high-energy X-ray beamline (HEMS at DESY, Hamburg, Germany). PC 1 is for reading data from area detector, PC 2 is for motor control, and PC 3 is connected to an *in situ* device.

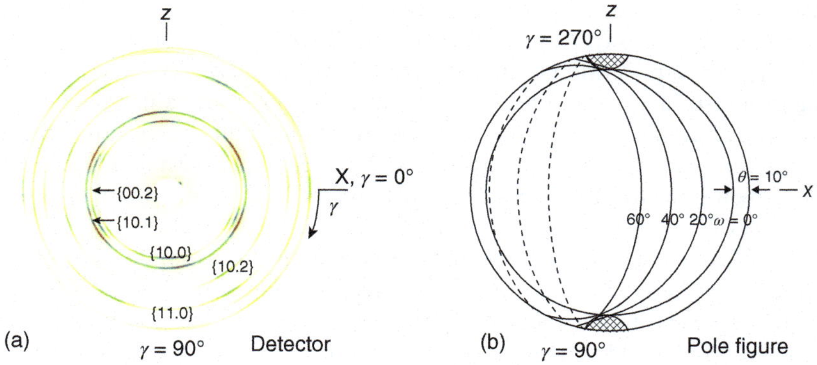

Figure 10.9 (a) Diffraction image from a strongly textured Mg sample on an area detector (transmission geometry). (b) Pole figure coverage by Debye–Scherrer rings obtained at different ω-positions. Solid lines in the pole figure are corresponding to Debye–Scherrer rings within the range from $\gamma = 90°$ to 270°. For emphasizing the blind area effect (shaded area) $\theta = 10°$ and equal area projection are applied to this pole figure.

of the twentieth century using X-ray films as the only available area detector at that time. The main difference is the wavelength of the incident beam on which both the range of the diffraction profile recorded at the detector as well as the scanning routine depend (Figure 10.9). For a complete texture measurement, the sample is rotated around the ω-axis with a fixed step-size. At each ω-position, the sample is irradiated and the occurring Debye–Scherrer rings are collected at the area detector (discrete scan). Another measuring scheme employs a continuous revolution of the sample, while Debye–Scherrer rings are collected within a specific range of the sample rotation (continuous scan). By using continuous scan, a data-set with higher grain statistics can be obtained, as the orientation information without a gap between fixed ω-positions is recorded.

The intensity along the circumference of Debye–Scherrer rings is proportional to pole density on the orientation sphere. Therefore, a textured sample shows an intensity variation along the Debye–Scherrer ring, γ angle, Figure 10.9a. A Debye–Scherrer ring recorded at a certain ω-sample position covers a circle in the corresponding $\{hkl\}$ pole figure and the full pole figure coverage is completed by the ω-rotation. The data gathered from the above steps are expressed in laboratory coordinates as the angles describing the sample position and diffraction geometry, $\{\chi, \varphi, \omega\}$ and $\{\gamma, \theta\}$, respectively. The laboratory coordinates are converted into the pole figure angles $\{\alpha, \beta\}$ for calculating the ODF according to:

$$\begin{bmatrix} \sin\alpha \cdot \cos\beta \\ \cos\alpha \\ \sin\alpha \cdot \sin\beta \end{bmatrix} = \begin{bmatrix} \cos\omega\cos\varphi - \sin\omega\sin\varphi\cos\chi & \sin\omega\cos\varphi + \cos\omega\sin\varphi\cos\chi & \sin\varphi\sin\chi \\ -\cos\omega\sin\varphi - \sin\omega\cos\varphi\cos\chi & -\sin\omega\sin\varphi + \cos\omega\cos\varphi\cos\chi & \cos\varphi\sin\chi \\ \sin\omega\sin\chi & -\cos\omega\sin\chi & \cos\chi \end{bmatrix} \cdot \begin{bmatrix} -\sin\theta \\ \cos\theta\cos\gamma \\ \cos\theta\sin\gamma \end{bmatrix} \tag{10.1}$$

where α stands for the pole distance and β for the azimuth angle in a pole figure. φ, χ represent the rotation and tilting angles of the Eulerian cradle and θ the Bragg angle.

As shown in Figure 10.9 there is a blind area in the pole figure, which originates from the diffraction geometry. Due to the low wavelength, however, this unmeasurable area, which corresponds to the diffraction angle θ, appears only in a small area. In case of a beam energy of 100 keV (wavelength of 0.124 Å) the diffraction rings that are of most interest locate at θ ranges less than 3°. The pole figure intensities measured on a regular $\{\omega, \gamma\}$ grid are irregularly distributed with respect to the $\{\alpha, \beta\}$ coordinates. For the commonly available ODF calculation programs based on series expansion or the WIMV method, the interpolation of the measured pole figures into regular $\{\alpha, \beta\}$ grid is required. The recently developed texture analysis programs allow the ODF estimation using pole figure data in arbitrary format. For example, ODF estimation is available from measured pole figures with an irregular $\{\alpha, \beta\}$ grid and incomplete pole figure coverage by using the MTEX software [4]. Some selected examples of the texture analysis using the synchrotron X-rays are presented in the following sections.

10.2.3 Examples of Texture Analyses Using Neutrons and Synchrotron X-rays

10.2.3.1 Local Texture Measurement of a Friction Welded Rod

Rotary friction welding (RFW), a solid state joining technique in which the welding process is conducted at a temperature below the melting point of the base material, is widely applied to weld dissimilar materials. To study the local texture variation in the welded rod, a nondestructive measurement was required with spatial resolution. Therefore, neutron diffraction was used with the wavelength of 1.69 Å from Ge (311) monochromator at STRESS-SPEC, FRMII at Garching (see Figures 10.5 and 10.6). Conventional Eulerian cradles are limited in rotation and tilt of a larger sample. Moreover, a combined analysis of texture and strain needs always a nondestructive method, so that the Stäubli robot RX160 installed at STRESS-SPEC was used. The present example shows the texture variation in an RFW of AA7020 alloy joined with 316L austenitic steel, having a sample dimension of Ø25 mm × 160 mm. The texture along the central line inside the welded rod was measured with a gauge volume of $2 \times 2 \times 2\,\text{mm}^3$ as a function of the distance to the weld. The gauge volume was obtained by employing a high precision slit system.

The RFW sample and the pole figures measured at different positions of the AA7020 part are presented in Figure 10.10. The texture of the base material (position C) is composed of typical β-fiber of rolled Al which includes the ideal texture components, like copper, S, and brass components. The thickness of the welded zone in the examined sample is less than 1 mm, which can be seen also in the minor influence on the pole figures at position B. The only influence is a reduction of the texture sharpness, which results from moderate short-time heating during welding. As the welded zone is small, the position B averages over the welded zone, the heat affected zone, and even a small part of the base material. The pole figures, particularly the (200) pole figure, show these overlap with an ideal ⟨100⟩ fiber component representing the rotation symmetry of the welding process and the recrystallization with the former deformation texture (position A). Due to the overlap the average texture is weaker than that in the base material.

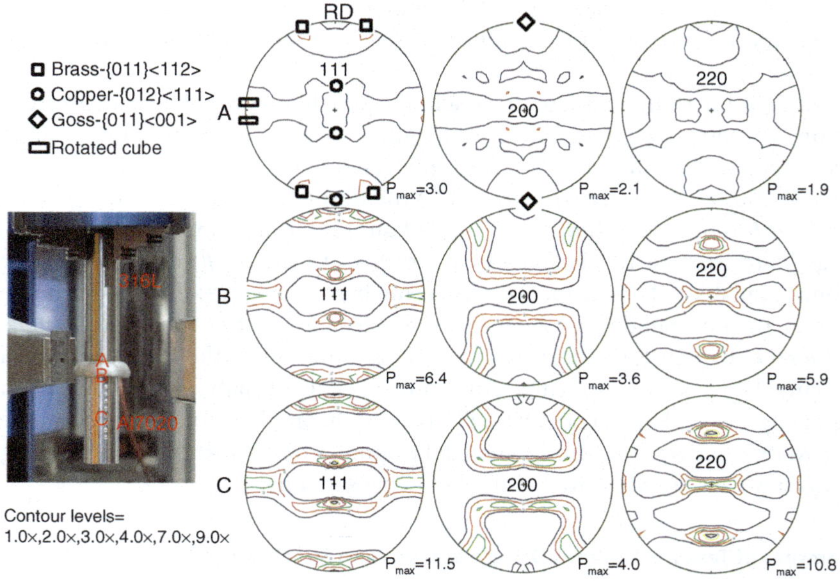

Figure 10.10 Pole figures at different points of AA7020 side in the RFW sample marked with A, B, and C, respectively.

10.2.3.2 Global Texture in Cu Wire

In this example, the determination of the global texture in a Cu wire with a thickness of 0.12 mm is shown [27]. Texture measurement in such a thin wire using other techniques, for example, electron or neutron diffraction, suffers from grain statistics and low diffraction intensity, respectively. In addition, a conventional X-ray measurement needs an intensity correction according to the wire geometry [28]. By employing a specially designed wire holder (Figure 10.11a), which extends slightly the wire bundle, the global texture of the thin wire was successfully obtained at the high-energy X-ray beam line BW5, HASYLAB at DESY, Germany. For increasing grain statistics, the measurements

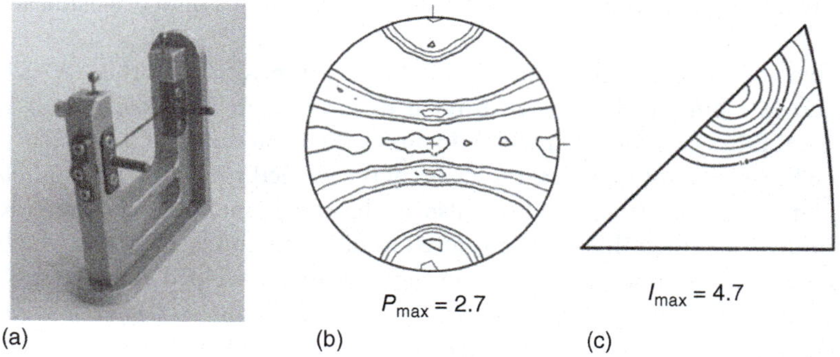

Figure 10.11 (a) Wire sample holder with 10 softly stretched Cu wires. (b) The average of six measured (111) pole figures; wire axis is in the center of the pole figure. (c) Inverse pole figure for the wire direction. (Brokmeier 1991 [26]. Reproduced with permission of Trans Tech Publications Ltd.)

were carried out at six different sample positions using a large beam size of $4 \times 1\,mm^2$ with the beam energy of 100 keV. Thereafter, the pole figures were averaged into one pole figure. The averaged (111) pole figure (Figure 10.11b) and the inverse pole figure in wire direction (Figure 10.11c) show a strong ⟨112⟩ fiber texture component parallel to the wire direction.

10.2.3.3 *In situ* Texture Measurement of Steel at Elevated Temperature

Texture transformation, that is, texture inheritance during phase transformation, was investigated by heating a steel sample [29]. The *in situ* heating experiments were conducted with high-energy X-rays with 100 keV at the high-energy X-ray beamline BW5 at DESY. The texture of the plain steel sample (carbon content: 0.2 mass%), having bcc crystal structure (α-ferrite), was measured at room temperature. Thereafter, the sample was heated until the phase transformation to fcc structure (γ-austenite) was completed. The texture measurement of the γ-austenite was carried out at elevated temperatures. Figure 10.12a shows the vacuum furnace used for the *in situ* texture measurement.

The furnace can be installed on the ω-rotation table and causes no additional diffraction peaks. Because of grain growth effects, a short measuring time is essential for this *in situ* measurement. For reducing the measuring time, the ω-rotation was conducted up to 45° with 5° steps, which covers only a quarter of the pole figure. For extracting the complete pole figures from the restricted data, the Rietveld texture analysis program MAUD [2], which is based on whole diffraction profile analysis, was employed. The calculated pole figures of the α-ferrite and the γ-austenite are shown in Figure 10.12b,c, respectively. The texture results match well with the Kurdjumov–Sachs orientation relationship for bcc ↔ fcc transformation: (111)fcc || (011)bcc and [101]fcc || [111]bcc.

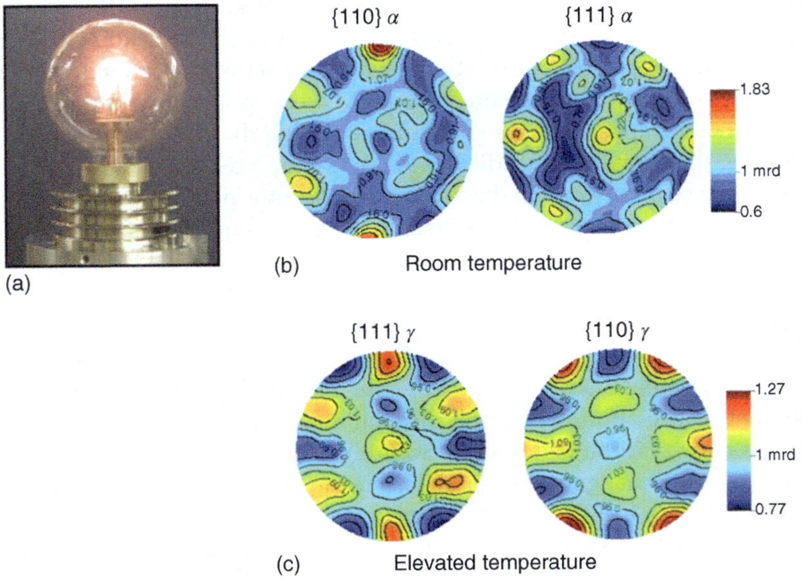

Figure 10.12 (a) Vacuum furnace with glass wall and graphite heating element. Pole figures of (b) α-ferrite and (c) γ-austenite calculated using the MAUD software.

10.2.3.4 *In situ* Texture Measurement under Loading

In situ texture measurement under mechanical loading provides an effective tool to study the interrelationships among texture evolution, activation of various deformation systems, and mechanical properties. This provides a method for tracking the texture evolution experimentally during the whole deformation process. *In situ* measurements guarantee to study the texture evolution without the influence originating from the usage of many different samples. Mechanical properties and the texture evolution during deformation are strongly dependent on the microstructural features of the initial material, for example, grain size, initial texture, and its homogeneity. This dependency becomes more pronounced in magnesium alloys having a hexagonal crystal structure, because the deformation modes available at low temperature are limited and their activities are determined by the grain orientations. Furthermore, deformation twinning is known to contribute to a marked yield surface asymmetry in many alloys with a hexagonal crystal structure. Conversely, the texture development itself is influenced by the activation of certain deformation modes.

For a basic understanding of the relationship between deformation mechanisms and texture evolution of a Mg alloy, texture was measured *in situ* during uni-axial compression of an extruded profile of AZ31(3Al-1Zn-0.4Mn-Mg, in mass%) alloy [26]. A universal testing machine (UTM), by which a sample can be loaded up to 20 kN, was installed on the ω-rotation table, Figure 10.13a. During the irradiation of the sample for the texture measurement, the loading was stopped, but not released, and the sample was rotated together with the UTM around the ω-axis through 96° in 3° steps, taking about 900 s. The texture measuring points were selected on the stress–strain curve and each of them corresponds to a characteristic point; after yielding, during strain hardening and at quasi-steady state plastic deformation. Figure 10.13b presents the mechanical response of the alloy and the texture variation during compression. A slight decrease in stress value is observed at each texture measuring point, due to stress relaxation during the texture measurement. Because of the rotational symmetry of the as-extruded texture and loading nature, the texture is represented using an inverse pole figure in the loading direction. A marked change in texture is observed during the compression. Most grains initially having their c-axes perpendicular to the loading axis, that is, $\langle 10.0 \rangle$ fiber texture component, are re-oriented such that their c-axes become parallel to the loading direction. This fast reorientation results in a formation of very strong $\langle 00.1 \rangle$ fiber texture component at fracture.

Texture simulation based on a viscoplastic self-consistent (VPSC) model [30] was performed using the time-resolved information from the *in situ* measurement. The experimental results were used to determine the fitting parameters. Using the simulations fitted to the experimental stress-strain response and *in situ* texture results, the activities of different deformation modes and their influence on the texture evolution were analyzed. It is clear that the high tensile twinning activity (80% of total activity) leads to the marked texture change at the initial deformation stage, that is, rotation into the loading direction, Figure 10.13c. With the ensuing decrease in the twinning activity, basal $\langle a \rangle$ slip accommodates the imposed deformation. At a certain stress level, the activity of $\langle c + a \rangle$ slip increases so that the relative low strain hardening is achieved.

As shown in the above examples, *in situ* texture measurement, also in combination with the numerical simulation, provides an effective tool to understand the activity of different deformation modes and anisotropic behavior as a function of initial texture and

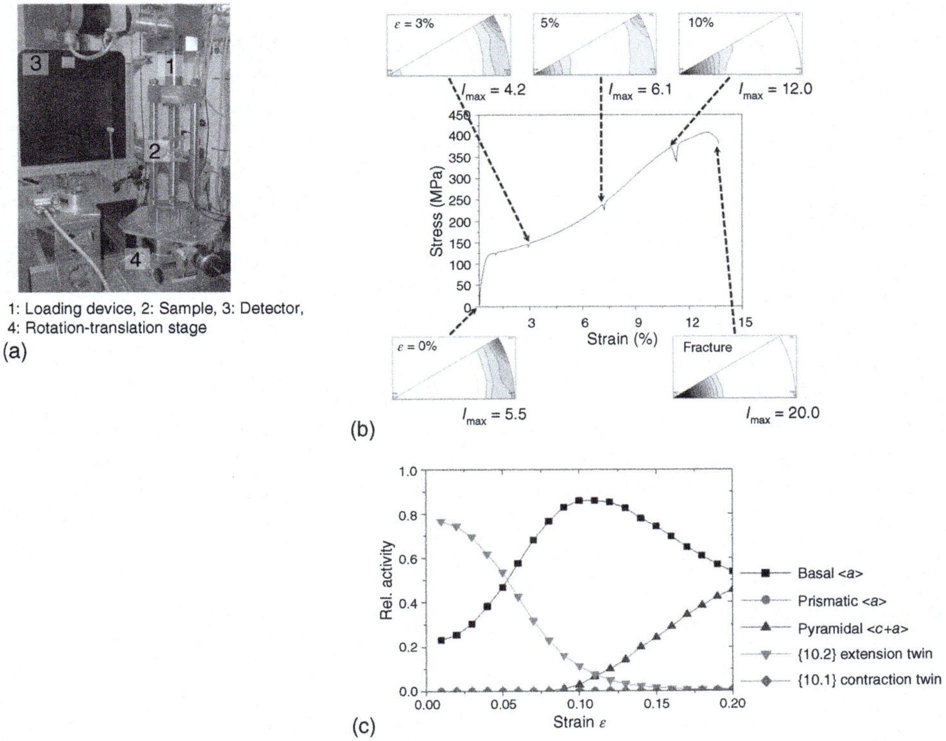

Figure 10.13 (a) UTM installed on the ω-rotation table. (b) Flow curve during compression of AZ31 extruded bar and the inverse pole figures in the loading direction at different loading stages. (c) Relative activity of each deformation mode as a function of the equivalent strain, calculated from VPSC texture simulation.

deformation degree [31]. Considering the fact that thermo-mechanical processes are mostly conducted at an elevated temperature, high-temperature *in situ* texture measurement has an important meaning to study the complex microstructural changes during high temperature deformation, such as dynamic recrystallization, deformation mechanisms, and phase transition. To perform a real time *in situ* texture measurement, it is necessary to fulfill some requirements of experimental setup, for example, a heating module providing an opening of the beam path for texture measurement, a stable and homogeneous temperature control over the sample volume, and a short measuring time to minimize the effect of static phenomena like grain growth.

10.3 Conclusion

Both neutrons and synchrotron radiation have sufficient penetration power for nondestructive analysis of condensed matter, which makes them attractive for global as well as for local texture investigations on advanced engineering materials [32]. In summary, a guideline to a more effective use of these novel techniques is obtained by understanding of the specific characteristics of each radiation source. Neutron diffraction is favored for a large sample volume and coarse-grained materials, for averaging over texture gradients for a direct correlation to macroscopic materials behavior as well as for multi-phase gradient materials. Synchrotron diffraction is favored for global textures on a small volume scale, for investigating of texture gradients, and for fast *in situ* experiments.

References

1 Bunge, H.J. (1993) *Texture Analysis in Materials Science*, Cuvillier Verlag, Göttingen.
2 MAUD http://maud.radiographema.eu/ (accessed 30 September 2016).
3 Bachmann, F., Hielscher, R., and Schaeben, H. (2010) *Solid State Phenom.*, **160**, 63–68.
4 mtex https://code.google.com/p/mtex/ (accessed 28 June 2016).
5 Matthies, S., Wenk, H.-R., and Vinel, G.W. (1988) *J. Appl. Crystallogr.*, **21**, 285–304.
6 Dahms, M. and Bunge, H.J. (1989) *J. Appl. Crystallogr.*, **22**, 439.
7 Helming, K., Schwarzer, R.A., Rauschenbach, B., Geier, S., Leiss, B., Wenk, H.-R., Ullemeier, K., and Heinitz, J. (1994) *Z. Metallkunde*, **85**, 545–553.
8 Kocks, U.F., Tome, C.N., and Wenk, H.-R. (2000) *Texture and Anisotropy*, Cambridge University Press, Cambridge.
9 Bunge, H.J. (1986) *Experimental Techniques of Texture Analysis*, DGM Informationsgesellschaft Verlag.
10 Williams, D.B. and Carter, C.B. (1996) *Transmission Electron Microscopy*, Plenum Press.
11 Zaefferer, S. (2003) *Adv. Eng. Mater.*, **5**, 745–752.
12 Engler, O. and Randle, V. (2009) *Introduction to Texture Analysis*, CRC Press.
13 Schwartz, A.J., Kumar, M., and Adams, B.L. (2000) *Electron Backscatter Diffraction in Materials Science*, Kluwer Academic.
14 Rauch, E.R. and Duft, A. (2005) *Mater. Sci. Forum*, **495–497**, 197–202.
15 Brokmeier, H.-G. (1999) *Textures Microstruct.*, **33**, 13–34.
16 Brokmeier, H.-G., Gan, W.M., Randau, C., Voeller, M., Rebelo-Kornmeier, J., and Hofmann, M. (2011) *Nucl. Instrum. Methods Phys. Res.*, **A642**, 87–92.
17 Hofmann, M., Schneider, R., Seidl, G.A., Rebelo-Kornmeier, J., Wimpory, R.C., Garbe, U., and Brokmeier, H.-G. (2006) *Physica B*, **385–386**, 1035–1037.
18 Brokmeier, H.-G., Randau, C., Tekouo, W., Hofmann, M., Gan, W., Müller, M., Schreyer, A., and Petry, W. (2010) *Mater. Sci. Forum*, **652**, 197–201.
19 Ullemeyer, K., Spalthoff, P., Heinitz, J., Isakov, N.N., Nikitin, A.N., and Weber, K. (1998) *Nucl. Instrum. Methods Phys. Res.*, **412** (**1**), 80–88.
20 Vogel, S.C., Hartig, C., Lutterotti, L., Von Dreele, R.B., Wenk, H.-R., and Williams, D.J. (2004) *Adv. X-Ray Anal.*, **47**, 431–436.
21 Lonardelli, I., Wenk, H.-R., Lutterotti, L., and Goodwin, M. (2005) *J. Synchrotron Radiat.*, **12**, 354–360.
22 Wenk, H.R. and Grigull, S. (2003) *J. Appl. Crystallogr.*, **36**, 1040–1049.
23 Bunge, H.J., Wcislak, L., Klein, H., Garbe, U., and Schneider, J.R. (2002) *Adv. Eng. Mater.*, **4**, 300–305.
24 Martins, R.V., Lienert, U., Margulies, L., and Pyzalla, A. (2005) *Mater. Sci. Eng.*, **A402**, 278–287.

25 Genzel, C., Denks, I.A., Gibmeier, J., Klaus, M., and Wagener, G. (2007) *Nucl. Instrum. Methods Phys. Res.*, **A578**, 23–33.

26 Schell, N., King, A., Beckmann, F., Fischer, T., Müller, M., and Schreyer, A. (2014) *Mater. Sci. Forum*, **772**, 57–61.

27 Brokmeier, H.-G., Weiss, B., Yi, S., Ye, W., Liss, K.D., and Lippmann, T. (2005) *Mater. Sci. Forum*, **495–497**, 131–136; *Textures Microstruct.* (1991), **14**, 567.

28 Montesin, T., Heinzmann, J.J., and Vadon, A. (1991) *Textures Microstruct.*, **14**, 567.

29 Brokmeier, H.-G., Yi, S., and Homeyer, J. (2009) in *Microstructure and Texture in Steels* (eds A. Haldar, S. Suwas, and D. Bhattacharjee), Springer, pp. 307–321.

30 Lebensohn, R.A. and Tome, C.N. (1993) *Acta Metall. Mater.*, **41**, 2611–2624.

31 Yi, S., Davies, C.H.J., Brokmeier, H.-G., Bolmaro, R.E., Kainer, K.U., and Homeyer, J. (2006) *Acta Mater.*, **54**, 549–562.

32 Brokmeier, H.-G. (2006) *Physica B*, **385–386**, 623–625.

11

Basics of Small-Angle Scattering Methods

Philipp K. Pranzas

11.1 Introduction

Small-angle scattering (SAS) is the elastic scattering by density fluctuations in a sample at angles mainly below 1° providing information about size, shape, composition, and orientation of the scattering structures. A set of SAS techniques using the different sources such as light, X-rays, and neutrons have been established with a wide range of applications. Small-angle neutron scattering (SANS) and small-angle X-ray scattering (SAXS) allow the characterization of structures with sizes between 1 and 100 nm, a size range which is particularly interesting in materials science. These methods are complemented by the ultra-small-angle scattering techniques USANS and USAXS for the investigation of larger structures up to several micrometers. Anomalous small-angle X-ray scattering (ASAXS) is used to obtain information about the distribution of a single element in the sample by measuring at different energies near the K-absorption edge of the investigated element. SAS using the grazing incidence geometry (GISANS/GISAXS) is applied for the characterization of surface structures.

In this chapter, basic SAS instrumentation and theory is briefly summarized. For further reading, it is referred to general SAS articles in the literature [1–7]. Examples for the use of neutrons and synchrotron radiation in materials science follow in Chapters 12 and 13.

11.2 Common Features of a SAS Instrument

In a SAS experiment a monochromatic, focused beam is scattered by inhomogeneities in the sample. In Figure 11.1, a schematic setup of a typical SANS instrument is shown exemplarily. A velocity selector (neutrons) or focusing mirrors (X-rays) are used as monochromators. A velocity selector consists of a rotor with absorbing lamellae. Only neutrons in a small wavelength (speed) distribution (e.g., $\Delta\lambda/\lambda = 10\%$) can pass according to the rotation speed of the rotor. The focused beam is obtained by an alignment of apertures, the collimation line (neutrons), or by a slit system (X-rays). The scattering pattern is recorded by a position-sensitive area or linear gas detector (neutrons and X-rays) or by CCD/image plate detectors (X-rays).

In this chapter, the fundamental scattering parameters that are essential to derive information about structures in the sample from the scattering signal are introduced.

Neutrons and Synchrotron Radiation in Engineering Materials Science: From Fundamentals to Applications, Second Edition. Edited by Peter Staron, Andreas Schreyer, Helmut Clemens, and Svea Mayer.

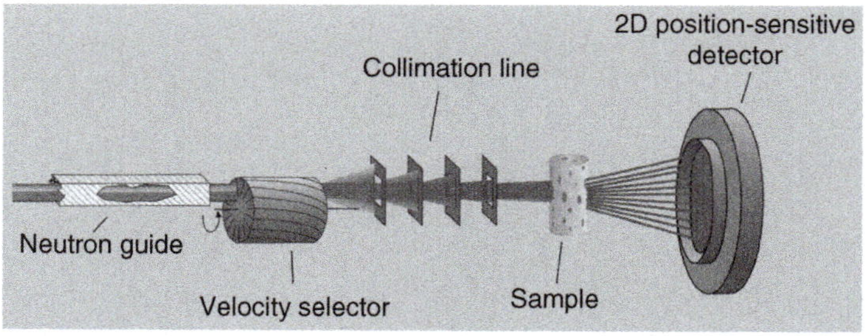

Figure 11.1 Main components of a SANS instrument.

11.3 Contrast

The key parameter that determines if neutrons or X-rays can see a certain structure is the contrast between the scattering body and the surrounding matrix. The different interactions of neutrons and X-rays with the scattering atom as well as the X-ray atomic scattering factors and the neutron scattering lengths/cross-sections were already introduced in Chapter 4.

The scattering length density η_s of a scattering body consisting of i different elements is calculated by

$$\eta_s = \sum_i b_i \cdot \frac{\delta \cdot N_A}{M_W} \tag{11.1}$$

with

δ : bulk density ($g \cdot cm^{-3}$)
N_A : Avogadro constant ($6.022 \times 10^{23}\ mol^{-1}$)
M_W : molecular weight or atomic mass of the scattering body ($g \cdot mol^{-1}$)
b : coherent scattering length (neutrons) or classical electron radius × atomic scattering factor f (X-rays)

The contrast is the scattering length density difference (SLDD) $\Delta\eta$ between the matrix and scattering body. In neutron scattering of biological samples it is common to substitute hydrogen atoms in a macromolecule by deuterium or to use the so-called contrast variation by changing the H_2O/D_2O ratio in the solution. At a certain D_2O concentration – the matching point – a part of or the whole molecule is "matched out"; it is "invisible" for the neutron because η is the same as that in the matrix.

11.4 Scattering Curve

The neutron is scattered at structures, which have a certain size and contrast to the matrix. Neutrons are scattered by structures in the sample following the reciprocal principle that larger structures lead to a scattering to smaller angles and vice versa. The scattering vector or momentum transfer $\vec{q}$ is the vector between the wave vector $\vec{k}_0$ of the incoming neutron and the wave vector $\vec{k}$ of the neutron scattered under the angle 2θ (Figure 11.2).

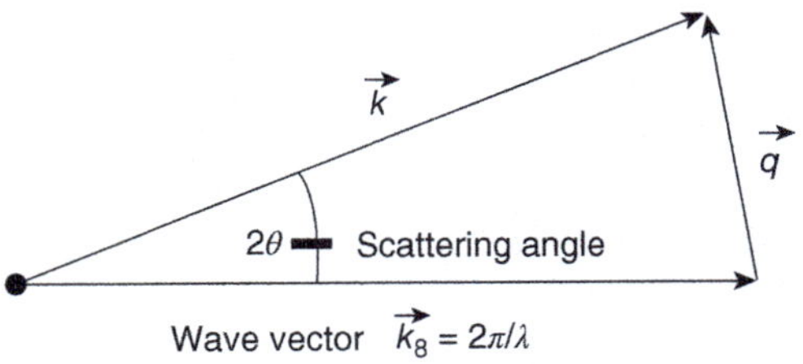

Figure 11.2 Definition of the scattering vector $\vec{q}$.

From Figure 11.2 the modulus of the scattering vector can be calculated:

$$|\vec{q}| = \frac{4\pi}{\lambda} \sin\theta \text{ (nm}^{-1} \text{ or Å}^{-1}) \quad (11.2)$$

q is commonly used because it combines the scattering angle with the wavelength λ, so that results obtained at different instruments measured at varying wavelengths can be compared directly. In the X-ray community, it is also common to use the symbols s or h, sometimes differing from q by a factor of 2π. The dimension d of the structures scattering at a certain scattering vector can be estimated by

$$d = \frac{2\pi}{q} \quad (11.3)$$

The resolution of a SAS experiment – the smallest dimension d_{min} – is, therefore, defined by q_{max}.

The three-dimensional information of the scattering pattern measured with a position-sensitive area detector (Figure 11.3a) is reduced by radial integration to two dimensions. A plot of the scattering intensity $I(q)$ over q – the scattering curve – is obtained (Figure 11.3b).

In the scattering pattern the intensity is smeared out, which results, for example, in a peak broadening, due to smearing effects such as the limited detector resolution, the width of the wavelength distribution of the monochromator, and the beam divergence

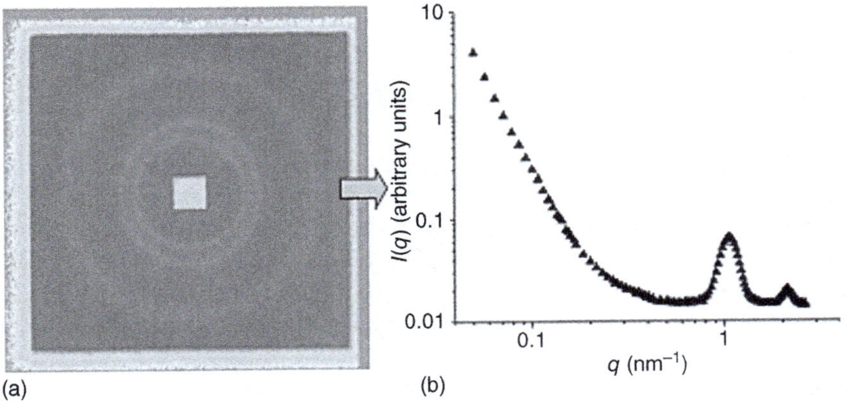

Figure 11.3 (a) Scattering pattern of silver behenate measured with SANS using a position sensitive area gas detector with 256 × 256 pixels (pixel size 2.2 mm) at a distance of 1 m from the sample. (b) Scattering curve calculated by radial integration from the scattering patterns measured at 1, 3, and 7 m from the sample.

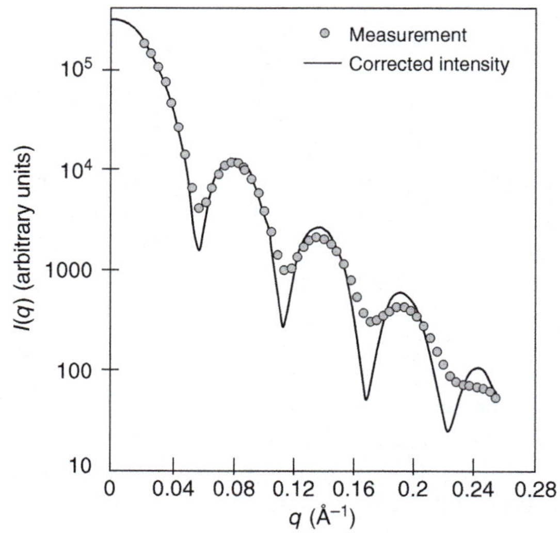

Figure 11.4 Measured and corrected scattering curve of the protein apoferritin [8].

depending on the slit/aperture (collimation) alignment and size. In Figure 11.4, this is demonstrated showing the measured and theoretically corrected ("desmeared") scattering curve of an apoferritin solution, an almost ideal monodisperse system of spheres [8]. For the comparison between measurement and theory and for the model calculation of size distributions the measured curve has to be desmeared or the theoretical curve has to be smeared out. Specific examples for scattering curves will be presented in Chapter 12.

Before analyzing the coherent scattering of a SANS curve, the background from the incoherent scattering, which is constant (isotropic) in the whole solid angle (4π), has to be subtracted first. Some elements such as hydrogen have a high incoherent scattering length when compared to the coherent scattering. A high content of such an element may lead to a bad signal/noise ratio. In SAXS, the isotropic background (e.g., fluorescence) is much lower.

11.5 Power Law/Scattering by Fractal Systems

A first characterization of complex systems is possible by analyzing the slope(s) of the scattering curve, known as *power law* behavior or scattering by fractal systems [9]. If the decay of the scattering intensity follows a q^{-D} behavior, with the fractal dimension D, an elongated (rod-like) structure in the sample can be assumed for $D = 1$ and a flattened (disk-like) structure for $D = 2$. The Porod scattering of sharp interfaces leads to $D = 4$ ("q^{-4} decay"). A schematic scattering curve of an elongated structure is shown in Figure 11.5.

In general, a smaller fractal dimension points to a more open structure. A value of D between 1 and 3 indicates the existence of a mass fractal – a fractal system of small units building larger aggregates with a self-similar structure over a large size range. Surface fractals – fractal systems with a dense core and with fractal structures on the surface – usually show a D value between 3 and 4.

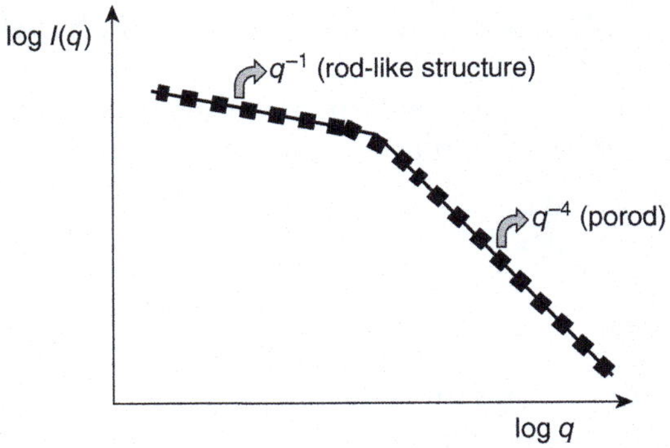

Figure 11.5 An ideal scattering curve of a rod-like structure.

11.6 Guinier and Porod Approximations

The Guinier equation (11.4) describes the intensity as a function of q [10]. It is valid only in the region of small q values ($q \cdot R_g < 1$).

$$I(q) \approx I(0)\,e^{-(1/3)q^2R_g^2} \tag{11.4}$$

$I(0)$ is the intensity extrapolated to $q = 0$. It is used, for example, for the determination of the molecular weight of macromolecules in diluted solutions. The radius of gyration R_g is in analogy to mechanics the "weight average" of all radii present in the sample. Depending on the model used for describing the structures in the sample, the average radius of the model structures can be calculated from R_g. The formulas are listed in the literature [3, 6], for example, the radius of spheres: $R^2 = 5/3R_g{}^2$. In a linear fit of the Guinier region in the so-called Guinier plot in which the logarithm of $I(q)$ is plotted versus q^2 (Figure 11.6a), the two parameters R_g and $I(0)$ can be determined from slope and ordinate. A non-existing linear range indicates the presence of very large structures, which scatter at low q values, perhaps outside the accessible q range, and the Guinier approximation should not be applied.

In the Porod plot $q^4 \cdot I(q)$ is plotted versus q^4 (Figure 11.6b) [11]. The linear fit $I(q) = C + A/q^4$ at large q values is convenient to determine the incoherent background (constant C).

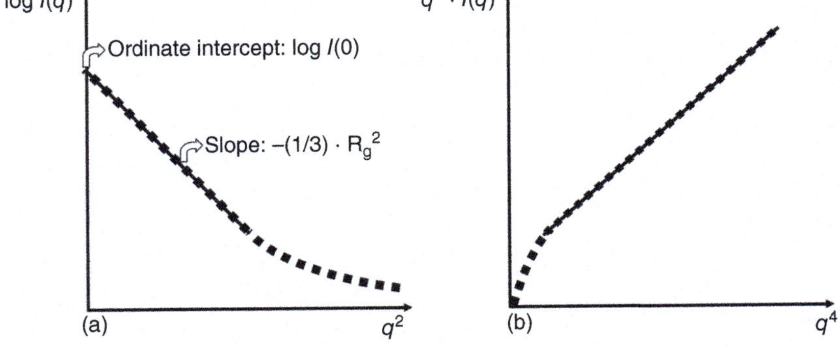

Figure 11.6 Schematic scattering curves: (a) Guinier plot and (b) Porod plot.

11.7 Macroscopic Differential Scattering Cross-section

For the calibration from the relative intensity to “absolute units,” the macroscopic differential scattering cross-section $\frac{d\Sigma}{d\Omega}$ is calculated as follows [4]:

$$\frac{d\Sigma}{d\Omega}(q) = \frac{I(q)}{M \cdot \left(\frac{I_0}{M}\right) \cdot D \cdot T \cdot \varepsilon \cdot \Delta\Omega} \quad (\text{cm}^{-1}) \tag{11.5}$$

with

$I(q)$	:	measured intensity
M	:	monitor counts
I_0	:	intensity of primary beam
I_0/M	:	primary beam/monitor ratio (calibrated by a standard with known scattering cross-section for example vanadium or H_2O)
D	:	sample thickness
T	:	transmission of sample ($= e^{-\Sigma D}$)
ε	:	detector efficiency/response
$\Delta\Omega$	:	solid angle element ($=dx \cdot dy/L^2$ with sample–detector distance L)

From the macroscopic differential scattering cross-section the form factor (size and shape) and the structure factor (degree of local order) of the scattering structures can be calculated using Eq. (11.6).

$$\frac{\partial\Sigma}{\partial\Omega}(q) = N_P \cdot V_P^2 \cdot (\Delta\eta)^2 \cdot P(q) \cdot S(q) + B_{inc} \tag{11.6}$$

with

N_P	:	number density of scattering bodies
V_P	:	volume of one scattering body
$\Delta\eta$	:	contrast or scattering length density difference
$P(q)$	:	form (shape) factor (also $F(q)$)
$S(q)$	:	interparticle structure factor, dependent on degree of local order, interaction between scattering bodies, for N_P approaching 0, $S(Q)$ becomes 1 (diluted system)
B_{inc}	:	incoherent background (isotropic)

11.8 Model Calculation of Size Distributions

The information about both, the size and shape of the scattering structures cannot be derived at the same time from the scattering curve. A first impression about sizes present in the sample can be derived by Fourier transformation of the scattering curve. The distance distribution function is obtained, showing the frequency distribution of distances between scattering centers. This distribution can be illustrated by all possible cuts through the scattering structures. The distance distribution is model independent and does not include shape information; therefore, a careful interpretation is obligatory.

For the calculation of the size distribution including the shape of the scattering structures, a model is chosen using preliminary information and additional results from other methods, for example, electron microscopy. With this model, the theoretical scattering curve is calculated and (least-squares) fitted to the measured curve, taking into account polydispersity and instrumental smearing. Commonly used model shapes of the scattering structures are spheres, ellipsoids, shells (hollow spheres), elliptic/hollow cylinders, or flattened/disk-like structures. The form factors of these simple bodies are listed in the literature [4, 6]. The form factor of spheres with the radius R is shown in Eq. (11.7):

$$P(q,R) = \left(3\frac{\sin(qR) - qR\cos(qR)}{(qR)^3}\right)^2 \tag{11.7}$$

In real systems, even a "monodisperse" distribution of structures has a certain width. For a model calculation including the determination of the size distribution Eq. (11.6) changes to

$$\frac{\mathrm{d}\Sigma}{\mathrm{d}\Omega}(q) = (\Delta\eta)^2 \cdot \int_0^{\infty} n(R) \cdot V(R)^2 \cdot P(q,R) \cdot S(q, R_{\mathrm{HS}}, f_{\mathrm{HS}}) dR \tag{11.8}$$

with

$n(R)dR$	:	number density of particles with sizes between R and $R + dR$
$S(q, \mathrm{R}_{\mathrm{HS}}, f_{\mathrm{HS}})$	:	structure factor describing the interparticle interference effect
R_{HS}	:	hard-sphere radius of a particle with radius R
f_{HS}	:	volume fraction of hard spheres

Normal or lognormal distributions or a certain number of B-splines [11] are used as distribution functions. A lognormal distribution with the width σ of a number of spheres n_0, with the position of the maximum R_0, is expressed by

$$n(R) = \frac{n_0}{\sqrt{2\pi}\sigma R_0} \exp\left(-\frac{(\ln(R/R_0))^2}{2\sigma^2}\right) \tag{11.9}$$

11.9 Magnetic Structures

For the characterization of magnetic structures, SANS measurements with applied magnetic field can be performed. Besides the nuclear scattering, additional magnetic scattering of the magnetic scattering lengths in the sample arises, which follows a $\sin^2\alpha$ behavior, where α is the angle between the scattering vector $\vec{q}$ and the direction of the magnetic field. The resulting scattering pattern is anisotropic with only the nuclear scattering parallel ($\alpha = 0°$) and the nuclear plus magnetic scattering perpendicular ($\alpha = 90°$) to the magnetic field (Figure 11.7). To obtain information about nuclear and magnetic scattering, a $\sin^2\alpha$ analysis is performed for the calculation of the scattering curves or horizontal and vertical sectors with a width of about 10° are used.

Polarized neutrons are ideal for the investigation of magnetic structures, because the magnetic moment due to the spin of the neutron is used. Only neutrons with a spin direction parallel to the magnetic field are chosen by a polarization device, for example, a super mirror with polarization ≥95%. These so-called polarized neutrons interact with the magnetic structures in the sample, while interference between nuclear and magnetic scattering takes place. With a spin flipper, for example a radio frequency flipper,

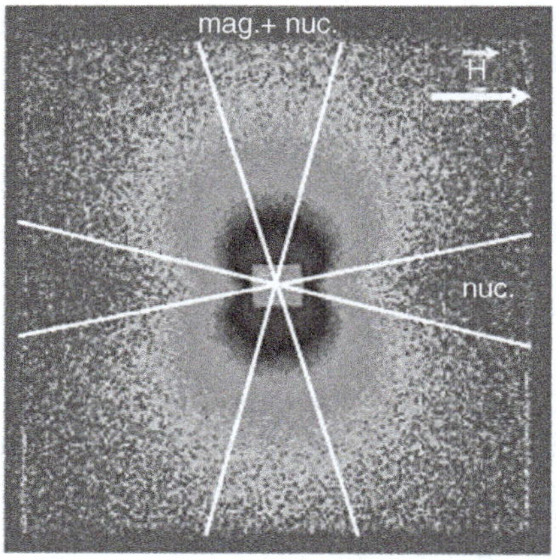

Figure 11.7 Magnetic scattering pattern of a Fe-1wt.% Cu model alloy.

the inverse (antiparallel) polarization state can also be realized. Using the scattering difference of these two states, the magnetic scattering is determined with high precision and without any background.

Synchrotron radiation can also be used to study magnetism in solids, utilizing the large resonant enhancement of the magnetic cross-section at the absorption edges of the metals in the sample [12]. The advantage of this method is its element specificity, which enables one to distinguish between the magnetic contributions of different elements. However, for many of the magnetic elements, the specific absorption edges that give rise to a large enhancement in the X-ray magnetic cross-section are in the low energy regime and, therefore, limit the applicable sample thickness.

References

1 Gunier, A. and Fournet, G. (1955) *Small Angle Scattering of X-rays*, John Wiley & Sons, Inc., New York.

2 Glatter, O. and Kratky, O. (eds) (1982) *Small Angle X-ray Scattering*, Academic Press, New York.

3 Feigin, L.A. and Svergun, D.I. (1987) in *Structure Analysis by Small-Angle X-ray and Neutron Scattering* (ed. G.W. Taylor), Plenum Press, New York, https://www.embl-hamburg.de/biosaxs/reprints/feigin_svergun_1987.pdf (accessed October 2016).

4 Kostorz, G. (1982) Small-angle scattering and its applications to materials science, in *Treatise on Materials Science and Technology*, vol. **15** (ed. G. Kostorz), Academic Press, New York, p. 227.

5 Lindner, P. and Zemb, T. (eds) (1991) *Neutron, X-ray and Light Scattering: Introduction to an Investigative Tool for Colloidal and Polymeric Systems*, North-Holland, Amsterdam.

6 May, R.P. (2002) Small-angle scattering, in *Neutron Data Booklet* (eds A.-J. Dianoux and G. Lander), ILL, p. 2.1-1, ftp://ftp.ill.fr//pub/cs/workshops/mdanse/data/NeutronDataBooklet.pdf (accessed October 2016), ILL Neutron Data Booklet (pdf, 9 Mb).

7 King, S.M. (1999) Small angle neutron scattering, in *Modern Techniques for Polymer Characterisation* (eds R.A. Pethrick and J.V. Dawkins), John Wiley & Sons, Inc., New York, http://www.isis.stfc.ac.uk/groups/large-scale-structures/publications/Introduction-to-SANS14599.pdf (accessed October 2016).
8 Zhao, J. (1995) *Bestimmung der in situ Struktur der Proteine L3 und L4 in der ribosomalen 50S Untereinheit mit Hilfe der Kernspinkontrastvariation*, Dissertation, Department of Physics, University of Hamburg, p. 31.
9 Teixeira, J. (1988) *J. Appl. Crystallogr.*, **21**, 781.
10 Guinier, A. (1939) *Ann. Phys.*, **12**, 161.
11 Porod, G. (1951) *Kolloid Z.*, **124**, 83.
12 Hannon, J.P., Trammell, G.T., Blume, M., and Gibbs, D. (1988) *Phys. Rev. Lett.*, **61**, 1245.

12

Small-Angle Neutron Scattering

Philipp K. Pranzas and André Heinemann

12.1 Introduction

Small-angle neutron scattering (SANS) is the ideal tool for a large number of investigations in the field of materials science, such as the characterization of precipitates, clusters, interfaces, grain sizes, and (magnetic) structures in materials such as metal alloys, ceramics, and polymers. SANS is also commonly used in molecular biology and soft matter science to obtain structural information about, for example, macromolecules, colloids, and membranes.

Due to the special properties of neutrons, described in Chapter 4, SANS is particularly suitable for the analysis of magnetic structures and of samples with light elements, especially after labeling with deuterium or using the H/D-contrast variation method. With SANS, thick samples of about 1 mm up to several centimeters – depending on the material – can be investigated. Therefore, measurements at low and high temperatures in a cryostat or furnace, or in a special vessel like a pressure cell or a shear apparatus, are possible. Scattering at large (backward) angles can also be measured with neutrons due to the angle-independent nuclear scattering. Furthermore, SANS is the method of choice for samples that are sensitive to radiation damage, for example, from X-rays.

In Figure 12.1, the SANS-1 instrument at the research reactor FRM II of the Heinz Maier-Leibnitz-Zentrum (MLZ) in Garching is shown schematically in its final stage of expansion [1]. One velocity selector out of two (with high (5%) or low (10%) $\Delta\lambda/\lambda$ resolution) is selectable to produce monochromatic neutrons. In the time of flight mode [2], where the full spectrum is used, a combination of a neutron guide and primary chopper is used to feed the so-called collimation with a series of neutron pulses. To perform time-resolved measurements with monochromatic neutrons, a special Time-Involved Small-Angle Neutron Experiment (TISANE) chopper can be moved into the beam [3, 4]. The divergence of the beam is adjusted in the collimation line. Collimators are an alignment of apertures at a certain distance from each other. As a rule of thumb, a collimation length similar to the distance between the sample and the detector should be chosen. Tunable apertures allow the adjustment of divergence and discrimination of background scattering. The position-sensitive area detector is moving in the evacuated scattering tube with typical sample–detector distances between 1 and 20 m. A second detector is useful for parallel measurements of different q-ranges or for high-resolution measurements at very small angles.

Neutrons and Synchrotron Radiation in Engineering Materials Science: From Fundamentals to Applications, Second Edition. Edited by Peter Staron, Andreas Schreyer, Helmut Clemens, and Svea Mayer.

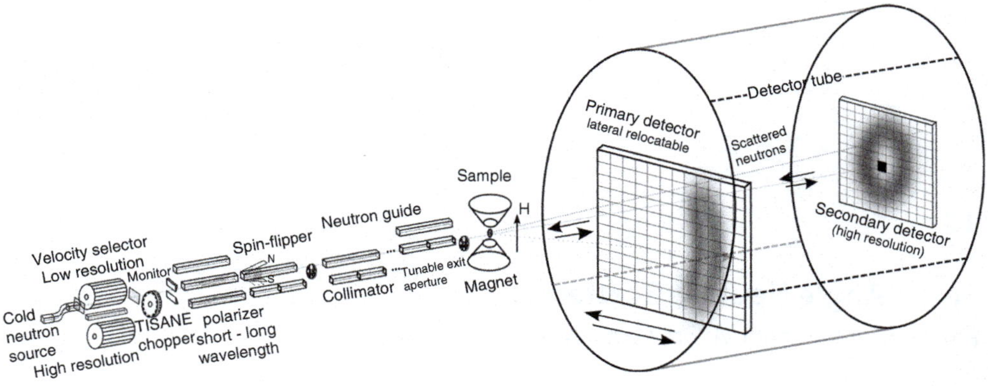

Figure 12.1 Special features of a SANS instrument (SANS-1@FRM II/MLZ) [1].

For the use of polarized neutrons, two V-shaped polarizers (e.g., V-shaped transmission polarizers optimized for different wavelength ranges) and a spin flipper (e.g., radio frequency flipper) are necessary to obtain polarized neutrons with a spin parallel to the direction of the magnetic field and to flip the neutron spin to the direction antiparallel to the field. A large variety of instruments, such as magnets, refrigerator cryostats, and furnaces, are commonly used at the sample position as well as linear translation, rotary, tilting, and lift stages for the adjustment of the sample. The typical range of the wave vector $\mathbf{q}$ is between 0.001 and about 10 nm^{-1}, corresponding to an accessible size range of about 1–1000 nm. In the following text, four different examples of SANS applications are presented to demonstrate the various possibilities of this method in the fields of engineering materials science and soft matter.

12.2 Nanocrystalline Magnesium Hydride for the Reversible Storage of Hydrogen

Hydrogen is one of the most interesting renewable energy sources regarding storage, transport, and conversion of energy in the future [5] (see also Section 19.3.2.4). Light metal hydrides offer a safe alternative for the storage of hydrogen in compressed or liquid form. Prototypes of cars, motor bikes, and boats using metal hydrides already exist. The usage of metal hydrides in notebooks and video cameras is considered as well. Magnesium hydride is interesting for industrial use because it has high hydrogen storage capacity of up to 7.6 wt.%. Absorption and desorption kinetics of hydrogen on MgH_2 as well as the hydride's cycling behavior and thermal stability are well known. High-energy ball milling is used to obtain a nanocrystalline material – nanoparticles consisting of crystallites separated by grain boundaries – with fast absorption/desorption kinetics. SANS in combination with USANS on a double crystal diffractometer (DCD) is the ideal tool to investigate the influence of microstructure, for example, crystallite and particle sizes, on the sorption properties and to characterize structural changes due to absorption and desorption of hydrogen over a very wide range of particle sizes [6, 7].

Model MgH_x samples with varying hydrogen content were prepared at 300 °C in a hydrogen titration apparatus after 20 h of high-energy ball milling in several sorption and desorption cycles [6]. The samples were measured with SANS and DCD in quartz cuvettes with a thickness of 1 mm. SANS measurements were performed at the instrument SANS-2 at the Geesthacht Neutron Facility (GeNF[1]) using distances between the sample and the detector of 1, 3, 9 m ($\lambda = 0.58$ nm) and 21 m ($\lambda = 0.58$ and 1.16 nm, $\Delta\lambda/\lambda = 0.1$). These five measurements covered the range of scattering vector $\mathbf{q}$ between 10^{-2} and 3 nm^{-1}, offering sufficiently large overlaps between the particular angular ranges. The SANS curves were obtained by radial integration of the two-dimensional scattering patterns (see also Chapter 11). Scattering data were normalized by monitor counts, corrected for sample transmission and detector response. The calibration to the absolute scale (macroscopic differential scattering cross section over q) was obtained at this time by an additional vanadium measurement. The scattering cross-section of vanadium – predominantly incoherent scattering in the whole room angle – is known. USANS experiments were carried out at the DCD instrument at GeNF [8] with a wavelength of 0.443 nm resulting in an accessible q-range from 10^{-4} to 10^{-2} nm^{-1}. The scattering curves in absolute values were obtained after de-smearing and correction of multiple scattering according to the model calculation procedure described in [9]. In Figure 12.2, the combined USANS–SANS scattering curves of several MgH_x samples are shown. The error bars of the scattering cross-section in DCD and SANS curves are in the size range of the used symbols with a mean error of about 5–10%. In DCD evaluation, the error is obtained by varying the fit parameters of the model calculation [9]. The error bars of the SANS curves are obtained from the intensity variation of the 2D scattering pattern during radial integration.

A first characterization of the scattering curves starts at large q values: the hydrogen content of the samples can be calculated from the different heights of the curves caused by the incoherent scattering of hydrogen. As the next step, the power law behavior of the scattering curves is analyzed. Two regions with a $q^{-2.5}$ decay are marked in Figure 12.2 at small and large q, separated by a Porod q^{-4} region. This indicates the presence of a bimodal particle size distribution. The two $q^{-2.5}$ regimes show the mass fractal behavior of the nanocrystalline sample.

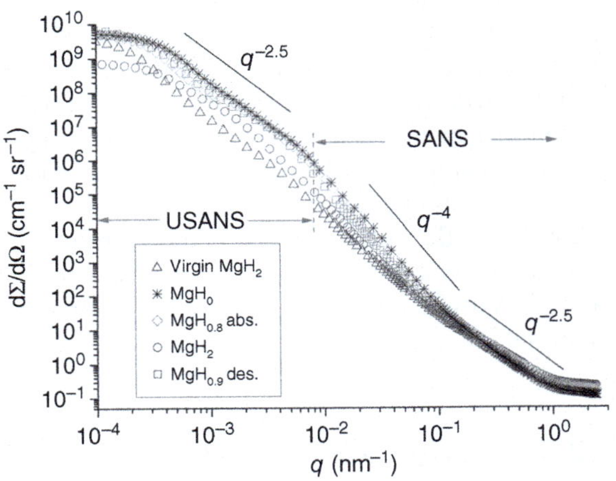

Figure 12.2 Combined USANS–SANS scattering curves of nanocrystalline MgH_x samples with different hydrogen content and cycling history. (Adapted from Pranzas 2006 [6]. Reproduced with permission of Elsevier.)

1 GeNF with the research reactor FRG-1 was operated until 2010 and subsequently replaced by the German Engineering Materials Science Centre (GEMS; http://gems.hzg.de) with two outstations at DESY and MLZ.

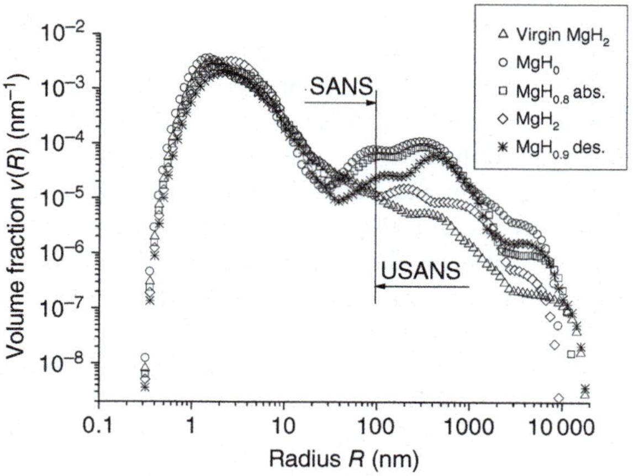

Figure 12.3 Volume fraction distributions of MgH_x samples calculated from the scattering curves shown in Figure 12.2. (Adapted from Pranzas 2006 [6]. Reproduced with permission of Elsevier.)

As the first step towards a quantitative structural characterization, size distributions were calculated from the SANS/USANS curves using as first approximation a model of hard spheres in a homogenous matrix (two-phase model) and constant scattering length density (SLD) differences $\Delta\eta_s$ between MgH_2 and Mg of 3×10^{10} cm^{-2}. The obtained volume fraction distributions are shown in Figure 12.3.

The small differences between the curves up to a radius of about 100 nm correspond to structural changes in the size range of the crystallites inside the MgH_x particles. The peak at 2–3 nm broadens and shifts to larger radii with proceeding cycling, which indicates a coarsening of the crystallites during cycling at 300 °C, which is in agreement with literature. The similar heights of these peaks indicate that there is a constant contrast of the crystallite structures within the particles independent of the hydrogen content. The volume fractions with diameters larger than about 400 nm characterize the particle size of the large MgH_x particles. In this size range, the contrast between particle and the surrounding atmosphere is of course expected to change with the hydrogen content. The comparison of the volume fractions of the samples with similar hydrogen concentrations like virgin MgH_2 and MgH_2 after one hydrogenation cycle indicates that particles of radii larger than 10 μm break-up and smaller particles form during the cycling process.

The results demonstrate the potential of SANS/USANS to characterize structures in metal hydrides over a large size range. For a further evaluation, additional information from other methods, such as electron microscopy or X-ray diffraction, is necessary to use more complex models, for example, fractal models. This is a good basis to study the influence of milling parameters, such as milling time and milling tool materials (steel or ceramic), on the microstructure of the hydrides. This structural characterization leads to a better understanding of sorption mechanisms in nanocrystalline light metal hydrides and gives useful information for the development of this new type of hydrogen storage material.

12.3 Precipitates in Steel

An excellent example for using SANS in engineering materials science is the investigation of the precipitation behavior in maraging steels [10]. Ni-containing maraging steels belong to a class of high-strength steels with very low carbon content. The strength and toughness of maraging steels make them a preferred choice for use in many technically ambitious areas, for example, for tools, medical instruments, light weight constructions, or military applications. The mechanical properties are mainly controlled by intermetallic precipitates in the nanometer size range dispersed in a rather soft martensitic matrix. For a further understanding of the influence of composition and heat treatment on the microstructures, TEM and SANS studies were carried out. The samples were solution heat treated for 1 h at 1273 K, air cooled, and subsequently aged for 0.25, 12, and 100 h at 748 K and for 16 h at 798 K, respectively [10].

A typical two-dimensional scattering pattern of a steel sample saturated in a vertical magnetic field (e.g., in the setup in Figure 12.1) of about 2 T is shown in Figure 12.4a. The magnetic scattering cross-sections calculated from sectors of the scattering patterns (the scattering cross-section of the vertical sector is subtracted from the cross-section of the horizontal sector, see Section 11.9) are presented in

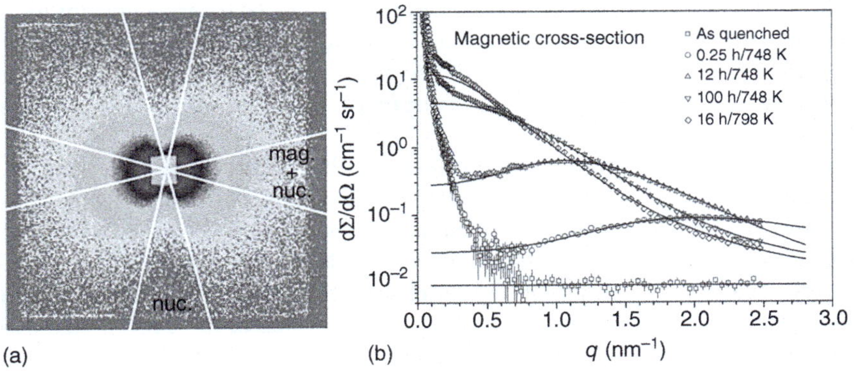

Figure 12.4 (a) Two-dimensional scattering pattern of a steel sample saturated in a vertical magnetic field of about 2 T and (b) magnetic scattering cross-sections calculated from sectors (horizontal–vertical) of the scattering patterns. The solid lines are fits obtained for the calculation of the size distributions shown in Figure 12.5. (Staron 2003 [10]. Reproduced with permission of International Union of Crystallography.)

Figure 12.5 Size distributions $\varphi(R) = n(R)V(R)$ of spherical precipitates with diameter D obtained from the magnetic SANS cross-sections shown in Figure 12.4. (Staron 2003 [10]. Reproduced with permission of International Union of Crystallography.)

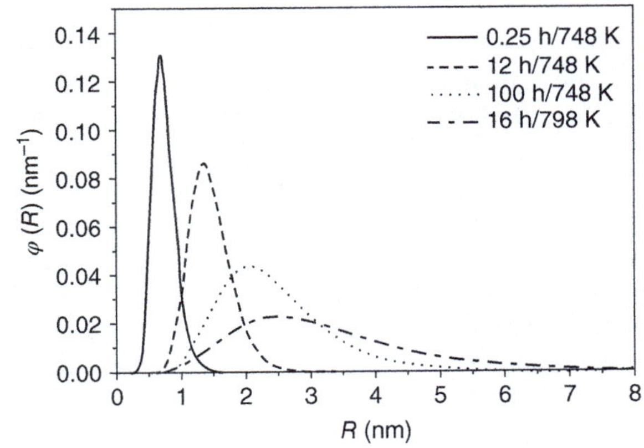

Figure 12.4b together with the fits (solid lines) obtained during the calculation of the size distributions (Figure 12.5). The SANS measurements were performed using a wavelength of 0.57 nm ($\Delta\lambda/\lambda = 0.1$) and again four detector distances. The error in the absolute scale of the calculated volume fractions obtained by the calibration with vanadium was approximately 5–10% [10].

The magnetic cross-section (Figure 12.4b) originates from precipitates that were formed during the aging process. The precipitates are assumed to be nonmagnetic and can, therefore, be treated as "magnetic holes" in the ferromagnetic matrix with saturated magnetization in the magnetic field of 2 T. As the magnetic SLD of the precipitates is assumed to be zero, the magnetic scattering length density difference (SLDD or "contrast") of the precipitates in the matrix is equal to the SLD of the matrix. Using this assumption the volume fraction of the precipitates can be calculated from the scattering curves.

A maximum is observed for the two samples aged at 748 K for 0.25 and 12 h. This peak originates from the interference of a large number of precipitates distributed in the matrix with short distances between each other. This interparticle interference effect is accounted for by introducing the hard sphere factor C_{HS}. This factor is a constant used to calculate the hard-sphere radius of a particle $R_{HS} = C_{HS} \cdot R$ with a radius R including a depleted zone around the particle [11]. For the aging time of 100 h at 748 K, larger but fewer precipitates were found with longer distances, which means less interference (smaller C_{HS} value).

In this work, a special evaluation procedure was used: as a direct calculation of the size distribution from the scattering curve could lead to several mathematical solutions providing fits with a similar quality, the shape and sizes of the precipitates were first determined by the analysis of TEM images. Afterward the size distribution was calculated using these parameters obtaining the number of the precipitates with a certain size. Following this evaluation procedure, the local information of the TEM image was combined with the statistics of the complete SANS sample volume. In Figure 12.5, the obtained volume fraction distributions of spherical precipitates with diameter D are shown.

The size distributions show the growth of the precipitates with increasing aging time. Precipitates with mean radii between 0.74 and 2.43 nm were formed at a temperature of 748 K. Number densities and volume fractions were determined, which indicate fast precipitation kinetics. The heat treatment that produces the maximum hardness in this alloy is 12 h/748 K.

The ratio A of magnetic to nuclear scattering intensity, which depends on the chemical composition of the precipitates, can be calculated from the measuring curves and compared with theoretical values [12]. Following this procedure, an intermetallic G-phase ($Ti_6Si_7Ni_{16}$) was concluded for the precipitates.

As SANS provides information about size distribution and composition of precipitates, it is the ideal method for the characterization of steels. Besides maraging steels other steel classes were analyzed by SANS as well [13–15]. For the investigation of magnetic structures, the use of polarized neutrons is often advantageous because the determination of the magnetic cross section can be performed with high precision and without any background [16]. For additional information the reader is referred to Chapter 21 and the references cited therein.

12.4 SiO_2 Nanoparticles in a Polymer Matrix – An Industrial Application

SANS is often used for the determination of size and volume fraction of components in composite materials. A typical example is the characterization of SiO_2 nanoparticles in aqueous or organic solution. SANS is an outstanding method for such investigations as large sample volumes can be measured, which leads to a good statistics, the contrast between the SiO_2 particles and the matrix is high, and – last but not least – no sample preparation is necessary and, therefore, no artifacts are formed.

Using an innovative, modified sol–gel process, Evonik Hanse GmbH (former hanse chemie AG), location Geesthacht, Germany, manufactures spherical, monodisperse, and nonagglomerated silica nanoparticles that are used for coatings or varnishes and as additives for a wide variety of other applications, for example, in adhesives. The small particle size and the narrow width of the size distribution and thereby the degree of agglomeration have significant influence on the properties of this new class of materials. The important step in the production process is the transfer of the monodisperse SiO_2 nanoparticles prepared from aqueous sodium silicate into a polymer or epoxy resin matrix.

Measurements at the SANS-2 instrument at GeNF were performed using again a wavelength of 0.58 nm and four detector distances. The evaluation of the resulting SANS curves showed that a very narrow particle size distribution of silica nanoparticles with sizes between 5 and 30 nm was obtained by the manufacturing process. It could easily be verified that the size distribution of the SiO_2 particles does not change significantly during the transfer from aqueous to polymer matrix as shown in Figure 12.6.

Interacting hard spheres were chosen as model system. The relatively sharp maximum in the SANS curve of the SiO_2 nanoparticle sample in aqueous solution originates from interparticle interference of the large number of particles in the concentrated solution of 40%. Therefore, the hard sphere constant factor C_{HS} [11] was used again for the calculation of the size distributions. The obtained distributions of the two samples in Figure 12.6 are almost identical. This is proof that the nanoparticles do not agglomerate during the transfer from aqueous to polymer matrix. The results demonstrate that with SANS size distributions of SiO_2 nanoparticles in concentrated and diluted solutions as well as in aqueous and polymer media can be easily determined. On the basis of these results, Evonik Hanse GmbH/hanse chemie AG applied for a patent for this special production process.

12.5 Green Surfactants

Glycolipids are amphiphilic molecules consisting of saccharide residues linked to an alkyl chain. They are applied in biology as biosurfactants and in environmental technology in wastewater treatment [17]. They are less environmentally damaging than synthetic surfactants and can be produced from renewable sources; therefore, they are usually referred to as *natural* or *green* surfactants. To study the phase and aggregation behavior of glycolipids, especially the formation of micelles, SANS and SAXS investigations were performed [18]. The SANS experiments were carried out at the SANS-1

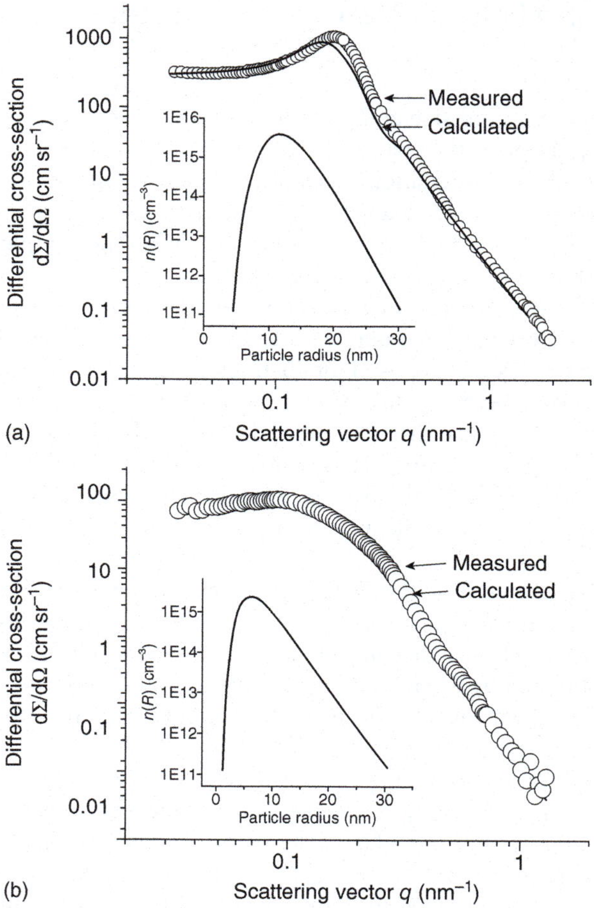

Figure 12.6 SANS curves and corresponding particle size distributions of (a) 40% SiO_2 nanoparticles in water and (b) 5% SiO_2 nanoparticles in cycloaliphatic bisepoxide. The solid lines are the obtained fits during the calculation of the size distributions.

instrument at GeNF. A wavelength of 0.85 nm ($\Delta\lambda/\lambda = 0.1$) and four sample-to-detector distances (from 0.7 to 7 m) were used to cover the q-range from 0.07 to 2.5 nm^{-1}. As an example surfactant, *N*-octyl-*b*-maltopyranoside (OM) was used. SANS and SAXS curves of OM are shown in Figure 12.7a, and the shell model and SLD profiles for neutrons and X-rays of bulk solvent, hydrophilic shell, and micelle core in Figure 12.7b.

The SANS and SAXS curves are obviously different due to the different SLDDs. The scattering-length density ρ (or η) is calculated by $\rho = \sum_i^n b_i/v$, where v is the volume of the head or tail group and b_i is the scattering length of the atoms. For neutron scattering in D_2O, seven H atoms of the OH groups in the head group were assumed to exchange with D atoms in the solvent.

The monodisperse spherical shell model shown in Figure 12.7b (hydration water in the shell of the micelle is not considered in this figure), which describes SANS and SAXS data, was applied to determine the radii of core and whole micelle as well as the contrast ratio between the core and hydrophilic shell. The resulting fits are the solid lines in Figure 12.7a. Using this model, it was shown that the micelle size is constant over a large range of OM concentrations and it is not sensitive to temperatures from 10 to 50 °C [18].

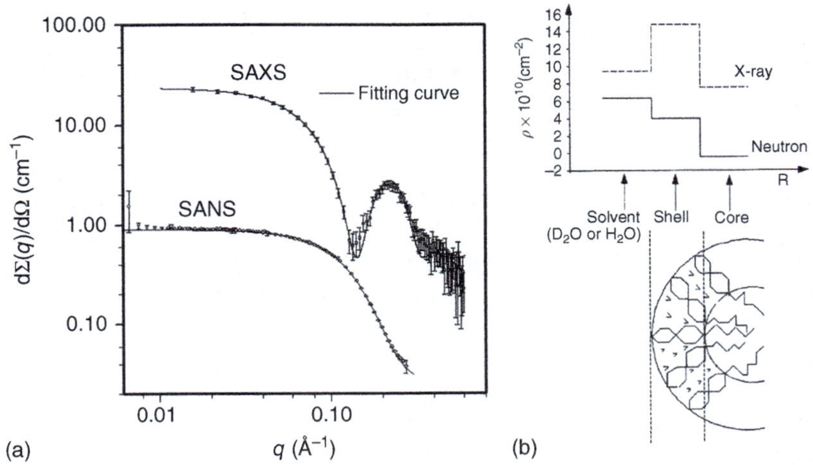

Figure 12.7 (a) SANS and SAXS curves of OM measured in D_2O. (b) Scattering length densities for neutrons and X-rays of bulk solvent, hydrophilic shell and micelle core, and shell model. (He 2002 [18]. Reproduced with permission of American Chemical Society.)

This example from the field of soft matter science shows the complementary information, which can be obtained by SANS and SAXS. The micellar structure was solved in this case using the combination of both methods. Additional results from other methods are often useful for completing the information obtained by SANS.

Acknowledgments

Valuable contributions of our colleagues Regine Willumeit-Römer, Helmut Eckerlebe, Peter Staron, and Vasyl Haramus are gratefully acknowledged.

References

1 Gilles, R., Ostermann, A., and Petry, W. (2007) Monte Carlo simulations of the new small-angle neutron scattering instrument SANS-1 at the Heinz Maier-Leibnitz Forschungsneutronenquelle. *J. Appl. Crystallogr.*, **40**, s428–s432, http://www.mlz-garching.de/sans-1 (accessed October 2016).

2 Dewhurst, C.D. (2008) D33—a third small-angle neutron scattering instrument at the Institut Laue Langevin. *Meas. Sci. Technol.*, **19**, 034007.

3 Gähler, R. and Golub, R. (1984) Time-dependent neutron optics - Quantum-mechanical effects on beam chopping and a new type of high-resolution neutron spectrometer (FOTOF). *Z. Phys. B*, **56**, 5.

4 Wiedenmann, A., Keiderling, U., Habicht, K., Russina, M., and Gähler, R. (2006) Dynamics of field-induced ordering in magnetic colloids studied by new time-resolved small-angle neutron-scattering techniques. *Phys. Rev. Lett.*, **97**, 057202.

5 Schlapbach, L. and Zuettel, A. (2001) Hydrogen-storage materials for mobile applications. *Nature*, **414**, 353–358.

6 Pranzas, P.K., Dornheim, M., Bellmann, D., Aguey-Zinsou, K.-F., Klassen, T., and Schreyer, A. (2006) SANS/USANS investigations of nanocrystalline MgH_2 for reversible storage of hydrogen. *Physica B*, **385-386**, 630–632.
7 Pranzas, P.K., Bösenberg, U., Karimi, F., Münning, M., Metz, O., Bonatto Minella, C., Schmitz, H.-W., Beckmann, F., Vainio, U., Zajac, D., Welter, E., Jensen, T.R., Cerenius, Y., Bormann, R., Klassen, T., Dornheim, M., and Schreyer, A. (2011) Characterization of hydrogen storage materials and systems with photons and neutrons. *Adv. Eng. Mater.*, **13**, 730–736.
8 Bellmann, D., Staron, P., and Becker, P. (2000) Performance of a double crystal diffractometer with different channel-cut perfect Si crystals. *Physica B*, **276**, 124–125.
9 Staron, P. and Bellmann, D. (2002) Analysis of neutron double-crystal diffractometer scattering curves including multiple scattering. *J. Appl. Crystallogr.*, **35**, 75–81.
10 Staron, P., Jamnig, B., Leitner, H., Ebner, R., and Clemens, H. (2003) Small-angle neutron scattering analysis of the precipitation behaviour in a maraging steel. *J. Appl. Crystallogr.*, **36**, 415–419.
11 Kinning, D.J. and Thomas, E.L. (1984) Hard-sphere interactions between spherical domains in diblock copolymers. *Macromolecules*, **17**, 1712–1718.
12 Große, M., Gokhman, A., and Böhmert, J. (2000) Dependence of the ratio between magnetic and nuclear small angle neutron scattering on the size of the heterogeneities. *Nucl. Instrum. Methods Phys. Res., Sect. B*, **160**, 515–520.
13 Leitner, H., Staron, P., Bischof, M., Clemens, H., Kozeschnik, E., Svoboda, J., and Fischer, F.D. (2004) Coarsening of secondary hardening carbides in a hot-work tool steel: experiments and simulation. MS&T 2004 Conference Proceedings 2004, pp. 623–632.
14 Leitner, H., Staron, P., Clemens, H., Marsoner, S., and Warbichler, P. (2005) Analysis of the precipitation behaviour in a high-speed steel by means of small-angle neutron scattering. *Mater. Sci. Eng., A*, **398**, 323–331.
15 Bischof, M., Erlach, S., Staron, P., Leitner, H., Scheu, C., and Clemens, H. (2005) Combining complementary techniques to study precipitates in steels. *Z. Metallk.*, **96**, 1074–1080.
16 Coppola, R., Kampmann, R., Magnani, M., and Staron, P. (1998) Microstructural investigation using polarized neutron scattering of a martensitic steel for fusion reactors. *Acta Mater.*, **46**, 5447–5456.
17 Lin, S.C. (1996) Biosurfactants: recent advances. *J. Chem. Technol. Biotechnol.*, **66**, 109–120.
18 He, L., Garamus, V.M., Funari, S.S., Malfois, M., Willumeit, R., and Niemeyer, B. (2002) Comparison of small-angle scattering methods for the structural analysis of octyl-, maltopyranoside micelles. *J. Phys. Chem. B*, **106**, 7596–7604.

13

Anomalous Small-Angle X-ray Scattering

Ulla Vainio

13.1 Introduction

Anomalous small-angle X-ray scattering (ASAXS) is a method suitable for studying the nanoscale structures of materials having complex elemental compositions. It is a contrast variation method, that is, an enhanced version of the common small-angle X-ray scattering (SAXS) method. While contrast variation in small-angle neutron scattering (SANS) is achieved by measuring samples with different isotope substitutions, in ASAXS, the contrast variation is achieved by measuring SAXS at different X-ray energies using only one sample. The first ASAXS experiments were made in the beginning of the 1980s by Stuhrmann [1] at the DORIS III synchrotron storage ring.

The need for an element-sensitive SAXS method becomes obvious when multicomponent materials are studied. In materials science, ASAXS has been used, for example, to study the nanostructure of catalysts [2, 3], hydrogen storage materials [4, 5], carbides in steel [6, 7], nanostructured glasses [8], and solid oxide fuel cells [9]. ASAXS is one of the few methods that can give statistically relevant structural information in the 1 – 100 nm size scale from complex noncrystalline materials.

To illustrate the usefulness of ASAXS when studying multicomponent systems, Figure 13.1 shows a scattering simulation of a two-dimensional random structure. In this simulation, fast Fourier transform (FFT) of real space images gives intensity patterns that are averaged over the azimuth angle to obtain corresponding intensity curves in the inverse space as a function of scattering vector magnitude q. The simulation shows a structure of pores (case a), a multicomponent structure of the pores and chemical inhomogeneities (case b), and the chemical inhomogeneities without other components (case c). The SAXS signal from the multicomponent system (B) is dominated by the signal from the pores and analysis of the inhomogeneities from this signal would hardly be possible. In ASAXS, the different contrast, Δb, at different energies, E_1 and E_2, leads to different scattering intensities $I(q)$, and we are able to separate the scattering coming from chemical inhomogeneities from the total scattering. This separated intensity can then be analyzed by using the standard methods of SAS (see Chapter 11).

Neutrons and Synchrotron Radiation in Engineering Materials Science: From Fundamentals to Applications, Second Edition. Edited by Peter Staron, Andreas Schreyer, Helmut Clemens, and Svea Mayer.

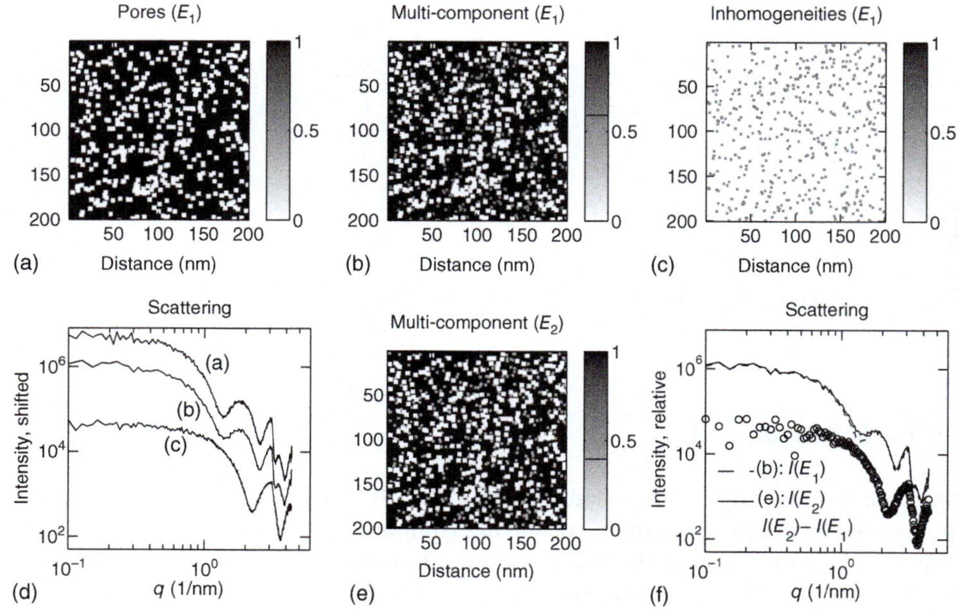

Figure 13.1 Two-dimensional simulations of randomly distributed pores and chemical inhomogeneities (a), (b), (c), and (e) and their small-angle X-ray scattering curves (d) and (f). The white "pores" and the gray quadratic inhomogeneities have average edge lengths of 5 and 3 nm, respectively. The color bars indicate a relative scale for the scattering length density b of the material, and the embedded two horizontal lines indicate the b of the inhomogeneities at energies E_1 and E_2. Here, E_1 is an energy far below the elements absorption edge and E_2 is an energy close to the edge.

13.2 Theory

13.2.1 Scattering Power of Elements

Anomalous scattering methods rely on the resonant processes taking place at the absorption edges of elements. When the X-ray energy is close to the energy that is sufficient enough to excite an electron from the K-, L-, or other shell of an atom, the scattering power of that atom decreases substantially. The scattering power is given by the atomic scattering factor

$$f(q,E) = f_0(q) + f'(q,E) + if''(q,E) \tag{13.1}$$

where the real part has been divided into two terms, the energy-independent term $f_0(q)$, which is the Fourier transform of the electronic charge density of the atom, and the energy-dependent term $f'(q,E)$. The imaginary component $f''(q,E)$ is related to the total cross-section of photoelectric absorption of X-rays, and together, $f'(q,E)$ and $f''(q,E)$ are called the *anomalous dispersion, anomalous scattering coefficients*, or *anomalous scattering factors* (see Chapter 4). There is actually nothing anomalous about these coefficients and they are well known and understood. Therefore, some attempts have been made to call the method resonant small-angle X-ray scattering rather than ASAXS.

For small scattering angles, we may approximate

$$f(E) = Z + f'(E) + if''(E) \tag{13.2}$$

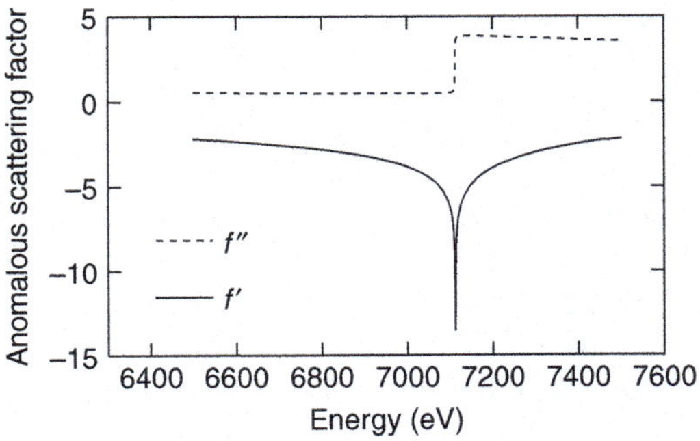

Figure 13.2 Theoretical scattering factor of iron around its K absorption edge of X-rays.

where the atomic number Z now represents the nonanomalous part of the scattering factor. For $f'(E)$ and $f''(E)$, the theoretical values calculated by Cromer and Liberman [10, 11] for neutral isolated atoms can be used, because they are readily available in popular software packages [12, 13]. In reality, the atoms have chemical environment, and experimental values for $f''(E)$ for a specific sample can be measured from the absorption (or fluorescence) spectra taken across the absorption edge of the element of interest, a technique called X-ray absorption near-edge spectroscopy (XANES). The real part of the dispersion, $f'(E)$ can then be calculated from the $f''(E)$ using the Kramers–Kronig transformation [14]. Figure 13.2 shows as an example the theoretical anomalous scattering factors of iron [13].

13.2.2 Contrast Variation

When the energy of X-rays is tuned far and close to the absorption edge of an element, the change in the SAXS intensity is typically a few percentage, as shown in Figure 13.1. However, when all the phases in the material have nearly similar scattering length densities, the change in intensity is stronger. This is seen, for example, in ASAXS studies of iron carbides [7] and oxyfluoride glass ceramics [8]. Notably, the L3 absorption edge gives a stronger anomalous effect than the K absorption edge.

13.2.3 Partial Structure Factor Formalism

The main goal of the ASAXS experiment is to separate the total scattering to its components and study the individual phase or element of interest. We may choose one of the many formalisms that have been developed to present the components of the total scattering intensity. In the following text, we present the partial structure factor formalism that has been adopted from anomalous wide-angle X-ray scattering. Despite their name, the partial structure factors in ASAXS contain not only the structure factor but also the form factor. Other formalisms include, for example, the Stuhrmann method [1] in which the intensity is divided into a resonant term, a nonresonant term, and a cross-term. The resonant term contains the energy-dependent part of the intensity while the nonresonant term represents the scattering measured far away from the absorption edge. For a comprehensive review on this topic, see [15].

The Debye formula can be used as a simple demonstration tool for understanding the partial structure factor formalism. For an isotropic system, the Debye formula gives the orientationally averaged intensity of a group of atoms [16]

$$I(q,E) = \sum_{i=1}^{N}\sum_{j=1}^{N} f_i^*(E)f_j(E)\frac{\sin\left(qr_{ij}\right)}{qr_{ij}} \tag{13.3}$$

here f_i and f_j are the atomic scattering factors of atoms i and j and the sums are over all the N atoms in the irradiated sample volume. The distance between the two atoms is r_{ij}. The magnitude of the scattering vector q is defined in Eq. (11.2).

Let us assume a material C in which particles of type A and B are present. For this three-phase system, we can rewrite Eq. 13.3 so that the correlations between the different phases are explicitly written out as six terms

$$\begin{aligned}
I(q,E) &= \left(f_{\mathrm{A}}^*(E)f_{\mathrm{B}}(E) + f_{\mathrm{B}}^*(E)f_{\mathrm{A}}(E)\right)\sum_{A}\sum_{B}\frac{\sin(qr_{\mathrm{AB}})}{qr_{\mathrm{AB}}} \\
&+ \left(f_{\mathrm{A}}^*(E)f_{\mathrm{C}}(E) + f_{\mathrm{C}}^*(E)f_{\mathrm{A}}(E)\right)\sum_{A}\sum_{C}\frac{\sin(qr_{\mathrm{AC}})}{qr_{\mathrm{AC}}} + \left(f_{\mathrm{B}}^*(E)f_{\mathrm{C}}(E)\right. \\
&\left. + f_{\mathrm{C}}^*(E)f_{\mathrm{B}}(E)\right)\sum_{\mathrm{B}}\sum_{\mathrm{C}}\frac{\sin(qr_{\mathrm{BC}})}{qr_{\mathrm{BC}}} \\
&+ 2\left|f_{\mathrm{A}}(E)\right|^2\sum_{\mathrm{A}}\sum_{\mathrm{A}}\frac{\sin(qr_{\mathrm{AA}})}{qr_{\mathrm{AA}}} + 2\left|f_{\mathrm{B}}(E)\right|^2\sum_{\mathrm{B}}\sum_{\mathrm{B}}\frac{\sin(qr_{\mathrm{BB}})}{qr_{\mathrm{BB}}} \\
&+ 2\left|f_{\mathrm{C}}(E)\right|^2\sum_{\mathrm{C}}\sum_{\mathrm{C}}\frac{\sin(qr_{\mathrm{CC}})}{qr_{\mathrm{CC}}}
\end{aligned} \tag{13.4}$$

Let us now introduce the partial structure factors $S(q)$ that contain all the energy-independent structural information

$$\begin{aligned}
I(q,E) &= \left(f_{\mathrm{A}}^*(E)f_{\mathrm{B}}(E) + f_{\mathrm{B}}^*(E)f_{\mathrm{A}}(E)\right)S_{\mathrm{AB}}(q) + \left(f_{\mathrm{A}}^*(E)f_{\mathrm{C}}(E) + f_{\mathrm{C}}^*(E)f_{\mathrm{A}}(E)\right)S_{\mathrm{AC}}(q) \\
&+ \left(f_{\mathrm{B}}^*(E)f_{\mathrm{C}}(E) + f_{\mathrm{C}}^*(E)f_{\mathrm{B}}(E)\right)S_{\mathrm{BC}}(q) \\
&+ 2\left|f_{\mathrm{A}}(E)\right|^2 S_{\mathrm{AA}}(q) + 2\left|f_{\mathrm{B}}(E)\right|^2 S_{\mathrm{BB}}(q) + 2\left|f_{\mathrm{C}}(E)\right|^2 S_{\mathrm{CC}}(q)
\end{aligned} \tag{13.5}$$

The same equations apply for multi-atom phases, in which case f_i can be replaced by the mean scattering length density b of a mixture of atoms in phase i, see Eq. (11.1). In addition, the properties of Fourier transforms allow us to add or subtract any constant from all the scattering length densities without introducing any change in the scattering intensity because the Fourier transform of a constant is a delta function at $q = 0$ and cannot be observed. Hence, we may subtract the scattering length density of phase C from the others. Using this trick, a three-phase system is conveniently reduced to a two-phase system.

By measuring SAXS intensities at three or more energies, we can further write a set of equations, for example, for two phases, A and B, at each q value in matrix form

$$\begin{pmatrix} I(q,E_1) \\ I(q,E_2) \\ I(q,E_3) \\ \vdots \end{pmatrix} = \begin{pmatrix} 2\left|b_{\mathrm{A}}(E_1)\right|^2 & b_{\mathrm{A}}^*(E_1)b_{\mathrm{B}}(E_1) + b_{\mathrm{B}}^*(E_1)b_{\mathrm{A}}(E_1) & 2\left|b_{\mathrm{BB}}(E_1)\right|^2 \\ 2\left|b_{\mathrm{A}}(E_2)\right|^2 & b_{\mathrm{A}}^*(E_2)b_{\mathrm{B}}(E_2) + b_{\mathrm{B}}^*(E_2)b_{\mathrm{A}}(E_2) & 2\left|b_{\mathrm{BB}}(E_2)\right|^2 \\ 2\left|b_{\mathrm{A}}(E_3)\right|^2 & b_{\mathrm{A}}^*(E_3)b_{\mathrm{B}}(E_3) + b_{\mathrm{B}}^*(E_3)b_{\mathrm{A}}(E_3) & 2\left|b_{\mathrm{BB}}(E_3)\right|^2 \\ \vdots & \vdots & \vdots \end{pmatrix} \begin{pmatrix} S_{\mathrm{AA}}(q) \\ S_{\mathrm{AB}}(q) \\ S_{\mathrm{BB}}(q) \end{pmatrix}$$

$$\mathbf{I}(q,E) = \mathcal{M}(E)\mathbf{S}(q) \tag{13.6}$$

In principle, by solving this set of equations, we can find out the vector of partial structure factors $\mathbf{S}(q)$ for each q value. Unfortunately, this inverse problem is usually ill-conditioned. The condition number k of the matrix $\mathcal{M}$, defined as $k = \| \mathcal{M}^{-1} \| \cdot \| \mathcal{M} \|$ for square matrices, tells how sensitive the solution is to errors in the intensities. Inverting the matrix is not recommended because of round-up errors, but results obtained by the classical Gaussian elimination or other methods that are implemented in mathematical software packages are equally affected by the poor condition of the matrix.

One way to test the reliability of the solved partial structure factors is by a Monte Carlo simulation of errors: solving the matrix equation many times while introducing artificial error to the measured intensities and the anomalous coefficients. If the solution stays stable, the result is trustworthy. When simulating ASAXS data, one should avoid the so-called inverse crimes that allow the simulation to produce too optimistic results [17].

Simon *et al.* [18] have shown for the case of an AlZnAg ternary alloy that, while the set of equations for that alloy is ill-conditioned for three measurement energies, the situation can be improved by measuring SAXS at more X-ray energies. Sometimes, regularization methods may be used to stabilize the solution [17, 19].

13.2.4 Model-Dependent ASAXS

A way to avoid the problem of solving the ill-conditioned set of equations is to fit the whole data set of intensities with a model of the complete structure [8, 20]. This can be done by least-squares minimization

$$\chi^2 = \sum_q \sum_E \frac{\left[\frac{d\Sigma}{d\Omega}_{\exp}(q,E) - \frac{d\Sigma}{d\Omega}_{\text{model}}(q,E)\right]^2}{\sigma^2_{\exp}} \tag{13.7}$$

here $\frac{d\Sigma}{d\Omega_{\exp}}(q,E)$ are the measured SAXS intensities in absolute units, while $\frac{d\Sigma}{d\Omega_{\text{model}}}(q,E)$ are the calculated intensities for a model of one's choice. Variance of the experimental intensities $\sigma^2_{\exp}$ weights the sum which is taken over the measured q values and energies E.

13.2.5 Subtraction Method

A common way to check for the quality of ASAXS data is to subtract two intensities measured at different X-ray energies from each other. The resulting intensity curve should come mostly from the anomalous element. For simple systems with only one anomalous phase, the shape of the subtracted curve should be independent of the selected energies. Quantitatively, this method will only work for a two-phase system for which the cross-term (see Eq. 13.5) is close to zero at the studied q-range. Even though this method is sometimes erroneously used for systems where the cross-term cannot be ignored, there are structures for which it can produce sufficiently accurate results, when the anomalous phase and other phases are not correlated. The applicability should be tested, for example, with simulations [21]. Figure 13.3 shows the subtracted curve and the partial structure factors of a simulated iron-carbon dumbbell system. In this case, the use of subtraction method would give fairly good results for particle size, but no quantitative fit of the form factor would be possible. The subtraction method has been used, for example, in the study of Pt-Ni catalyst nanoparticles [2].

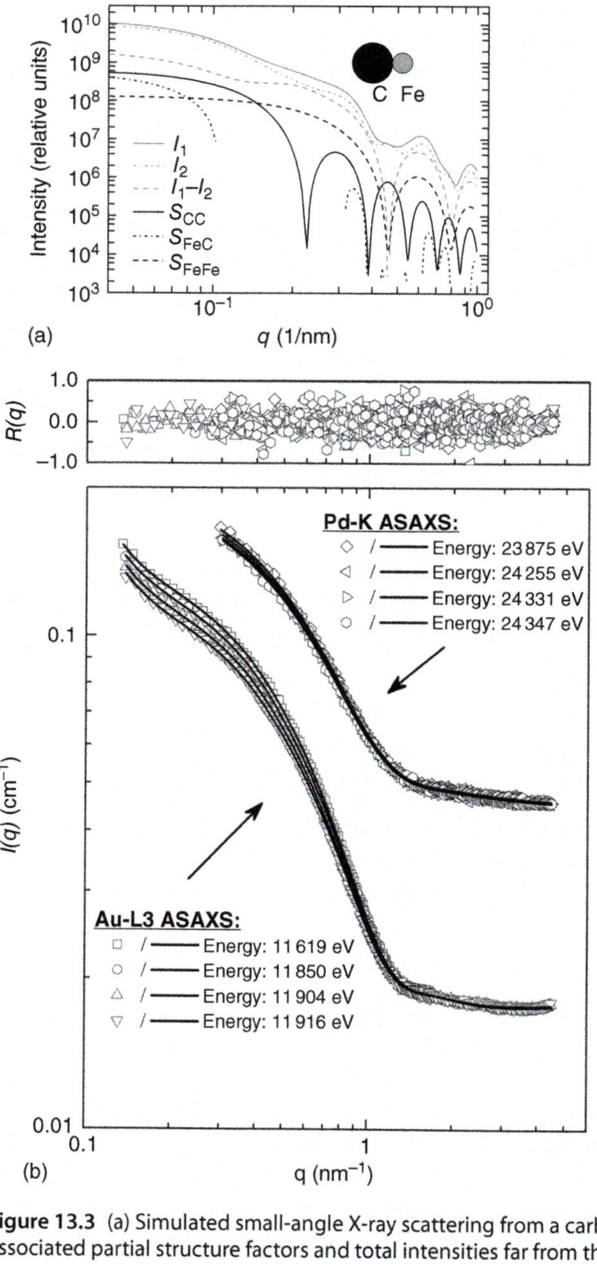

Figure 13.3 (a) Simulated small-angle X-ray scattering from a carbon-iron dumbell structure and the associated partial structure factors and total intensities far from the edge (E_1 = 6600 eV) and close to the edge (E_2 = 7110 eV). The cross-term, $S_{FeC}(q)$, is partially negative. The carbon and iron sphere radii are 20 and 10 nm, respectively. The simulation was made using the Debye formula, Eq. 13.3. (b) Experimental ASAXS intensities, marked with symbols, from a Au/Pd nanoparticle sample at Au L3 and Pd K absorption edges. Fitted curves are marked with lines. For residuals, $R(q) = (I_{exp}(q) - I_{fit}(q))^2/\sigma^2_{exp}$ [22]. (Courtesy of Sylvio Haas.)

13.3 Experiments

At dedicated ASAXS beamlines, special care has been taken to ensure that the X-ray beam size and position on the sample do not change when the energy of the X rays is changed. The energy resolution of the monochromator at ASAXS beamlines is typically $\frac{\Delta E}{E} \lesssim 10^{-4}$. Due to the changing energy, the probed q range will be different in scattering images taken at different energies and the data must be interpolated or re-binned to same q range. To ensure that energy-dependent effects from intensity monitors are not misinterpreted as the ASAXS effect, the SAXS intensities should be calibrated onto the absolute scale. Further, the energy-dependent background coming from fluorescence and resonant Raman scattering close to the edge must be subtracted from all the curves prior to analysis, except in the case of direct model fitting, where the background can be a fitted.

13.4 Example: ASAXS on Catalyst Nanoparticles

Au/Pd nanoparticles can be used as catalysts in different applications, such as hydrogen peroxide synthesis and aldehyde generation. To study the nanochemistry of these catalyst particles, Haas *et al.* [22] synthesized Au clusters in solution. Addition of a small amount of Pd onto the Au clusters increased the catalytic activity of the clusters tremendously but adding more Pd gave no further increase. Based on the synthesis method, one would expect Pd to form a shell on top of the Au nanoparticles, and hence ASAXS measurements were made below Au and Pd absorption edges to quantify the core–shell structure at the 7T-MPW-SAXS beamline at BESSY II in Berlin. However, a simultaneous model fitting of the SAXS curves showed that core–shell models had to be excluded, and only a model with a homogeneous mixture of Pd and Au inside the particles gave a good fit to all six SAXS curves (Figure 13.3b). The model fitting allowed for the estimation of the mean radius of the particles, the polydispersity of the radius, and the scattering contrast of the particles. From the contrast, the number density of the particles in solution, the molar fraction of gold and palladium within the particles, and the mass density of the particles could be calculated. The ASAXS results were crucial in explaining the origin of the high catalytic activity of the bimetallic Au/Pd clusters.

13.5 Summary and Outlook

ASAXS completes the normal SAXS method by allowing us to probe the elemental distribution within nanostructures. In the future, ever more complex structures need to be characterized with ASAXS. With the possibility to tune polarization of X rays nowadays fairly effortlessly at synchrotrons, magnetic ASAXS [23] could be used to gain both magnetic and chemical information from nanostructures. ASAXS measurements will also be rendered easier once energy-sensitive two-dimensional detectors capable of rejecting the energy-dependent background become commercially available.

References

1 Stuhrmann, H.B. (1981) Anomalous small-angle scattering. *Q. Rev. Biophys.*, **14**, 433–462.

2 Tuaev, X., Rudi, S., Petkov, V., Hoell, A., and Strasser, P. (2013) In situ study of atomic structure transformations of Pt-Ni nanoparticle catalysts during electrochemical potential cycling. *ACS Nano*, **7**, 5666–5674.

3 Høydalsvik, K., Fløystad, J.B., Voronov, A., Voss, G.J.B., Esmaeli, M., Kehres, J., Granlund, H., Vainio, U., Andreasen, J.W., Rønning, M., and Breiby, D.W. (2014) Morphology changes of Co catalyst nanoparticles at the onset of Fischer–Tropsch synthesis. *J. Phys. Chem. C*, **118**, 2399–2407.

4 Boesenberg, U., Vainio, U., Pranzas, P.K., Bellosta von Colbe, J.M., Goerigk, G., Welter, E., Dornheim, M., Schreyer, A., and Bormann, R. (2009) On the chemical state and distribution of Zr- and V-based additives in reactive hydride composites. *Nanotechnology*, **20**, 204003.

5 Karimi, F., Pranzas, P.K., Hoell, A., Vainio, U., Welter, E., Raghuwanshi, V.S., Pistidda, C., Dornheim, M., Klassen, T., and Schreyer, A. (2014) Structural analysis of calcium reactive hydride composite for solid state hydrogen storage. *J. Appl. Crystallogr.*, **47**, 67–75.

6 Jemian, P.R., Weertman, J.R., Long, G.G., and Spal, R.D. (1991) Characterization of 9Cr-1MoVNb steel by anomalous small-angle X-ray scattering. *Acta Metall. Mater.*, **39**, 2477–2487.

7 Song, W., Vainio, U., and Bleck, W. (2014) Nano precipitates in high carbon bearing steel 100Cr6. Proceedings of the 29. Aachener Stahlkolloquium 2014, pp. 75–85.

8 Haas, S., Hoell, A., Wurth, R., Rüssel, C., Boesecke, P., and Vainio, U. (2010) Analysis of nanostructure and nanochemistry by ASAXS: accessing phase composition of oxyfluoride glass ceramics doped with Er^{3+}/Yb^{3+}. *Phys. Rev. B*, **81**, 184207.

9 Allen, A.J., Ilavsky, J., Jemian, P.R., and Braun, A. (2014) Evolution of electrochemical interfaces in solid oxide fuel cells (SOFC): a Ni and Zr resonant anomalous ultra-small-angle X-ray scattering study with elemental and spatial resolution across the cell assembly. *RSC Adv.*, **4**, 4676–4690.

10 Cromer, D.T. (1965) Anomalous dispersion corrections computed from self-consistent field relativistic Dirac-Slater wave functions. *Acta Crystallogr.*, **18**, 17–23.

11 Cromer, D.T. and Liberman, D. (1970) Relativistic calculation of anomalous scattering factors for X-rays. *J. Chem. Phys.*, **53**, 1891–1898.

12 Brennan, S. and Cowan, P.L. (1992) A suite of programs for calculating X-ray absorption, reflection, and diffraction performance for variety of materials at arbitrary wavelengths. *Rev. Sci. Instrum.*, **63**, 850–853.

13 Evans, G. and Pettifer, R.F. (2000) CHOOCH: a program for deriving anomalous-scattering factors from X-ray fluorescence spectra. *Appl. Crystallogr.*, **34**, 82–86.

14 Ravel, B. and Newville, M. (2005) ATHENA, ARTEMIS, HEPHAESTUS: data analysis for X-ray absorption spectroscopy using IFEFFIT. *J. Synchrotron Radiat.*, **12**, 537–541.

15 Tatchev, D. (2008) Structure analysis of multiphase systems by anomalous small-angle X-ray scattering. *Philos. Mag.*, **88**, 1751–1772.

16 Guinier, A. (1994) *X-Ray Diffraction in Crystals, Imperfect Crystals, and Amorphous Bodies*, Dover Publications, New York, p. 49.

17 Kaipio, J. and Somersalo, E. (2004) *Statistical and Computational Inverse Problems*, Applied Mathematical Sciences, vol. **160**, Springer-Verlag, New York.

18 Simon, J.P., Lyon, O., and de Fontaine, D. (1985) A comparison of the merits of isotropic substitution in neutron small-angle scattering and anomalous X-ray scattering for the evaluation of partial structure functions in a ternary alloy. *J. Appl. Crystallogr.*, **18**, 230–236.

19 Pirkkalainen, K., Vainio, U., Kisko, K., Elbra, T., Kohout, T., Kotelnikova, N.E., and Serimaa, R. (2006) Structure of nickel nanoparticles in a microcrystalline cellulose matrix studied using anomalous small-angle X-ray scattering. *J. Appl. Crystallogr.*, **40**, s489–s494.

20 Vainio, U., Lauten, R.A., Haas, S., Svedström, K., Veiga, L.S.I., Hoell, A., and Serimaa, R. (2012) Distribution of counterions around lignosulfonate macromolecules in different polar solvent mixtures. *Langmuir*, **28**, 2465–2475.

21 Vainio, U., Pirkkalainen, K., Kisko, K., Goerigk, G., Kotelnikova, N.E., and Serimaa, R. (2007) Copper and copper oxide nanoparticles in a cellulose support studied using anomalous small-angle X-ray scattering. *Eur. Phys. J. D*, **42**, 93–101.

22 Haas, S., Fenger, R., Fertitta, E., and Rademann, K. (2013) Cascade catalysis of highly active bimetallic Au/Pd nanoclusters: structure-function relationship investigation using anomalous small-angle X-ray scattering and UV–Vis spectroscopy. *J. Appl. Crystallogr.*, **46**, 1353–1360.

23 Fischer, P., Zeller, R., Schütz, G., Goerigk, G., Haubold, H.-G., Pruegl, K., and Bayreuther, G. (1998) Probing the magnetic microstructure of an amorphous GdFe system with magnetic anomalous small angle X-ray scattering. *J. Appl. Crystallogr.*, **83**, 7088–7090.

14

Imaging

Wolfgang Treimer

14.1 Radiography

14.1.1 Fundamentals

A radiograph is the two-dimensional attenuation distribution of a path-integrated projection from a three-dimensional object. In this context, films in the case of X-rays and gas counters in the case of neutrons have been historically established. For neutron radiography also film-converter foil systems were developed and used. In the past decades more and more electronically based systems (image foils) and electronic recording systems (CCD cameras) became available with improved image properties and qualities.

Currently, film imaging is used only in very special cases; much more common are CCD cameras that supply a hardware resolution (pixel size) down to some micrometers which is of the same order of films. The advantages of electronic recording are the possibilities of easy image processing, high bit depth (optical density), real-time imaging, and information handling (delivery, storage). The next chapter will give all interesting details about the recording cameras; here we will summarize only some necessary facts that are important for our considerations.

The method radiography seems to be a simple technique, which needs only a few parameters to work with. First, some definitions must be established, which are also important for the discussion of tomography. The step from radiography = projection to tomography = many projections is then rather small. Only some mathematical tools have to be applied to get a 3D image from a series of radiographs (projections).

For a neutron or X-ray radiography, one must have in general a beam, an object, and a detector system for imaging. The beam is specified by the

- flux $\phi = \frac{\text{number of photons (neutrons)}}{\text{unit time} \times \text{unit area}}$
- wavelength distribution ($\Delta\lambda/\lambda$)
- divergence ϕ, and
- polarization P (or spin state in the case of neutrons, here not discussed).

For radiography and tomography, Φ, $\Delta\lambda/\lambda$, and ϕ are of principal interest; in the case of polarized neutron radiography one has to use polarized neutrons in front of the sample and a spin analyzer in front of the sample and/or in front of the detector (film). The flux used for a radiograph is a quantity, which strongly depends on the

Neutrons and Synchrotron Radiation in Engineering Materials Science: From Fundamentals to Applications,
Second Edition. Edited by Peter Staron, Andreas Schreyer, Helmut Clemens, and Svea Mayer.

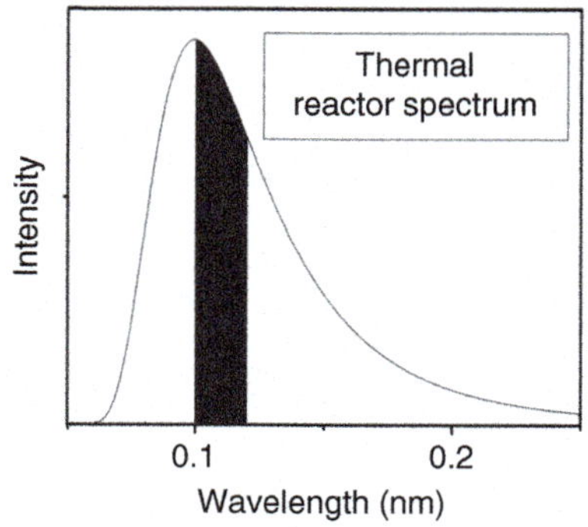

Figure 14.1 Cold reactor spectrum: black bar denotes a certain $\Delta\lambda/\lambda$-band in the spectrum, the area of the band is proportional to the intensity of monochromatic radiation.

monochromaticity of the beam, that is, its wavelength distribution ($\Delta\lambda/\lambda$) and divergence ϕ. In order to illustrate the dependence of the available intensity consider a source S emitting radiation with a wavelength spectrum as shown in Figure 14.1.

The same holds for X-rays. The Bremsstrahlung spectrum has a characteristic shape that is nearly independent of the target material (the kinetic energy of the accelerated electrons is much larger than the binding energy of the bound electrons). In the case of X-rays the maximum photon energy is determined by the maximum acceleration voltage, yielding the shortest wavelength in the spectrum. It is given by (h = Planck constant, c = speed of light, ν_{max} = maximum frequency)

$$\lambda_{min} = \frac{hc}{h\upsilon_{max}} \tag{14.1}$$

One can see from Eq. (14.1) that the wavelength is of the order of 0.2 nm and less and the energy $h\nu$ ranges from approximately 10 up to 250 keV. In the case of neutrons the kinetic energy is given by

$$E_{kin} = \frac{m \cdot v^2}{2} \tag{14.2}$$

with the corresponding neutron velocity calculated by $v = \frac{h}{m\lambda} = 3.95603 \times 10^3\,\mathrm{m\,s^{-1}}$ (λ in 0.1 nm or Å). The neutron energy E_{kin} is then given by $E_{kin} = \frac{m \cdot v^2}{2} \rightarrow E(\lambda) = 0.081\,804\,\left[\frac{eV}{\lambda}\right]$. That yields for $\lambda = 0.1$ nm ~0.081 804 eV (with m = neutron mass = $1.6\,749\,286(10)10^{-27}$ kg, 1 (eV) = $6.24\,149\,596 \times 10^{18}$ (J)).

The neutron flux Φ from a research reactor has usually the shape of a Maxwell spectrum due to the thermal equilibrium of neutrons with the moderator. The spectrum is different in the case of neutron guides, because of their different curvatures and coatings. Neutron radiography (and tomography) uses mainly the whole spectrum, accepting both a very intense neutron beam and an integrated absorption behavior of the sample.

14.1.2 Interactions of Neutrons with Matter

There are three possibilities that neutrons can interact with matter (Figure 14.2).

Commonly, all interactions are described by so-called cross-sections, referring to the particular interaction the neutrons undergo. The interactions of neutrons with matter can be distinguished into macroscopic Σ and microscopic cross-sections σ. One can

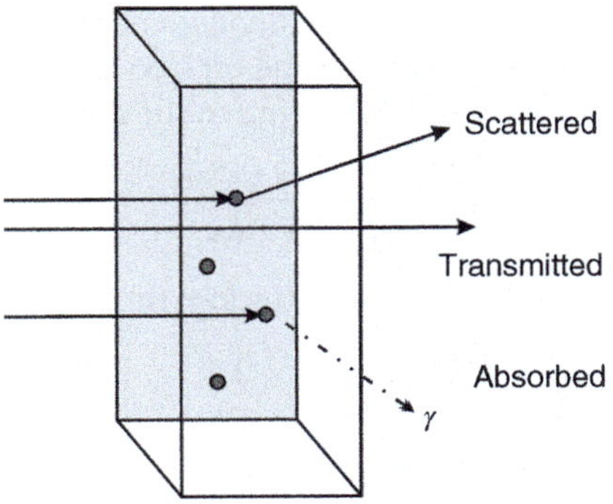

Figure 14.2 Possible interactions of a neutron with matter.

define σ and Σ as follows (definition after [1, 2]). The cross-section σ is called *microscopic* if it refers to a single nucleus. The dimension $[\sigma] = (\text{cm}^2)$; unit = 1 barn = 10^{-24} cm^2. The differential cross-section

$$\frac{d\sigma}{d\Omega} = \frac{\text{number of interacting particles/unit time} \times \text{unit cone } d\Omega}{\text{number of incident particles/unit time} \times \text{unit area} \times \text{unit cone } d\Omega}$$
$$= [\text{area}]$$

The cross-section Σ is called *macroscopic* if $\Sigma = N \times \sigma$ and N denotes the number of nuclei per cubic centimeter. The dimension of Σ is (cm^{-1}). From the change of the intensity I (or neutron flux Φ)

$$-\frac{dI}{I} = \Sigma \times dz \tag{14.3}$$

one gets the probability $\Sigma \times dz$ that neutrons will interact with matter along the path dz. With A_0 (Loschmidt or Avogadro number = 6.02×10^{23}) the number of atoms per cubic centimeter N is given by

$$N = \frac{\rho}{A} \times A_0 \tag{14.4}$$

if A is the atomic weight and ρ the density. The macroscopic cross-section then becomes

$$\Sigma = N \times \sigma = \frac{\rho}{A} \times A_0 \times \sigma \tag{14.5}$$

Example Determine the macroscopic cross-section of iron.

Solution
Iron has a density of 7.86 g cm^{-3}, the microscopic cross-section for absorption of iron is 2.56 barns, and the gram atomic weight is 55.847 g. Thus, $\Sigma = 0.217$ cm^{-1}.

For a material consisting of j-different atoms, Σ becomes the weighted sum of the cross-sections of the constituents

$$\Sigma = \sum_j N_j \sigma_j \tag{14.6}$$

where N_i is the density of the nuclei of kind i and σ_i the corresponding microscopic cross-section of the nuclei. From this equation the attenuation law can be expressed as

$$I(z) = I_0 \cdot \exp\left(-\int_{\text{path}} \Sigma(z)\, dz\right) \tag{14.7}$$

if $\Sigma(z)$ means the z-dependent cross-section along the path of the neutron through the sample. Note that the reciprocal value, $1/\Sigma = \lambda_{\text{free}}$, is the mean free path of neutrons in the material. Referring to the absorption law one can define a "half-thickness" $z_{1/2}$ as

$$z_{1/2} = \frac{\ln(2)}{\int_{\text{path}} \Sigma(z)\,\mathrm{d}z} \tag{14.8}$$

In the case of scattering, the macroscopic scattering cross-section is

$$\Sigma_s(z) = \sum_j N_j(z) \times \sigma_{s,j} \tag{14.9}$$

that is, the weighted sum of all contributing nuclei ($\sigma_{s,j}$) along the z-direction. For neutron radiography, the absorption cross-section is of major importance, that is, $\Sigma = N_j\sigma$ is replaced by the absorption cross-section $\Sigma_a = N_{j,a}.\sigma_{j,a}$ if the dominant interaction is absorption:

$$\Sigma_a(z) = \sum_j N_j(z) \times \sigma_{a,j} \tag{14.10}$$

The total macroscopic cross-section $\Sigma_t(z)$ is then defined as

$$\Sigma_t(z) = \Sigma_a(z) + \Sigma_s(z) \tag{14.11}$$

The scattering process plays an important role in radiography and tomography because scattering contributions usually result in a decrease of sharpness in the picture. (However, scattering processes can also be used as imaging signals.) Figure 14.3 shows approximately the relationship between X-ray and neutron cross-sections and points out the complementary nature of both radiations (see Chapter 4).

14.1.3 Geometries

For radiography, radiation is usually collimated by slits or a collimator system that reduces the propagation directions of the radiation and, therefore, the intensity. To estimate the intensity in front of a sample that is situated at a certain distance from the source one can deduce the flux as follows. If a source emits radiation into 4π, the source flux I_0 decreases at a certain distance R. Within a solid angle $\Delta\Omega = (\phi_h \times \phi_v)$ ($\phi_{h,v}$ is the horizontal and vertical divergence) the intensity I is proportional to the area of the surface of the sphere $\Delta A = ds \times ds$. The element ds can be determined as $(\phi \times R)$, where ϕ is the one-dimensional divergence. The intensity will decrease as $1/R^2$. Figure 14.4 demonstrates the $1/R^2$-law for a given distance d, $2d$, and $3d$ from a point source. A given intensity I at a distance d is distributed over a four times larger area at $2d$ and, therefore, the density is decreased by a factor 4.

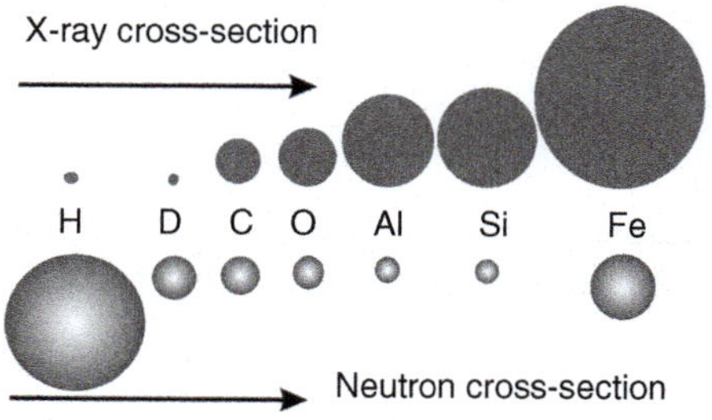

Figure 14.3 For X-rays the "amount or strength" of interaction increases with the atomic number (and number of electrons), which does not hold for neutrons.

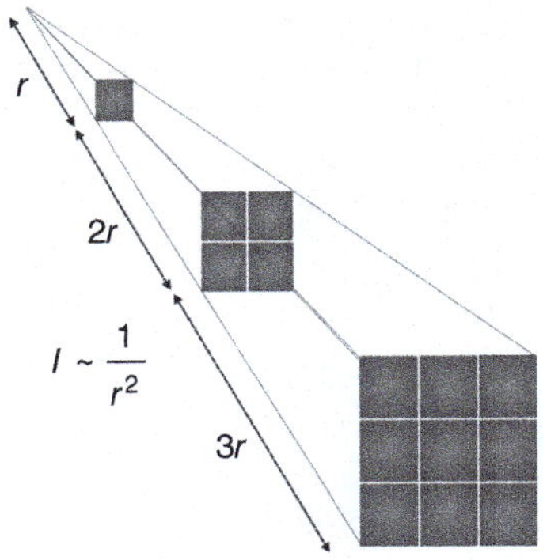

Figure 14.4 The number of photons or neutrons/unit area decreases as $1/d^2$ ("$1/R^2$-law"; $d = R$ = distance from the source).

For a known divergence ϕ (which can be measured), the available intensity I in front of a sample can be calculated from a given source flux I_0. The intensity (=neutrons/area × time) at a distance d decreases as shown in Figure 14.4. Moreover, an area in an object A_{object} is enlarged and blurred to an area A_{image} depending on the following parameter:

1) L/D ratio ~ divergence
2) distance object–detector (film, CCD camera)
3) blurring due to secondary detection.

Ad (1 and ad (2). The (effective) source size D, the distance L of the object from the source, and the distance l of the object from the detector play important roles in radiography and determine – together with other parameters – the quality of a radiograph. Every point in a sample is scaled up to an area with the diameter d on the detector (film, CCD camera) as

$$d = \frac{l}{L/D} \tag{14.12}$$

L/D defines the divergence ϕ, as can be seen in Figure 14.5.

Note: A large L/D number means a low divergence and neutron flux and vice versa. Depending on the sample problem to be imaged, L/D must be adapted to the boundary conditions (neutron flux, spatial resolution of the detector, etc.) of the experimental setup.

In radiography and tomography, one uses either a pinhole or a collimator. In the case of a pinhole, D is the diameter of the pinhole. In Table 14.1, L/D and divergence ϕ are displayed for given pinholes in millimeters if $L = 10$ m.

Example For an $L/D = 250$ a point in a sample, 50 mm apart from the detector, is imaged as a circle with a diameter of 200 µm; if $l = 100$ mm, the circle becomes 400 µm.

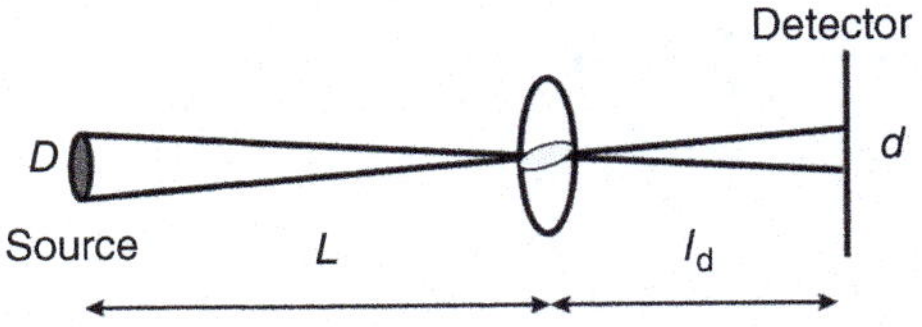

Figure 14.5 Definition of the L/D-ratio: D = size of the (effective) source, L = distance source–object (point), and l = distance object–detector.

Table 14.1 Geometric parameter for a given distance source–object $L = 10$ m and distance object–point detector $l = 50$ mm, the pinhole diameter is varied from 10 to 50 mm.

Pinhole diameter (mm)	10	20	30	40	50
L/D	1000	500	333	250	200
Divergence ϕ (°)	0.057	0.114	0.171	0.228	0.285
$L = 50$ mm: point spread (mm)	0.05	0.1	0.15	0.2	0.25

Note: The larger the L/D and the shorter the distance object–detector (film) l (Figure 14.5) the sharper is the attenuation radiograph. In the case of phase radiography, the second condition (the shorter the distance object–detector) does not hold.

Ad (3). Neutrons can be detected only by means of (n,α), (n,γ), or (n,β) – reactions to measure an electronic signal, that is, one needs nuclear reactions to convert neutrons into photons. Due to the finite thickness of the converter and the emission cone of produced photons, an additional uncertainty of localization of neutrons is produced (Figure 14.6).

The capture of a neutron leads to isotropic emission of secondary radiation. A film or CCD camera focused on the converter screen will therefore measure – for a well-defined incident direction – on the converter different spots on the image screen (which forms an area spot instead of a point spot). A point or a line is enlarged simply by the converter process, despite the fact of the assumption of a perfect parallel beam (which does not exist!). One recognizes the role of the geometries and the converter foil thickness that determine sharpness and resolution of an image.

14.1.4 Resolution Functions

To take this blurring (mentioned above) into account one can define a resulting secondary radiation probability density function represented by an edge or line spread function LSF as [2]:

$$\mathrm{LSF}(x, x_0) = \frac{\delta/\pi}{1 + \delta^2 \times (x - x_0)^2}$$

$$\delta = \frac{1}{d_{\mathrm{conv}}} \tag{14.13}$$

where d_{conv} (in the direction of z) is the distance of the converter foil to the image plane and x_0 is the center of the incident beam. Usually, d_{conv} can be taken as the thickness of

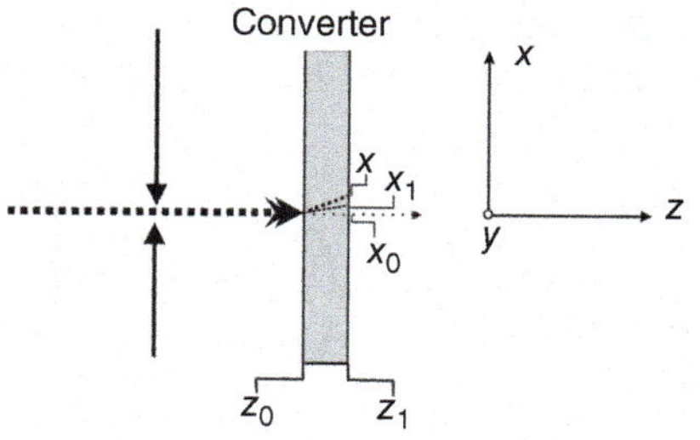

Figure 14.6 A neutron propagating into the z-direction and spatially defined by a fine slit impinges on a converter foil with a thickness $\Delta z = z_1 - z_0$ and produces photons scattered into the directions (x_0,y_0), (x_1,y_1), …, (x,y). y_0, y_1, (x,y) are not shown.

the converter if the film is attached or the CCD camera is focused on it. LSF(x,x_0) is also called the LSF for this system. Figure 14.7 shows three different LSF for three different converter thicknesses.

The point spread function (PSF) can be derived from LSF by rotating around the z-axis which was kept constant in the case of the LSF. For $x_0 = 0$, $y_0 = 0$ the PSF can be written as [2]

$$\mathrm{PSF}(x, y) = \frac{\delta^2/2\pi}{[1 + \delta^2\{x^2 + y^2\}]^{3/2}} \tag{14.14}$$

In Figure 14.8 three PSF images are calculated for different converter thicknesses ((a) = 50 μm, (b) = 100 μm, and (c) = 300 μm).

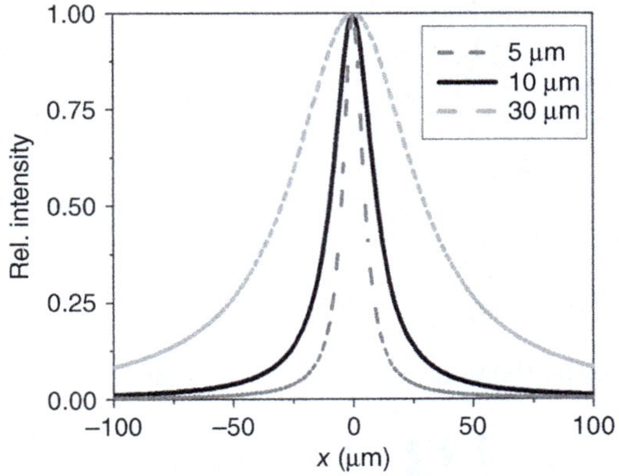

Figure 14.7 Line spread function (LSF) as a function of converter thickness. 10 mm^{-1} means $d_{conv} = 0.1$ mm (x_0 was set zero).

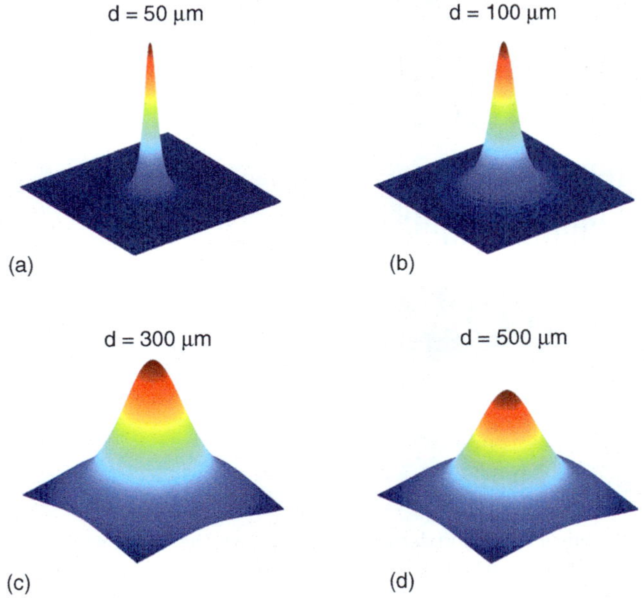

Figure 14.8 Four PSF images calculated for converter thicknesses d = 50 μm, 100 μm, 300 μm and 500 μm.

Note: Every action on the picture (measurement, restrictions, filtering, etc.) decreases the information. To determine the blurring in an image one must calculate the 2D-convolution of the (instrumental) PSF with the original object function.

The modulation transfer function (MTF) is known as the modulus of the optical transfer function (OTF) which is the Fourier transform (FT) of the LSF. It characterizes a transfer system such as a radiography or a tomography setup. The question to be answered is how does the contrast in an image depend on the spatial resolution of the transfer system. The MTF can simply be determined by measuring the contrast C given by $(I_{max} - I_{min})/(I_{max} + I_{min})$ for different line pairs per millimeter and plotting C as a function of the spatial frequency. For this procedure, one uses imaging gratings having different line pairs per millimeter (see Figure 14.9). Another grating looks like the one shown in Figure 14.10.

The MTF can also be derived from the FT of the first derivative of the LSF as $|\text{MTF}| = \text{FT}\{\text{LSF}\}$ [3].

In Figure 14.11, a perfect edge $E^*(x)$ as well as a measured one $E(x)$ are shown. Due to divergence and other effects (blurring due to converter foil, finite pixel element of a CCD camera, etc.) a point is spread out and decreases the sharpness of the image. The measured edge $E(x)$ can be approximated either by a function

$$E(x) = \frac{1}{2} + \frac{1}{\pi}\tan^{-1}(\delta \times x) \tag{14.15}$$

or by the error function erf(x) (Figure 14.12).

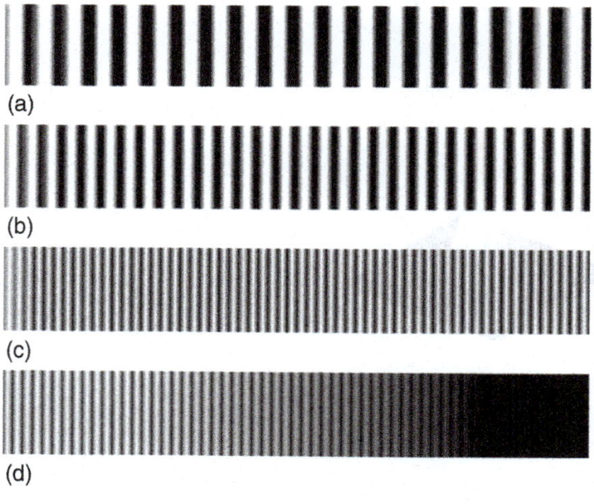

Figure 14.9 Line gratings with different spacing (a–c) and (d) shows for a constant line spacing from left to right with decreasing contrast.

Figure 14.10 From left to right: decreasing line spacing decreases the contrast. The function which describes the contrast as a function of line spacing (line pairs per millimeter) is called the *modulation transfer function (MTF)*.

14.1.5 Image Degradation

The image degradation is based on a number of experimental components that can be summarized as follows:

- beam divergence ϕ
- large distance object–detector
- motion of the sample during an exposure
- converter unsharpness (roughness of the surface + finite thickness of the converter)
- object scattering (neutrons are scattered in the object, lose their incident direction, which leads to a reduction of image contrast and artifacts).

The contrast of an image (neighboring pixels or edges in a picture) is given by

$$C = \left| \frac{\mathrm{gr}_1 - \mathrm{gr}_2}{\mathrm{gr}_1 + \mathrm{gr}_2} \right| \tag{14.16}$$

gr_i = gray value or pixel value of a pixel. The final contrast C of an image is given by its smallest value.

Note: The difference between two gray levels does not determine C uniquely; C changes within the range of gray levels within an image. For the same difference $\mathrm{gr}_2 - \mathrm{gr}_1 = 10$, one gets different contrasts: if $\mathrm{gr}_1 = 10$ and $\mathrm{gr}_2 = 20$ one gets a contrast $C = 33\%$; for $\mathrm{gr}_1 = 100$ and $\mathrm{gr}_2 = 110$ one yields a contrast $C \sim 4.8\%$!

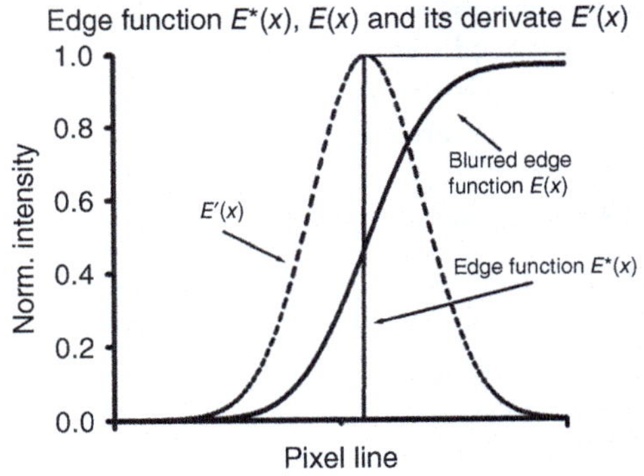

Figure 14.11 Edge functions: a perfect edge $E^*(x)$ is measured as $E(x)$; its first derivative has the shape of a Gaussian function.

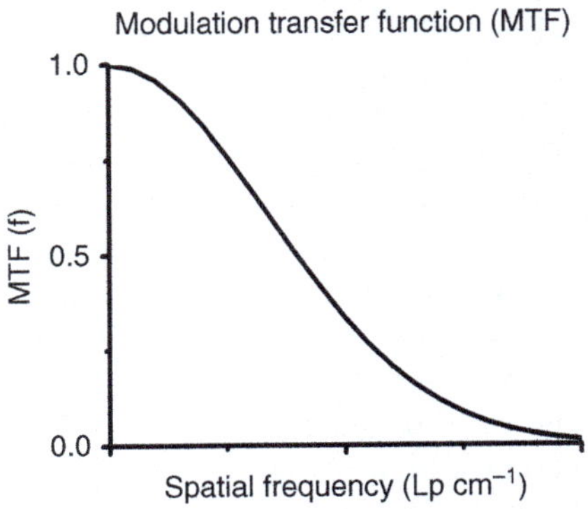

Figure 14.12 MTF of the edge function of Figure 14.11.

The image contrast is also influenced by the following variables:

- absorption differences in the specimen
- wavelength (band, spectrum) of the primary radiation
- beam hardening (the spectrum alters to shorter wavelengths along the path through the sample)
- scattering in the converter.

For a film-converter system, the X-ray film itself must fulfill some conditions, that is, the film contrast itself is determined by the following items:

- grain size or type of film
- film processing chemicals
- concentrations of film processing chemicals
- time of development
- temperature of development
- degree of mechanical agitation (physical motion).

Modern CCD cameras have nearly superseded the film-converter technique, because the digital image allows a number of image processing actions. All image processing techniques known from medical imaging can also be applied to neutron radiography and tomography. There is a large number of applications of neutron radiography, but also some "new" developments that must be shortly mentioned. Due to better experimental conditions and the need for industrial applications neutron radiography has experienced new attention, and developments such as energy-dependent radiography, (rapid) real-time radiography, and phase contrast radiography have increased the interest in this method.

14.1.6 Other Imaging Techniques

In this chapter we summarize some techniques that become more and more important due to better detectors (higher efficiency) and dedicated radiography stations that are optimized for these techniques.

14.1.6.1 Energy Dispersive Radiography

A neutron spectrum from $\lambda = 0.05$ to 1 nm covers a large range of lattice constants of crystals which are parts of materials. Therefore, it can be used for a special imaging technique that will be described as "energy dispersive radiography" (EDR). The transmission of neutrons through a material depends on many factors such as the composition of the material itself, the particular absorption (and scattering) cross-sections, the wavelength (energy) of the neutrons, temperature, isotopes, and, the most important factor, the crystalline structure and its orientation. If certain regions or parts in a material contain, for example, aluminum (Al), copper (Cu), or iron (Fe) and brass, then one can observe the following effect in a radiography (Figure 14.13).

Consider the behavior of the attenuation cross-section of, for example, Al, Cu, brass, and Fe in the close neighborhood of 4 Å or 0.4 nm (see Figure 14.13). These measurements were performed at CONRAD at the BER II reactor (HMI), a radiography, and tomography instrument for cold neutrons. Using a double monochromator system of two graphite crystals the energy and, therefore, the wavelength could be tuned within

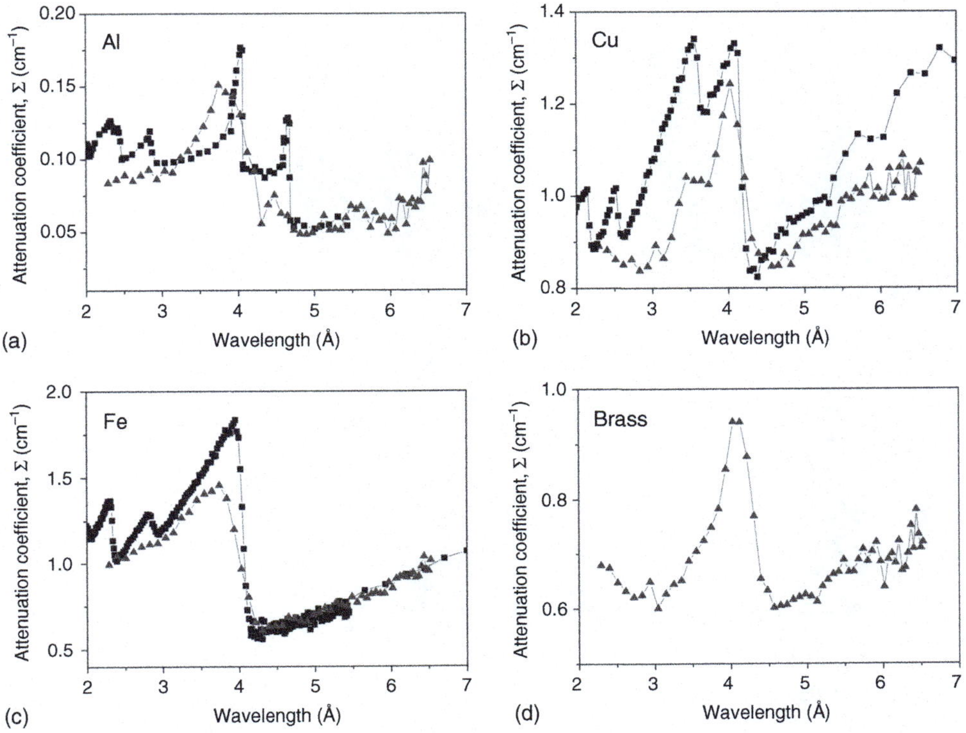

Figure 14.13 Attenuation coefficients of (a) Al, (b) Cu, (c) Fe, and (d) brass as a function of the neutron wavelength. Data points represented by triangles are own measurements at the HMI in 2005/2006.

a range from 0.28 up to 0.62 nm. The wavelength dependence enables sophisticated radiographies as shown below because the attenuation can be the same at different energies (different wavelengths).

To show this, an object (a set of step wedges) was imaged by means of neutron radiography at different neutron energies. Theses radiographies require the possibility of using different neutron energies E_1 and E_2, selected out of the spectrum (e.g., by a chopper or by monochromator crystals). The incident intensities I_{01} and I_{02} in front of the sample are different (due to the spectrum) as well as the transmitted intensities I_1 and I_2. Figure 14.14 shows five different materials imaged at different energies. The step wedges had thicknesses of 5, 10, 15, and 20 mm.

To use these features for radiography one must normalize the radiographs according to Eq. (14.17). All parts having the same attenuation coefficient will be suppressed. The ratio of the normalized intensities yields an exponential expression,

$$\frac{\frac{I_1}{I_{01}}}{\frac{I_2}{I_{02}}} = \frac{I_1 \cdot I_{02}}{I_2 \cdot I_{01}} = \frac{\mathrm{e}^{-\Sigma(\lambda_1)d}}{\mathrm{e}^{-\Sigma(\lambda_2)d}} = e^{[\Sigma(\lambda_2)-\Sigma(\lambda_1)]d} \tag{14.17}$$

which decreases or increases. The contribution to attenuation of some materials is also shown in Figure 14.15.

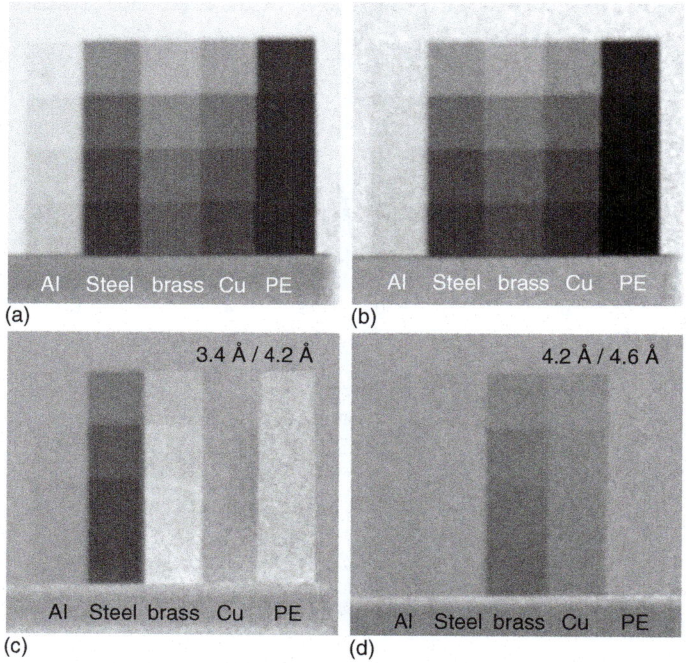

Figure 14.14 Radiographs of wedges of different materials (Al = aluminum, Cu = copper, PE = polyethylene) at (a) $\lambda = 0.34$ nm and (b) $\lambda = 0.643$ nm. Panels (c)) and (d) show normalized radiographs. Note the different visibilities of the materials for those energies.

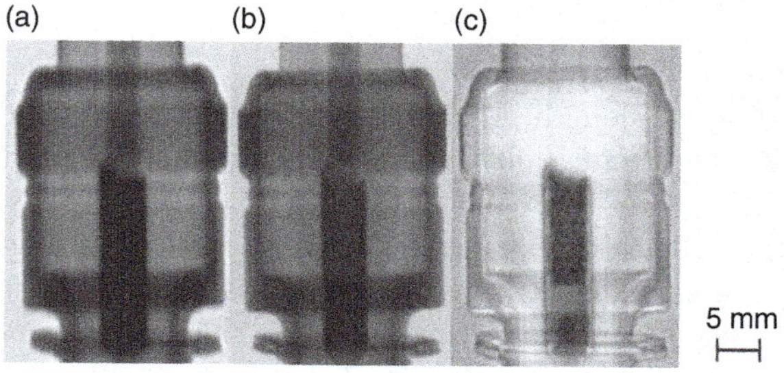

Figure 14.15 Radiographs of a spark plug at 6.9 Å (a) and 3.2 Å (b). The division of both images yields image (c). (Kardjilov 2003 [4]. Reproduced with permission of Elsevier.) Due to application of Eq. (14.17) the surrounding part becomes transparent and the inner part becomes sharper visible.

14.1.6.2 Real-Time Radiography

To visualize rapid processes one has to distinguish between oscillating and simple emerging (nonoscillating) processes. In the latter case, each shot (or frame) must image (by absorption) within a certain time interval Δt an instantaneous progress of the process. The process within Δt itself can be visualized by registering as many frames as possible. In the case of an oscillating piston of a model aircraft engine, the motion was triggered, so that different positions of the piston could be imaged at different times. The sharpness of the pictures and, therefore, of the produced film depends on the number of neutrons that can be collected within one pixel (Δx) within Δt (per frame) and how well defined the triggering can be managed. Therefore, the exposure time Δt should be (much) smaller than the velocity of the process within a certain Δx ($x + \Delta x$), that is, any detail of the oscillation process should move less than half of a pixel within the exposure time Δt [5, 6].

14.1.6.3 Phase Contrast Radiography

The idea of phase contrast imaging is the following: consider a (neutron) wave which is coherent in a region given by the first Fresnel zone (FZ). The size of the first FZ A_{FZ} is

$$A_{FZ} = \frac{R \cdot r}{R + r} \cdot \pi \cdot \lambda \cdot \frac{1}{\sin(\theta)} \qquad (14.18)$$

with R = distance point source–object, r = distance object–detector, and θ the angle between the ray propagation vector and the plane of observation (usually $\theta = 90°$). This size A_{FZ} determines the coherent source for any interference effects. For standard geometries (e.g., $R \sim 500$ cm, $r \sim 1$ cm, $\lambda \sim 0.5$ nm) A_{FZ} is approximately 1.6×10^{-7} cm^2 and the diameter of the coherent source ~4 μm. Therefore, one has to create a "point-like" source by a small pinhole that determines, together with the divergence of the beam the size, the coherent area d_{coh} of a real neutron beam. The coherent size of a source d_{coh} can also be estimated by simple geometric considerations as given by Treimer *et al.* [5]

$$d_{coh} = \lambda \cdot \frac{1}{2}\frac{R}{D} \qquad (14.19)$$

where d_{coh} is the size of the monochromatic source, R/D the common known "L/D ratio," and D denotes the distance object–detector. A short calculation demonstrates that D must be kept very small, of the order some hundred microns or even less, to achieve a reasonable size of the coherent "source area." In the case of phase contrast, the coherent part of the wave propagates through an object and experiences position-dependent phase shifts, that is, the propagating wave front is modulated by different indices of refractions in the sample. The modulation occurs quite rapidly in the so-called near field region ($z < d^2/\lambda$), z = distance object–detector, d = diameter of the object and it broadens with increasing z [6, 7]. The fringe modulation caused, for example, by an edge is washed out in the case of a too large divergence and/or source size. To observe the strongest phase contrast, one has to operate in the near field region, where $z < d^2/\lambda$, where the modulation of the amplitude is best. It can be shown that the optimal contrast occurs at $z = d^2/2\lambda$. This was (partially) proven by the investigations of contrast to realize experimentally phase images of steel needles at different detector distances and ratios L/D [8]. One approach to realize phase contrast was performed by neutron radiography some years ago. The achieved contrast, however, was not as brilliant as it should be. This was due to a number of experimental conditions, mainly due to the small intensity (several hours exposure time). First experiments based on phase effects (refraction) were already performed in 1989 [9], and the first neutron radiography based on phase contrast was performed in 2000 [10]. However, phase contrast imaging is more often used by synchrotron radiation due to the higher brilliance of the source.

One can calculate the contrast, knowing that the FT of the structure is measured. The phase development close behind the sample yields strong intensity oscillations that were measured at different distances from the object. However, one can also use simple ray-tracing, adding up all amplitudes emerging from the structure, which causes the phase gradient that contribute to a measuring point. Figure 14.16 shows the result of such calculations for a phase boundary consisting of Al and Si, which have very similar index of refraction. Therefore, for equal thickness, no intensity variation occurs (both 50 μm). In the calculation it was assumed that both slabs were illuminated by a plane wave. This explains the small oscillations that appear approximately more than 200 μm off the boundary at 600 μm.

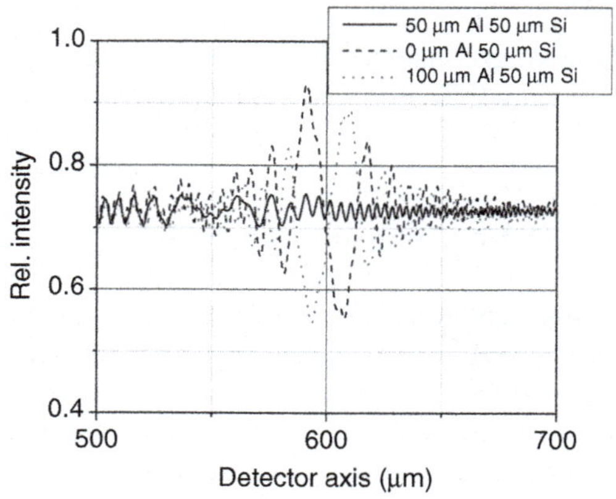

Figure 14.16 Phase boundary between two slabs (Al and Si) having nearly the same index of refraction, but different thickness. At the position of 600 μm, Al and Si stick together. (Treimer 2005 [5]. Reproduced with permission of Springer.)

14.2 Tomography

14.2.1 Mathematical Introduction

The image procedure described in the previous chapter yields the path-integrated information onto a 2D detector (film or electronic device). However, it is often additionally important to have a 2D slice image or a 3D image of the object. To do this in the simplest way, the object can be parted in one direction (e.g., z-direction, perpendicular to the (x,y) plane) into small slices. Each slice, having a finite thickness Δz, is considered as a 2D function $f = f(x,y)$ which describes, for example, the position-dependent attenuation coefficient $\mu = \mu(x,y)$. This slice is scanned (cf. Figures 14.17 and 14.18) under equidistant angles from 0° to 180° (360°) and so-called projections $P_\theta(t)$ are measured. From these projections 2D images of the slice can be reconstructed and stacked together to a 3D image reconstruction of the object.

It must be pointed out that any function f can be taken into account, which is bounded and finite in a given region and zero outside this region. This is easy to realize for solid or liquid samples or gas in a box, but it is not so easy for electric and magnetic fields.

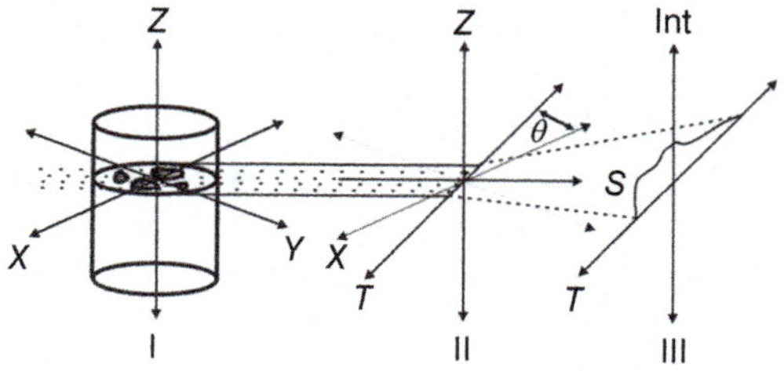

Figure 14.17 CT-geometry to scan an object. Position (I) is the object location with object coordinates (*x*, *y*, *z*), (II) shows the coordinate system for a certain (rotated) orientation of the object with respect to the 2D detector system, and (III) shows the measured projection for this rotated orientation of the sample, *T* = translation direction. (Kharfi http://www.intechopen.com/books/imaging-and-radioanalytical-techniques-in-interdisciplinary-research-fundamentals-and-cutting-edge-applications/mathematics-and-physics-of-computed-tomography-ct-demonstrations-and-practical-examples. Used under CC-BY 3.0 license http://creativecommons.org/licenses/by/3.0/.)

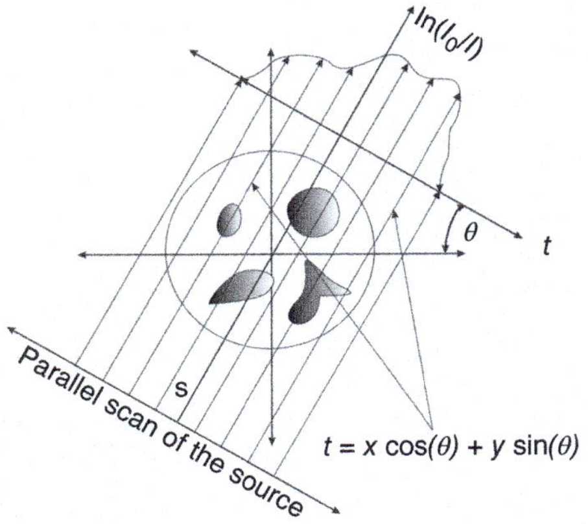

Figure 14.18 A sample is scanned under the angle θ. The straight line across the slice is given by $t = x\cos(\theta) + y\sin(\theta)$. t is the scanning direction perpendicular to the ray direction s. (Kharfi http://www.intechopen.com/books/imaging-and-radioanalytical-techniques-in-interdisciplinary-research-fundamentals-and-cutting-edge-applications/mathematics-and-physics-of-computed-tomography-ct-demonstrations-and-practical-examples. Used under CC-BY 3.0 license http://creativecommons.org/licenses/by/3.0/.)

The sample is scanned under angles θ with $0 < \theta < 180°$ (360°) and the transmitted intensity is registered as a function of the translation t. To get a 2D image of the slice, we follow the theoretical description given by Kak and Slaney [11], and Rosenfeld and Kak [12]. A set of pencil rays traverses the sample as shown in Figure 14.19 and the transmitted intensity is measured as a function of the translation t. The transmitted intensity is given by Lambert's law as

$$I(x, y) = I_0 \exp\left(-\int_{\text{ray}} \mu(x, y)\right) \tag{14.20}$$

where I_0 is the intensity of the incident neutron beam. Now one defines a new rectangular coordinate system (t,s) that expresses the rotated detector system (t,s) with respect to the fixed sample system (x,y) (or vice versa, if the sample is rotated as it is done in neutron tomography). The path across the sample is given in terms of t and θ by eliminating s,

$$t = x \cdot \cos(\theta) + y \cdot \sin(\theta) \tag{14.21}$$

From Eq. (14.20) one can define a so-called projection $P_\theta(t)$ as

$$P_\theta(t) = \ln\left(\frac{I_0}{I}\right) = \int_{\text{path}} \mu(x, y)\,\mathrm{d}s \tag{14.22}$$

The set of all projections $P_\theta(t)$ of $\mu(x,y)$ is called the *Radon transform* of $\mu(x,y)$. From these projections a 2D image can be reconstructed using the "Fourier slice theorem."

14.2.2 Slice Theorem, Shannon Theorem

The basic idea of image reconstruction from projections is contained in the Fourier slice theorem. It states that the FT of a projection $P_\theta(t)$ of $\mu(x,y)$ is a subset of the FT of $\mu(x,y)$:

$$\mathrm{FT}\{P_\theta(t)\} = S_\theta(w) \subset \mathrm{FT}\{\mu(x, y)\} = M(u, v) \tag{14.23}$$

To demonstrate this relation, we follow again the treatise in [10, 11]. We assume $\mu(x,y)$ to be bounded and finite, so $\mu(x,y)$ has a FT $M(u,v)$;

$$M(u, v) = \int_{-\infty}^{\infty}\int_{-\infty}^{\infty} \mu(x, y) \cdot \mathrm{e}^{2\pi \cdot i \cdot (ux+yv)} \cdot \mathrm{d}x \cdot \mathrm{d}y \tag{14.24}$$

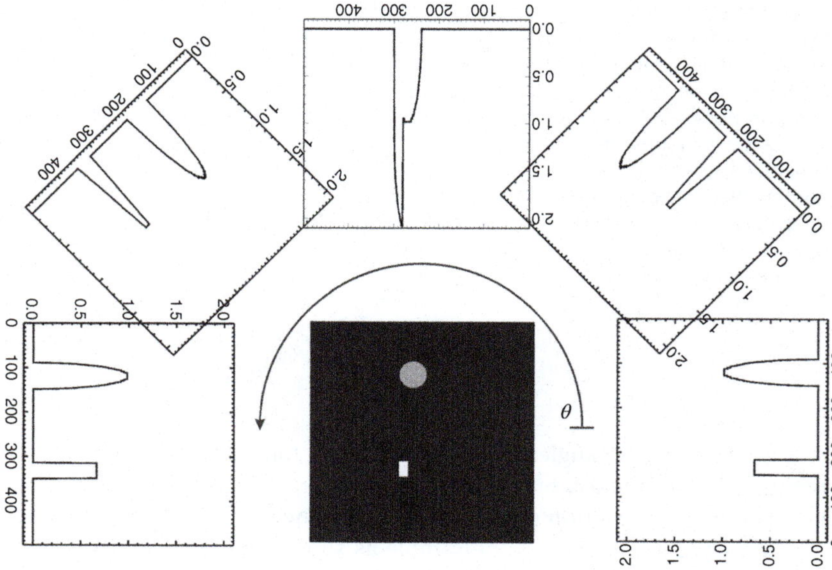

Figure 14.19 Scanning of a simple 2D object (square and circle) and corresponding projections. (Kharfi http://www.intechopen.com/books/imaging-and-radioanalytical-techniques-in-interdisciplinary-research-fundamentals-and-cutting-edge-applications/mathematics-and-physics-of-computed-tomography-ct-demonstrations-and-practical-examples. Used under CC-BY 3.0 license http://creativecommons.org/licenses/by/3.0/.)

To prove the Slice theorem one can consider the simple case of $\theta = 0$. In this case, the two coordinate systems (x,y) and (u,v) coincide, that is, x is parallel to u and y is parallel to v. A projection $P_\theta(t)$ in this case is simply the integral

$$P_{\theta=0}(t) = \int_{-\infty}^{\infty} \mu(x,y) \cdot \mathrm{d}y \tag{14.25}$$

If we look at the FT of $\mu(x,y)$ in the case of $\theta = 0$, then the FT of the object becomes (due to $v = 0$)

$$\begin{aligned} M(u,0) &= \int_{-\infty}^{\infty}\int_{-\infty}^{\infty} \mu(x,y) \cdot \mathrm{e}^{-2\pi \cdot i \cdot (ux)} \cdot \mathrm{d}x \cdot \mathrm{d}y \\ &= \int_{-\infty}^{\infty} \left\{ \int_{-\infty}^{\infty} \mu(x,y) \cdot \mathrm{d}y \right\} \mathrm{e}^{-2\pi \cdot i \cdot (ux)} \cdot \mathrm{d}x \end{aligned} \tag{14.26}$$

Now, one can substitute Eq. (14.23) into Eq. (14.24) and get

$$M(u,0) = \int_{-\infty}^{\infty} P_{\theta=0}(x) \cdot \mathrm{e}^{2\pi \cdot i \cdot ux} \cdot \mathrm{d}x \tag{14.27}$$

If the FT of $P_\theta(t)$ is given by

$$S_\theta(w) = \int_{-\infty}^{\infty} P_\theta(t) \cdot \mathrm{e}^{-2\pi \cdot i \cdot w \cdot t} \cdot \mathrm{d}t \tag{14.28}$$

then $S_{\theta=0}(w)$ equals $M(u,0)$, the FT of $\mu(x,y)$ for $\theta = 0$. Because the orientation of the object in the (u,v) system is arbitrary with respect to the (x,y) system this relation enables us to use it for all angles θ keeping in mind that the FT conserves the rotation taken in real space. The FT of $P_\theta(t)$ yields the values of the FT of $\mu(x,y)$. A dense set of values in the Fourier space (given in polar coordinates due to the polar scanning of the object) can be approximated to rectangular coordinates and back transformed into real space.

An important question must be answered: How many projections yield the best result? This is answered by the so-called Shannon theorem which states that a unique reconstruction is given if the image is sampled with a frequency that is twice higher than the highest frequency in the Fourier transformed image. For the parallel mode scanning using N steps for one projection, one can easily derive that M projections (angles θ) are required to fulfill the Shannon theorem.

If D is the diameter of the object that shall be scanned, and Δx is the difference between two scanning points, then the number of scanning points/projection is

$$N = \frac{D}{\Delta x} \tag{14.29}$$

and Δx fulfills the Shannon condition. The rotation of the sample around 180° means a path length y of an edge point $\pi D/2$. We require that $\Delta x = \Delta y$, so that the number projections M must be

$$M = \frac{\frac{D}{2}\pi}{\Delta y} = \frac{D \cdot \pi}{2 \cdot \Delta y}$$

Using now Eq. (14.27) we get ($\Delta x = \Delta y$)

$$M \geq \frac{\pi}{2} \cdot N \tag{14.30}$$

N and M determine the quality of the reconstruction. Note that Eq. (14.28) is only one of some other requirements. If M does not fulfill the inequality Eq. (14.30), that is, if less than M projections are registered, one can then calculate the reduced $D = D^*$ that fulfills the Shannon condition with Eqs. (14.29) and (14.30).

14.2.3 Image Reconstruction

As mentioned in the previous chapter, a 2D reconstruction of the slice $\mu = \mu(x,y)$ can be yielded by the back transformation (inverse Fourier transformation) of the FT $S_\theta(w)$. However, the more efficient (and elegant) way is the so-called filtered backprojection (FBP), which is described in detail in [10, 11]. The basic idea is embedded in the scanning mode.

The sample (slice) is scanned from 0° to 180°, which involves that $P_\theta(t) = P_{\theta+180°}(-t)$ and suggests to use a polar coordinate system rather than a rectangular one. The projections are registered in polar coordinates, so that the Fourier transformed projections of $\mu(x,y)$, the $S_\theta(w)$ data, are discrete polar function values. So it seems to be apparent to write $\mu(x,y)$ as a FT representation in polar coordinates. The function $\mu(x,y)$ written as the Fourier representation with $M(u,v)$ in rectangular coordinates is

$$\mu(x,y) = \int_{-\infty}^{\infty}\int_{-\infty}^{\infty} M(u,v)\,\mathrm{e}^{2\pi\cdot i(ux+yv)}\,\mathrm{d}x\,\mathrm{d}y \tag{14.31}$$

The same representation in polar coordinates is

$$\mu(x,y) = \int_{0}^{2\pi}\int_{-\infty}^{\infty} M(w,\theta)\cdot \mathrm{e}^{+2\pi\cdot i\cdot w\cdot(x\cos(\theta)+y\sin(\theta))}w\cdot \mathrm{d}w\cdot \mathrm{d}\theta \tag{14.32}$$

The transformation of the (x,y) system into the (w,θ) system uses the relations

$$\begin{aligned} u &= w \cdot \cos(\theta) \\ v &= w \cdot \sin(\theta) \\ du \cdot dv &= w \cdot dw \cdot d\theta \end{aligned} \tag{14.33}$$

Substituting these formulas in Eq. (14.32) we get (with the slice theorem)

$$\begin{aligned} \mu(x,y) &= \int_0^{\pi} \int_{-\infty}^{\infty} [M(w,\theta) \cdot e^{+2\pi \cdot i \cdot w \cdot t} |w| \cdot dw] \cdot d\theta \\ &= \int_0^{\pi} \left[\int_{-\infty}^{\infty} S_{\theta}(w) \cdot e^{+2\pi \cdot i \cdot w \cdot t} |w| \cdot dw \right] \cdot d\theta \end{aligned} \tag{14.34}$$

The integral given in brackets can be considered as the FT of $P_{\theta}(t)$, however, multiplied by the $|w|$-function which is a special filtering functions of the frequencies in the frequency space:

$$Q_{\theta}(t) = \int_{-\infty}^{\infty} S_{\theta}(w) \cdot |w| \cdot e^{+2\pi \cdot i \cdot w \cdot t} \cdot dw \tag{14.35}$$

With $t = x \cdot \cos(\theta) + y \cdot \sin(\theta)$, $\mu(x,y)$ simply becomes

$$\mu(x,y) = \int_0^{\pi} Q_{\theta}(x \cos(\theta) + y \sin(\theta)) \cdot d\theta \tag{14.36}$$

A product in Fourier space inverse Fourier transformed is the convolution of the inverse Fourier transformed functions, that is, we use the relation

$$\mathrm{FT}^{-1}\{S_{\theta}(w) \cdot |w|\} = P_{\theta}(t) \otimes \mathrm{FT}^{-1}\{|w|\} \tag{14.37}$$

The convolution operation is denoted by $\otimes$. The function $|w|$ is not a square integrable function, so it has no inverse FT; however, $\mathrm{FT}^{-1}\{|w|\}$ can be approximated by several filter response functions (convolution kernels, filter functions). The reconstruction of $\mu(x,y)$ from projections $P_{\theta}(t)$ works now in the following way: each $P_{\theta}(t)$ is convoluted with a proper filter function (e.g., Shepp–Logan) and the resulting values are "smeared" over the (x,y)-planes as demonstrated in Figure 14.19.

In Figure 14.20, four different reconstructions are shown. The (x,y) plane looks like the first image if only one projection is used, then like the second image if two projections, the $\theta = 0$ and 90° projections, are added, and so on. Summing up 1000 projection yields image (d) in Figure 14.20. Note that the star artifacts are nearly suppressed.

To illustrate one of the filtered projection, a part of the filtered function (circle) is extracted and shown in Figure 14.21.

14.3 New Developments in Neutron Tomography

The examples above identify another question: What can be done if attenuation of neutrons (or X-rays) as an imaging signal does not work? Or to state the physical problem: How can structures in objects that do not contribute to absorption be visualized?

Theorem (without proof): Any interaction that can be measured can be used as imaging signal.

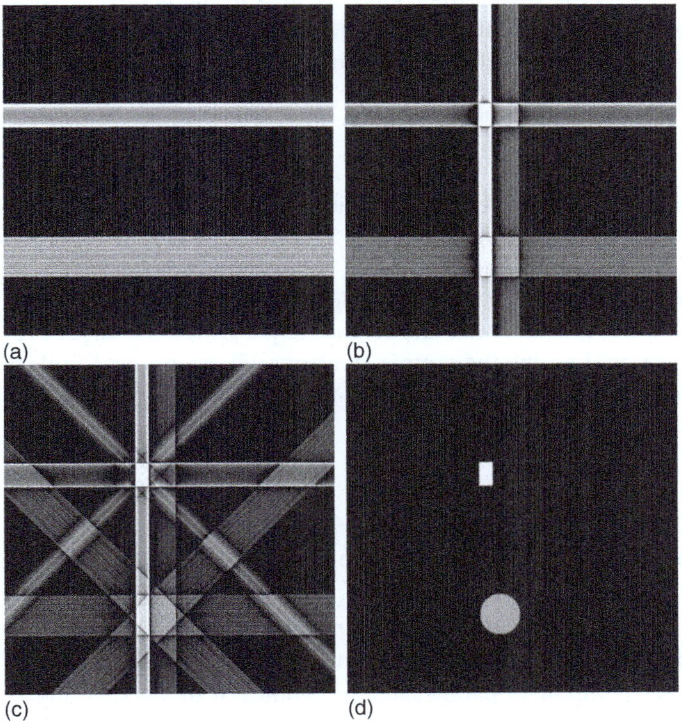

Figure 14.20 The convoluted projections are smeared back (backprojected) over and summed up in the (x, y) plane (Kharfi http://www.intechopen.com/books/imaging-and-radioanalytical-techniques-in-interdisciplinary-research-fundamentals-and-cutting-edge-applications/mathematics-and-physics-of-computed-tomography-ct-demonstrations-and-practical-examples. Used under CC-BY 3.0 license http://creativecommons.org/licenses/by/3.0/.)

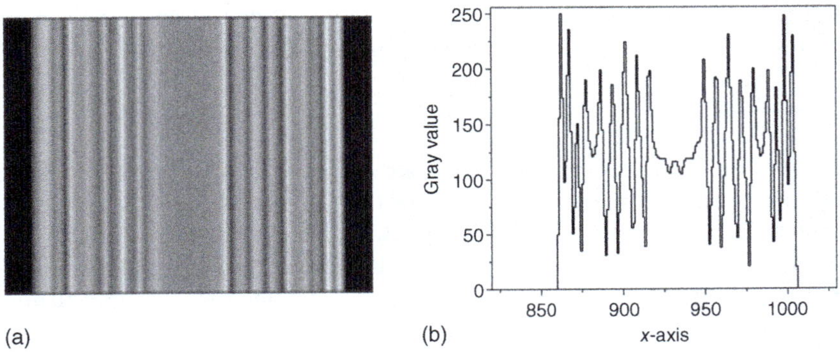

Figure 14.21 (a) Part of the image "1 projection" (see Figure 14.21), rotated by 90° and (b) the corresponding filtered function (=convolution of the projection P with a filter function) which is smeared back (backprojected) over the (x, y) plane which yields the left part of the figure.

As was shown in Section 14.3 the size of the coherence area of the neutron (X-ray) wave determines the effect: if the coherent width (lateral coherence length) is smaller than the object, then refraction occurs. If the object is smaller than the coherent width, small angle scattering occurs.

14.3.1 Refraction

We consider a function $n(x,y)$ that describes the position-dependent index of refraction: $n(x,y) \neq 1$ inside the sample and $n(x,y) = 1$ outside. From the Snellius law one gets $n_1 \sin(\varepsilon_1) = n_2 \sin(\varepsilon_2)$, with ε_1 the corresponding angle of incidence and ε_2 the refraction angle; n_1 and n_2 are the corresponding indices of refraction of medium 1 and medium 2. From simple geometrical optics it is also well known that larger the ε_1 the larger is ε_2 up to the case of total reflection. Leaving the sample, neutrons will have a final angle of deflection Δ, which is the sum of all individual deflections δ_l ($\delta_l = \overline{\varepsilon}_l - \varepsilon_{l+1}$) along the path through the sample:

$$\Delta = \sum_{i=1}^{n} \delta_i \tag{14.38}$$

A deflection δ_l from the preceding direction is only present at the adjacent volumes with different indices of refraction and if $\overline{k} \cdot \nabla n(x,y) \sim 0$ ($\overline{k}$ is the neutron wave vector). One can determine $\nabla n(x,y)$ and by that $n(x,y)$ by measuring Δ of each path (line integral, Eq. (14.38)) and varying the orientation θ of the sample from 0° to 180°. If $\overline{k} = \overline{k}_{\parallel}, \overline{k}_{\perp}$ is represented by a component parallel and perpendicular to $\nabla n(x,y)$, and $\overline{k}_{\perp}$ is the component of $\overline{k}$ parallel to $\nabla n(x,y)$, then a point $P_\theta(t)$ of a projection is given by

$$P_\theta(t) = P(t,\theta) = \int_{\text{path}} \nabla n(x,y) \cdot \overline{k}_{\perp} \cdot \mathrm{d}s \tag{14.39}$$

If the incident direction is known, Δ_t and $P_\theta(t)$ can be determined from the experiment. The total set of functions $P_\theta(t)$ is then another form of the Radon transform of $n(x,y)$. The reconstruction of $n = n(x,y)$ from projections can follow the well-known mathematical procedures. An estimation of the total angle of deflection Δ is given by the calculation of the index of refraction n, which is given for neutrons by

$$n = 1 - \frac{\lambda^2 N b_c}{2\pi}$$

With $\lambda = 4.76 \times 10^{-10}$ (m), N = number of unit cells per cubic meter $\sim 5 \times 10^{28}$, and b_c = coherent scattering length $\sim 10^{-15}$ (m) of the sample volume of interaction; $1 - n \sim 10^{-6}$ gives the size of the angle of refraction which has to be measured. This can only be done with a special double crystal instrument as described in [13, 14]. A simple calculation of Δ_t for a given orientation θ for a brass sample (Figure 14.22, ∅ = 10 mm four holes with ∅ = 1, 2, 3, and 4 mm) is shown in Figure 14.22. Figure 14.23 shows the tomograms of a brass cylinder, ∅ 10 mm, ∅ holes: 1, 2, 3, and 4 mm.

A 3D-refraction tomography could also be realized [15]. In worse cases, there is no absorption contrast; only refraction contrast yields a good reconstruction as demonstrated in Figure 14.24.

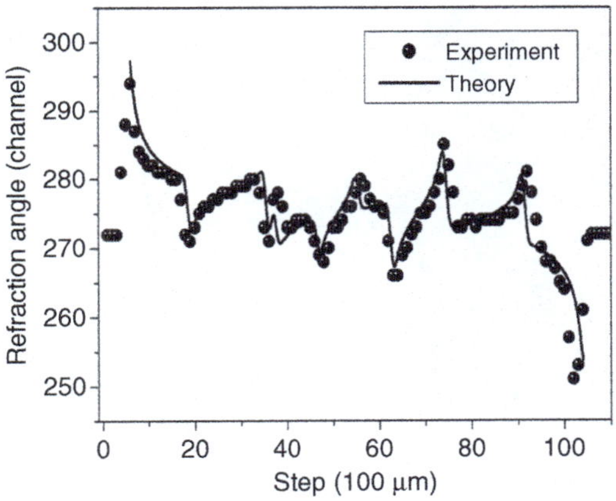

Figure 14.22 Calculated and measured peak shift for a certain orientation θ of a brass sample.

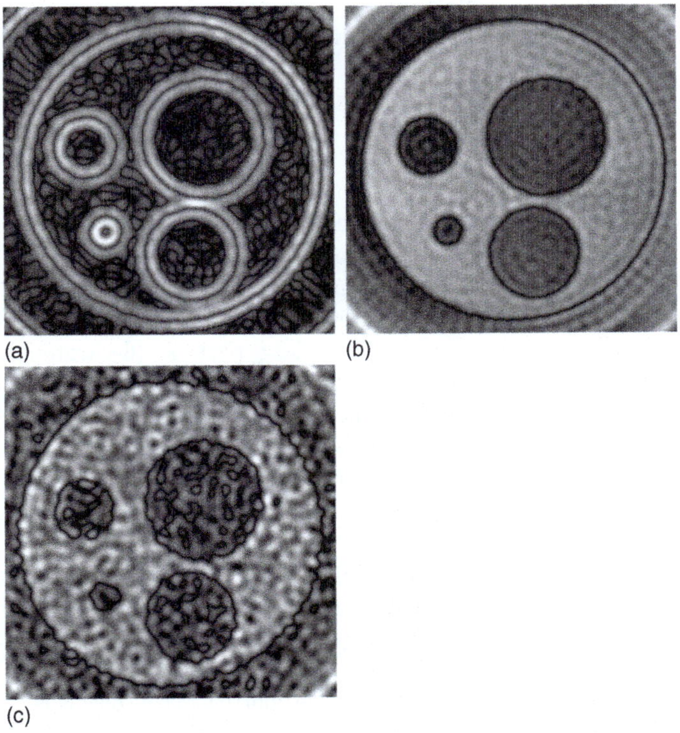

Figure 14.23 Reconstruction using (a) refraction data only, (b) integrated refraction data, or (c) absorption data only.

14.3.2 Ultra-Small Angle Neutron Tomography

In the case of USANS (ultra-small angle neutron scattering) tomography $\mu(x,y)$ has to be replaced by a suitable function that represents the USANS effects in the sample. The broadening B (nm^{-1}) of a scattering function due to (ultra) small angle scattering can

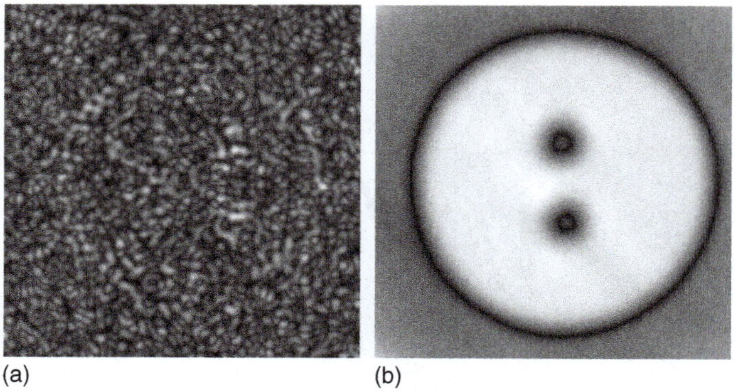

Figure 14.24 Al-sample: diameter 6 mm; two holes, each 2 mm. (a) Attenuation data and (b) reconstruction with refraction data only.

be (roughly) approximated by a Gaussian distribution and – involving multiple scattering – written as [16]

$$B_\theta(t) = \sqrt{\int_{\text{path}} \frac{\sigma(x,y) \cdot N(x,y)}{R(x,y)^2} \cdot \mathrm{d}s} \tag{14.40}$$

where σ is the scattering cross-section, N is the particle density, and R is a parameter with the dimension of a length specifying an average size or correlation length in the scattering object (corresponding to the Gaussian approximation). Now one can use B, B^2, as imaging signals to use ultra-small angle scattering for image reconstruction. Therefore, the detection of clusters of approximately 150 nm large particles (ß-carotene), dissolved in D_2O (~5–12%), could be realized. In this experiment refraction, USANS and absorption were measured simultaneously with the double crystal diffractometer [13, 14]: the shift of the rocking curve yields the final deviation and determines the refraction, the broadening the small angle scattering, and the decrease of the integral intensity the absorption of the neutron beam. All these data were derived from a single rocking curve.

Figure 14.25 shows the first successful results. Combining USANS and refraction data yields a complete reconstruction. With pure attenuation, refraction, or USANS data only incomplete, imperfect information is reconstructed (Figure 14.26).

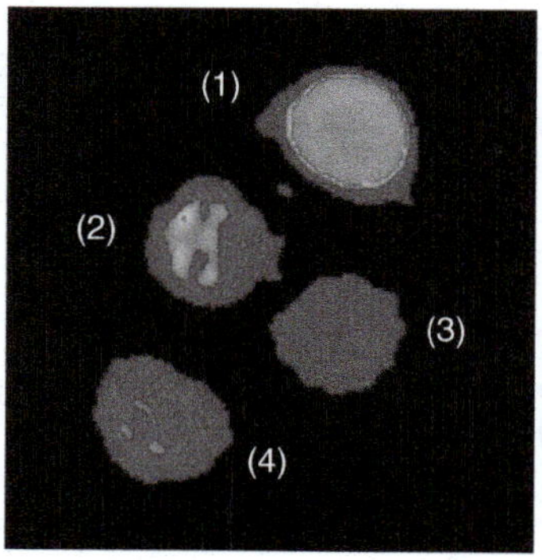

Figure 14.25 2D reconstruction from USANS data only. Blue ~3.8 wt%, green ~5.8 wt%, red ~12 wt%; particle size: ~150 nm (β-carotene). (Strobl 2004 [14]. Reproduced with permission of Elsevier.)

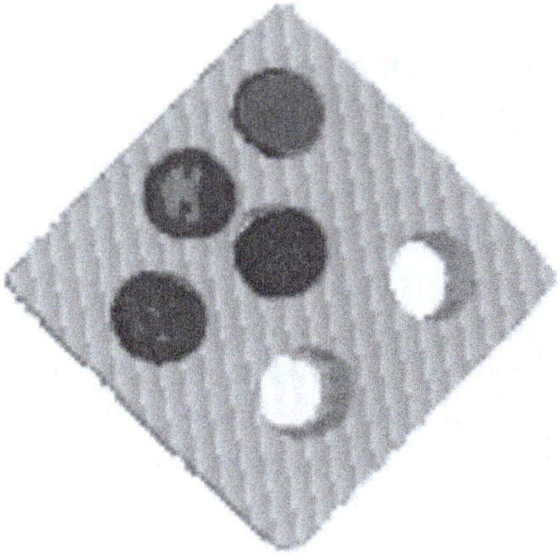

Figure 14.26 Reconstruction of an aluminum box having six holes, four were filled with different weight percentages (wt%) of D_2O and β-carotene (see also Figure 14.25). (Strobl 2004 [14]. Reproduced with permission of Elsevier.)

14.3.3 Radiography and Tomography with Polarized Neutrons

In [7] and some years later [17, 18] it was demonstrated that radiography and tomography with polarized neutrons and their applications show the large potential of CT in material research.

Neutrons have a spin 1/2 and interact with magnetic fields. The spin rotates around a magnetic field with the so-called Larmor frequency $\omega_L = \gamma_L \cdot B$; γ_L is the gyromagnetic ratio of the neutron, $\gamma_L = g\cdot\mu_N/\hbar = -1.83247\times10^8$ rad s^{-1} ET^{-1}. With $\omega_L = \phi/t$ the angle of rotation ϕ is given by $\omega_L\cdot t$, and t is the time the spin moves in the magnetic field B. The number of spin rotations in a field B is determined by the length of the path "s" of the neutron in B, that is by the path integral

$$\int_{\text{path}} B\cdot \mathrm{ds}\cdot\phi = \omega_L\cdot t = \gamma_L\cdot B\cdot t = \frac{\gamma_L}{v}\int B\cdot \mathrm{ds} = \frac{\gamma_L\cdot m}{h}B\cdot s\cdot\lambda \tag{14.41}$$

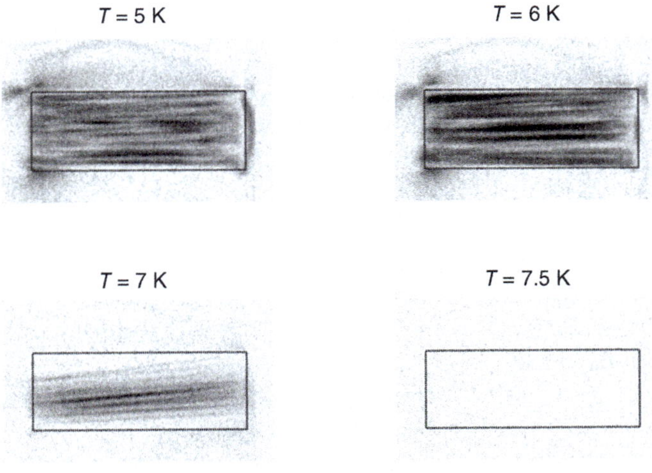

Figure 14.27 Observation of temperature-dependent flux trap in a Pb cylinder (diameter 12 mm; length = 30 mm). Fringes are locations of constant depolarization due to spin interaction with the trapped field, fringes vanish if $T > T_c = 7.2$ K.

v is velocity of the neutron, B is the magnetic field, m is the neutron mass, s is the path length in the magnetic field B, λ is the wavelength, and h is the Planck constant. The velocity and the wavelength are related to each other (de Broglie) as $\lambda = h/p$, $p = m \cdot v$, $m =$ mass of the neutron, $v =$ velocity. $h/m = 3.956\,034 \times 10^{-7}$ ($m^2\,s^{-1}$), for $v = 3956\,m\,s^{-1}$, the wavelength $\lambda = 1 \times 10^{-10}$ (m) $= 0.1$ (nm) $= 1$ Å. With a polarized neutron beam that can be spin-analyzed behind a sample a new field of neutron imaging is opened. Especially in the field of superconductivity a number of new results could be already obtained by imaging the bulk of samples when they are in the Meissner phase or intermediate state [19–21], see Figure 14.27.

These methods offer new possibilities for a dedicated computerized tomography, especially for nondestructive testing and material research. Coherence properties, phase-based radiography, and imaging with polarized neutrons will all contribute to a much more detailed investigation of solid-state physics problems.

References

1 Glasstone, S.G. and Edlund, M.C. (1961) *Kernreaktortheorie*, Springer-Verlag, Wien, p. 30.

2 Harms, A.A. and Wyman, D.R. (1986) *Mathematics and Physics of Neutron Radiography*, Reidel, Dordrecht, Holland.

3 Krestel, E. (1980) *Bildgebende Systeme für die medizinische Diagnostik, Grundlagen*, Siemens Aktiengesellschaft, Berlin, München.

4 Kardjilov, N., Baechler, S., Bastürk, M., Dierick, M., Jolie, J., Lehmann, E., Materna, T., Schillinger, B., and Vontobel, P. (2003) New features in cold neutron radiography and tomography: part II. Applied energy-selective neutron radiography and tomography. *Nucl. Instrum. Methods Phys. Res., Sect. A*, **501**, 536–546.

5 Treimer, W., Kardjilov, N., Feye-Treimer, U., Hilger, A., Manke, I., and Strobl, M. (2005) Absorption- and phase-based imaging signals for neutron tomography. *Adv. Solid State Phys.*, **45**, 407.

6 Wu, X. and Liu, H. (2003) Clinical implementation of x-ray phase-contrast imaging: theoretical foundations and design considerations. *Med. Phys.*, **30** (**8**), 2169.

7 Treimer, W., Hilger, A., Kardjilov, N., and Strobl, M. (2005) Review about old and new imaging signals for neutron computerized tomography. *Nucl. Instrum. Methods Phys. Res., Sect. A*, **542**, 367.

8 Pogany, A., Gao, D., and Wilkins, S.W. (1997) Contrast and resolution in imaging with a microfocus x-ray source. *Rev. Sci. Instrum.*, **68**, 2774.

9 Podurets, K.M., Somenkov, V.A., and Shilstein, S.S. (1989) Neutron radiography with refraction contrast. *Physica B*, **156**, 691–693.

10 Allman, B.E., McMahon, P.J., Nugent, K.A., Paganin, D., Jacobson, D.L., Arif, M., and Werner, S.A. (2000) Phase radiography with neutrons. *Nature*, **408**, 158.

11 Kak, A.C. and Slaney, M. (1999) *Principles of Computerized Tomographic Imaging*, IEEE Press, New York.

12 Rosenfeld, A. and Kak, A.C. (1982) *Digital Picture Processing, Computer Science and Applied Mathematics*, Academic Press, New York.

13 Treimer, W., Strobl, M., Hilger, A., Seifert, C., and Feye-Treimer, U. (2003) Refraction as imaging signal for computerized (neutron) tomography. *Appl. Phys. Lett.*, **83**, 2.

14 Strobl, M., Treimer, W., Hilger, A., and Feye-Treimer, U. (2004) Neutron tomography in double crystal diffractometers. *Physica B*, **350**, 155.

15 Strobl, M., Treimer, W., and Hilger, A. (2004) First realisation of a three-dimensional refraction contrast computerised neutron tomography. *Nucl. Instrum. Methods Phys. Res., Sect. B*, **222**, 653.

16 Strobl, M., Treimer, W., and Hilger, A. (2004) Small angle scattering signals for (neutron) computerized tomography. *Appl. Phys. Lett.*, **85** (**3**), 488.

17 Pfeiffer, F., Weitkamp, T., Bunk, O., and David, C. (2006) Phase radiography with neutrons. *Nat. Phys.*, **2**, 258.

18 Treimer, W., Hilger, A., Strobl, M., Kardjilov, N., and Manke, I. (2006) Wavelength tunable device for neutron radiography and tomography. *Appl. Phys. Lett.*, **89**, 203504.

19 Treimer, W., Ebrahimi, O., Karakas, N., and Prozorov, R. (2012) Polarized neutron imaging and 3D calculation of magnetic flux trapping in bulk of superconductors. *Phys. Rev. B*, **85**, 184522-1–184522-9.

20 Treimer, W., Ebrahimi, O., and Karakas, N. (2012) Observation of partial Meissner effect and flux pinning in superconducting lead containing non-superconducting parts. *Appl. Phys. Lett.*, **101**, 162603–1–162603–4.

21 Treimer, W. (2014) Radiography and tomography with polarized neutrons. *J. Magn. Magn. Mater.*, **350**, 188–198.

15

Neutron and Synchrotron-Radiation-Based Imaging for Applications in Materials Science – From Macro- to Nanotomography

Felix Beckmann

In recent years, 2D and 3D imaging has become a standard tool for scientific and industrial applications in materials science. Employing the high-intensity sources at neutron facilities and the highly brilliant and collimated X-ray sources at second and third-generation synchrotron radiation (SR) facilities, specialized contrast techniques were developed. Attenuation-contrast and phase-contrast techniques were optimized to analyze specimens used in basic research studies as well as structural components used in industrial applications.

The image formation at neutron sources was described in Chapter 14. Here, a short introduction to the image formation at SR sources is given. The concepts presented here are also valid for investigations with neutrons. After a short discussion of image creation and quality in tomographic applications, the instrumentations operated by the Helmholtz-Zentrum Geesthacht, Germany, for computerized tomography using neutrons (NCT) and microtomography using synchrotron radiation (SRμCT) is presented.

15.1 Introduction

The interaction of neutrons with matter is described in detail in Chapter 14. Imaging at second- and third-generation SR facilities makes use of the low divergence and high intensity of the source to record parallel 2D projections of samples using monochromatic X-rays. The principle setups and the concept of image formation are presented in Figure 15.1.

15.1.1 Attenuation-Contrast Projections

To obtain attenuation-contrast projections, radiograms are taken at minimal detector–sample distance. A single projection is obtained by recording the primary intensity $I_0(x, z)$ and the attenuated intensity $I(x, z)$, where x denotes the horizontal and z the vertical dimension. The attenuation projection $P(x, z)$ can be determined by

$$P(x,z) = \int \mu(x,y,z)\,\mathrm{d}y = -\ln\left[\frac{I(x,z)}{I_0(x,z)}\right] \tag{15.1}$$

Neutrons and Synchrotron Radiation in Engineering Materials Science: From Fundamentals to Applications, Second Edition. Edited by Peter Staron, Andreas Schreyer, Helmut Clemens, and Svea Mayer.

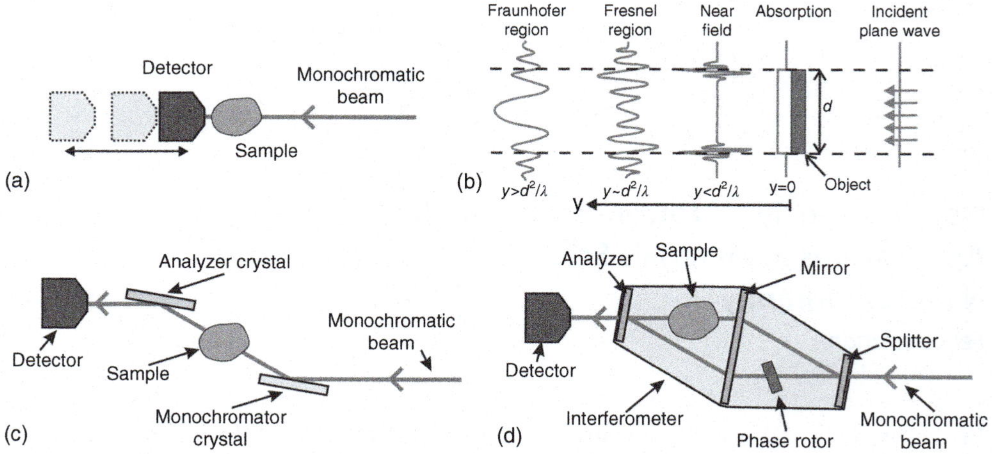

Figure 15.1 (a) Schematic setup used for imaging at different sample–detector distances. (b) Coherent image formation with partially coherent beam depending on the object–detector distance *y*, sample dimension *d*, and wavelength λ. (c) Concept for phase outline contrast applying diffraction-enhanced imaging. A monochromator and analyzer crystal is used to record images at different positions of the rocking curve of the analyzer crystal. (d) Experimental setup for interferometric phase contrast. A monolithic X-ray interferometer is used to divide the monochromatic beam into two coherent beams. The phase rotor introduces a defined phase shift and the direct phase projection can be obtained.

Using the atomic number Z, photon energy E, and mass density ρ, the attenuation coefficient μ except close to absorption edges can be estimated by

$$\mu \propto \frac{Z^4}{E^3}\rho \tag{15.2}$$

The photon energy E is matched with the sample characteristics in order to obtain sufficient contrast in the radiogram. Therefore, only the structure of the most absorbing material of the specimen can be investigated well. To investigate samples consisting mainly of light elements, the more sensitive phase contrast has to be used [1].

15.1.2 Phase-Contrast Projections

X-rays and neutrons are not only attenuated by the specimen, they are also shifted in phase. This behavior is described by the complex index of refraction

$$n = 1 - \delta - i\beta \tag{15.3}$$

where the real part δ corresponds to the phase shift due to refraction and the imaginary part β belongs to absorption. Depending on the type of the source different phase-contrast imaging techniques were developed, which are suited for different types of sample composition. At incoherent X-ray sources of second-generation SR facilities, multiple crystal reflections or X-ray interferometers have to be used. At third-generation sources, phase-enhancement can be achieved by varying the sample–detector distance due to the partial coherent X-ray beam [2, 3]. In Figure 15.1b, the coherent image formation of a transmission image is shown. Using sample dimension d and wavelength λ four regions of sample–detector distances y can be distinguished [4]:

$y=0$	absorption	the image is a pure absorption image
$y<d^2/\lambda$	near field	contrast is given by sharp changes in the refractive index, that is, at interfaces
$y\sim d^2/\lambda$	Fresnel region	the image loses more and more resemblance with the object
$y>d^2/\lambda$	Fraunhofer region	the image intensity is the Fourier transform of the object transmission function

15.1.3 Phase-Enhanced Projections

Using a partially coherent beam, images are taken in the near field. Rotating the sample and performing a tomographical scan emphasizes the sharp changes in the refractive index of the sample. Then, this is called *phase outline contrast* [5, 6]. A more sensitive technique makes use of multiple crystal reflections. In Figure 15.1c, the schematic setup is shown. Images at different positions of the rocking curve of the two crystals are recorded. From these images, an absorption image and a phase-enhanced image can be obtained [7, 8]. Using this as the contrast method in tomography again results in mapping of borders of changes in the refractive index.

15.1.4 Direct Phase-Contrast Projections

In order to obtain the refractive index in the tomogram, phase projections of the sample have to be measured. A direct technique makes use of an X-ray interferometer, which allows for the measurement of a phase projection modulo 2π. The principle setup is shown in Figure 15.1d. For one phase projection, interference patterns – with and without the sample – are obtained at different overall phase shift introduced by the phase shifter [9–11]. Using the electron density σ, the phase shift ϕ is given by

$$\phi \propto \frac{\sigma}{E} \tag{15.4}$$

15.1.5 Indirect Phase-Contrast Projections

An indirect method to measure the phase projection makes use of the partial coherence of third-generation SR beams. By varying the sample–detector distance and recording images, different spatial frequencies in the projected phase images vanish (Talbot effect). The so-called holographic reconstruction procedure then determines the quantitative phase projection [12]. The combination of the obtained images at several detector distances and the data reconstruction however is sophisticated and is, therefore, not offered as a normal user experiment at synchrotron sources. A fast method, called propagation-based phase contrast, is using one detector distance, which allows for high spatial resolution. Due to the contrast principle the method is only feasible for defined sample systems where β/δ is constant. Furthermore, the tomograms are not representing quantitative values [13, 14]. The only quantitative method currently offered as a user experiment is the grating-based phase contrast. It is established at SR facilities for spatial resolutions up to 5 μm [15–17]. The experimental setup used at beamline for providing grating-based phase-contrast microtomography at P07/PETRA III is shown in Figure 15.14 [18].

15.2 Parallel-Beam Tomography

The concept of 3D imaging using parallel-beam geometry is based on recording a set of 2D projections of a specimen at different sample rotations [19]. In Figure 15.2, the principle setup used for parallel-beam tomography is given. Here, a virtual test object consisting of several spheres and cubes is used to simulate the tomographic measurement, reconstruction data evaluation, and visualization process.

The detector size is 501 × 501 pixels. The basic volume of the sample consists of 501 × 501 × 501 voxels. The sample is defined by the combination of the following objects presented in Table 15.1. The position is given in percentage of the maximal size of 501. Furthermore, two boxes are tilted in the volume. The maximum projected attenuation of the sample is about 2. For the shown simulated measurement projections were calculated from 0° to 180° with an angular increment of 0.5°.

15.2.1 Measurement and Reconstruction

In Figure 15.2, the measurement of a single projection is shown. If the rotation axis is perpendicular to the beam, a single slice perpendicular to the axis is projected onto a single line on the 2D detector for all sample rotations. Therefore, measurement and

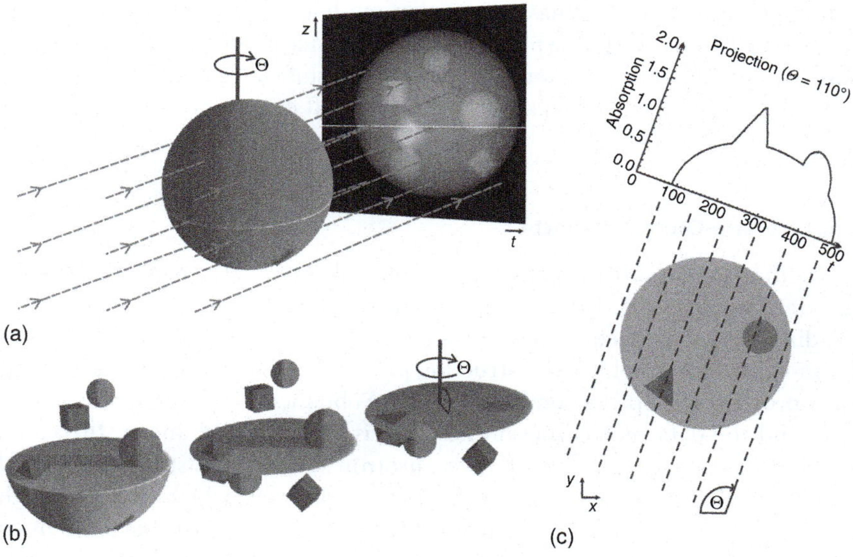

Figure 15.2 Schematic setup used for recording of 2D projections in parallel-beam geometry (a). The sample consists of several spheres and cubes. Different colors represent different attenuation (b). If the rotation axis of the sample is perpendicular to the beam, the slice perpendicular to the axis is projected onto the same line of the 2D detector for all sample rotations (c).

Table 15.1 Composition of the virtual sample used for simulation.

Shape	x (%)	y (%)	z (%)	Radius/size (%)	Attenuation	Color
Sphere	45	42	50	40	0.002595	Yellow
Sphere	20	45	50	10	0.006986	Red
Sphere	60	40	25	11.8	0.006986	Red
Sphere	40	50	75	6.67	0.006986	Red
Box	65	22	40	12.5	0.009980	Blue
Box	58	62	60	11.1	0.009980	Blue
Box	26	26	37	10	0.009980	Blue

reconstruction can be reduced to a 2D slice with 1D projections. The mathematical description of this tomographic measurement for a single slice $f(x, y)$ is given by the Radon transform [20]

$$P_\theta(t) = \iint f(x, y)\delta(x\cos\theta + y\sin\theta - t)\,dx\,dy \tag{15.5}$$

The tomographic reconstruction is the implementation of the inverse Radon transform. A schematic view of the measurement and reconstruction process is given in Figure 15.3. The complete measurement for a single slice $f(x, y)$ is the so-called sinogram

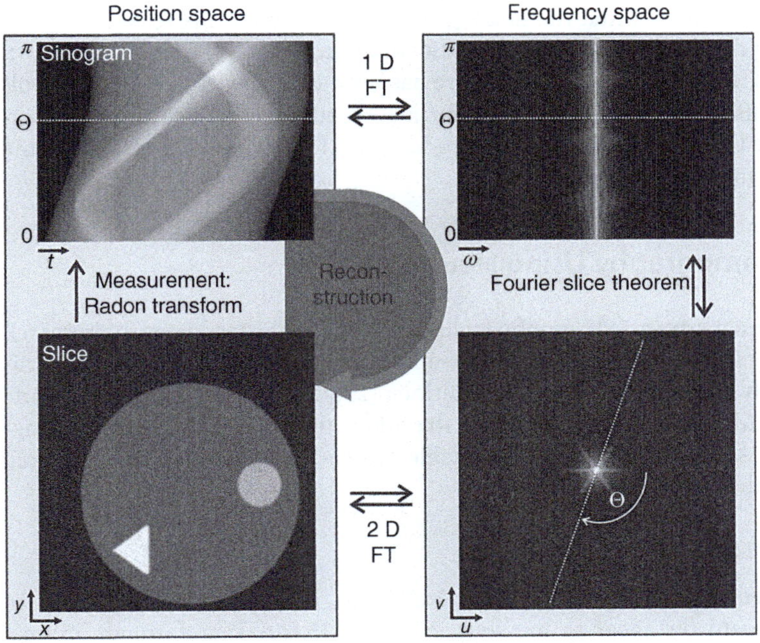

Figure 15.3 Scheme of the measurement and reconstruction process for parallel-beam tomography. On the left side in the position space the tomographic measurement of a single slice is represented by the sinogram. The tomographic reconstruction can be performed as the sinogram and the slice are connected by the Fourier slice theorem in the frequency space shown on the right side.

$P_\theta(t)$. A 1D Fourier transformation of the projection at the angle θ represents a single line in the 2D Fourier transformation of the slice $f(x, y)$. This is described by the so-called Fourier slice theorem. An implementation of this reconstruction is the back-projection of filtered projection algorithm presented in Chapter 14.

15.2.2 Density Resolution and Detector Quality

The spatial resolution in a tomogram can be measured by obtaining the modulation transfer function (MTF), which was discussed in Chapter 14. Here, special emphasis will be given to the quality of the density resolution in tomography. Assuming an implementation of an optimal measurement, which should, therefore, result in an artifact-free reconstruction, a simulation of a standard tomographic measurement and reconstruction was performed for detector systems, which differ in the number of distinguishable analog digital units (ADUs). In Figure 15.4, a resulting single reconstructed slice, the histogram of all slices and the obtained Gauss fits to the histogram are given. It can clearly be seen that for low-statistical qualities of the used detector the different absorbing materials can hardly be separated.

15.2.3 Data Evaluation and Visualization

For further data evaluation very often thresholds are used to distinguish different materials in the tomograms. For this purpose, the intersections of the Gauss peaks are calculated. It is very important to weigh the Gauss peaks to equal mass as otherwise the thresholds will be distorted as shown in Figure 15.5. For 3D visualization of the total reconstructed volume most often the floating point representation is scaled to a byte value. By applying digital filters, the noise in the tomograms can be reduced as shown in Figure 15.6. In Figure 15.7, a special software has been used to virtually slice the volume, to colorize different material, and to perform 3D renderings. The software VGStudioMax of Volume Graphics GmbH Heidelberg was used for the digital filtering and the 3D volume rendering [21].

15.3 Macrotomography Using Neutrons

As an example for a neutron radiography facility the GENRA-3 facility at the research reactor Geesthacht FRG-1 will be shown. The reactor was shut down in 2010. The instrument was routinely used for industrial applications [22]. At the Heinz Maier-Leibnitz Zentrum at the FRM II in Garching, Germany, the cold neutron radiography and tomography station ANTARES became recently available. It is currently the most sophisticated neutron tomography station in Europe [23].

15.3.1 Experimental Setup

The setup for neutron tomography was installed in the irradiation chamber of GENRA-3 at different distances to the source (Figure 15.8a). Therefore, an optimal value of L/D with respect to the size of the sample was selected. This is required to obtain a good spatial resolution in the tomogram. The neutron beam parameters for different sample–collimator distances P at GENRA-3 are given in Table 15.2 and compared to the newly available ANTARES beamline in Table 15.3.

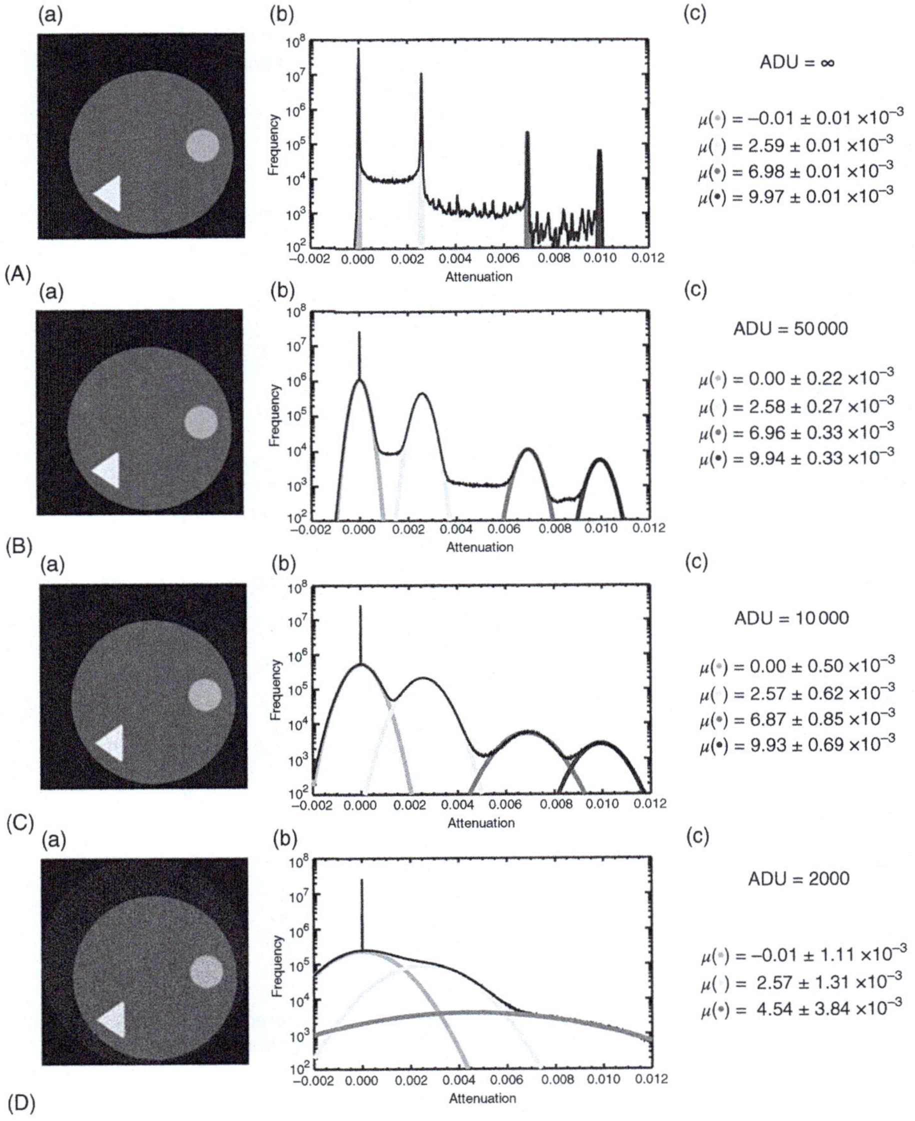

Figure 15.4 Simulated measurement and reconstruction for the sample presented in Figure 15.2. (a) reconstructed slice;(b) histogram of all slices representing the frequency of different attenuation values; and (c) obtained parameter for the shown Gauss peaks. The quality of the detector decreases from (A) to (D), resulting in a noisy tomogram where different absorptions used for spheres and cubes cannot be distinguished.

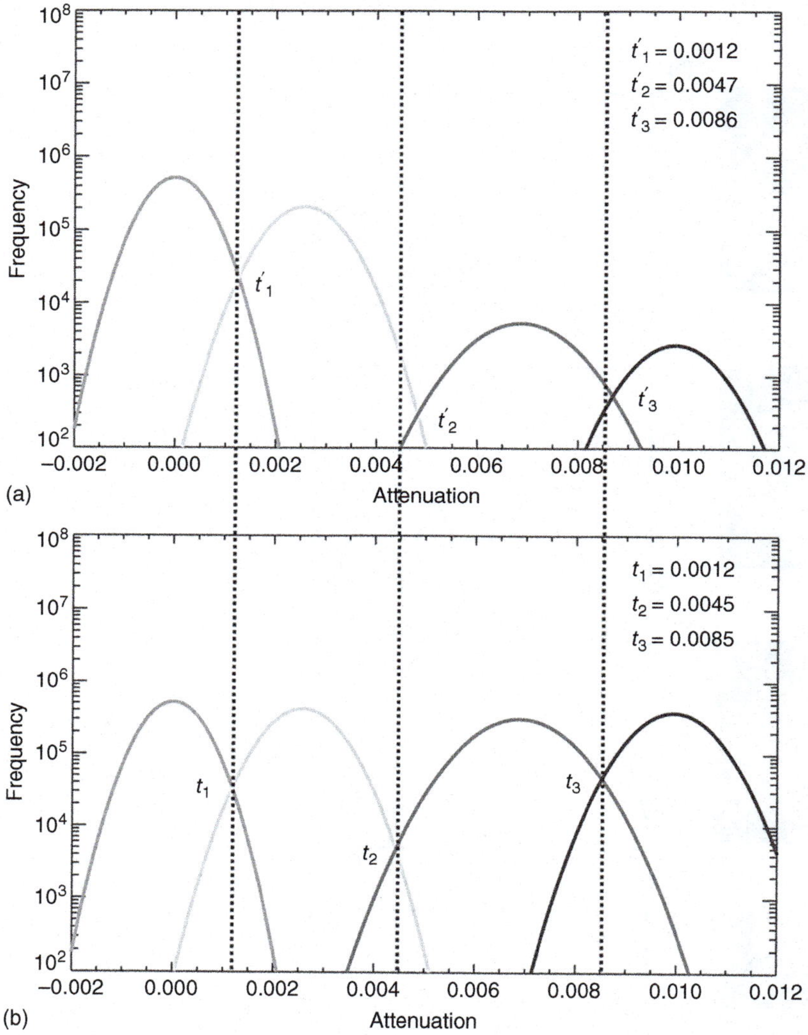

Figure 15.5 Calculation of the thresholds used for visualization. (a) Distorted thresholds of the original peaks. (b) Correct thresholds of peaks weighted to equal mass. The used simulated scan is presented in Figure 15.4C.

In Figure 15.9, the setup for tomography installed at GENRA-3 is shown. It consists of a 2D neutron detector and a high-precision sample manipulator. The principle setup is given in the sketch in Figure 15.8b. The main beam parameter are given in Table 15.4.

15.3.2 Measurements and Results

In order to conduct the tomographical experiment at GENRA-3, we chose as a sample a water tap consisting of different types of metal and high-absorbing rubber components (see Figure 15.9b). We set up the sample axis at 0.5 m distance from the collimator

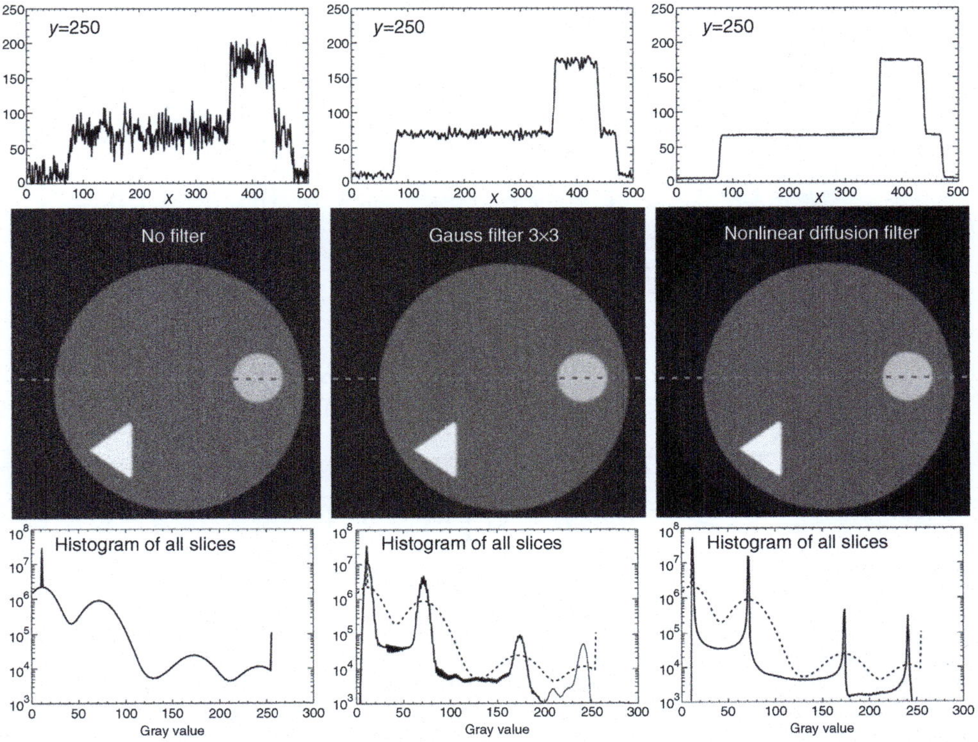

Figure 15.6 Demonstration of different digital filters to reduce the noise in the tomogram. The used simulated scan is presented in Figure 15.4C. The slices were scaled from −0.0005 to 0.0105 to byte 0 to 255. The resulting thresholds presented in Figure 15.5 are $t_1 = 39.6$, $t_2 = 116.4$, and $t_3 = 209.5$, respectively.

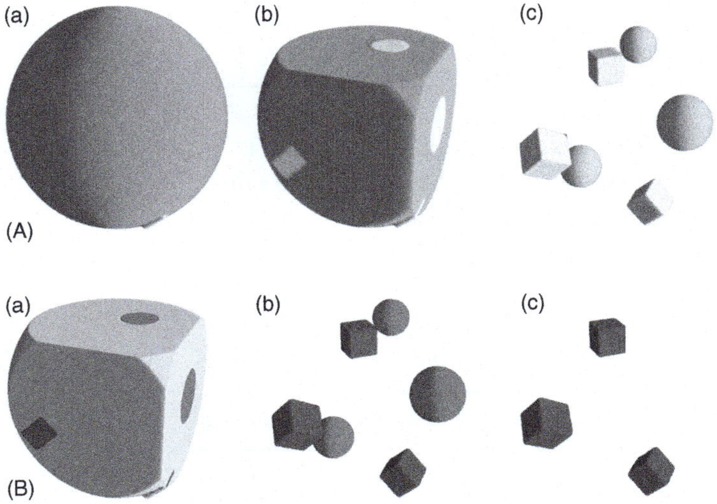

Figure 15.7 Volume renderings of the simulated scan presented in Figure 15.4C. (A) The unfiltered volume (a), virtually sliced (b), and the visualization of the inner spheres and boxes (c). By applying the digital filter presented in Figure 15.6c, the different material is segmented and can be separately visualized (B).

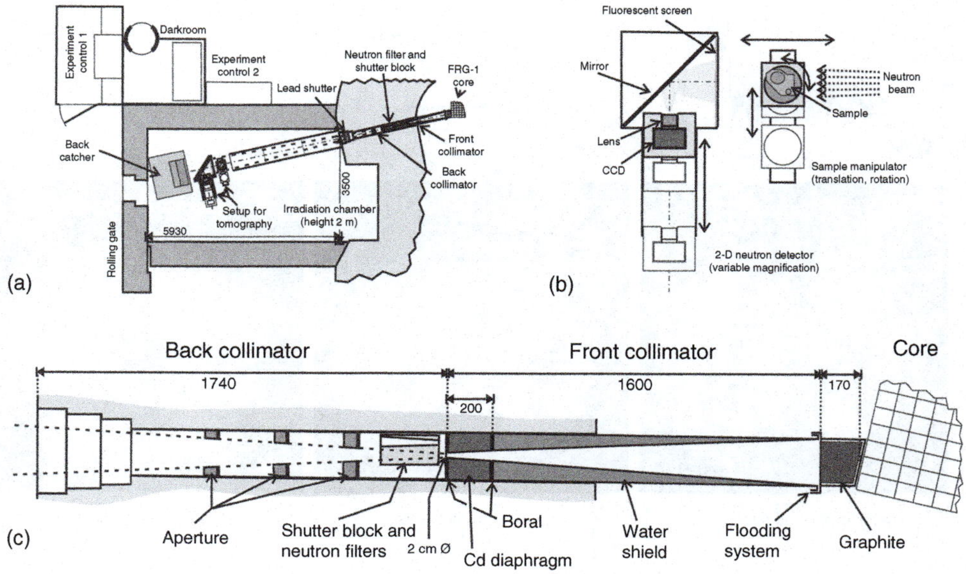

Figure 15.8 (a) Schematic view of the neutron radiography facility GENRA-3 at the research reactor Geesthacht FRG-1. The tomographic apparatus can be installed at different distances to the source resulting in *L/D* from 100 up to 300. The distance *L* is the distance to the aperture installed in the collimator with diameter $D = 2$ cm. The *L/D* fraction represents the divergence of the beam and is used as a quality value in neutron radiography. (b) Experimental setup for neutron tomography. The optical magnification of the detector can be varied from a field of view of 10 cm × 10 cm to 40 cm × 40 cm. (c) Collimator for GENRA-3 installed at FRG-1. The shutter block can cover different neutron filters to characterize the spectrum of the neutron beam. A 4 cm Bi filter is used for a thermal and a 0.2 Cd filter for an epithermal neutron beam.

Table 15.2 Neutron beam parameters at GENRA-3.

P (m)	0.5	1.5	2.5	4.0
Size (cm^2)	16 × 16	24 × 24	32 × 32	47 × 47
Flux (n cm^{-2} s^{-1})	7.4×10^6	3.2×10^6	1.7×10^6	6.8×10^5
L/D	100	150	200	300

Table 15.3 Neutron beam parameters at ANTARES. Beam size up to 35 × 35 cm^2.

Flux (n cm^{-2}s^{-1})	4×10^8	1×10^8	2.6×10^7	2.6×10^5
L/D	200	400	800	8000

(a) (b) (c)

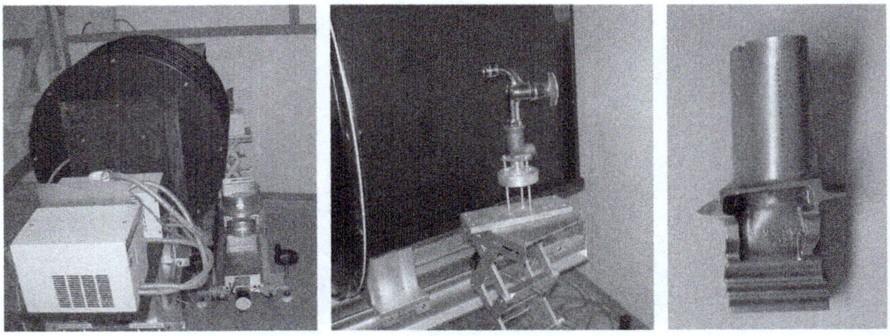

Figure 15.9 Experimental setup used for NCT (a). Investigation of a water tap (b). A viewgraph of a turbine blade made of nickel-based alloy (c).

Table 15.4 Components used for NCT setup at GENRA-3.

CCD camera	Kodak KAF-1400, 1320 × 1035 pixels, 6.8 μm × 6.8 μm, 12 bit digitization
Optical lens	Nikkor 35 mm $f/1.4$, Nikon, Tokyo, Japan
Optical mirror	Plane mirror, optical surface Al covered by SiO, LINOS Photonics, Göttingen, Germany
Fluorescent screen	$^{6}LiF/ZnS$: Ag, 40 cm × 40 cm, 250 μm thick, ND, green emitting, substrate 1 mm aluminum, Applied Scintillation Technologies, Harlow, UK

and sample–detector distance to 0.13 m (see Figure 15.3). The tomographical scan was performed by taking 30 flat field images (images without sample) and 720 projections at different sample rotations equally stepped between 0 and 2π. The exposure time of each image was 10 s. The total scan lasted 4.5 h. To reconstruct the sample, the axis of rotation was carefully aligned by using software to be perpendicular to the lines of the projections. These aligned projections then were reconstructed, resulting in a 3D dataset consisting of 498 × 498 × 611 voxels representing the size of 130 mm × 130 mm × 160 mm. The visualization of the 3D dataset is presented in Figure 15.10. By using the software VGStudioMax 3D volume renderings and virtual cuts are created. Furthermore, the software is able to extract and colorize the high absorbing sealing rings of the water tap.

A further tomographical scan was performed on a turbine blade made out of a nickel-based alloy (Inconel). The aim of the study is the investigation of the internal cooling system. We recorded 720 projections with an exposure time of 15 s. The total scan lasts 5 h. In Figure 15.9c, an optical viewgraph of the sample is given. The reconstructed data set consists of 437 × 437 × 433 voxel representing a volume of 114 mm × 114 mm × 113 mm. Renderings of the total and sliced volume are presented in Figure 15.11a–c. By using the software VGStudioMax, the voids in the blade representing the cooling canal system are segmented and visualized in blue in Figure 15.11d.

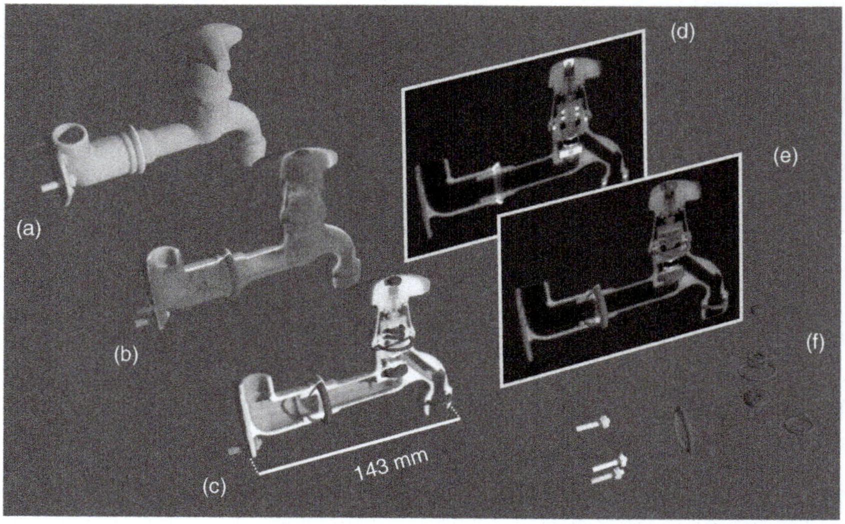

Figure 15.10 Volume rendering of the total reconstructed volume of the water tap (a). The high-absorbing sealing rings out of rubber material are shown as white regions in the vertical cut in (d). Using the software VGStudioMax these regions are segmented and colored in red in the 3D volume. At the position of the vertical cut, half of the metal material is visualized semitransparent (b) and transparent (c). The 3D arrangement of the sealing rings is shown in combination with the vertical cut (e) and the segmented screws (f).

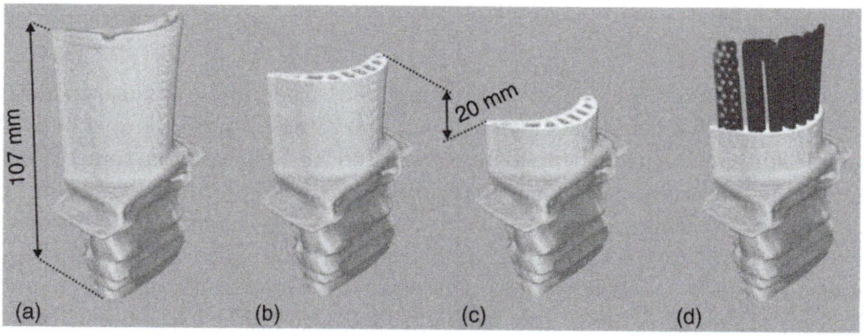

Figure 15.11 Volume renderings of the reconstructed data set of a turbine blade made of a nickel-based alloy (Inconel). A volume rendering of the total data set is given in (a). Then the volume is sliced at 20 mm (b) and 40 mm (c) from the top. The internal canal system is clearly visible. These voids then are segmented and colored in dark gray. A combination of the cooling system (dark gray) and the remaining nickel-based alloy matrix (gray) is visualized in (d).

15.4 Microtomography Using Synchrotron Radiation

As an example for a microtomography beamline, the beamline W2 of the storage ring DORIS III at DESY is presented. This beamline HARWI-II was specialized for the use of high-energy X-rays from 20 to 250 keV dedicated to texture, strain, and imaging measurements for materials science applications [24]. Special attention was given to the optics concept to obtain a high-energy high-flux X-ray beam with a large field of view optimized for imaging materials science applications [25]. The storage ring DORIS III was shut down in 2013. Therefore, the current experimental setup for microtomography used at the third generation SR storage ring PETRA III is presented in Figure 15.17.

15.4.1 Beamline Optics

The main components for the beamline optics are the HARWI-II wiggler, the front-end filter, the monochromator, the beam stop, and the diagnostics table (Figure 15.12). The design and features of the wiggler installed in the DORIS-III storage ring are presented in detail elsewhere [26]. The main parameters of the wiggler are given in Table 15.5.

To reduce the heat load on the first crystal of the monochromator, a 3 mm carbon was permanently installed as a high-pass filter. For the use of high-photon energies, a 7 mm carbon filter and a 1 or 2 mm copper filter could additionally be placed into the beam. The sketch of the filter system is given in the inset of Figure 15.12. A detailed description of the system located in the DORIS-DESY injection tunnel 36 m downstream of the HARWI-II wiggler is given in [27].

The resulting flux behind the filter system is presented in Figure 15.13. The remaining total power of the white beam (70 mm × 10 mm) at minimum gap and minimum filter

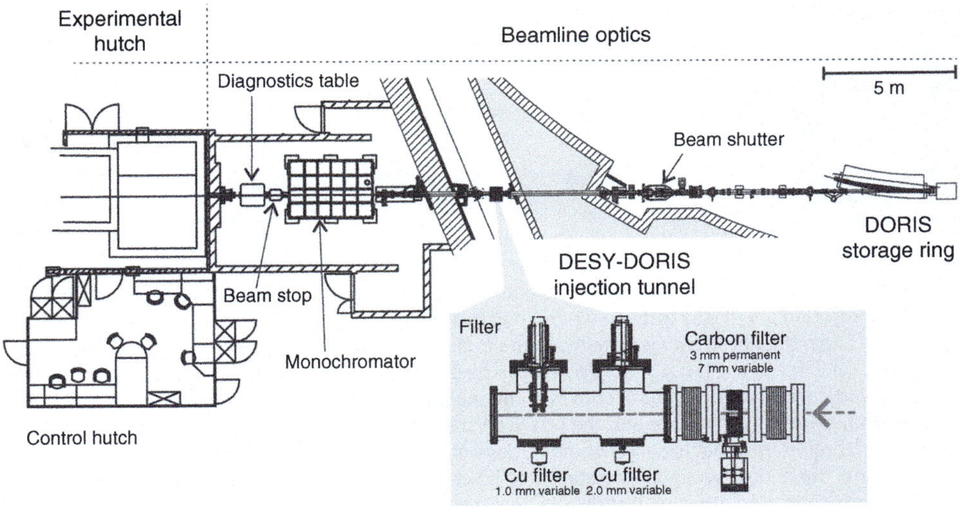

Figure 15.12 Sketch of the HARWI-II beamline showing the beamline optics. The permanent and optional high-pass filter installed in the DORIS-DESY injection tunnel is shown in the inset. A permanent 3 mm-thick carbon filter can be extended by an optional filter (7 mm carbon and 1 or 2 mm copper).

Table 15.5 Parameters of the HARWI-II wiggler.

Total length	**4 m**	**Peak field**	**1.91 T**
Period length	110 mm	K	20.3
Minimal gap	14 mm	Critical energy	26.7 keV

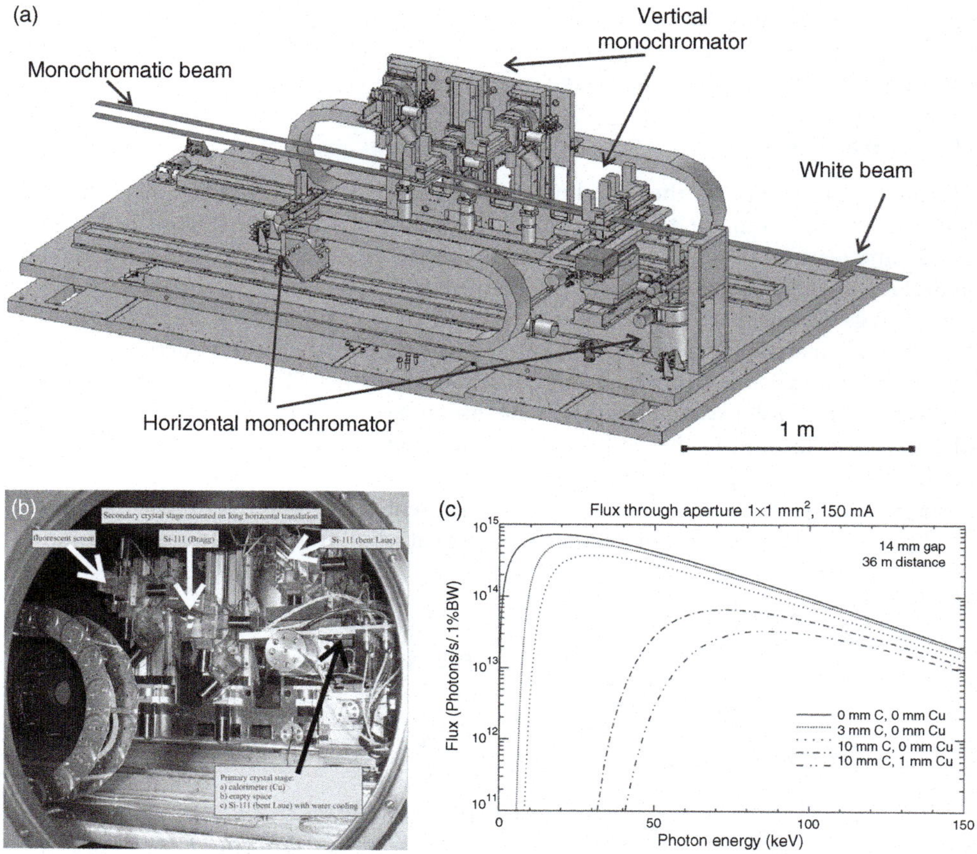

Figure 15.13 (a) Sketch of the monochromator setup. The user can switch between two different monochromators. The horizontal monochromator in the foreground and the vertical monochromator in the background are placed into the white X-ray beam by moving the upper plate perpendicular to the beam. The monochromator is installed in the monochromator tank in the optics hutch. The tank is of dimension 3 m × 2 m × 1 m (*L* × *W* × *H*) and can be evacuated to below 10^{-6} mbar. (b) View on the installed vertical monochromator as shown in the sketch (top). The primary crystal stage (right foreground) carries a calorimeter and the first Si crystal. The secondary stage (left background) consists of three stages that allow for combinations of different secondary crystal stages with the first crystal. Here, the middle secondary crystal stage is set to 36 keV photon energy (Si-111). The fixed exit monochromator has an offset of 40 mm. (c) Comparison of the flux of the HARWI-II wiggler at minimal gap of different filter settings with the unfiltered spectrum at the position of the high-pass filter (36 m distance from the source).

(3 mm carbon) is about 12.6 kW. To further reduce the heat load on the first crystal the wiggler gap can be increased. By use of different combinations of the filters and wiggler gaps the optimized conditions with respect to photon energy and beam size can be found.

The concept of the monochromator setup is presented in Figure 15.13. Two different monochromator systems, one reflecting in the horizontal plane, and one reflecting in the vertical plane, are installed. The horizontal monochromator is mainly used for diffraction applications and provides an X-ray beam with a size of 10 mm × 10 mm from

50 to 250 keV. The vertical monochromator is optimized for imaging applications and will deliver an X-ray beam with a size of 70 mm × 10 mm and a photon energy range of 15–200 keV. Different pairs of crystals in Laue–Laue and Laue–Bragg geometry will be installed. Furthermore, bent-Laue crystals are used to increase the energy band path of the reflected X-ray beam.

A fluorescent screen, a Si-111 (Bragg), and Si-111 (bent-Laue) crystal were installed on the secondary crystal stage. The primary crystal stage was equipped with a calorimeter and a water-cooled bent-Laue crystal. Figure 15.13 shows a view into the monochromator tank revealing the installed monochromator system. Both, sketch and viewgraph show the same settings used for 36 keV photon energy.

15.4.2 Experimental Setup

The dedicated setup used for attenuation-contrast microtomography is presented in Figure 15.14. It consists of a 2D X-ray detector and a sample manipulator stage. The sample manipulator provides both for the rotation and for the lateral positioning of the specimen. The incident X-rays are converted into visible light which then is projected onto a CCD camera by an optical lens system. The used components are given in Table 15.6.

This setup was operated as a user experiment for attenuation-contrast microtomography at beamline HARWI-II using photon energies from 20 to 200 keV. An attenuation projection is calculated from an image with and without the sample.

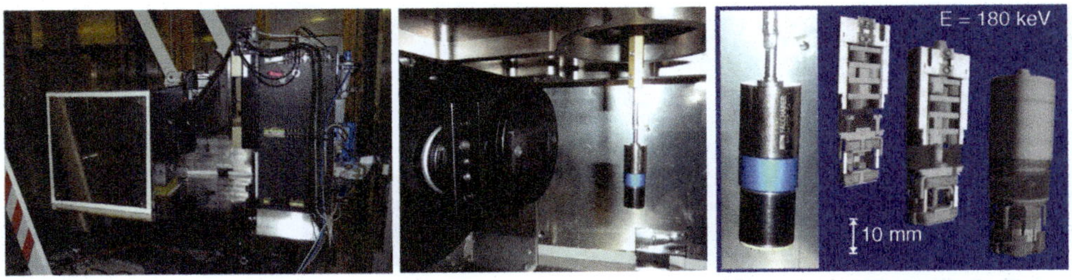

(a) (b) (c)

Figure 15.14 (a) Setup for microtomography installed at the HARWI-II beamline. The experiment is mounted on a lift table, which can be lowered by 1.5 m. The pit will then be covered by a roof. The length of the total system is about 1.7 m. (b) Close-up showing a stepper motor installed as sample in the microtomography system. The main components of the apparatus are a high-precision sample manipulator and a 2D X-ray detector. The sample manipulator consists of a rotational axis with integrated *x*–*y*–*z* translation of the sample. The 2D X-ray detector is built from a single crystal luminescent screen, a lens and a CCD camera. The optical magnification can be adjusted to the size of the sample. (c) Result performing microtomography using a photon energy of about 180 keV. The internal structure of the stepper motor can clearly be visualized.

Table 15.6 Components used for SRμCT at DORIS III

CCD camera	PL09000, Finger Lakes Instrumentation; 16 bit digitalization at 8 MHz, 3056 × 3056 pixels; each 12 μm × 12 μm
Optical lens	Camera lens, Nikon Inc., 35 or 50 mm focal length
Fluorescent screen	$CdWO_4$ single crystal, thickness 200–1000 μm

For microtomography typically a set of 900 radiograms at different sample rotations, equally stepped between 0 and π, were taken. The tomographical reconstructions were calculated by using "back-projection of filtered projection" algorithm. For this reconstruction method, the center of rotation is determined within a fraction of pixel done by an automatized algorithm [28].

For performing microtomography at HARWI-II, the microtomography apparatus was installed onto a lift table in the experimental hutch (Figure 15.14). From 2006 to 2012 the beamline was in routine user operation to provide attenuation-constant and phase-contrast microtomography using the photon energy range from 20 to 200 keV.

A few selected examples of materials science applications obtained at the HARWI-II beamline are presented in Figures 15.15 and 15.16 [29–32].

In Figures 15.17 and 15.18 the new instrumentation for microtomography at the storage ring PETRA-III is presented. To cover the full energy range, two similar experiments are installed at the beamline P05 (7–50 keV) and P07 (35–150 keV). In Figure 15.17 the overall concept is shown. The system consists of three main components: the substructure, the X-ray detector and the rotation axis. The setup used for grating based phase contrast is shown in Figure 15.17c. First results performing phase-contrast microtomography at high photon energies can be found in [18]. The main features of the setup in comparison to the HARWI-II beamline are given in Table 15.7.

To make full use of the beamtime and to increase the sample throughput a standardized sample changing system was built. The system installed at the P07 beamline at PETRA III is presented in Figure 15.18.

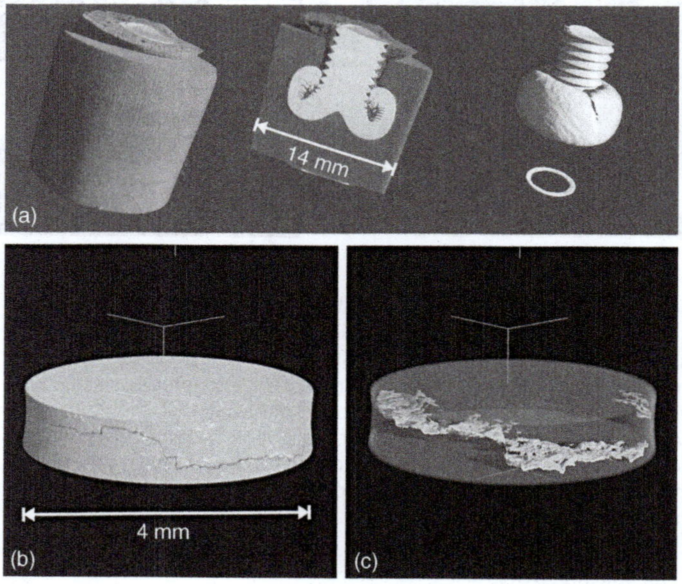

Figure 15.15 (a) Investigation of an Al-sample studying the material flow in a friction rivet sample. This new joining technique is used for polymers and lightweight metal alloys (here Al). The sample is scanned using 36 keV photon energy. Volume renderings are presented of the total sample (a, left), sliced sample (a, middle), and the Al-rivet, that is, transparent polymer, (a, right). (b,c) Investigation of damage mechanisms and crack extension in an AA2024 aluminum alloy using 28 keV photon energy at HARWI-II. Cut of one specimen, as a full solid (b), semivisible with the cracks appearing in gray (c).

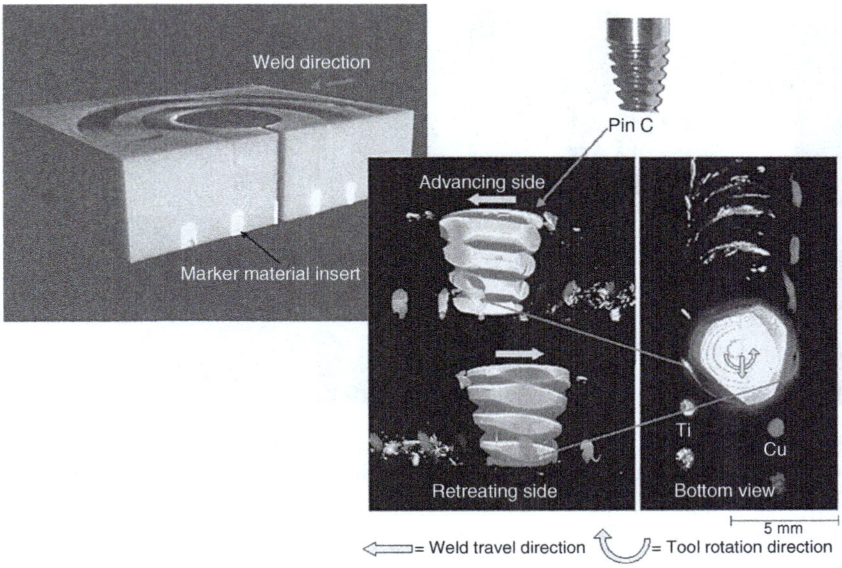

Figure 15.16 Simultaneous marker material displacement (Cu and Ti) around the aluminum workpiece and the friction stir welding tool (Pin C – conical and threaded with three equally spaced flats along entire thread length). Left: Volume rendering showing the aluminum sample (5 mm thickness) and the introduced marker material. On the right, the aluminum is set transparent to reveal the 3D distribution of the marker material around the pin. The tomographical scans were performed at beamline HARWI-II using photon energy of about 64 keV.

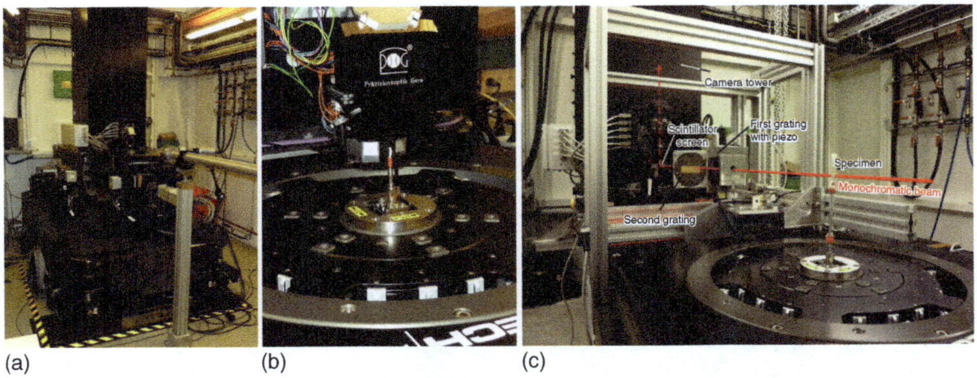

Figure 15.17 (a) Instrument for microtomography installed at the beamline P07/PETRA III. (b) Sample position in attenuation contrast mode. (c) Setup used for grating-based phase-contrast imaging.

Two examples demonstrating results obtained with the new instrumentation for microtomography at the beamlines P05 and P07 are presented in Figures 15.19 and 15.20. First results using SR-μCT at P05 for the investigation of biodegradable bone screws consisting of novel Mg alloys [33] are presented in Figure 15.19.

In Figure 15.20, high-energy SR-μCT at P07 was applied on a hybrid joint of aluminum and polymer composite (PEI with carbon fibers). The sample was produced by Friction-based Injection Clinching Joining (F-ICJ) as part of an exploratory study [34].

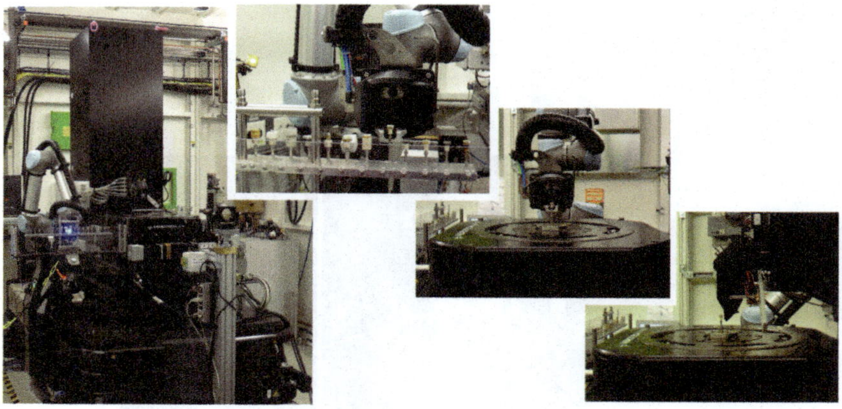

Figure 15.18 Sample changer for high-throughput tomography at beamline P07/PETRA III. (a) Sample changer scanning the sample ids, (b) selecting one sample, (c) installation into rotation axis, and (d) sample changer in rest position for scanning the sample.

Table 15.7 Parameters of SRμCT at DORIS III and PETRA III.

	DORIS III	PETRA III
Photon energy (keV)	7–200	7–150
Sample height (mm)	3.5	2
Sample width (mm)	70	7
Spatial resolution (μm)	2	1

Figure 15.19 SRμCT sample measured at beamline P05/PETRA III. Rendered view representing the bone formation around a medical implant. The bone screw consists of a novel Mg alloy. (Sample diameter 3 mm.)

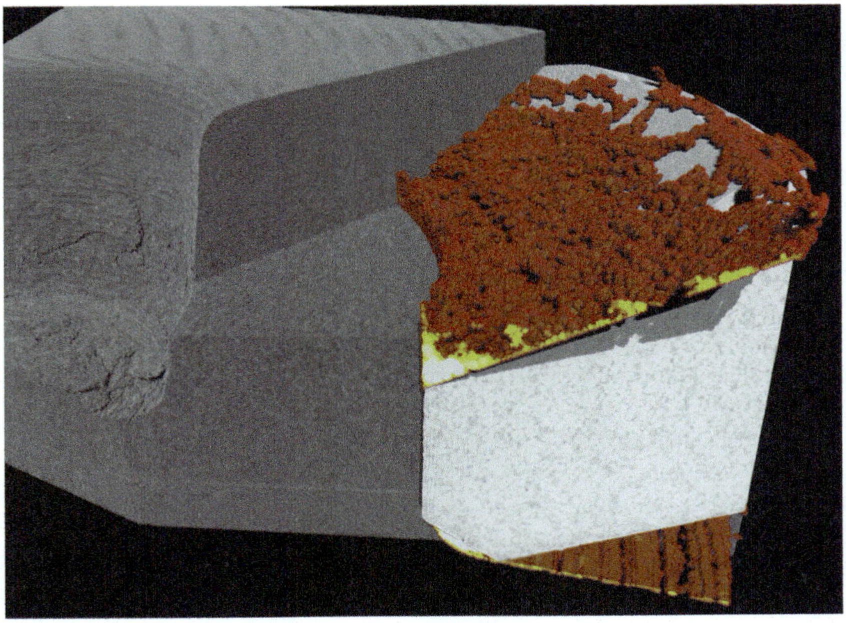

Figure 15.20 SRμCT sample measured at beamline P07/PETRA III. Volume rendering of a hybrid joint of an aluminum and polymer composite (PEI with carbon fibers). To visualize the microstructure of the voids (red) produced by Friction-based Injection Clinching Joining (F-ICJ) the polymer composite (gray) and the aluminum (white) is partly removed by software. The shown sample height is 2 mm.

SR-μCT was used to analyze the material flow in the polymer and the volume of voids as part of the microstructural investigation.

15.5 Summary and Outlook

Different setups for tomography and microtomography at a neutron source and at a SR facility are presented. The experiments are designed to operate in user mode. Complementary access to both the techniques, neutron tomography, and SR tomography, allows the development of new techniques and will give new insight in the 3D behavior of samples especially in materials science.

Exploiting the source characteristics of the new third-generation SR facility PETRA III at DESY will allow extending the spatial resolution into the nanometer regime. The standardization of the different imaging techniques will allow for optimal investigations of samples consisting of low-absorbing and absorbing components.

References

1 Bonse, U. and Busch, F. (1996) X-ray computed microtomography (CT) using synchrotron radiation (SR). *Prog. Biophys. Mol. Biol.*, **65**, 133–169.

2 Snigirev, A., Snigireva, I., Kohn, V., Kuznetsov, S., and Schelokov, I. (1995) On the possibilities of X-ray phase contrast microimaging by coherent high-energy synchrotron radiation. *Rev. Sci. Instrum.*, **66** (**12**), 5486–5492.

3 Cloetens, P., Barrett, R., Baruchel, J., Guigay, J.-P., and Schlenker, M. (1996) Phase objects in synchrotron radiation hard X-ray imaging. *J. Phys. D: Appl. Phys.*, **29**, 133–146.

4 https://en.wikipedia.org/wiki/Phase-contrast_X-ray_imaging

5 Raven, C., Snigirev, A., Snigireva, I., Spanne, P., Souvorov, A., and Kohn, V. (1996) Phase-contrast microtomography with coherent high-energy synchrotron X-rays. *Appl. Phys. Lett.*, **69** (**13**), 1826–1828.

6 Spanne, P., Raven, C., Snigireva, I., and Snigirev, A. (1999) In-line holography and phase-contrast microtomography with high energy X-rays. *Phys. Med. Biol.*, **44**, 741–749.

7 Bravin, A. (2003) Exploiting the X-ray refraction contrast with an analyser: the state of the art. *J. Phys. D: Appl. Phys.*, **36**, A24–A29.

8 Pagot, E., Cloetens, P., Fiedler, S., Bravin, A., Coan, P., Baruchel, J., and Härtwig, J. (2003) A method to extract quantitative information in analyser-based X-ray phase contrast imaging. *Appl. Phys. Lett.*, **82** (**20**), 3421–3423.

9 Momose, A. (1995) Demonstration of phase-contrast X-ray computed tomography using an X-ray interferometer. *Nucl. Instrum. Methods Phys. Res., Sect. A*, **352**, 622–628.

10 Beckmann, F., Bonse, U., Busch, F., and Günnewig, O. (1997) X-ray microtomography (μCT) using phase contrast for the investigation of organic matter. *J. Comput. Assisted Tomogr.*, **21** (**4**), 539–553.

11 Momose, A. (2002) Phase-contrast X-ray imaging based on interferometry. *J. Synchrotron Radiat.*, **9**, 136–142.

12 Cloetens, P., Ludwig, W., Baruchel, J., Van Dyck, D., Van Landuyt, J., Guigay, J.P., and Schlenker, M. (1999) Holotomography: quantitative phase tomography with micrometer resolution using hard synchrotron radiation X-rays. *Appl. Phys. Lett.*, **75** (**19**), 2912–2914.

13 Paganin, D., Mayo, S.C., Gureyev, T.E., Miller, P.R., and Wilkins, S.W. (2002) Simultaneous phase and amplitude extraction from a single defocused image of a homogeneous object. *J. Microsc.*, **206**, 33–40.

14 Weitkamp, T., Haas, D., Wegrzynek, D., and Rack, A. (2011) *ANKAphase:* software for single-distance phase-retrieval from inline X-ray phase contrast radiographs. *J. Synchrotron Radiat.*, **18**, 617–629.

15 Weitkamp, T., Diaz, A., David, C., Pfeiffer, F., Stampanoni, M., Cloetens, P., and Ziegler, E. (2005) X-ray pase imaging with grating interferometer. *Opt. Express*, **12** (**16**), 6296–6304.

16 Pfeiffer, F., Weitkamp, T., Bunk, O., and David, C. (2006) Phase retrieval and differential phase-contrast imaging with low-brilliance X-ray sources. *Nat. Phys.*, **2**, 258–261.

17 Herzen, J., Donath, T., Beckmann, F., Ogurreck, M., David, C., Mohr, J., Pfeiffer, F., and Schreyer, A. (2011) X-ray grating interferometer for materials-science imaging at a low-coherent wiggler source. *Rev. Sci. Instrum.*, **82** (**11**), 113711.

18 Hipp, A., Beckmann, F., Lytaev, P., Greving, I., Lottermoser, L., Dose, T., Kirchhof, R., Burmester, H., Schreyer, A., and Herzen, J. (2014) Grating-based X-ray phase-contrast imaging at PETRA III. *Proc. SPIE*, **9212**, 921206.
19 Kak, A.C. and Slaney, M. (1988) *Principles of Computerized Tomographic Imaging*, IEEE Press, New York.
20 Radon, J. (1927) Über die Bestimmung von Funktionen durch ihre Integralwerte längs gewisser Mannigfaltigkeiten. Berichte über die Verhandlungen der Königlich Sächsischen Gesellschaft der Wissenschaften zu Leipzig. *Math-Phys. Kl*, **69**, 262–277.
21 Volumegraphics http://www.volumegraphics.com (accessed 28 June 2016).
22 Beckmann, F., Vollbrandt, J., Donath, T., Schmitz, H.W., and Schreyer, A. (2005) Neutron and synchrotron radiation tomography: new tools for materials science at the GKSS Research Center. *Nucl. Instrum. Methods Phys. Res., Sect. A*, **542**, 279–282.
23 MLZ http://www.mlz-garching.de/antares (accessed 28 June 2016).
24 Beckmann, F., Dose, T., Lippmann, T., Martins, R.V., and Schreyer, A. (2007) The new materials science beamline HARWI-II at DESY. AIP Conference Proceedings, SRI 2006, Daegu, Korea, 2007, CP879, pp. 746–749.
25 Beckmann, F., Donath, T., Fischer, J., Dose, T., Lippmann, T., Lottermoser, L., Martins, R.V., and Schreyer, A. (2006) New developments for synchrotron radiation-based microtomography at DESY. *Proc. SPIE*, **6318**, 631810.
26 Tischer, M., Gumprecht, L., Pflüger, J., and Vielitz, T. (2007) A new hard X-ray wiggler for DORIS III. AIP Conference Proceedings, SRI 2006, Daegu, Korea, 2007, CP879, pp. 339–342.
27 Schulte-Schrepping, H. and Hahn, U. (2007) Hard X-ray wiggler front end filter design. AIP Conference Proceedings, SRI 2006, Daegu, Korea, 2007, CP879, pp. 1042–1045.
28 Donath, T., Beckmann, F., and Schreyer, A. (2006) Automated determination of the center of rotation in tomography data. *J. Opt. Soc. Am. A:*, **23** (**5**), 1048–1057.
29 Amancio, S., dos Santos, J.F., Beckmann, F., and Schreyer, A. (2006) Computer Microtomographic Investigation of Polyetherimide/Al 2024-T351 Friction Riveted Joints, DESY Annual Report, pp. 485–486.
30 Braun, R., Steglich, D., and Beckmann, F. (2006) Damage Mechanisms and LCF Crack Extension in Al2024, DESY Annual Report, pp. 497–498.
31 Amancio-Filho, S.T., Roeder, J., Nunes, S.P., dos Santos, J.F., and Beckmann, F. (2008) Thermal degradation of polyetherimide joined by friction riveting (FricRiveting). Part I: influence of rotation speed. *Polym. Degrad. Stab.*, **93**, 1529–1538.
32 Zettler, R., dos Santos, J.F., Donath, T., Herzen, J., Beckmann, F., and Lohwasser, D. (2006) Removing the Guess Work from Material Flow and Deformation Mechanics Associated with the FSW of Aluminium and Its Alloys through Application of Micro-Computer Tomography and Dedicated Metallurgical Techniques, DESY Annual Report, pp. 667–668.
33 Galli, S., Szakacs, G., Lukac, F., Vlcek, M., Jimbo, R., Naito, Y., Wennerberg, A., Herzen, J., Hammel, J., and Willumeit, R. (2014) Osseointegration of resorbable magnesium screws – A SRμCT study. *Eur. Cells Mater.*, **28** (**3**), 64.
34 Abibe, A. (2015) Friction-based Injection Clinching Joining (F-ICJ): a new joining method for hybrid lightweight structures, PhD thesis. Technical University Hamburg-Harburg.

16

μ-Tomography of Engineering Materials

Astrid Haibel and Julia Herzen

16.1 Introduction

Since the early 1970ss computed tomography (CT) has been available for medical diagnostics. It was developed in 1972 by Sir Godfrey Newbold Hounsfield, one of the pioneers in the field of CT. CT machines were the first imaging devices for providing detailed visualization of the internal three-dimensional anatomy of living creatures. The use of CT also for industrial applications started in the 1980s. In particular, the construction of third-generation synchrotron sources combined with the improvement of data acquisition in speed and memory since the 1990s has led to a boom in high-end synchrotron tomography also in materials science with a wide range of different applications, for example, the visualization and quantitative analysis of material components, microporosities, cracks, precipitations, or grains.

16.2 Advantage of Synchrotron Tomography

Synchrotron radiation is electromagnetic radiation, generated by the acceleration of ultrarelativistic electrons through magnetic fields. It was discovered in a General Electric synchrotron accelerator in 1947 [1]. The wavelength range of this radiation includes the infrared, optical, ultraviolet, and X-ray light spectrum.

The high brilliance, exceeding other natural and artificial light sources by many orders of magnitude, and the continuous wavelength spectrum of this radiation combined with suitable narrow band pass monochromators make synchrotron radiation a unique method for experiments requiring continuously tunable sources from the infrared up to the X-ray regions. Further unique properties of synchrotron radiation are high collimation, low emittance, and high level of polarization.

In case of synchrotron tomography, the monochromatic radiation allows us to resolve and visualize different sample components even if they have similar absorption coefficients with a spatial resolution in the sub-micrometer range within short measurement times. It is possible to adjust the X-ray energy to the absorption properties of the samples. Moreover, the monochromatic and highly collimated beam can be used for phase-contrast imaging, based on the high coherence of synchrotron sources. Furthermore, the possibility of monochromizing the beam can be used, for example, for

Neutrons and Synchrotron Radiation in Engineering Materials Science: From Fundamentals to Applications, Second Edition. Edited by Peter Staron, Andreas Schreyer, Helmut Clemens, and Svea Mayer.

absorption edge tomography. Due to the monochromatic and almost parallel beam, the images are widely free of artifacts known from laboratory tomography.

16.3 Applications and 3D Image Analysis

16.3.1 Discharging Processes in Alkaline Cells

Alkaline batteries have been used since about 100 years. They show an excellent performance characteristic, a favorable cost structure, and are extremely reliable in many industrial and household applications [2–5].

The lifetime of such batteries is strongly correlated with the dissolution process of the zinc particles in the anode material [6]. The zinc utilization in alkaline batteries in order to continue the improvement of the cell performance is still an important topic under discussion [7].

In this application example for absorption synchrotron tomography, the visualization and analysis of the discharging processes of a battery by an *in situ* experiment is shown. The complete process including shrinking and disintegrating of zinc particles in the anode electrolyte material, that is, the morphology and spatial distribution of zinc and zinc oxide during discharging, is observed [8].

For the measurement, a battery was located on a sample holder in the synchrotron beam during the whole experiment, that is, during the tomographic measurements as well as during the discharging. The same part of the battery was observed for all the five different discharging states (see Figure 16.1a–e). A slice of a fully discharged battery is shown in Figure 16.1f. The discharging times between the tomographic measurements were 15, 45, 90, 180, and 540 min.

The battery consists of zinc gel as anode (inner granular structure, 67% zinc and 33% electrolyte) and manganese oxide as cathode (outer green ring) isolated by a separation fleece (blue ring between the anode and cathode) and covered by a steel tube (outer red ring). The electrolyte is made of 36% potassium hydroxide, 60% water, and 4% zinc oxide.

As a first qualitative result we see that the amount of small particles in the middle of the images in Figure 16.1 decreases in favor of the surrounding areas, that is, the zinc particles dissolve from outside and zinc oxide accrues. Also the outer ring in the figures increases, that is, the manganese oxide layer swells. As a consequence the separation fleece is being impacted. Equation (16.1) describes the chemical reaction at the cathode and at the anode:

$$\begin{aligned} &\text{Anode}: \mathrm{Zn} + 2\mathrm{OH}^- \rightarrow \mathrm{ZnO} + \mathrm{H_2O} + 2\mathrm{e}^- \\ &\text{Cathode}: 2\mathrm{MnO_2} + 2\mathrm{H_2O} + 2\mathrm{e}^- \rightarrow 2\mathrm{MnOOH} + 2\mathrm{OH}^- \end{aligned} \tag{16.1}$$

For the quantitative 3D image analysis of the zinc particle dissolving process, the particles were separated from the other components in Boolean images by choosing suitable gray value thresholds. The closeness of zinc particles in the battery inhibits a spatial separation of each particle in the Boolean records. Therefore, a 3D watershed algorithm combined with a Euclidian distance transformation was used to separate the particles in the Boolean image. Then the particles were identified by a labeling algorithm and color coded [9, 10]. The result for the zinc particles in the fully charged state of the battery is displayed in Figure 16.2.

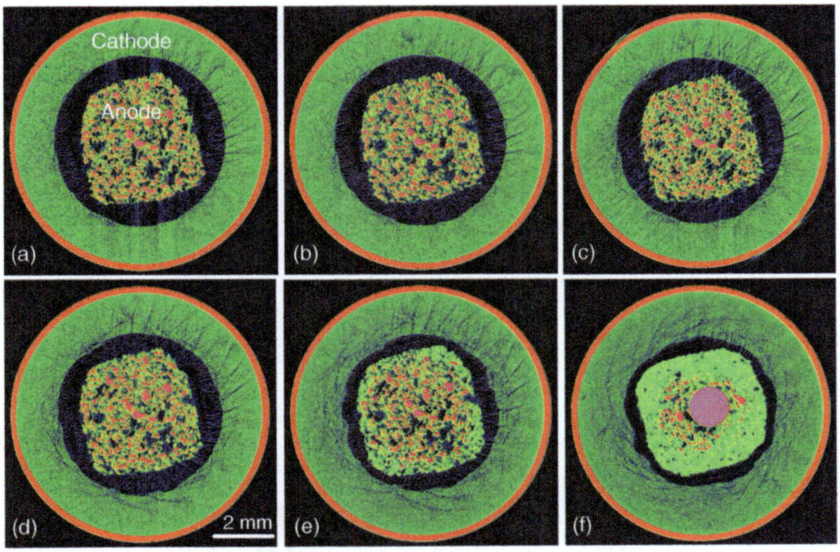

Figure 16.1 Six different discharging states of the investigated battery. (a) Fully charged state with voltage $U = 1.59$ V. (b) State after 15 min discharging time with voltage $U = 1.47$ V. (c) State after 45 min. discharging time with voltage $U = 1.42$ V. (d) State after 90 min discharging time with voltage $U = 1.37$ V. (e) State after 3 h discharging time with voltage $U = 1.10$ V. (f) A second, structurally identical battery in the almost discharged state (9 h discharging time, $U = 0.5$ V; here the arrester is visible in the middle of the image). (Haibel 2010 [8]. Reproduced with permission of The Electrochemical Society.)

Figure 16.2 Labeled and color-coded zinc particles. They act as anode in alkaline batteries. (Haibel 2010 [8]. Reproduced with permission of The Electrochemical Society.)

Based on this particle separation algorithm the decrease of the zinc particle fraction for each discharging step and the shape factor of each of these particles could be quantified. The shape factor describes the surface-to-volume ratio of the particles, normalized to 1 for spheres. The results are shown in Figure 16.3a,b.

The diagrams show that during discharging the mean diameter and the shape of the particles are widely unchanged; only the particle number decreases. This confirms the

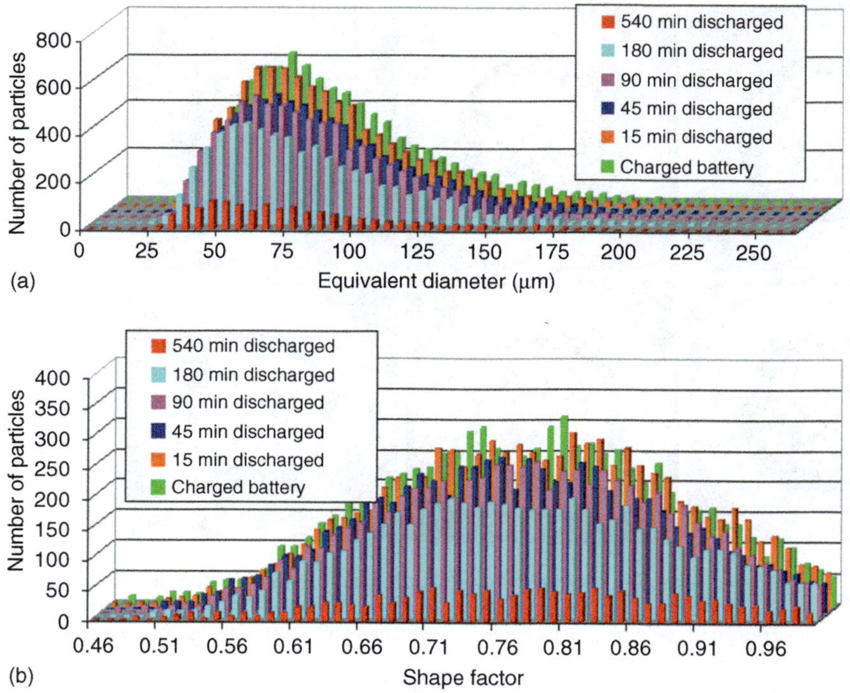

Figure 16.3 *In situ* tomography on alkaline batteries: (a) zinc particle number versus diameter for each discharging step. (b) Zinc particle number versus particle shape for each discharging step (from 0 (planes) to 1 (spheres)).

visual impression of Figure 16.1, which indicates that the particle dissolution takes place from the outside of the zinc gel. Only the outer particles disappear whereas the inner particles remain unchanged.

16.3.2 Microstructural Investigations of Nb_3Sn Multi-filamentary Superconductor Wires

The low-T_C superconductor Nb_3Sn is most commonly used to achieve huge magnetic fields higher than 10 T [11, 12]. But, as for the high-T_C materials, the intermetallic Nb_3Sn-phase is very brittle, too. Thus, the superconducting wires cannot be bent to form the magnet coils in a simple production step.

To enable coil forming, the synthesis of the superconducting Nb_3Sn phase within the wires has to be done after coiling. In a first step, the Nb, Cu, and Sn sub-elements that are assembled inside a tantalum diffusion barrier and a copper can are tubular extruded (see Figure 16.4). Then the strand size is reduced to a cross-section less than 1 mm by swaging and drawing.

Subsequently, the strands are wired to magnet solenoids. Finally, a heat treatment is applied to transform the constituents into the superconductive phase. The formation reaction occurs through a solid state diffusion reaction at high temperatures in a protecting atmosphere. Absorption tomography has been applied to observe the microstructure during the heat treatment [13].

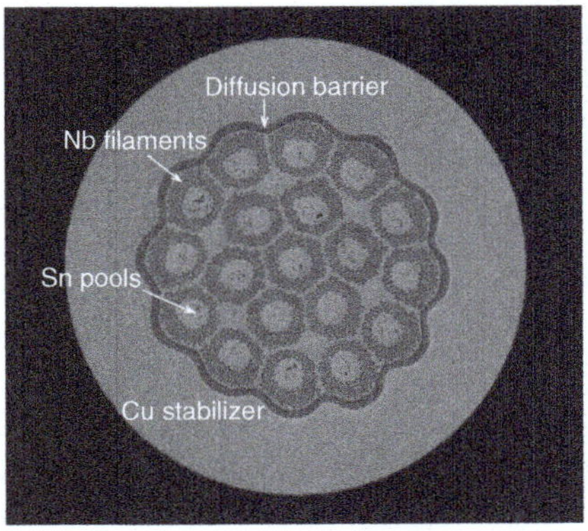

Figure 16.4 Light microscopic image of a Nb_3Sn precursor wire cross-section. The diameter of the wire is 0.7 mm.

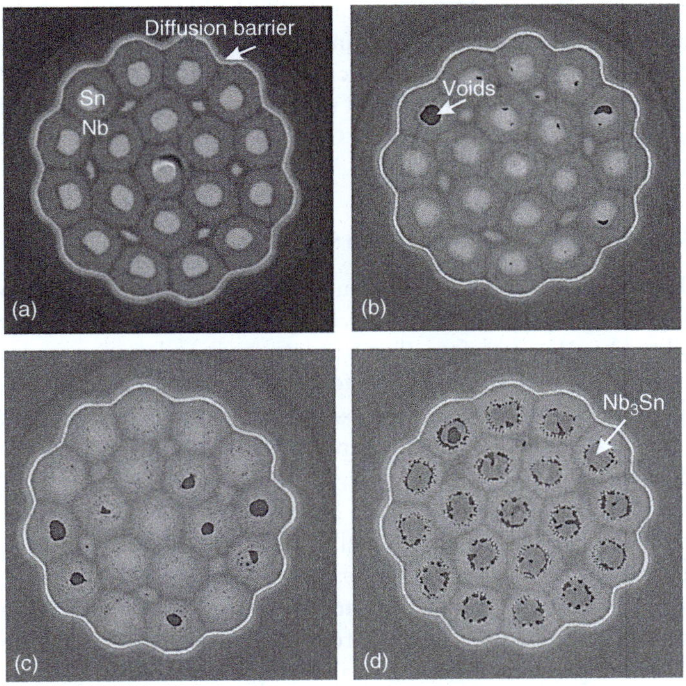

Figure 16.5 2D slices of tomograms of Nb_3Sn wires (diameter 0.7 mm): (a) non-heat-treated, ductile state. (b) 24 h heat treated at $T = 220\,°C$. (c) 24 h heat treated at $T = 310\,°C$. (d) 24 h heat treated at $T = 760\,°C$.

In Figure 16.5 the cross-sections of four internal Nb_3Sn strands are shown after different thermal treatments reconstructed from the synchrotron absorption contrast. It is obvious that the pure Sn in the non-heat-treated strand section (Figure 16.5a) is successively transformed into the superconducting phase Nb_3Sn (Figure 16.5b–d). The linear absorption coefficients μ at the used X-ray energy $E = 50\,keV$ are for pure Sn $\mu = 78.12\,cm^{-1}$, for pure Nb $\mu = 56.81\,cm^{-1}$, for pure Cu $\mu = 23.42\,cm^{-1}$, and for Nb_3Sn $\mu = 64.85\,cm^{-1}$. Hence, in the original state of the wire the gray values in the image for the Sn pools are much brighter than the gray values of the surrounding Nb. After the reaction treatment the gray values of the new formed Nb_3Sn are darker than the remaining Nb.

During the thermal treatment, the formation of voids in the strand matrix can be observed [14]. These voids may cause a degradation of the superconductor critical current. Therefore, it is desirable to avoid or at least to better control the void growth during the strand heat treatment, for example, through optimized strand designs or heat treatment cycles. Taking a glance at Figure 16.5 two types of voids are visible. One void type is formed below the melting temperature of Sn in the Sn diffusion centers of the strands (see Figure 16.5b). The cross-section of these voids can approach that of an entire Sn pool. The formation of these voids can be explained by a Sn volume increase upon β-Sn ($\rho = 7.26\,g\,cm^{-3}$ [15]) to γ-Sn transformation ($\rho = 6.6\,g\,cm^{-3}$ [16]), which starts at $T = 162\,°C$ [16, 17]. Tetragonal β-Sn is transformed into the rhombic γ-Sn, which causes a strong volume expansion of about 10% and, therefore, a plastic deformation of the wire. After the cooling, when γ-Sn is transformed back to β-Sn, voids evolve.

A second type of void arises in the reacted Nb_3Sn strands above the melting temperature of Sn, which has comparatively small (about 1 μm^2) cross-sections and is distributed around the superconductor filaments. The formation of these voids has been attributed to differences in the diffusion rates of Cu in Sn and Sn in Cu and to the different densities of the formed phases [18, 19].

16.3.3 Influence of the Foaming Agent on Metallic Foam Structures

Metallic foams are highly porous materials with unique properties. They distinguish themselves from other metallic materials by low density, high specific stiffness, and high energy-absorption capability [20, 21]. Therefore, they become more and more popular for industrial applications. In order to produce metallic foams with reproducible macroscopic properties, the precise knowledge of their microscopic structure is needed, that is, the knowledge of the physical processes during foaming is important for the optimization of the process parameter.

The metal foam presented here was manufactured by admixing 0.5 wt% titanium hydride powder acting as blowing agent to aluminum alloy powder [22]. Subsequently, the powder was pressed to the precursor material. By heating up the precursor to above both the decomposition temperature of the blowing agent and the melting temperature of the metal, hydrogen is released in the melt and a porous structure is generated. The influence of a blowing agent heat pre-treatment and of a varying blowing agent particle size on the foam structure was investigated by means of synchrotron tomography [23]. Figure 16.6 shows four different prepared foams in the early foaming state.

The treated TiH_2 powder (Figure 16.6a) causes a hydrogen release at higher temperatures (decomposition temperature $T = 520\,°C$) than the untreated powder ($T = 400\,°C$) (Figure 16.6b), which coincides much better with the melting temperature of the alloy ($T = 525\,°C$) [24]. Therefore, most hydrogen can be used for forming regular pores, whereas the hydrogen release in the sample prepared with untreated foaming agent leads to crack formation in the still solid metal.

The sample prepared with the small TiH_2 particles (Figure 16.6c) shows a small pore size distribution of a high number of small pores after foaming. For the sample prepared with a broad TiH_2 size distribution (Figure 16.6d) the pore size distribution is also widened.

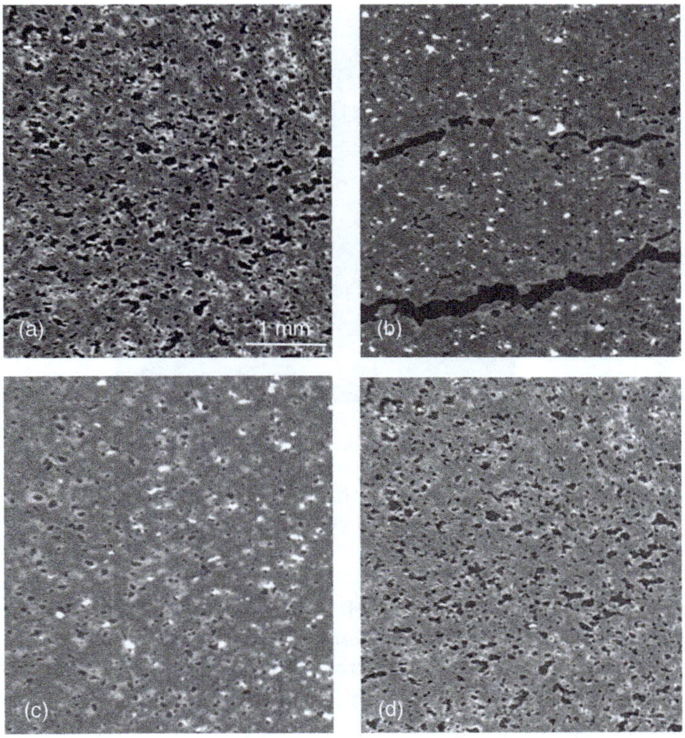

Figure 16.6 Tomographic slices of four differently prepared aluminum foams in the early foaming stage. (a) Sample prepared with pre-heat-treated TiH_2 (180 min/480 °C, decomposition temperature $T = 520$ °C). (b) Sample prepared with untreated TiH_2 (decomposition temperature $T = 400$ °C). (c) Sample prepared with sieved TiH_2 (particle size <20 μm). (d) Sample prepared with unsieved TiH_2 (particle size 20–180 μm).

Hence, the absorption tomographic experiments have shown that by varying the blowing agent parameter (size and/or heat treatment) the metallic foam structure is controllable [25].

16.3.4 *Ex vivo* Grating-Based Phase Contrast Imaging of Human Carotid Arteries

For samples consisting mainly of light elements where the attenuation contrast decreases strongly or for samples containing elements with almost the same absorption coefficients, more sensitive phase-contrast imaging methods are an interesting measurement alternative (see Chapter 15).

The precondition for phase contrast is (partially) coherent X-ray radiation. Insertion devices, like undulators in synchrotron storage rings, produce partially coherent X-rays. Therefore, phase-contrast X-ray imaging methods have become standard methods for tomography at third generation synchrotrons (see, e.g., [26–36]).

As an example for the potential of phase-contrast imaging in Figure 16.7 an absorption tomographic image of a human carotid artery is compared with a tomographic phase-contrast image of the same sample measured with a grating interferometer [37].

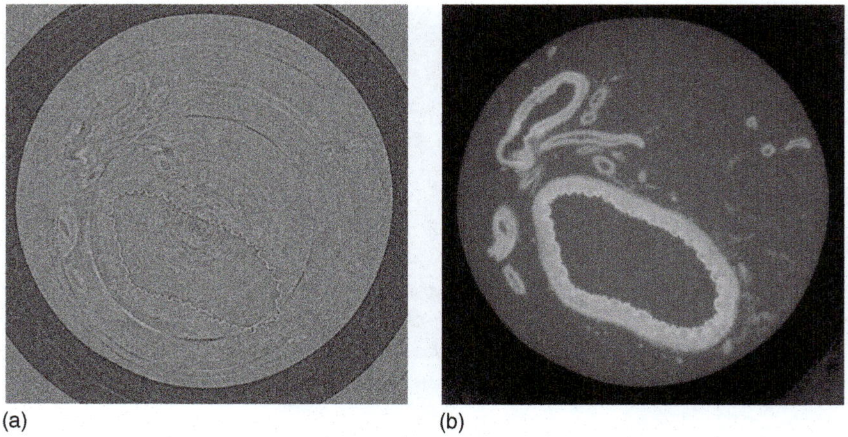

Figure 16.7 Reconstructed tomographic slices through a near-normal human carotid artery in conventional attenuation-based contrast (a) and grating-based phase-contrast (b). The artery was measured fixated in formalin with a pixel size of 5 μm at the beamline ID19 at (ESRF, Grenoble, France) [37].

The attenuation in the left image suffers from the expected low contrast for the vessel walls, while the phase contrast image reveals very small vessels with high contrast. Figure 16.8 shows the 3D rendering of the phase contrast tomographic image. Due to the excellent soft tissue contrast, several vasa vasorum (see arrows in Figure 16.8) can be identified without the use of contrast dye.

16.4 Image Artifacts

Tomographic images often contain different kinds of artifacts. A choice of common artifacts is shown in this section.

16.4.1 Ring Artifacts

Tomographic images are often superimposed by so-called ring artifacts. Ring artifacts are concentric rings in the images around the center of rotation of the tomographic setup caused, for example, by differences in the individual pixel response of the detector or by impurities on the scintillator crystals. They complicate the post-processing of the data, that is, the segmentation of individual image information.

Figure 16.9a shows a tomographic slice of a superconductor strand of Nb_3Sn superimposed by many ring artifacts [38].

A common approach to reduce ring artifacts is the flat field correction [39]. Thereby, images of the background without sample before and after the data acquisition and in certain intervals during the data acquisition are measured. The resulting flat fields include the nonuniform sensitivity of the CCD camera pixels, the nonuniform response of the scintillator screen, as well as the inhomogeneities in the incident X-ray beam. The measurement data are corrected by dividing the normalized radiographic images of the samples and the flat fields ($I_{corr} = I_{measurement}/I_{flat\ field}$). However, ring artifacts are not removed completely by this correction when different camera elements

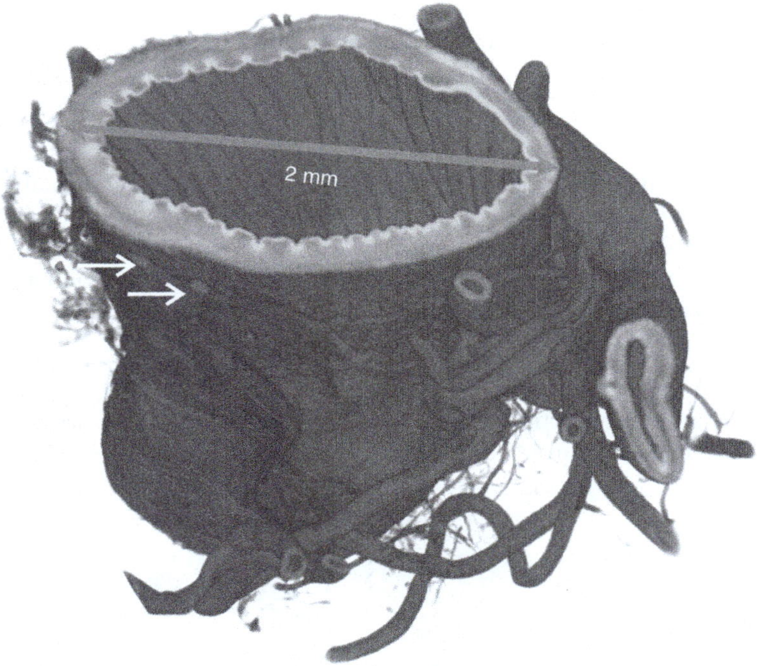

Figure 16.8 3D phase-contrast image of the near-normal carotid artery (measured at ID19, ESRF, 23 keV, 5 μm pixel size). The arrows points to some vasa vasorum [37].

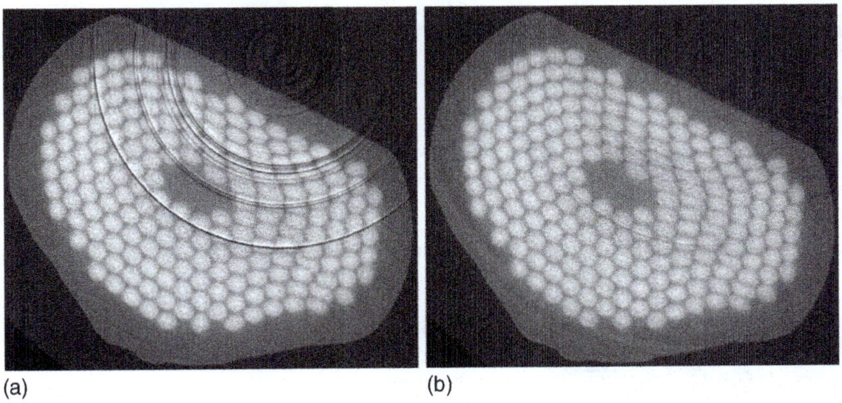

Figure 16.9 Binary powder-in-tube superconductor strand of Nb_3Sn and copper (sample diameter ≈ 0.7 mm). (a) Tomographic image superimposed by ring artifacts. (b) The same tomographic slice after using a sinogram correction algorithm [38].

have intensity-dependent, nonlinear response functions or the incident beam has time-dependent nonuniformities.

In addition, ring artifacts can be avoided by moving the sample or the detector during the measurement in defined horizontal and vertical steps [40, 41]. Then the characteristic of all detector elements is averaged, which leads to significantly reduced ring artifacts.

A third way to reduce ring artifact is the post-processing of the images, for example, size and shape filtering after the reconstruction or sinogram processing during the reconstruction (see Figure 16.9b, [42–44].

16.4.2 Image Noise

The quality of CT images can be degraded by image noise resulting from an unfavorable conversion rate of X-ray photons to visible light photons especially for higher X-ray energies (see Figure 16.10a) or from high-energy spikes. The use of median filtering in the reconstruction process allows both noise reduction and edge preservation and preserves useful details in the image (Figure 16.10b) [45].

16.4.3 Edge Artifacts

Edge enhancement is an adverse artifact for absorption tomography. These artifacts appear as aliasing patterns and as overshoots in the areas of sharp density transitions (see Figure 16.11).

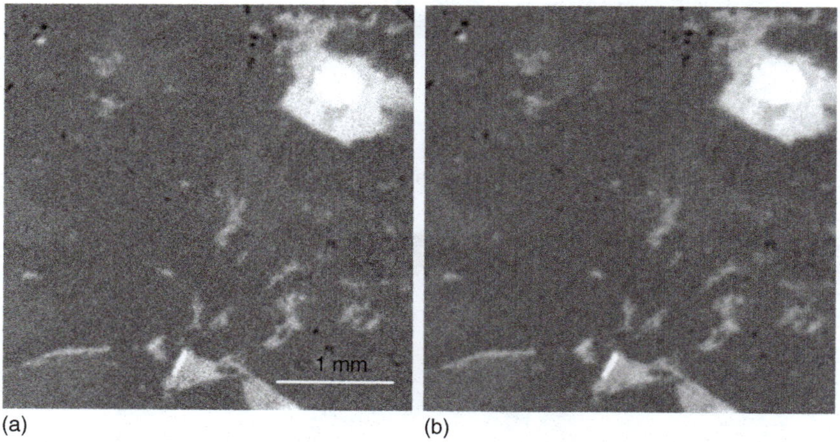

Figure 16.10 2D slice of a tomogram of granite rock. (a) Image noise. (b) Median filtered image.

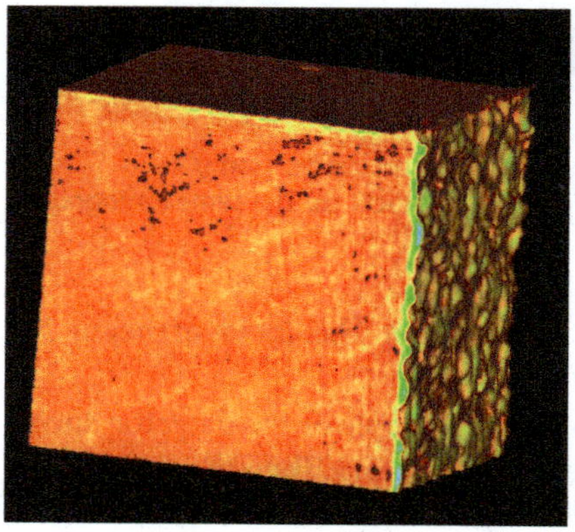

Figure 16.11 Tomographic slice of a homogeneous aluminum cube with a strong edge enhancement (sample size $2 \times 2 \times 2\ mm^3$).

Due to the edge enhancement, a precise preprocessing of the data, for example, the separation of the several components of a sample, is hampered. The effect is caused by scattering of X-rays on materials surfaces, where the absorption coefficients change strongly. As shown in Figure 16.11 the effect increases with the finish roughness. The three polished surfaces show a lower effect than the rougher one. To avoid this artifact, the distance between the sample and detector have to be minimized, which reduces the angle of beam spread.

16.4.4 Motion Artifacts

Motion artifacts occur either due to the movement of the sample during the measurement, for example, tilting, shaking, or shrinking, or due to the inexact movement of the step motors during the data acquisition. The sample in Figure 16.12 was intentionally shifted during the data acquisition to some horizontal steps to demonstrate the influence of this motion on the reconstructed 3D image. The Figure 16.12A(a) shows one sinogram of the measurement. A sinogram is the set of all projections, plotted versus the measured angular range (0 – 180°). The shift during the measurement can be seen clearly in the sinogram as discontinuities in the sinusoids (Figure 16.12A(a)). In Figure 16.12A(b) the associated tomographic slice is displayed.

The sample movement can be corrected only in a poor way, whereas the correction of the inexact movement of the motors is possible. Here, the discrepancy can be corrected by a translational displacement of each sinogram line at the optimal sinus curve (see Figure 16.12B(a)) [46].

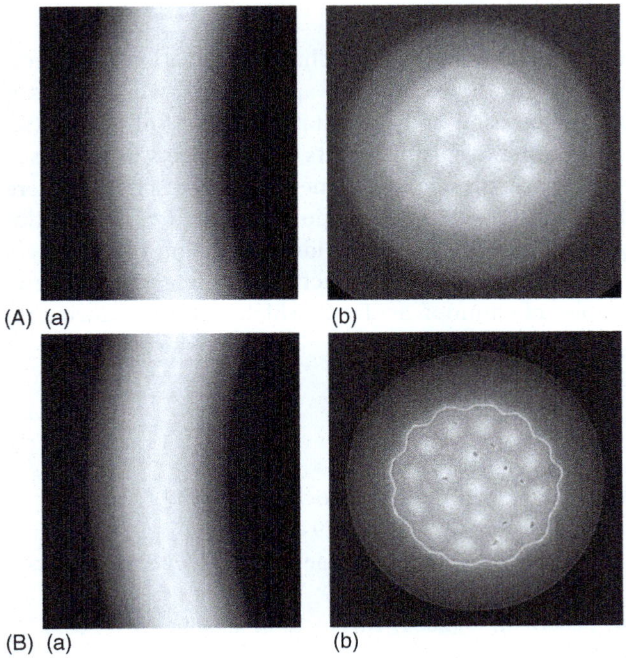

Figure 16.12 Tomographic slice of a superconductor wire cross-section Nb_3Sn (diameter 0.7 mm). (A) (a) Inexact movement of the step motor. (b) Unsharp image of the sample. (B) (a) The corrected sinogram. (b) The same image as above but without motion artifacts.

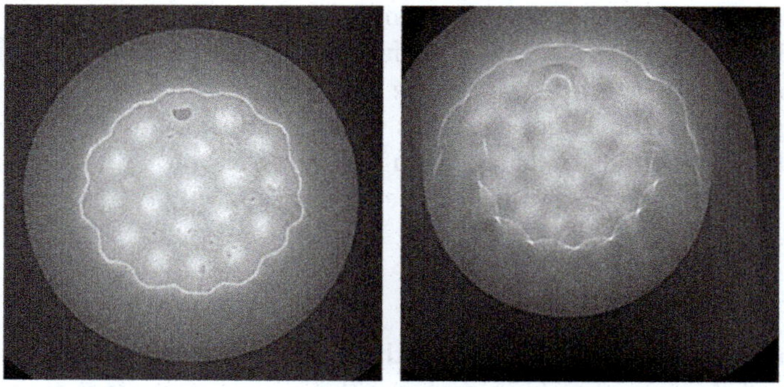

(a) (b)

Figure 16.13 Tomographic slice of a superconductor Nb_3Sn cross-section with a diameter of 0.7 mm. (a) The corrected image. (b) The rotation axis of the image differs from the correct axis by 30 pixels.

16.4.5 Centering Errors of the Rotation Axis

A deviation of the rotation axis during the measurements create semicircle-shaped artifacts for a 180° tomographic scan as shown in Figure 16.13b. These artifacts can be corrected easily by an adjustment of the parameter of the wrong rotation axis during the reconstruction calculation [47, 48]. Figure 16.13a shows the result of this correction.

16.5 Summary

In this chapter, we have described the advantages of synchrotron radiation for tomography. We have given an overview of the wide range of applications, which starts in biology and medicine with the visualization and quantitative analysis of tissues (see Section 16.3.4), teeth, structure of bones, implants, plants, small insects down to single cells followed by applications in materials science and neighboring disciplines. Here, the application field includes the investigation of defects, porous structures (see Section 16.3.3), granular structures (see Section 16.3.1), cracks, voids (see Section 16.3.2), fibers, precipitations or inclusions, small components, inner structures of stones, and soil. Further, we have shown some examples of common artifacts, which can occur during the measurement or reconstruction.

References

1 Elder, F.R., Gurewitsch, A.M., Langmuir, R.V., and Pollock, H.C. (1947) Radiation from electrons in a synchrotron. *Phys. Rev.*, **71** (**11**), 829–830.

2 Dell, R.M. (2000) Batteries, fifty years of materials development. *Solid State Ionics*, **134**, 139–158.

3 Köhler, U., Antonius, C., and Bäuerlein, P. (2004) Advances in alkaline batteries. *J. Power Sources*, **127**, 45–52.

4 de Souza, C.C.B.M., de Oliveira, D.C., and Tenorio, J.A.S. (2001) Characterization of used alkaline batteries powder and analysis of zinc recovery by acid leaching. *J. Power Sources*, **103**, 120–126.

5 Ohms, D., Kohlhase, K., Benczur-Ürmössy, G., and Schädlich, G. (2002) New development on high power alkaline batteries for industrial applications. *J. Power Sources*, **105**, 127–133.
6 Horn, Q.C. and Shao-Horn, Y. (2003) Morphology and spatial distribution of ZnO formed in discharged alkaline Zn/MnO_2 AA cells. *J. Electrochem. Soc.*, **150** (**5**), A652–A658.
7 Huot, J.-Y. and Malservisi, M. (2001) High-rate of zinc anodes in alkaline primary cells. *J. Power Sources*, **6**, 133–139.
8 Haibel, A., Manke, I., Melzer, A., and Banhart, J. (2010) In situ microtomographic monitoring of discharging processes in alkaline cells. *J. Electrochem. Soc.*, **157** (**4**), A387–A391.
9 Lohmann, G. (1998) *Volumetric Image Analysis*, John Wiley & Sons, Ltd., Chichester.
10 Ohser, J. and Mücklich, F. (2000) *Statistical Analysis of Microstructures in Materials Science*, John Wiley & Sons, Inc., New York.
11 Devred, A. *et al.* (2006) Overview and status of the next European dipole joint research activity. *Supercond. Sci. Technol.*, **19** (**3**), 67–83.
12 Goedeke, A. (2005) Performance boundaries on Nb_3Sn superconductors. PhD thesis. University of Twente, Enschede, The Netherlands.
13 Haibel, A. and Scheuerlein, C. (2007) Synchrotron tomography for the study of void formation in internal tin Nb3Sn superconductors. *IEEE Trans. Appl. Supercond.*, **17** (**1**), 34–39.
14 Naus, M.T., Lee, P.J., and Larbalestier, D.C. (2000) The interdiffusion of Cu and Sn in internal Sn Nb_3Sn superconductors. *IEEE Trans. Appl. Supercond.*, **10** (**1**), 983.
15 Lide, D.R. (ed.) (1995) *Handbook of Chemistry and Physics*, 76th edn, CRC Press, Boca Raton, FL.
16 Khitrova, V.I. (1988) Crystal structure of γ-Sn. *Sov. Phys. Crystallogr.*, **33** (**1**), 139–141.
17 Kubiak, R. (1986) Evidence for the existence of the γ form of tin. *J. Less-Common Met.*, **116**, 307–311.
18 Paul, A. (2004) The Kirkendall effect in solid state diffusion. PhD thesis. Technische Universiteit Eindhoven.
19 Scheuerlein, C., Gasser, P., Jacob, P., Leroy, D., Oberli, L., and Taborelli, M. (2005) The effect of CuSn intermetallics on the interstrand contact resistance in superconducting cables for the large hadron collider. *J. Appl. Phys.*, **97** (**3**), 033909-1–033909-7.
20 Ashby, M. *et al.* (2000) *Metal Foams: A Design Guide*, Butterworth-Heinemann, Boston, MA.
21 Banhart, J., Fleck, N., and Mortensen, A. (eds) (2003) *Cellular Metals: Manufacture, Properties, Applications*, MetFoam 2003, MIT-Verlag, Berlin.
22 Banhart, J. (2001) Manufacture, characterisation and application of cellular metal foams. *Prog. Mater Sci.*, **46**, 559–632.
23 Bütow, A. (2004) Strukturuntersuchungen an Metallschäumen in verschiedenen Entwicklungsstadien. Diploma thesis. Hahn-Meitner-Institut Berlin, SF3.
24 Matijasevic, B., Fiechter, S., Zizak, I., Görke, O., Wanderka, N., Schubert-Bischoff, P., and Banhart, J. (2004) Decomposition behaviour of as-received and oxidized TiH_2 powder. Powder Metallurgy World Congress, Conference Book 2004.

25 Haibel, A., Bütow, A., Rack, A., and Banhart, J. (2004) Quantitative analysis of pore-particle-correlations in metallic foams. WCNDT Congress, Conference Proceedings 2004.

26 Raven, C., Snigirev, A., Snigireva, I., Spanne, P., Sourvorov, A., and Kohn, V. (1996) Phase-contrast microtomography with coherent high-energy synchrotron X-rays. *Appl. Phys. Lett.*, **69** (**13**), 1826–1828.

27 Spanne, P., Raven, C., Snigireva, I., and Snigirev, A. (1999) In-line holography and phase-contrast microtomography with high energy X-rays. *Phys. Med. Biol.*, **44**, 741–749.

28 Snigirev, A., Snigireva, I., Kohn, V., Kuznetsov, S., and Schelokov, I. (1995) On the possibilities of X-ray phase contrast microimaging by coherent high-energy synchrotron radiation. *Rev. Sci. Instrum.*, **66** (**12**), 5486–5492.

29 Cloetens, P., Barrett, R., Baruchel, J., Guigay, J.-P., and Schlenker, M. (1996) Phase objects in synchrotron radiation hard X-ray imaging. *J. Phys. D*, **29**, 133–146.

30 Momose, A. (1995) Demonstrating of phase-contrast X-ray computed tomography using an X-ray interferometer. *Nucl. Instrum. Methods Phys. Res., Sect. A*, **352**, 622–628.

31 Beckmann, F., Bonse, U., Busch, F., and Günnewig, O. (1997) X-ray microtomography (μCT) using phase contrast for the investigation of organic matter. *J. Comput. Assisted Tomogr.*, **21** (**4**), 539–553.

32 Momose, A. (2002) Phase-contrast X-ray imaging based on interferometry. *J. Synchrotron Radiat.*, **9**, 136–142.

33 Momose, A. (2003) Phase-sensitive imaging and phase tomography using X-ray interferometers. *Opt. Express*, **11**, 2303.

34 Cloetens, P., Ludwig, W., Baruchel, J., Van Dyck, D., Van Landuyt, J., Guigay, J.P., and Schlenker, M. (1999) Holotomography: quantitative phase tomography with micrometer resolution using hard synchrotron radiation X-rays. *Appl. Phys. Lett.*, **75** (**19**), 2912–2914.

35 Weitkamp, T., Diaz, A., David, C., Pfeiffer, F., Stampanoni, M., Cloetens, P., and Ziegler, E. (2005) X-ray phase imaging with a grating interferometer. *Opt. Express*, **13** (**16**), 6296–6304.

36 Pfeiffer, F., Weitkamp, T., Bunk, O., and David, C. (2006) Phase retrieval and differential phase-contrast imaging with low-brilliance X-ray sources. *Nat. Phys.*, **2**, 258–261.

37 Saam, T., Herzen, J., Hetterich, H., Fill, S., Willner, M. *et al.* (2013) Translation of atherosclerotic plaque phase-contrast CT imaging from synchrotron radiation to a conventional lab-based X-ray source. *PLoS One*, **8** (**9**), e73513. doi: 10.1371/journal.pone.0073513

38 Boin, M. and Haibel, A. (2006) Compensation of ring artefacts in synchrotron tomographic images. *Opt. Express*, **14** (**25**), 12071–12075.

39 Herman, G.T. (1980) *Image Reconstruction from Projections – The Fundamentals of Computed Tomography*, Academic Press, New York.

40 Davis, R. and Elliott, J.C. (1997) X-ray micro tomography scanner using time-delay integration for elimination of ring artefacts in the reconstructed image. *Nucl. Instrum. Methods Phys. Res., Sect. A*, **394**, 157–162.

41 Görner, W., Hentschel, M.P., Müller, B.R., Riesemeier, H., Krumrey, M., Ulm, G., Diete, W., Klein, U., and Frahm, R. (2001) BAMline: the first hard X-ray beamline at BESSY II. *Nucl. Instrum. Methods Phys. Res., Sect. A*, **467–468**, 703–706.

42 Sijbers, J. and Postnov, A. (2004) Reduction of ring artifacts in high resolution micro-CT reconstructions. *Phys. Med. Biol.*, **49** (**14**), 247–253.

43 Raven, C. (1998) Numerical removal of ring artifacts in microtomography. *Rev. Sci. Instrum.*, **69**, 2978–2980.

44 Antoine, C., Nygard, P., Gregersen, O., Holmstad, R., Weitkamp, T., and Rau, C. (2002) 3D images of paper obtained by phase-contrast X-ray microtomography: image quality and binarisation. *Nucl. Instrum. Methods Phys. Res., Sect. A*, **490**, 392–402.

45 Marion, A. (1991) *An Introduction to Image Processing*, Chapman & Hall, London, p. 274.

46 Lu, W. *et al.* (2002) Tomographic motion detection and correction directly in sinogram space. *Phys. Med. Biol.*, **47**, 1267–1284.

47 https://arxiv.org/abs/1306.1392v1.

48 Donath, T., Beckmann, F., and Schreyer, A. (2006) Automated determination of the center of rotation in tomography data. *J. Opt. Soc. Am. A:*, **23**, 1048–1057.

Part III

New and Emerging Methods

17

3D X-ray Diffraction Microscope

Henning F. Poulsen, Wolfgang Ludwig, and Søren Schmidt

Three-dimensional X-ray diffraction (3DXRD) is a novel technique aimed at a fast and nondestructive characterization of the individual crystalline elements within millimmeter–centimeter-sized polycrystalline specimens. It is based on two principles: the use of highly penetrating hard X-rays from a synchrotron source (X-ray energies in the range 30–100 keV) and the application of "tomographic" reconstruction algorithms for the analysis of the diffraction data. In favorable cases, the position, morphology, phase, and crystallographic orientation can be derived for hundreds of elements simultaneously as well as their elastic strains. Furthermore, the dynamics of the individual elements can be monitored during typical processes such as deformation or annealing. Hence, for the first time, information on the interaction between elements can be obtained directly. The provision of such data is vital in order to extend beyond state-of-the-art structural models.

Notably, in a 3DXRD experiment one must prioritize between spatial, angular, and time resolution. This has led to a variety of 3DXRD strategies, which we in the following sections will summarize as four standard modes of operation. The first two modes enable fast measurements of the *average characteristics* of each grain – such as their center-of-mass (CMS) position, volume, average orientation, and/or average strain tensor – while the exact location of the grain boundaries is unknown. The two other modes enable full 3D mappings of grains and orientations.

In the following text, initially the 3DXRD setup will be presented and 3DXRD strategies are discussed, and subsequently, the standard modes of operation are presented in more detail with emphasis on reconstruction principles and with selected examples of use. For general reference, a mathematical description of the methodology and additional applications, see [1].

Alternative approaches to provision of 3D maps based on X-ray diffraction exist. These are based on inserting wires [2], slits [3], or collimators [4] between the sample and the detector and scanning the sample with respect to these elements. Not surprisingly such methods will be slower than the tomographic approach of 3DXRD, but they may be associated with other advantages, such as improved options for measuring the local elastic strain. In particular, we mention the technique of "differential-aperture X-ray microscopy" [2]. Using a polychromatic microbeam with energies of 8–20 keV and scanning a wire, sub-micrometer resolution in 3D has been demonstrated, see Chapter 18.

Neutrons and Synchrotron Radiation in Engineering Materials Science: From Fundamentals to Applications, Second Edition. Edited by Peter Staron, Andreas Schreyer, Helmut Clemens, and Svea Mayer.

17.1 Basic Setup and Strategy

The basic 3DXRD setup – sketched in Figure 17.1 – is quite similar to conventional tomography settings at synchrotrons. A (nearly) parallel monochromatic X-ray beam impinges on the sample as a uniform field. The sample is mounted on an ω rotation stage, where ω is the rotation around an axis perpendicular to the incoming beam. As an option x-, y-, and z-translations as well as additional rotations may be added.

Any part of the illuminated structure, which fulfills the Bragg condition, will generate a diffracted beam. This beam is transmitted through the sample and probed by a 2D detector. To probe the complete structure, and not just the part that happens to fulfill the Bragg condition, the sample is rotated. Hence, exposures are made for equi-angular settings of ω with a step of $\Delta\omega$. To provide a uniform sampling, the sample is rotated by $\Delta\omega$ during each exposure. Essential to 3DXRD is the idea to mimic a 3D detector by positioning several 2D detectors at different distances to the center-of-rotation, L, and exposing these either simultaneously (many detectors are semitransparent to hard X-rays) or subsequently.

Presently, typically two types of detectors are used: near-field detectors with a spatial resolution of ~5 μm in close proximity to the sample (L in the range 2–10 mm) and far-field detectors: conventional CCDs with a resolution of 100–200 μm (L in the range 20–50 cm). The former provide information on position and orientation degrees-of-freedom, while the latter probes strain and orientation. With the near-field detector, images may be acquired at several, typically three, distances as illustrated in Figure 17.1. This enables ray tracing of the diffracted beam, also known as *tracking*.

The incoming beam may illuminate the full sample or be focused in one direction to probe only a layer within the material (see Figure 17.1). In the following discussion, we shall refer to these as the 3D and 2D cases respectively. Notably, the 2D case is restricted to use at third-generation synchrotron sources, while the 3D case also applies to second-generation sources and in favorable cases even to work with laboratory

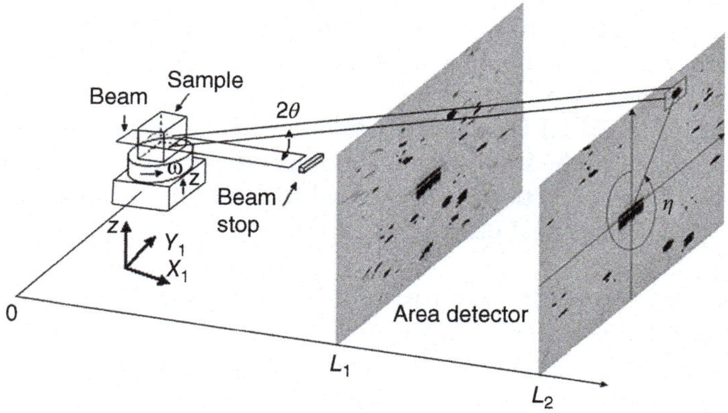

Figure 17.1 Sketch of the 3DXRD principle for the case of the incoming monochromatic beam being focused in one dimension. Two types of detectors are used: high-spatial-resolution detectors close to the sample and low-resolution detectors far away from the sample. The Bragg angle 2θ, the rotation angle ω and the azimuthal angle η are indicated for the diffracted beam arising from one grain of a coarse-grained specimen, and for two settings of the high-resolution detector. The axes for the laboratory co-ordinate system are also shown.

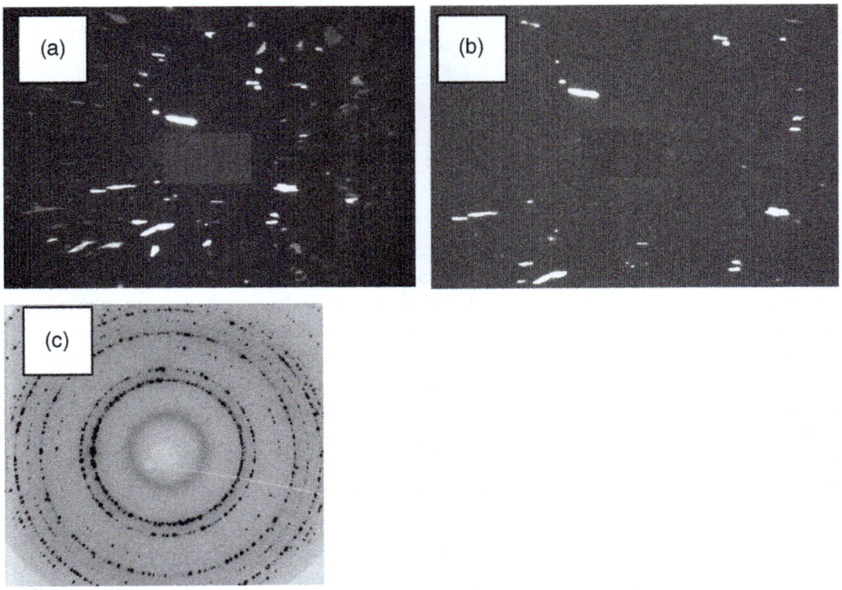

Figure 17.2 Typical 3DXRD data from a coarse-grained undeformed polycrystal. (a) and (b) are "near-field" exposures with the high-spatial-resolution detector (see Figure 17.1) at distances of $L = 4$ mm and $L = 8$ mm, respectively. The background is shown as black. Panel (c) is a corresponding "far-field" image made with the low-resolution detector at $L = 400$ mm and shown with an inverted gray scale to improve visibility. In the latter case the diffraction spots are positioned on Debye–Scherrer rings.

sources. In the 2D case, 3D information is generated simply by repeating data acquisition for a set of layers and stacking the resulting reconstructions. An example of a set of exposures obtained in the 2D case is provided in Figure 17.2.

A vital difference between the two types of detectors relates to data acquisition time: with the current technology the far-field detectors provide a time resolution that is several orders of magnitude better than the near-field ones. This makes it necessary to carefully consider the priority of time and space resolution in a 3DXRD experiment. In practice, we may distinguish between four modes of operation, as sketched in Figure 17.3.

Modes I and II are based (mainly) on far-field information. Mode II provides information on the phase, CMS position, volume, as well as average orientation and average strain tensor of each grain in the illuminated part of the specimen. Mode I is an option for very fast measurements, where only a subset of the illuminated grains are characterized and no spatial information is available. In addition, the orientation and strain characterization will not be complete. Experimentally, the main difference between the two modes is the ω range covered.

Modes III and IV aim at generation of complete 3D maps. In the case of an undeformed material, the orientation is constant within each grain, and the aim is to provide a *grain map*. In the case of a deformed material, the orientation varies locally. This makes it relevant to measure *orientation maps*, where – similar to EBSD – each voxel in the sample is associated with its own orientation. In both the cases, evidently a high-spatial-resolution detector is needed. The various modes will be detailed in further text.

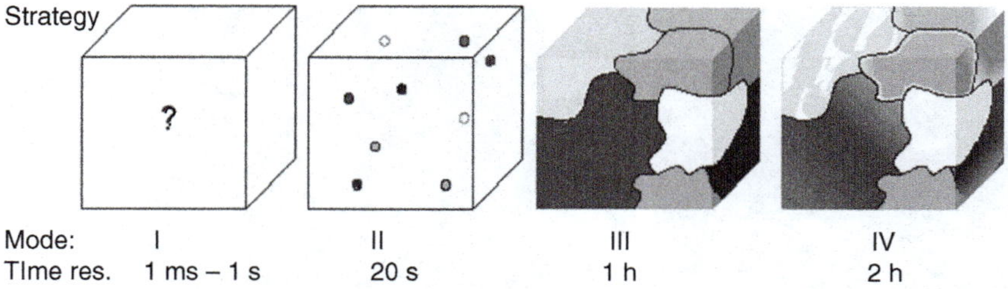

Figure 17.3 Sketch of four typical 3DXRD modes of operation, listed in direction of increasing spatial information. Mode I: statistical mode, with no spatial information and where not all grains within the illuminated volume are probed. Mode II: information of center-of-mass position, volume, average orientation, and average strain tensor for each grain. Mode III: 3D grain map for undeformed specimens. Mode IV: 3D orientation map for deformed specimens. In all cases, orientations are represented by a gray scale. Typical time resolutions for characterization of 100 grains are indicated. (Courtesy of E. Nielsen.)

In many experiments combinations of these modes are relevant. As an example, one may wish to start by mapping extended parts of a sample in order to identify sub-volumes of particular interest. Then one focuses on such parts and performs a fast CMS study on these parts during *in situ* processing. This procedure is then complemented by mapping extended parts again after the processing is completed.

17.1.1 The 3DXRD Microscope

Presently, one dedicated 3DXRD microscope exists. This was developed in collaboration between Risø National Laboratory and the ESRF, and installed at the Materials Science beamline at the ESRF. The optics enable work in the energy range 30–90 keV with a choice between parallel beam, vertical focusing only, and both vertical and horizontal focusing. The energy bandwidth can be varied from 10^{-4} to 10^{-2}, with the most popular setting being $\sim 10^{-3}$. The microscope can carry loads of up to 200 kg. It is equipped with a 25 kN stress rig and several furnaces with operational temperatures of up to 1300 °C. In addition, 3DXRD experiments have been performed at sector 1 at the APS. Currently, this setup is being converted into a dedicated experimental station.

17.2 Indexing and Characterization of Average Properties of Each Grain

A very important simplification arises in the case where the diffraction pattern is composed of a set of (primarily) nonoverlapping diffraction spots. Three examples of such datasets are shown in Figure 17.2. A large number of problems within polycrystal and powder research can be tailored to apply to this situation.

In this case, the diffraction spots can be sorted with respect to element of origin, say the grain, by means of a polycrystalline indexing scheme [5]. The starting point for such schemes is to apply image analysis routines that identify the spots and determine their CMS positions. Next, the direction of the corresponding diffracted X-rays are determined. In the case of near-field data – that is the tracking setup illustrated in Figure 17.1 – ray tracing is used. Having identified a set of spots at different detector-to-sample distances that belong to the same reflection, the best fit to a straight line through the CMSs of these spots is determined. In the case of far-field data, one may simply assume that all grains are positioned at the center-of-rotation and for any spot draw the line from its center on the detector to the rotation axis. In both the cases, once the direction of the ray is known, the corresponding scattering vector is readily determined.

The sorting can be based on three principles: orientation, position, or grain volume (the integrated intensity of any diffraction spot is proportional to the volume of the associated grain of origin). Sorting based on orientation and position has been implemented in the program GRAINDEX [5], which assumes the crystallographic space group to be known. The main limitation is in terms of the probability of spot overlap on the detector. The probability is determined by the number of grains, the texture, and the orientation spread of each grain. Simulations show that for samples with a weak texture and grains with a near-perfect lattice – that is with an orientation spread of order 0.1° or less – several thousand grains can be indexed simultaneously [1]. On the other hand, plastic deformation introduce orientation spread within the grains, which in practice prohibits indexing of embedded grains in materials that are deformed to more than 20%.

Once a grain has been indexed all the tools of the conventional single crystal diffraction analysis is available. Hence, based on the output of GRAINDEX, a comprehensive characterization of each grain immediately follows (mode II analysis). If only the far-field detector is used, the grain volume and average orientation can be characterized. Furthermore, as detailed in Chapter 8, one can determine the average values of all the components of the elastic strain tensor [6, 7]. The time resolution is often ~10 s, limited by the read-out time of the detector. The accuracy is of the order 0.3° for orientations and $\Delta\varepsilon/\varepsilon = 10^{-4}$ for the strain. Furthermore, it is possible to perform a full crystallographic analysis of each grain, for example, a structural refinement [8]. This mode is well suited for *in situ* studies of dynamic phenomena in materials science as it is fast and relatively straightforward.

In a variant of this mode also the CMS position of each grain is determined. At the expense of time resolution, this can be done by tracking using a high-spatial-resolution detector as outlined above. To determine positions with high accuracy, a nonlinear least square fit to the CMS positions of the spots on the detector is performed. The accuracy is currently ~5 μm.

For very fast data acquisitions, mode I is relevant. In this case, one simply repeats acquisitions with the far-field detector, oscillating around a given ω setting. Assuming that grains do not rotate, one can then monitor the change in the volume of the grains giving rise to diffraction spots at this ω setting. Several hundred so-called *growth curves* of deeply embedded grains can be obtained simultaneously in this way [9].

17.2.1 Application I: Nucleation and Growth Studies

Traditionally, nucleation and growth phenomena have been analyzed using ensemble average properties, such as the volume fraction of transformed material. However, the predictive power of average properties is limited by the neglect of heterogeneities. For example, nucleation may take place preferentially at specific sites, and the growth rate of nuclei may depend strongly on orientation, size, stoichiometry, or relationships with neighboring volumes. 3DXRD is an ideal tool to study the effect of heterogeneities,

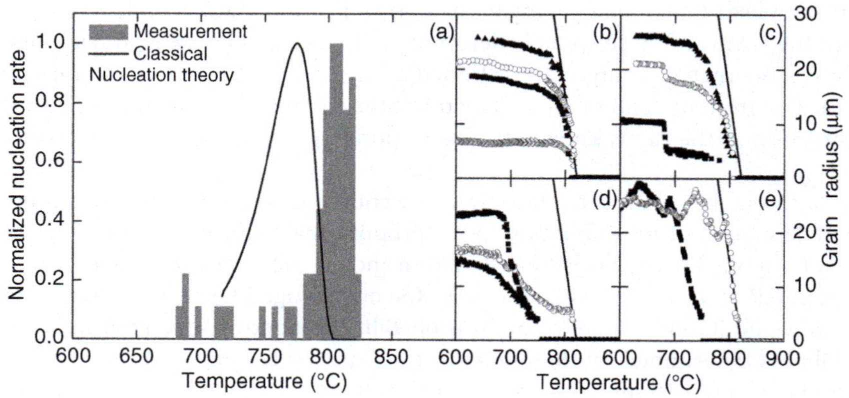

Figure 17.4 (a) Nucleation of ferrite grains during cooling of a carbon steel. The normalized observed nucleation rate (bars) is compared to theoretical prediction (line) based on a classical model [12]. (b–e) Examples of growth curves for individual ferrite grains. The lines are predictions based on the parabolic growth model of Zener. (Offerman 2002 [11]. Reproduced with permission of AAAS.)

and as such it has been used for a series of studies related to recrystallization [9], solidification [10], and phase transformations in steel [11], ceramics, and ferroelectrics. In all cases it was demonstrated that the ensemble average "Avrami-type" models are at best gross simplifications.

As an example of mode I work, Offerman *et al.* studied the phase transformation from the high-temperature austenite to the medium-temperature ferrite phase in carbon steel [11]. The results shown in Figure 17.4 relate to a study where the steel was cooled at a constant rate from 900 to 600 °C over 1 h. During this process the nucleation and growth of ferrite grains was inferred from the appearance and evolution of ~60 distinct diffraction spots originating from the ferrite. By counting the number of new spots appearing in a given time interval, the nucleation rate is determined as a function of temperature, a parameter that cannot be determined with high fidelity in any other way. In Figure 17.4a, the nucleation rate is compared to a prediction from a classical nucleation model [12]. From analysis of the data shown in the figure it is found that the predicted and measured activation energies differ by more than two orders of magnitude. Relevant revisions to the model are currently being discussed.

Shown in Figure 17.4b–e are examples of resulting growth curves for the nuclei. It appears that the growth of some grains, those in "b", followed closely the predictions of Zener, while others displayed large discrepancies. The latter was explained by several effects including local enrichment in the carbon content of the austenite.

17.2.2 Application II: Plastic Deformation

Polycrystal deformation is a topic of prime interest to both metallurgists and geoscientists. However, despite 70 years of effort, there is no consensus on how to approach the modeling. In particular, it is unclear to what extent deformation behavior is determined by the initial grain orientation, the grain–grain interactions, or by the emerging dislocation structures. 3DXRD provides *in situ* methods for addressing these issues, for example, through real-time observations of grain rotations in bulk materials during the deformation process [13, 14]. As an example of mode II work, Figure 17.5 provides a mapping of the rotation path of 95 grains in a 4 mm-thick polycrystalline Al sample

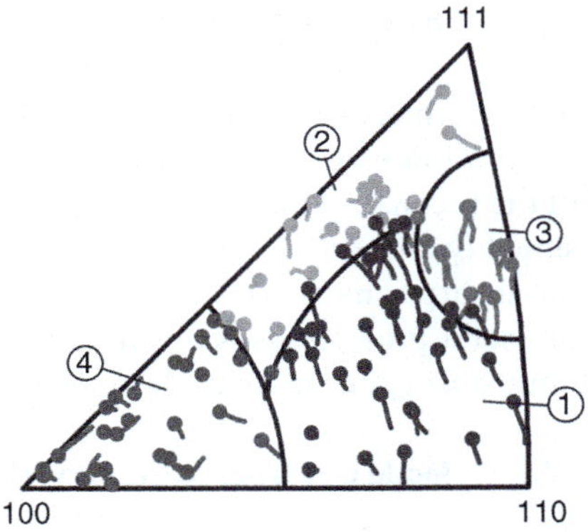

Figure 17.5 Rotation of the tensile axis of 95 grains within an Al specimen, as expressed in a stereographic triangle. The curves are the observed paths for the average orientation of each grain during tensile deformation from 0 to 6%. The final orientation is marked by a circle. The rotation behavior is interpreted in terms of four regions marked by solid lines. (Winther [15]. Reproduced with permission of Elsevier.)

strained in tension. These data provide a detailed combinatorial database for critical evaluation of models for polycrystalline deformation. In major parts of the orientation triangle, the paths exhibit a clear dependence on initial orientation. It is found that the grain responses do not match predictions from classical Taylor, Sachs, or self-consistent models, but can be described by subdividing orientation space in four regions with distinctly different rotation behavior [15].

17.2.3 Application III: Studies of Subgrains and Nanocrystalline Materials

Notably, 3DXRD studies are also relevant for crystalline structures with a grain size smaller than the spatial resolution. Evidently, in such cases, the generation of a grain map is not possible. However, by focusing the beam to say 10 μm × 10 μm, diffraction spots from individual grains as small as 70 nm can be detected. By tracing the integrated intensities and positions of such spots as function of time, one can infer changes in the volume, orientation, and strain of the grains of origin.

The main limitation for such studies is spot overlap. To overcome this problem two approaches have been pursued. The first approach is based on reducing the number of illuminated grains by investigating foils [16]. Provided the foil thickness is at least 10 times the grain size, for many annealing processes the grains at the center of the foil may be considered bulk grains. This methodology has been applied to a series of coarsening studies of subgrains in aluminum deformed to both medium and very high strains.

The second approach is to increase the angular resolution of the instrument by two orders of magnitude – to 0.004° in η and ω and 0.0005° in 2θ. Implemented at sector 1-ID at APS, this improvement is achieved at the expense of flux by use of a six-bounce monochromator and a detector positioned at a distance of 4 m from the sample [17]. This setup is designed specifically for studies of dislocation patterning. When deforming metals, typically the dislocations assemble into walls separating nearly perfect dislocation-free crystals, the subgrains, with dimensions of 100 nm to 2 μm. With the setup at APS, individual spots from millions of illuminated subgrains in a 300 μm thick Cu sample can be resolved. Full 3D reciprocal space maps are gathered continuously while deforming the specimen. Uniquely, from their time-dependence fundamental questions can be addressed in a direct way, such as "how and when do the dislocation structures form?" and "how do the structures subdivide as function of increasing strain?" In the case of tensile deformation of Cu, it is found that the dislocation structures form just after the plastic on-set and that the structures display intermittent dynamics – a surprising fact not predicted by theory [17].

17.3 Mapping of Grains and Orientations

3DXRD microscopy enables a relatively fast generation of large 3D grain maps and 3D orientation maps by use of tomographic reconstruction algorithms. The development of the reconstruction algorithms is nontrivial, as the complexity in terms of the dimensionality and sheer size of the reconstruction space is much larger than, for example, for absorption contrast tomography. Another difference is that in 3DXRD the number of useful projections is given by the number of observable reflections, which may be limited by experimental geometry to be as few as five.

17.3.1 Mode III: Mapping Grains in Undeformed Specimens

At this stage focus a range of approaches have been presented for the 2D case:

- *Simple backprojection.* In the 2D case, reconstruction algorithms are strictly speaking not needed, as the shape of any grain can be retrieved by back-projecting the periphery of one associated diffraction spot along the direction of the diffracted X-rays. This approach is simple and fast [6], but in practice is associated with a spatial resolution that is substantially worse than the approaches below.
- *Forward projection.* For each pixel in the layer of interest in the sample, a forward-projection algorithm scans orientation space to test which orientations match the diffraction patterns. Cross-talk between the pixels in the layer is neglected. Hence, one can only infer the set of *possible* orientations. However in practice, the constraint that all pixels belonging to a given grain must have the same orientation often implies that there is one and only one (possible) orientation for each pixel. If that is not the case, for example, in cases with substantial spot overlap, the ambiguous result may be used as input for inverse methods presented below. Two forward-projections programs have been developed by Suter and coworkers [18] and Schmidt [19].
- *Algebraic reconstruction algorithm.* The fact that there is a linear relationship between grain volume and the integrated intensity of an associated diffraction spot implies that algebraic approaches are possible. In [20], the so-called 2D-ART algorithm is detailed. It assumes that the orientations and approximate CMS positions of all grain sections are known *a priori*, for example, determined by GRAINDEX. The method attempts to reconstruct the boundary of each grain separately. For a specific grain, this is done by associating a "grain-density" to each pixel in the layer. Once the solution has converged, the grain boundary can be defined by setting a threshold. A complete grain map may be obtained by superposing the solutions, the boundaries, of the individual grains. However, such a map will not be space filling as boundaries from neighboring grains may overlap or leave "voids" in the map. To avoid this, one may choose for each pixel the grain where the reconstruction exhibited the largest density.

  First *experimental data* for 2D-ART is reproduced in Figure 17.6. The grain map resulted from an independent reconstruction of 27 grains based on five reflections each. From the overlap between grains and voids in Figure 17.6, the spatial resolution is estimated to be ~5 μm. The limitation is identified to be the instrumental point-spread-function. A few grains are obviously missing from the map – for the particular geometry used the number of independent reflections were too few to enable reconstruction.

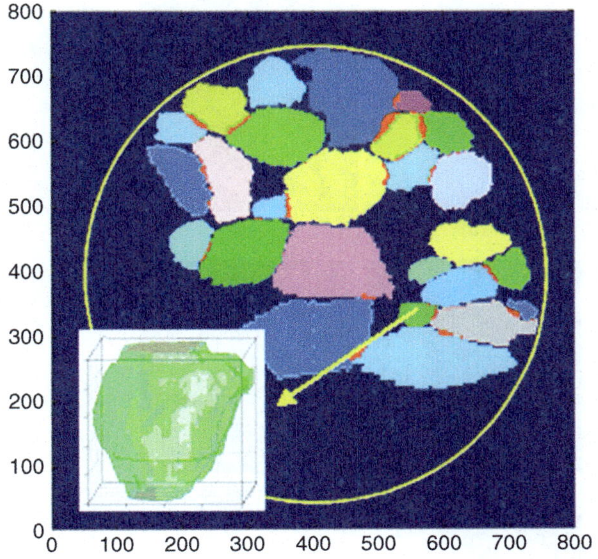

Figure 17.6 First reported partial grain map of one layer using 2D-ART. The map is in units of micrometer with the surface of the Al polycrystal indicated by a solid yellow circle. The grain sections were reconstructed independently. This data set was acquired in 10 min. By stacking layers 3D maps are obtained, an example of which is shown in the inset. Orientations are represented by a gray level scheme. (Fu 2003 [21]. Reproduced with permission of Society of Photo-Optical Instrumentation Engineers.)

- *Monte-Carlo-based reconstruction.* Stochastic approaches are attractive as they easily enable genuine simultaneous reconstructions of all grains rather than grain-by-grain reconstruction as 2D-ART above. On the other hand, they tend to be slow and "get stuck" in local minima if the configuration space is too large. Two approaches have been demonstrated: (i) A *restoration* approach, where a coarse or ambiguous grain map first is generated by a forward-projection algorithm [18] or by ART [22]. A Monte-Carlo-based routine is then used to "restore" the correct pixel affiliations in the grain boundary near regions. (ii) An *indexing* approach, in this approach there is no need for an initial grain map; instead the grain map is reconstructed based only on the output from GRAINDEX, that is the number of grains and their orientations [22].

Extensive simulations have been performed in order to estimate the merits of various approaches. The most critical parameter is found to be the number of reflections available for the analysis. Remarkably, for 2D-ART and Monte-Carlo-based reconstructions, reasonable quality maps can be gathered with as little as five projections. In Alpers *et al.* [22], several algorithms are compared with the result that the Monte-Carlo-based reconstructions are superior. Corresponding work on algorithms for the 3D case is currently undergoing (e.g., [23]).

17.3.2 Mode IV: Mapping Orientations in Deformed Specimens

For deformed specimens it is relevant to generate orientation maps, where each voxel in the specimen is associated with each own orientation, see Figure 17.3. Mathematically speaking, the task in this case is to reconstruct a vector-field $\underline{r}(\underline{x})$, where $\underline{r}$ symbolize orientations and $\underline{x}$ position in direct space. To the knowledge of authors, for such fields no transform or algebraic reconstruction algorithm is readily available. Stochastic approaches are possible, but here it is a concern that the configurations space is much larger for deformed case than for the un-deformed.

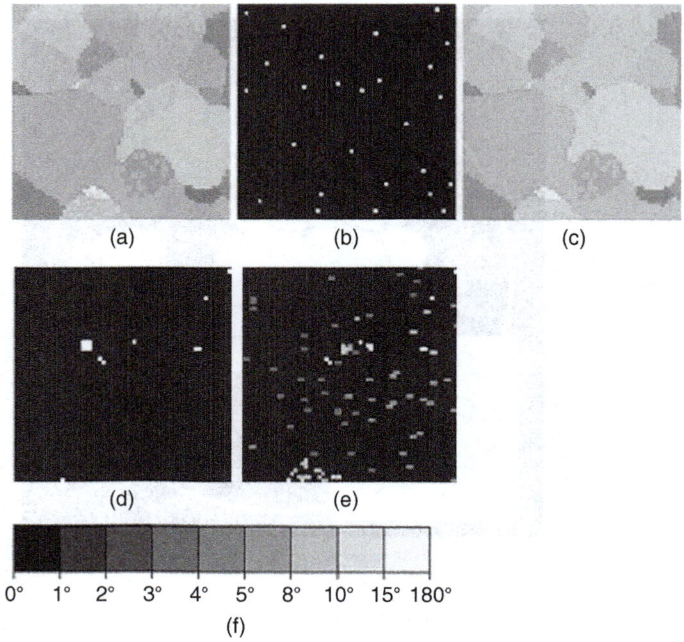

Figure 17.7 The reconstruction of test case representing a deformed microstructure. (a) Reference orientation map, (b) seeds, and (c) reconstructed orientation map. (d) Difference of the reference and the reconstructed grain maps. Black pixels denote identical grain labels, white pixels represent mismatching ones. (e) Difference of the reference and the reconstructed orientation maps. The gray level of the pixels is determined by the misorientation of corresponding orientation pairs, as shown in the key at (f). (Rodek 2007 [24]. Reproduced with permission of International Union of Crystallography.)

Recently, Rodek *et al.* presented a Monte-Carlo-based algorithm for moderately deformed specimens [24], where the size of configuration space has been reduced by orders of magnitude by using known properties of the microstructure. The routine assumes that prior information arising, for example, from a forward projection, makes it possible to identify the number of grains, and for each to determine an approximate average orientation and one pixel in the layer which belongs to this grain. These points function as seeds for the algorithm. This approach was tested on a set of phantoms representing microstructures typical of various degrees of deformation. The results are encouraging. As an example in Figure 17.7 the reconstruction of a microstructure with 22 grains and with an internal orientation spread within grains of up to 22° is illustrated. (In this reproduction of the two maps on top it is difficult to see the orientation gradient within the grains.) At low noise levels, the reconstructions were nearly perfect. At the time of writing, analysis on real data is pending.

17.3.3 Application I: Recrystallization

The first 3D movies of grain coarsening related to recrystallization [25], at least in the initial stage of the process, is relevant to visualize the growth of a single grain (nucleus). More specifically, nucleation in an aluminum single crystal deformed to 42% was stimulated by hardness indents. The crystal was subject to *in situ* annealing

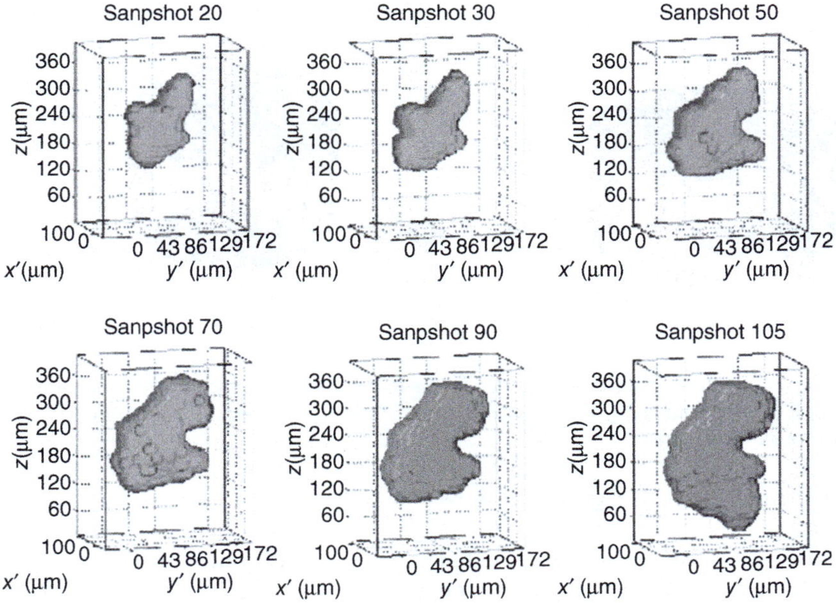

Figure 17.8 Storyboard showing the growth of one emerging nucleus in an Al single crystal deformed by 42% as function of annealing. (Schmidt 2004 [25]. Reproduced with permission of AAAS.)

for 30 h at temperatures between 270 and 310 °C. The mapping comprised 50 z-layers with a spacing of 6 μm. One emerging nucleus was identified at an early stage of the growth. Exposures were made around the ω setting of a specific reflections belonging to this grain. The nucleus was mapped continuously with a time resolution of at most 10 min. Snapshots of the resulting 3D movie are shown in Figure 17.8. The data quality is sufficient that two general features of the growth are revealed:

1) The growth is very heterogeneous, so the shape is, at times, remarkably irregular.
2) The growth does not occur smoothly with time, but is often jerky.

These unique observations were explained by the heterogeneous nature of the deformed microstructure.

17.3.4 Application II: Grain Growth

In this study, the evolution of the morphology of several hundred grains was monitored simultaneously for the first time [26]. The base material Al–0.1% Mn was cold rolled 80% and annealed for 8 min at 300 °C producing a fully recrystallized sample with grain sizes between 30 and 100 μm. Subsequently, a cylindrical sample with dimensions of 700 μm in diameter and 2 mm in height was selected for further investigations at the 3DXRD microscope. The sample was annealed in an external furnace. Before the first annealing and following each annealing step a cylindrical volume with a diameter of 700 μm and height of 350 μm was fully characterized by mapping out the sample in layers using a planar beam shape at 50 keV. The grain volumes were reconstructed using the Grain Sweeper [19] algorithm. In total, five annealing steps were made at annealing temperatures between 400 and 450 °C. Initially, 491 grains were present leaving only 49 grains

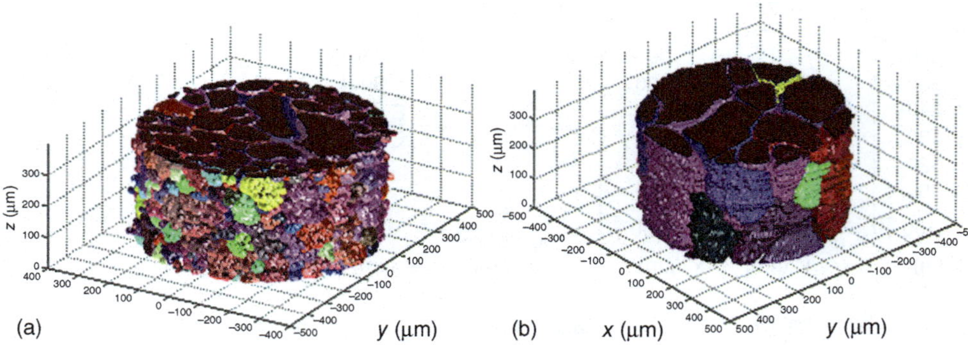

Figure 17.9 The fully reconstructed grain volumes: (a) the initial grain volume containing 491 grains and (b) final grain volume containing 49 grains. Gray scales are related to the crystallographic orientation of the individual grain.

after the final annealing step. Figure 17.9 shows the fully reconstructed initial and final grain volumes.

The information present in the reconstructed volumes, such as number of grain edges, and so on, is to be extracted. Such information may serve as a crucial constraint to any grain growth model.

17.4 Combining 3DXRD and Tomography

The 3DXRD setup sketched in Figure 17.1 is very similar to the setup typically used for parallel beam tomography, cf. Chapter 14. Hence, it is possible to probe the same specimen with both techniques either simultaneously or consecutively. More specific to 3DXRD, we note that the basic 3DXRD algorithms developed so far all apply to monophase materials only. A combined data set enables the methodology to be extended to multiphase materials, by first segmenting the volume to be reconstructed based on density and then applying the 3DXRD method to each phase at a time.

17.4.1 Grain Mapping by Tomography

Conventional wisdom tells us that attenuation and/or phase contrast tomography is not sensitive to orientations and as such cannot be used to map grains in a monophase material. Recently, two methods that overcome this basic limitation have been demonstrated by a group of people at ESRF and Risø.

1) *Diffraction contrast tomography (DCT).* The principle is illustrated in Figure 17.10: assume a grain is in the Bragg condition. Then there will be two contributions to attenuation of the direct beam: the absorption in the sample and the missing intensity due to diffraction. The latter additional contrast is clearly visible if the grains are near-perfect (mosaic spread well below 1°). Performing first a conventional reconstruction, a density map of the sample is obtained. Based on this one can simulate what the absorption data should look like for a noncrystalline sample. Subtracting the actual images from the simulated ones, one may sort the residual "grain contrast" images, based on a combination of real space (grain shape and position) and orientation space scanning procedures that compare simulated and measured angles.

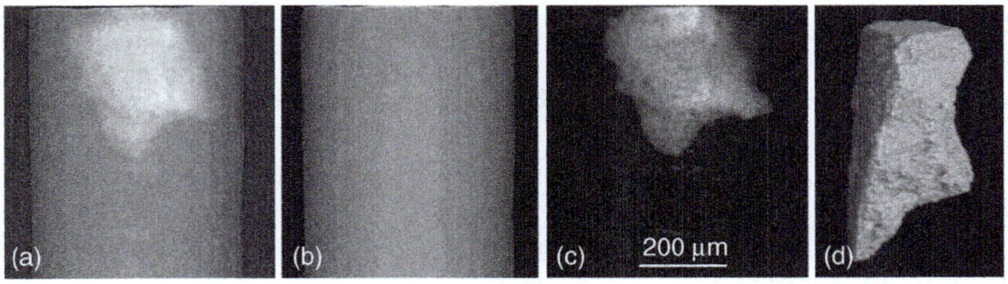

Figure 17.10 Diffraction contrast tomography: (a) experimental projection with one grain fulfilling the Bragg condition, (b) calculated absorption background, (c) extracted grain projection, in all cases with brightness in image proportional to attenuation, and (d) reconstruction of 3D grain shape by means of algebraic reconstruction techniques (different scale).

Having indexed a grain in this way, the 3D shape can be reconstructed based on the 3D-ART approach.

DCT and 3DXRD may be seen as the equivalent of "dark-field" and "bright-field" imaging in TEM. Combined data sets have been acquired in several experiments at ID19 at ESRF. The results demonstrate that this combined method produce grain maps of superior resolution to those based on 3DXRD only. The drawback is that the number of projections required needs to be much higher, of the order of a few 1000 images.

2) *Topo-tomography* is an alternative tomographic scanning procedure, which allows reconstructing a single grain, using a setup with additional tilt-stages [27]. For a selected grain, the scattering vector of a suited reflection is aligned parallel to the rotation axis of the tomographic setup. This particular setting assures that the diffraction condition is maintained while the sample is turned 360° around the rotation axis. Tomographic reconstruction can be performed from data acquired in direct and/or in diffracted beam by means of a standard (cone beam) filtered backprojection algorithm. In comparison to mode III reconstructions by 3DXRD above, the spatial resolution can be substantially better. The disadvantage is the restriction to one grain only.

17.5 Outlook

At the time of writing, the ID11 beamline is undergoing a major refurbishment. A new generation of hard X-ray optics, an advanced feedback system, and temperature control will be installed in order to generate a stable, full energy and bandwidth tunable beam, with dimensions as small as 50–100 nm. At the same time, work is ongoing to develop a new type of fluorescence screen – a structured screen – for the high-resolution detectors. This detector is specified to have a resolution of 1 μm and an efficiency in the excess of 25% (in comparison present detectors have a resolution of 5 μm and an efficiency of a few percentage).

For modes III and IV work, these developments are predicted to enable the generation of 3D maps with a spatial resolution of 1 μm or alternatively complete maps of 1-mm^3-sized materials with a resolution of 5 μm in a few minutes. For modes I and II work, it is predicted that crystalline structures as small as 5–10 nm can be studied during *in situ* processing.

References

1 Poulsen, H.F. (2004) *Three-Dimensional X-Ray Diffraction Microscopy*, Springer, Berlin.
2 Larson, B.C., Yang, W., Ice, G.E., Budai, J.D., and Tischler, T.Z. (2002) *Nature*, **415**, 887–890.
3 Bunge, H.J., Wcislak, L., Klein, H., Garbe, U., and Schneider, J.R. (2003) *J. Appl. Crystallogr.*, **36**, 1240–1255.
4 Wroblewski, T., Clauss, O., Crostack, H.-A., Ertel, A., Fandrich, F., Genzel, C., Hradil, K., Ternes, W., and Woldt, E. (1999) *Nucl. Instrum. Methods Phys. Res., Sect. A*, **428**, 570–582.
5 Lauridsen, E.M., Schmidt, S., Suter, R.M., and Poulsen, H.F. (2001) *J. Appl. Crystallogr.*, **34**, 744–750.
6 Poulsen, H.F., Nielsen, S.F., Lauridsen, E.M., Schmidt, S., Suter, R.M., Lienert, U., Margulies, L., Lorentzen, T., and Juul Jensen, D. (2001) *J. Appl. Crystallogr.*, **34**, 751–756.
7 Martins, R.V., Margulies, L., Schmidt, S., Poulsen, H.F., and Leffers, T. (2004) *Mater. Sci. Eng.*, **387–389**, 84–88.
8 Vaughan, G.B.M., Schmidt, S., and Poulsen, H.F. (2004) *Z. Kristallogr.*, **219**, 813.
9 Lauridsen, E.M., Poulsen, H.F., Nielsen, S.F., and Juul Jensen, D. (2003) *Acta Mater.*, **51**, 4423–4435.
10 Iqbal, N., van Dijk, N.H., Offerman, S.E., Moret, M.P., Katgerman, L., and Kearley, G.J. (2005) *Acta Mater.*, **53**, 2875–2880.
11 Offerman, S.E., van Dijk, N.H., Sietsma, J., Grigull, S., Lauridsen, E.M., Margulies, L., Poulsen, H.F., Rekveldt, M.T., and van der Zwaag, S. (2002) *Science*, **298**, 1003–1005.
12 Lange, W.F. III,, Enornoto, M., and Aaronson, H.I. (1988) *Metall. Trans. A*, **19**, 427–440.
13 Margulies, L., Winther, G., and Poulsen, H.F. (2001) *Science*, **291**, 2392–2394.
14 Poulsen, H.F., Margulies, L., Schmidt, S., and Winther, G. (2003) *Acta Mater.*, **51**, 3821–3830.
15 Winther, G., Margulies, L., Schmidt, S., and Poulsen, H.F. (2004) *Acta Mater.*, **52**, 2863.
16 Gundlach, C., Pantleon, W., Lauridsen, E.M., Margulies, L., Doherty, R., and Poulsen, H.F. (2004) *Scr. Mater.*, **50**, 477–481.
17 Jakobsen, B., Poulsen, H.F., Lienert, U., Almer, J., Shastri, S.D., Sørensen, H.O., Gundlach, C., and Pantleon, W. (2006) *Science*, **312**, 889–892.
18 Suter, R.M., Hennessy, D., Xiao, C., and Lienert, U. (2006) *Rev. Sci. Instrum.*, **77**, 123905.
19 West, S.S., Schmidt, S., Sørensen, H.O., Winther, G., Poulsen, H.F., Margulies, L., Gundlach, C., and Juul Jensen, D. (2009) *Scripta Mater.* **61**, 875–878.
20 Poulsen, H.F. and Fu, X. (2003) *J. Appl. Crystallogr.*, **36**, 1062–1068.
21 Poulsen, H.F., and Fu, X. (2003). *J. Appl. Crystallogr.*, **36**, 1062–1068
22 Alpers, A., Poulsen, H.F., Knudsen, E., and Herman, G.T. (2006) *J. Appl. Crystallogr.*, **39**, 582–588.
23 Markussen, T., Fu, X., Margulies, L., Lauridsen, E.M., Nielsen, S.F., Schmidt, S., and Poulsen, H.F. (2004) *J. Appl. Crystallogr.*, **37**, 96–102.

24 Rodek, L., Poulsen, H.F., Knudsen, E., and Herman, G.T. (2007) *J. Appl. Crystallogr.*, **40**, 313–321.
25 Schmidt, S., Nielsen, S.F., Gundlach, C., Margulies, L., Huang, X., and Juul Jensen, D. (2004) *Science*, **305**, 229–232.
26 Schmidt, S., Olsen, U.L., Poulsen, H.F., Sørensen, H.O., Lauridsen, E.M., Margulies, L., Maurice, C., and Juul Jensen, D. (2008). *Scripta Mater.* **59**, 491–494
27 Ludwig, W., Cloetens, P., Härtwig, J., Baruchel, J., Hamelin, B., and Bastie, P. (2001) *J. Appl. Crystallogr.*, **34**, 602–607.

18

3D Micron-Resolution Laue Diffraction

Gene E. Ice

18.1 Introduction

A fundamental question of materials science is how atomic-scale interactions self-organize atoms into mesoscopic structures. This simple question is important because the physical behavior of most materials is dominated by the mesoscale structures and dynamics. For example, self-organization is essential to understand grain-growth and deformation microstructure and to understand their effects on plasticity, strength, fracture, transport, and other materials properties.

To understand how mesostructures arise, and how they influence materials behavior, it is essential to map *local* elemental composition, crystal/local structure, and geometrical/chemical defect distributions. In materials, this information is mathematically approximated by three-dimensional (3D) tensor fields that are typically highly heterogeneous. For this reason, 3D quantitative probes are essential. X-ray microdiffraction is particularly interesting as it provides detailed *atomic-resolution* information about local crystalline structure correlated with the *mesoscale* (0.1 – 10 μm) real-space resolution of the probe. Furthermore, unlike almost any other probe, X-ray microbeams can nondestructively characterize materials properties in three-dimensions and can observe mesoscale evolution as a response to underlying driving forces (e.g., stress).

18.2 The Need for *Polychromatic* Microdiffraction

Although diffraction with micron-scaled beams can provide detailed structural information, *monochromatic* microdiffraction studies are complicated; to efficiently diffract microbeams from small volumes, samples must be simultaneously positioned with submicron precision and aligned with arcsecond angular accuracy. In polycrystalline materials, the penetration of X-rays into the sample further complicates data interpretation as more than one (or even none) of the grains intercepted by the incident beam can diffract (Figure 18.1a). If the sample is oriented to position one grain to satisfy a Bragg condition, the neighboring grains illuminated by the penetrating beam can be rotated into or out of the beam. This problem is exacerbated by sphere-of-confusion errors in rotations of the sample (Figure 18.1b). High-energy monochromatic beams that flood many grains can be used for mapping of crystal structures as described in the

Neutrons and Synchrotron Radiation in Engineering Materials Science: From Fundamentals to Applications, Second Edition. Edited by Peter Staron, Andreas Schreyer, Helmut Clemens, and Svea Mayer.

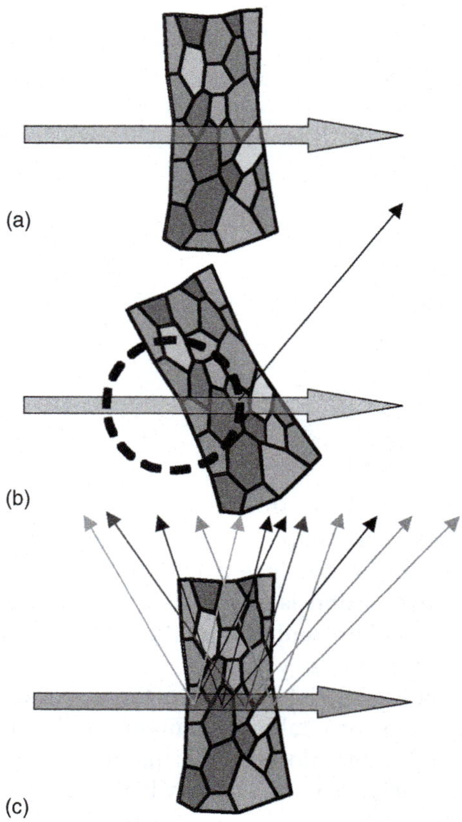

Figure 18.1 (a) A monochromatic X-ray beam has a low probability of diffracting from an arbitrary crystal grain orientation. (b) If the sample is rotated to orient a grain for diffraction the grain moves within the instrumental sphere-of-confusion even if it is centered on the goniometer rotation axis. Grains away from the rotation axis are in the incident beam path over an even smaller angular range. (c) With a polychromatic beam, the subgrain volumes that intercept the incident beam generate Laue patterns that overlap at the detector.

previous chapter (Chapter 17). Here, the spatial resolution is determined by the spatial resolution of the detector and by corrections for sphere-of-confusion in the sample rotations. Polychromatic beams however, have the advantage of requiring no sample rotations for diffraction studies. In this case, spatial resolution is determined to a large degree by the beam size, which can be submicron and even nanoscale. Furthermore, issues of strain or deformation on the measured diffraction volume have – at least in principle – a small effect.

With a polychromatic (Laue) approach however, there is a completely different problem: an over abundance of diffraction with each subgrain volume diffracting into overlapping Laue patterns on the detector (Figure 18.1c). The complexity of the overlapping Laue patterns is greatly reduced by the use of small X-ray beams that limits the number of subgrains illuminated and can be further simplified by an experimental approach called differential aperture microscopy, which resolves the single-crystal-like Laue diffraction from subgrain volumes along the illuminating beam. The theories behind differential aperture microscopy and polychromatic microdiffraction are described briefly below.

18.3 Theoretical Basis for Advanced Polychromatic Microdiffraction

18.3.1 Modified Ewald's Sphere Description of Laue Diffraction

Monochromatic diffraction is conceptually represented by an Ewald's sphere diagram, which elegantly illustrates the diffracting conditions relative to the sample orientation. *Laue diffraction* can be similarly modeled by a *modified* Ewald's sphere diagram as shown in Figure 18.2. Bragg reflections for a crystal are indicated in reciprocal space by a reciprocal-space lattice that is aligned to the real-space orientation of the crystal. A Bragg reflection is excited when the momentum transfer of an accessible diffraction direction for a particular wavelength matches a reciprocal space lattice co-ordinate. For a collimated monochromatic beam, possible momentum transfers in reciprocal space lie on the Ewald's sphere, and the sample must be precisely oriented to pass a reciprocal lattice point through the Ewald's sphere.

With Laue diffraction however, a finite bandpass creates a volume in reciprocal space between maximum and minimum Ewald's spheres (note that the centers of the limiting spheres lie along the incident beam direction but are displaced from each other). Bragg reflections that fall within this volume are excited and the directions of the Laue pattern peaks indicate the Bragg plane normals of the crystal. More exactly, the intensity in each pixel in a Laue image is proportional to a line integral through reciprocal space weighted by the incident beam spectral distribution [1].

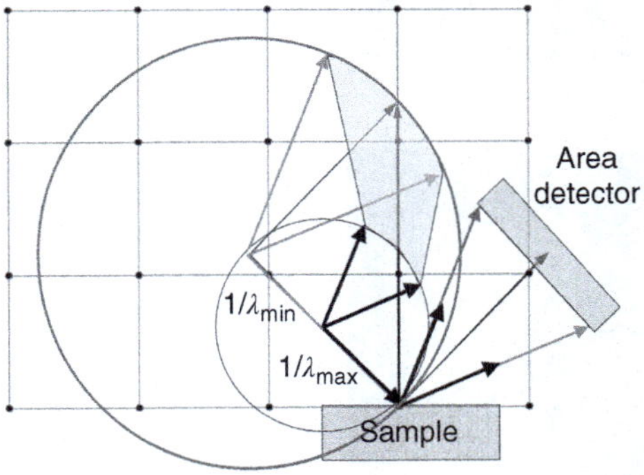

Figure 18.2 A two-dimensional modified Ewald's sphere diagram. In reciprocal space, the momentum of an X-ray beam depends on its direction and wavelength; for short wavelength, $1/\lambda_{min}$, the momentum vector magnitude is larger than that for large wavelengths, $1/\lambda_{max}$. The momentum transfer is simply the vector difference between the final and incident X-ray momentums. For a fixed wavelength, the possible momentum transfers lies on a sphere. The centers of the spheres for different wavelengths lie along the incident beam direction and the spheres touch at the 0, 0, 0 position (no momentum transferred). The Laue pattern observed by an area detector is made up of radial line integrals through reciprocal space. Lattice points (and other diffuse scattering) that lie within the bounds of the minimum and maximum Ewald's spheres contribute to the line integrals.

18.3.2 Qualitative Information: Phase, Texture, Elastic Strain, Dislocation Density

With the understanding that Laue images represent radial line integrals in reciprocal space, microLaue images can offer immediate qualitative information about the sample volume illuminated by the beam.

18.3.2.1 Phase

For example, as shown in Figure 18.3a, the image from a small high-symmetry unit cell looks qualitatively distinct from that of a large or low-symmetry unit cell (Figure 18.3b). Hence, phase distributions can often be determined by mapping the number of reflections emanating from subgrain volumes within the sample.

18.3.2.2 Texture

Similarly, microLaue patterns are highly sensitive to orientation and with good quality crystals orientations with uncertainties less than 0.001° can be measured [2]. Therefore,

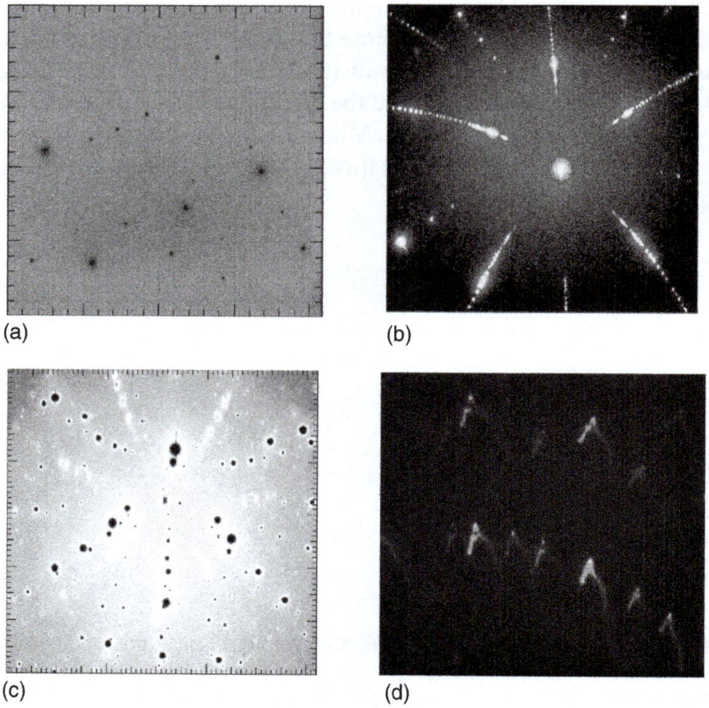

Figure 18.3 (a) Laue pattern from a $Ni_{80}Fe_{20}$ single crystal. The small unit cell and high symmetry face-center-cubic (FCC) real-space lattice results in a sparse body-center-cubic (BCC) reciprocal-space lattice and, consequently, only a few Laue spots are detected. (b) Overlapping Laue patterns from a GaN film on a SiC single crystal. The two patterns are superimposed because the GaN film is crystallographically aligned to the SiC substrate. (c) High temperature superconducting film with buffer layer on a Ni single crystal. There are actually three patterns in the Laue image: one from the substrate and two from the two film overlayers. The patterns are at slightly different orientations due to crystallographic tilt between the layers. (d) Deformed Ni crystal shows a dramatically redistribution of intensity near the Bragg peaks. This redistribution is due primarily to long-range rotations of the Bragg planes caused by unpaired dislocations.

measurements of local grain orientations (texture) are easily made with one to three orders of magnitude better angular resolution than with typical electron probe measurements. For example, in Figure 18.3c, the relative misorientation of crystal layers in an engineered high-temperature superconductor structure is easily determined. Furthermore, these orientation measurements can be made tens to hundreds of microns below the sample surface so that the local texture in thin films, multilayers, and true 3D samples can be studied.

18.3.2.3 Dislocation Tensor

In addition, dislocations are Krivoglaz defects of the second kind [3], which introduce long-range rotations in the Bragg planes [4] so that sharp Laue spots are smeared (Figure 18.3d). Single images can be used to model the dislocation density and distribution through the illuminated sample volume [5], or microLaue images can be resolved by differential aperture microscopy to determine the local Laue pattern within a submicron volume and the 3D distribution of lattice rotations can be used to unambiguously extract the local dislocation or Nye tensor [6, 7].

18.3.2.4 Elastic Strain Tensor

Finally, changes in the angles between reflections can be used to determine the local elastic stress. The angles between reflections can be used to determine distortions of the unit cell shape. These distortions are described by the local deviatoric strain tensor [8]. At least four indexed reflections are needed to determine the deviatoric strain tensor (distortion of the unit cell shape). If the energy of one reflection is determined, then not only the unit cell shape but also its volume can be determined and the full strain tensor can be found by comparison to the undeformed unit cell [8, 9].

18.4 Technical Developments for an Automated 3D Probe

Although Laue diffraction is the oldest X-ray diffraction method [10], the recent extension to 3D imaging of crystalline structure has relied on modern developments in X-ray sources, optics, and detectors [11]. To go beyond a conceptual tool to an automated 3D probe for mapping crystal structure, five elements have been assembled in the "first-of-its-class" 3D X-ray Crystal Microscope deployed on station 34-ID-E at the Advanced Photon Source (APS) (Figure 18.4). These are (i) an ultra-brilliant X-ray source, (ii) a nondispersive monochromator that can be cycled between monochromatic and polychromatic status, (iii) nondispersive focusing optics with submicron resolution, (iv) a scanning-wire differential aperture (Figure 18.5), and (v) a high-resolution area detector. The purpose and performance of each element is described briefly in the following text.

18.4.1 Source

An ultra-brilliant wide-bandpass X-ray source is the first essential element for automated polychromatic microdiffraction. The combination of small beam size and good collimation needed for simultaneously good real-space and momentum transfer resolution greatly restricts the phase space of the usable beam. This makes it important that

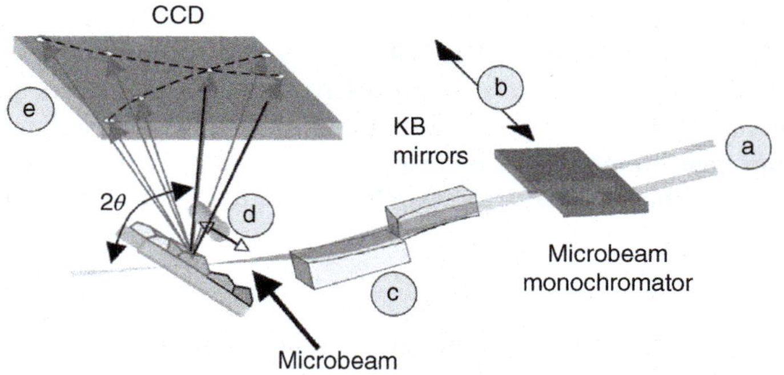

Figure 18.4 Schematic of the 3D X-ray crystal microscope an instrument optimized for polychromatic microdiffraction experiments. The instrument includes (a) an ultra-brilliant X-ray source; (b) a microbeam monochromator that can switch the beam from polychromatic to monochromatic radiation; (c) a nondispersive focusing Kirkpatrick–Baez mirror system; (d) a differential aperture for deconvolving the Laue patterns generated along the incident beam; and (e) an optimized X-ray sensitive area detector.

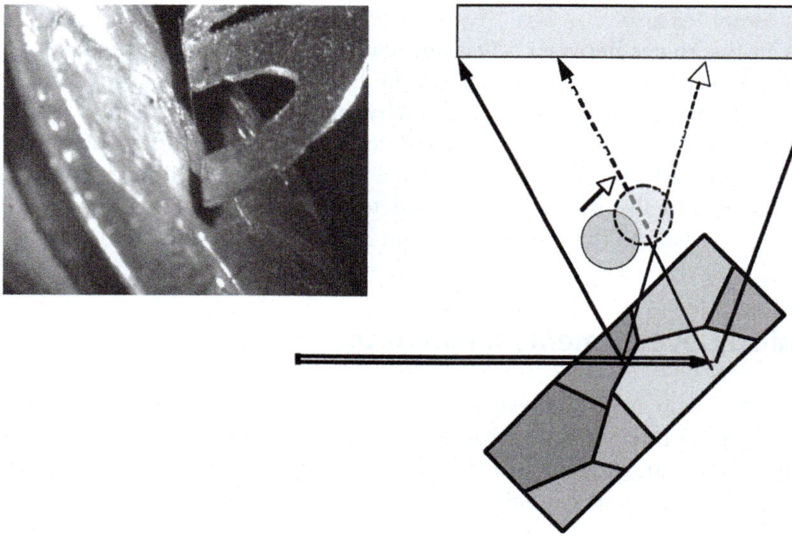

Figure 18.5 Differential aperture. The intensity is monitored at each pixel as the wire is moved through the diffracted beam. The intensity origin for each pixel is determined by ray tracing back from the pixel position past the edge of the wire and onto the incident beam path.

the source has high brilliance (photons/s/mm^2/mrad^2/eV) over the required bandpass, and favors small electron source emittance (size^2 × divergence^2). Small source size also allows for small focal spot size with modest demagnification. An undulator on a modern third generation synchrotron like Spring-8 in Japan or the National Synchrotron Radiation Research Center (NSRRC) of Taiwan have about 12 (4) orders of magnitude greater X-ray brilliance for monochromatic experiments and about 10 (2) orders of magnitude greater brilliance for wide-bandpass measurements than a laboratory (bend magnet synchrotron) source (Figure 18.6). This high brilliance allows for microLaue images to

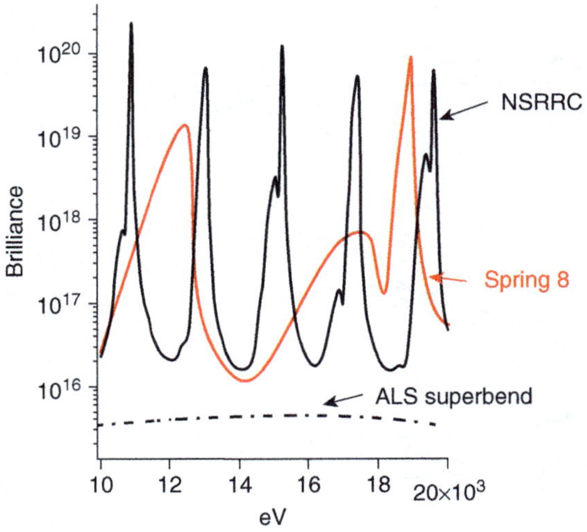

Figure 18.6 Third generation undulator sources such as Spring-8 and the NSRRC have about two orders of magnitude greater integrated brilliance than a bend magnet with large fluctuations in the spectra that complicate polychromatic microdiffraction. Special undulators and clever use of the spectral distribution off axis can be used to smoothen the fluctuations. For monochromatic experiments, the increased brilliance translates into approximately four or more orders of magnitude greater signal for microdiffraction experiments.

be collected in milliseconds from many samples. Even though the undulator source is intrinsically highly collimated with a relatively small source, typically only about 0.1% of the total beam emittance is used in polychromatic microdiffraction. This makes it possible to field many polychromatic beamlines on one undulator source.

New third generation sources like the NSRRC (Figure 18.6) can provide high brilliance in closely spaced harmonic peaks that span the 10–20 keV range. Such sources are highly desirable for monochromatic measurements but may require specially designed undulators for polychromatic experiments. Such undulators might include helical fields with variable pitch or rapidly tunable undulators to provide time-average coverage of a wide bandpass.

Bend magnet sources can also be successfully used for polychromatic microdiffraction despite their lower overall brilliance. Currently, the performance of polychromatic microdiffraction is influenced more by the speed of the detector readout than by the counting statistics in the Laue images. Sources that are only a factor of 100 lower in integrated brilliance can, therefore, do much of the science that is done with an undulator source and can be optimized for particular experiments to achieve very high performance.

18.4.2 Microbeam Monochromator

A microbeam monochromator is the second important element of a polychromatic microdiffraction system. The monochromator allows for measurements of the full strain tensor. As described above, Laue images record radial line integrals through reciprocal space. Strain is determined by comparing the measured unit cell shape to the shape of an undistorted unit cell. Laue images cannot however distinguish between unit cells with slightly different volumes but the same shape. Therefore, in order to determine the full strain tensor, it is important to measure the energy of one reflection. Of course, the deviatoric strain tensor, which is most sensitive to shear stains can be measured without an energy measurement. On station 34-ID-E at the APS the microbeam monochromator can be translated into or out of the beam so the beam incident on the focusing

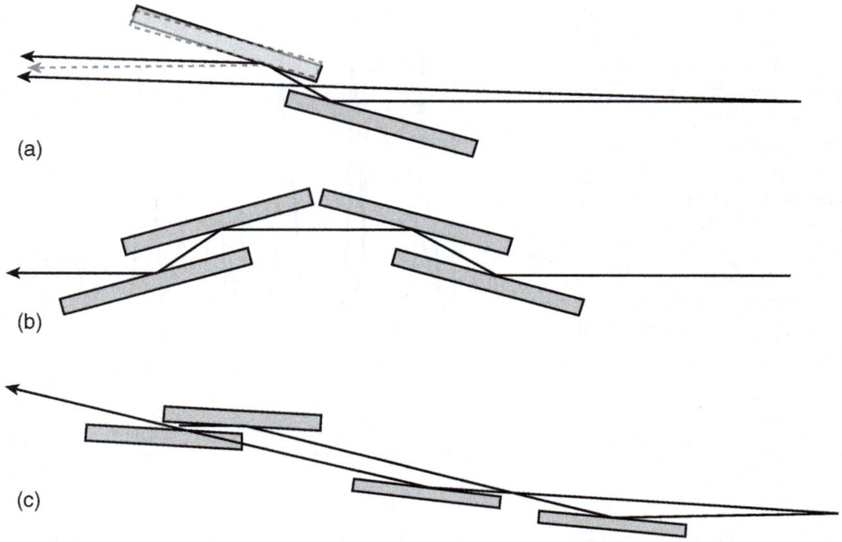

Figure 18.7 (a) Schematic of the microbeam monochromator on beamline 34-ID-E at the APS. The polychromatic beam is taken ~1 mm above the central beamline axis. This makes the polychromatic beam to have a more uniform spectral intensity throughout the useful bandpass of the optics. It also changes the angle of the beam slightly, which is compensated by a small temperature difference between the first and second crystals. (b) The micromonochromator design on beamline 12.3.2 at the ALS is a four crystal design that is intrinsically zero offset. (c) The design for the VESPERS beamline at the Canadian Light Source uses mirrors and a small displacement monochromator to alternate between co-linear polychromatic or monochromatic beams.

optics is either polychromatic or monochromatic. The design also maintains a fixed offset independent of X-ray energy or bandpass [12].

The 34-ID-E monochromator utilizes a two-crystal fixed exit design with a small (~1 mm) offset. The polychromatic beam is taken from ~1 mm above the beam axis and the monochromatic beam is taken from near the beam axis (Figure 18.7a). This optimizes the effective beam brilliance of the undulator source when used for monochromatic measurements and delivers a more uniform energy distribution for polychromatic experiments. Beamline 12.3.2 at the Advanced Light Source (ALS) uses a four-crystal design that is intrinsically zero offset (Figure 18.7b). The VESPERS beamline at the Canadian Light Source uses mirrors in addition to the crystals [13] to switch between co-axial polychromatic or monochromatic beams (Figure 18.7c). This design combines the simplicity, high-throughput and ease of use of a two-crystal monochromator with the advantages of an intrinsically zero offset dispersion-free design.

18.4.3 Nondispersive Focusing Optics

Nondispersive focusing optics are the essential elements for polychromatic microdiffraction. On 34-ID-E, total external reflection Kirkpatrick–Baez mirrors are used to focus to an $\sim 300 \times 300\ \text{nm}^2$ focal spot. The mirrors are fabricated by profile coating on ultra-smooth substrates [14, 15]. Kirkpatrick–Baez optics are rapidly developing with hard X-ray nanoprobe optics demonstrating properties suitable for polychromatic nano diffraction [15–17]. Compact nested designs are particularly appealing as they allow for larger numerical aperture, more stable optics, and shorter average focal lengths [18].

18.4.4 Area Detector

An area detector to record the Laue pattern is also an essential element of a polychromatic microdiffraction station. Unlike fluorescence microscopy it is important that the scattering is out of the ring plane to improve the reflectivity of the polarized incident beam. On 34-ID-E and other (2D) polychromatic microdiffraction beamlines the detector is typically aligned at 90° vertical to the incident beam. This geometry is a compromise that allows for simple sample alignment with reasonable reflectivity of X-rays between 10 and 20 keV (Figure 18.4). The nominal 90° scattering angle also optimizes the penetration depth of elastically scattered X-rays for a given absorption coefficient.

A detector optimized for polychromatic microdiffraction is quite distinct from area detectors used in protein crystallography and other more common X-ray applications. For example, extremely linear pixel spacing is important to measure strains to 0.01%. Similarly, large dynamic range, low readout noise, fast readout, and low point-spread functions are all important. Readout speed is particularly important to collect meaningful 3D images in a reasonable time. On 34-ID at the APS, amorphous Si detectors are used to record the images. A set of three detectors collects from a large solid angle, and the pixel spacing is highly linear which allows for strain measurements <0.01%. On beamline 12.3.2 at the ALS, a Pilatus detector with single photon sensitivity greatly extends the dynamic range of the experiments and is ideally suited for weak scattering near strong scattering signals.

18.4.5 Differential Aperture

The final element for a 3D X-ray Crystal Microscope is a differential aperture. The differential aperture deconvolves the Laue patterns from subgrain volumes along the incident beam [6]. The principle of differential aperture microscopy is indicated in Figure 18.5. A smooth high Z wire is passed near the surface of the sample (within ~200 μm). The wire is calibrated prior to the measurement so that its position is precisely known relative to the incident beam and the axis of the detector. As the wire moves along the sample surface, it occludes rays that are collected at the detector. The intensity as a function of wire position is observed for each pixel. The derivative of the measured intensity/wire step in each pixel is the sum of a positive and negative image of the intensity distribution along the wire that reaches that particular pixel separated by the diameter of the wire. By ray tracing back from the pixel and wire locations, and taking into account small effects due to the angle of the intercepted beam, the origin of the intensity into each pixel can be mapped with submicron precision along the incident beam. A picture of the 3D X-ray crystal microscope on 34-ID-E is shown in Figure 18.8 and a close up of the differential aperture is shown in Figure 18.5. Efforts are now underway to accelerate differential aperture microscopy with multiple wires and more intelligent step scans.

18.4.6 Software

Although Laue diffraction is a long-established technique, until recently, no software existed to allow for automated mapping of multiple grains, either for differential

Figure 18.8 The 3D X-ray crystal microscope on beamline 34-ID-E. The components are mounted on an optical table, which allows new optical approaches to be rapidly tested. The CCD detector directly above the sample stage has been replaced with three amorphous Si detectors that provide much more rapid readout and larger solid angle coverage, but with poorer signal-to-noise.

aperture microscopy or for precision determination of strain and modeling of deformation parameters. This software is still a work in progress, but is now the subject of an international effort to develop a standard package that will foster polychromatic microprobes on beamlines around the world. Existing software packages are described in various references [8, 19] and more advanced user-friendly software is under development. The XMAS program [19] is particularly user friendly for 2D data collection.

18.5 Research Examples

The mission of station 34-ID-E is centered on fundamental investigations of long-standing issues in materials science. These issues include grain growth and 3D organization in polycrystalline materials, 3D dynamics of mesoscale structures, and fundamental assumptions in materials physics modeling. Studies of 3D organization underway include measurements of colony structure in processed materials and studies of fractal misorientation [20] in deformed materials including the development of dislocation hierarchical structures. Other studies of 3D structures include measurements of percolation and diffusion behavior.

Studies of dynamics underway include 3D local growth of grains, measurements of ripening, and measurements of anomalous grain growth (i.e., Sn whiskers) [21]. Other measurements include studies of nucleation sites, the role of deformation energy and elastic stress in grain nucleation, and grain/subgrain rotations during deformation as a function of global and local environmental conditions.

Fundamental assumptions are being tested by experiments that include mapping of intra- and intergranular stresses to guide our understanding of the constitutive equations of materials near grain boundaries and other interfaces, measurements of grain boundary shear versus habit, and measurement of deformation scaling laws in three-dimensions.

18.5.1 3D Grain Boundary Networks

How grain boundary/polycrystalline networks interact to applied forces and environmental conditions is a central materials challenge of the twenty-first century. Modelers want to know the constitutive equations at grain boundaries, and how they change with boundary type. This information impacts efforts to model materials behavior and to engineer ideal microstructures for applications ranging from corrosion resistance to high strength. Similarly, synthesis and processing control of grain boundary distributions is an area of active research with particularly complex issues for nanoscale and advanced layered structures.

Polychromatic microdiffraction allows for unprecedented testing of longstanding assumptions regarding grain boundaries. For example, in coincident site lattice (CSL) theory, grain boundaries are considered to have lower energy if some of the lattice sites of the neighboring grains overlap [22]. The lattice is assigned a number, Σ, which is the inverse of the fraction of shared sites; if 1/3 of the sites are shared then $\Sigma = 3$. According to theory, high Σ boundaries can only have a slight misorientation because the energy minimum for high Σ boundaries is small and the subsequent energy well is shallow. The good angular precision of polychromatic microdiffraction allows for detailed tests of this and other predictions of CSL theory. Preliminary measurements actually indicate an opposite trend in the distribution of lattice misorientations near CSL boundary conditions; measured misorientations from CSL boundaries increase with Σ. Other interesting questions, which can be addressed, include the role of curvature and faceting in grain boundaries. Grain boundary normals should have lower energy in ideal CSL directions. However, curved surfaces can also lower energy. Here, the role of faceting may allow for energy minimization and studies are underway to study how 3D surfaces accommodate competing processes toward energy minimization. For example, Figure 18.9 shows a step in the surface of a polycrystalline Ni sample at a grain boundary triple junction as well as grain boundary grooving. Polychromatic microdiffraction is certain to shed new light on grain boundaries and on how grain boundaries affect materials behavior.

18.5.2 Deformation Behavior and Grain Boundaries

Most theories of deformation are based on the crystallographic orientation of grains with respect to the applied load; the orientation controls the critical-resolved shear stress along slip systems. These theories have now been tested for the first time with the use of high-energy X-ray diffraction as described in Chapter 17. However, local environments also affect deformation behavior. For example, Pang *et al.* [23] has recently

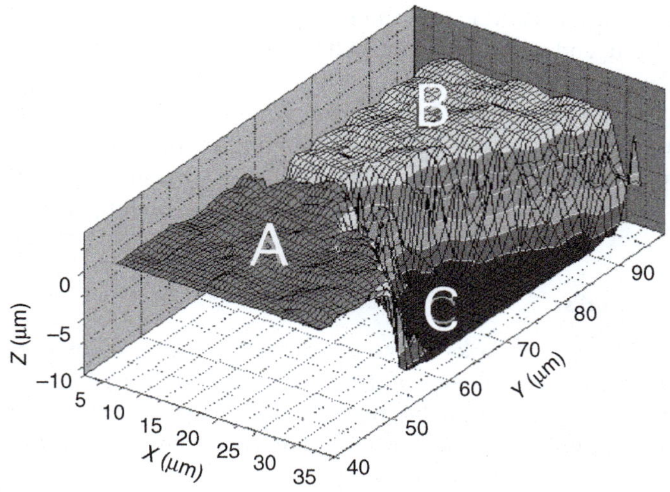

Figure 18.9 The surface of a Ni triple grain junction. The orientation of grains A, B, and C were mapped by polychromatic microdiffraction and show grain boundary grooving, and a step in one of the grains surfaces. Internal grain boundary surfaces are also recovered with these measurements.

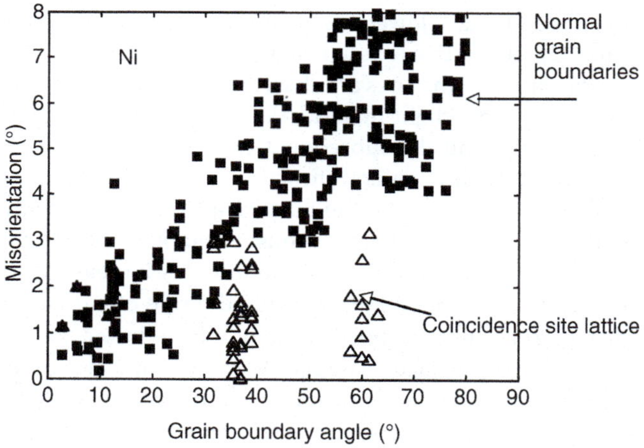

Figure 18.10 Grain boundary angle before deformation (*x*-axis) compared to the change in misorientation across a grain boundary after deformation (*y*-axis). There is a positive correlation between grain boundary misorientation before deformation and the change in misorientation after deformation. Grain boundary misorientations near coincidence site lattice boundaries however appear to have a distinctly different behavior from other boundaries.

studied rotations *across* grain boundaries before and after deformation. The findings show a strong correlation between grain boundary angle and misorientation magnitude, and observes that CSL boundaries have a much different behavior during deformation than other boundary types (Figure 18.10). Indeed, deformation behavior near surfaces and interfaces appears to be quite distinct from the behavior near the center of the grains [24].

18.5.3 Deformation in Single Crystals

Although the deformation of materials appears to have important complexity introduced by the presence of grain boundaries, there is strong evidence that even in single crystals our understanding of deformation is a work in progress. For example, dislocations tend to organize into mesoscale dislocation walls and cells with scaling laws that extend over many orders of magnitude of deformation. Measurements of single-crystal surface roughening have identified self-affine (fractal) behavior in point-to-point surface roughening of single crystals that extends over decades in length [25]. With polychromatic microdiffraction we can now measure misorientation and the development of 3D deformation structure to test theoretical predictions of deformation theories and to understand how bulk deformation compares to the surface behavior. Pang *et al.* [20] has tested misorientation *both* at the surface and in the bulk to determine whether the surface observations are indicative of the bulk behavior or are somehow restricted to the near surface region. Her measurements find that the Hurst coefficient is sensitive to both depth and crystallographic orientation/strain load (Figure 18.11). These studies are certain to provide new information about deformation in single crystals that can be directly compared to finite element and molecular dynamics models.

18.5.4 Grain Growth on Surfaces and in Three Dimensions

The penetrating nature of X-ray microdiffraction allows for precision measurements of thin-film growth that correlate to the underlying substrate. For example, Budai *et al.* [26] has explained the transition between "good" and "bad" growth of high temperature superconducting films on textured Ni substrates by demonstrating how the growth mechanism changes with temperature. Similarly, Barabash *et al.* [27, 28] has studied defects in GaN thin films grown on patterned Si and SiC substrates (Figure 18.12). The results clarify how defects are introduced as a response to thermal and lattice mismatch strains in these highly engineered materials.

More generally, the nondestructive nature of polychromatic microdiffraction enables a fundamentally new approach to the study of grain growth in three dimensions. Previous studies of grain growth have been limited to 2D growth, have depended on statistical information to understand how grains compete for material during grain growth, or have studied individual grains with little information about the local environment. Budai [29] has recently demonstrated that 3D images of grain structure can be measured and that grain growth can be observed before and after thermal annealing steps. The high angular precision of the measurements allows for a detailed study of how grain and subgrain boundaries are annealed out. Because the local environment – including the local stress state – can be measured, theories of grain growth can be directly compared to observed local movement of boundaries (Figure 18.13).

18.5.5 Anomalous Grain Growth

In addition to normal grain growth, there are very interesting and technologically important cases where anomalous grain growth occurs. For example, Sn films are important as a low toxicity alternative to Pb. However, Sn films can spontaneously grow whiskers that can short circuit electronics. The mechanism behind this phenomena has been the subject of intense speculation and study. In particular, stress, stress gradients, crystallographic orientation, intermetallic phase formation, and surface oxide fracture have

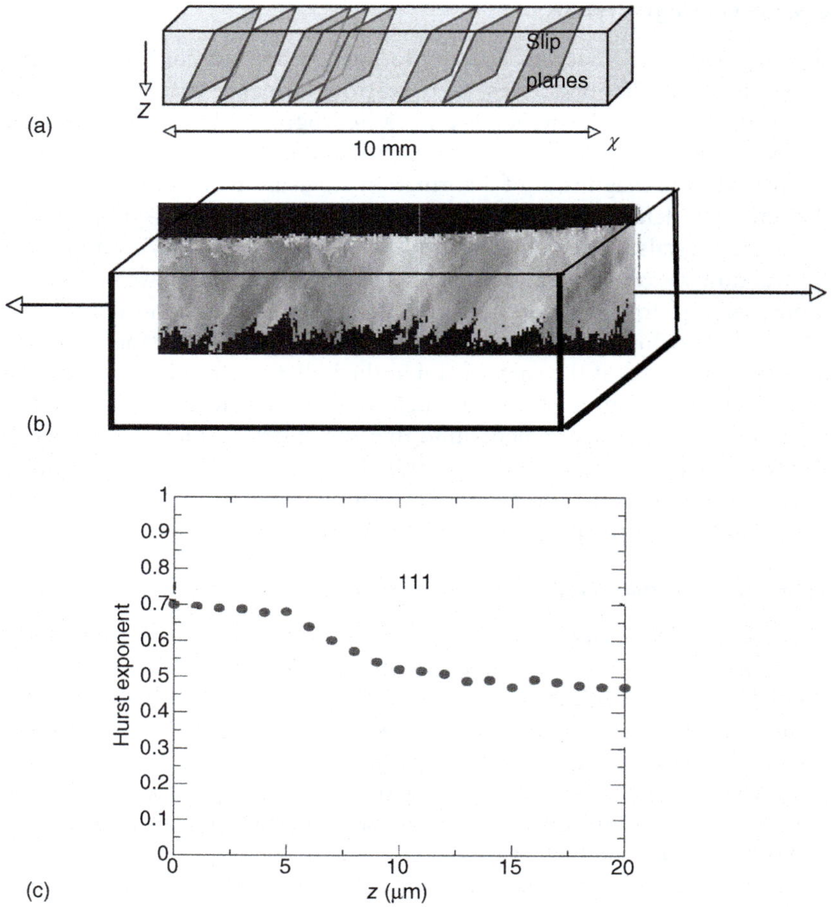

Figure 18.11 (a) Schematic of single crystal sample used to study deformation structure. Tensile specimens were pulled along various crystallographic orientations. Misorientations in points along the tensile specimen were studied as a function of depth and crystal orientation. A gray scale map (b) indicates the orientation distribution in an interior surface of the sample. Changes in the Hurst coefficient as a function of depth (c) indicate that both proximity to surface and crystal orientation effect the misorientation distribution of deformed single crystals.

all been invoked as root causes for Sn whisker development. Polychromatic microdiffraction enables unprecedented tests of these mechanisms. For example Sobiech *et al.* [21] used high-resolution synchrotron microdiffraction to study strain *gradients* near growing whiskers (Figure 18.14). The measurements support a theory of whisker formation that emphasize the role of strain gradients [30]. Other Sn whisker experiments with polychromatic microdiffraction are also underway to better understand the role of local grain boundary networks, crystal orientations, and strain fields generated by intermetallic growth. Results from these experiments are certain to provide essential new information about anomalous grain growth in Sn films and show the potential for addressing anomalous grain growth in a range of materials.

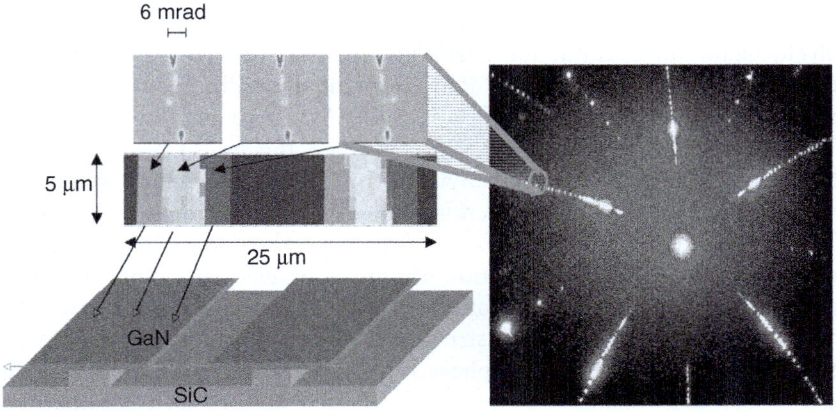

Figure 18.12 The overall Laue image from this thin GaN film on SiC looks like a single crystal because the film is crystallographically aligned to the SiC substrate. A detailed look at individual reflections find that some are streaked (SiC) due to the finite penetration of the X-ray beam into the substrate. Other reflections change the relative position depending on the location of the X-ray beam near pedestals etched into the substrate. These reflections are from the GaN film, which has wings tilted upward on each side of the pedestals.

Figure 18.13 Al grain structure at an interior sample surface before and after thermal annealing steps. The colors indicate crystal orientation and the thickness of the lines indicates the misorientations across the grain boundaries; the thicker the line the larger the misorientation.

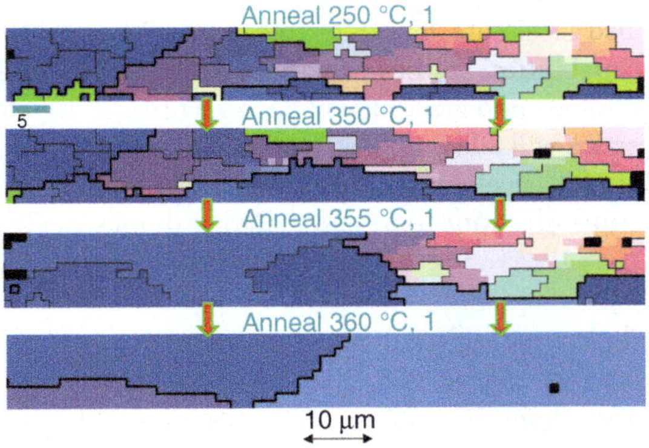

Figure 18.14 Maps of the out of plane strain tensor element (*zz*) and an in-plane element (*xx*) near a fresh Sn whisker. Strain gradients can be estimated from these measurements. (Ice 2011 [15–17]. Reproduced with permission of AAAS.)

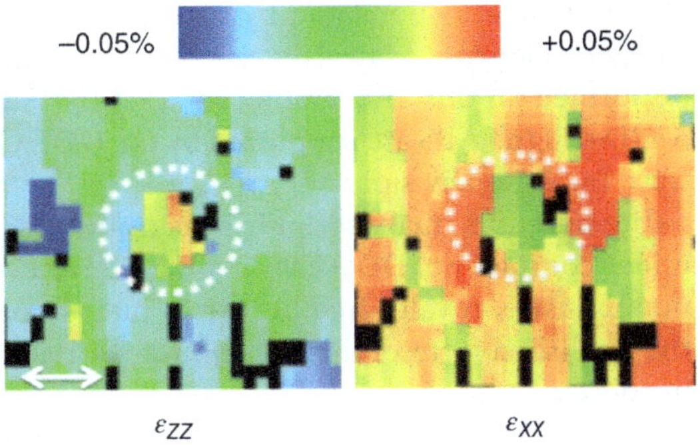

18.6 Future Prospects and Opportunities

Because polychromatic microdiffraction is still an emerging field, research opportunities currently swamp available facilities. Polychromatic microdiffraction now exists in some form at a number of facilities including the Advanced Photon Source, the Advanced Light Source, the Pohang Light Source, Spring-8, the Swiss Light Source, the Canadian Light Source, and the European Synchrotron Radiation Facility. Additional facilities are planned or are under construction at the Australian Light Source and the NSRRC.

Extensive studies of the electromigration response of electronic interconnects have revealed the importance of deformation in the electromigration process [31]. Similarly, measurements of domain wall structures, stresses in implanted materials, fracture, minor phases and precipitates, phase transitions, and interface phases will all contribute to our understanding of materials behavior. Recent measurements to identify tiny microcrystalline grains inside bulk amorphous substrates, demonstrates an ability to test how materials segregate and accommodate metastable equilibrium under rapid annealing [32].

As equipment and techniques improve, the range of possible experiments will also increase. Faster detectors, detectors with energy resolution, and smaller beams will all extend experimental possibilities. We envision in the near future, for example, a gain of at least a factor of 10 in data acquisition speed (10–100 frames s^{-1}) and an improvement of a factor of 10 in spot size (500–50 nm). Specially designed differential apertures can also accelerate data collection by an additional factor of 2–3. With existing and emerging capabilities, studies of nanocrystalline materials, and studies of grain boundary morphology and faceting will become routine. In addition, improved control of beam properties will allow for detailed studies of defects in micron-scaled volumes [33]. This will greatly accelerate diffuse measurements of defect properties and will allow point defect concentrations to be associated with mesoscale defects distributions.

Acknowledgment

This chapter would not have been possible without the efforts of many collaborators. These include, J.-S. Chung, W. Lowe, B.C. Larson, J.D. Budai, J.Z. Tischler, W. Liu, W. Yang, J.W.L. Pang, F.J. Walker, A. Khounsary, L. Assoufid, A. Macrander, and R.I. Barabash. Work sponsored by the U.S. Department of Energy (DOE), Office of Basic Energy Sciences, Materials Sciences and Engineering Division. Experiments were performed on Unicat beamline 34-ID at the Advanced Photon Source Argonne Il. Both 34-ID and the APS are supported by the DOE Office of Basic Energy Sciences, Scientific User Facilities Division.

The submitted manuscript has been authored by a contractor of the U.S. Government under contract no. DE-AC05-00OR22725. Accordingly, the U.S. Government retains a nonexclusive, royalty-free license to publish or reproduce the published or reproduce the published form of this contribution, or allow others to do so, for U.S. Government purposes.

References

1 Barabash, R., Ice, G.E., Larson, B.C., Pharr, G.M., Chung, K.-S., and Yang, W. (2001) *Appl. Phys. Lett.*, **79**, 749–751.

2 Yang, W.G., Larson, B.C., Ice, G.E., Tischler, J.Z., Budai, J.D., Chung, K.S., and Lowe, W. (2003) *Appl. Phys. Lett.*, **82**, 3856–3858.

3 Krivoglaz, M.A. and Glebov, O.A. (1966) *X-ray Diffraction in Nonideal Crystals*, Springer.

4 Barabash, R.I., Ice, G.E., and Pang, J.W.L. (2005) *Mater. Sci. Eng., A*, **400**, 125–131.

5 Barabash, R.I., Ice, G.E., and Walker, F.J. (2003) *J. Appl. Phys.*, **93**, 1457–1464.

6 Larson, B.C., Yang, W., Ice, G.E., Budai, J.D., and Tischler, J.Z. (2002) *Nature*, **415**, 887–891.

7 Larson, B.C., Tischler, J.Z., El-Azab, A., and Liu, W. (2008) *J. Eng. Mater. Technol.*, **130**, 021024-10.

8 Chung, J.S. and Ice, G.E. (1999) *J. Appl. Phys.*, **86**, 5249–5256.

9 Budiman, A.S., Shin, H.A.S., Kim, B.J., Hwang, S.H., Son, H.Y., Suh, M.S., Chung, Q.H., Byun, K.Y., Tamura, N., Kunz, M., and Joo, Y.C. (2012) *Microelectron. Reliab.*, **52**, 530–533.

10 von Laue, M. (1915) Concerning the detection of X-ray interferences. Nobel lecture, http://nobelprize.org/physics/laureates/1914/laue-lecture.pdf (accessed 28 June 2016).

11 Liu, W., Ice, G.E., Larson, B.C., Yang, W., Tischler, J., and Budai, J.D. (2004) *Metall. Mater. Trans. A*, **35**, 1963–1967.

12 Ice, G.E., Chung, J.-S., Lowe, W., Williams, E., and Edelman, J. (2000) *Rev. Sci. Instrum.*, **71**, 2001–2006.

13 Feng, R., Gerson, A., Ice, G., Reininger, R., Yates, B., and McIntyre, S. (2007) Synchrotron Radiation Instrumentation, Parts 1 and 2, AIP Conference Proceedings, vol. **879**, pp. 872–874.

14 Liu, C., Assoufid, L., Conley, R., Macrander, A.T., Ice, G.E., and Tischler, J.Z. (2003) *Opt. Eng.*, **42**, 3622–3628.

15 Liu, W., Ice, G.E., Tischler, J.Z., Khounsary, A., Liu, C., Assoufid, L., and Macrander, A.T. (2005) *Rev. Sci. Instrum.*, **76**, 113701-1–113701-6.

16 Mimura, H., Handa, S., Kimura, T., Yumoto, H., Yamakawa, D., Yokoyama, H., Matsuyama, S., Inagaki, K., Yammura, K., Sano, Y., Tamasaki, K., Nishino, Y., Yabashi, M., Ishikawa, T., and Yamauchi, K. (2010) *Nat. Phys.*, **6**, 122–125.

17 Ice, G.E., Budai, J.D., and Pang, J.W.L. (2011) *Science*, **334**, 1234–1239.

18 Liu, W.J., Ice, G.E., Assoufid, L., Liu, C.A., Shi, B., Khachatryan, R., Qian, J., Zschack, P., Tischler, J.Z., and Choi, J.Y. (2011) *J. Synchrotron Radiat.*, **18**, 575–579.

19 Tamura, N., Macdowell, A.A., Spolenak, R., Valek, B.C., Bravman, J.C., Brown, W.L., Celestre, R.S., Padmore, H.A., Batterman, B.W., and Patel, J.R. (2003) *J. Synchrotron Radiat.*, **10**, 137–143.

20 Pang, J.W.L., Ice, G.E., and Liu, W.J. (2010) *Mater. Sci. Eng., A*, **528**, 28–31.

21 Sobiech, M., Wohlschlogel, M., Welzel, U., Mittemeijer, E.J., Hugel, W., Seekamp, A., Liu, W., and Ice, G.E. (2009) *Appl. Phys. Lett.*, **94**, 221901.

22 Bollman, W. (1982) *Crystal Lattices, Interfaces, Matrices: An Extension of Crystallography*, published by the author.

23 Pang, J.W.L., Liu, W., Budai, J.D., and Ice, G.E. (2014) *Acta Mater.*, **65**, 393–399.

24 Pang, J.W.L., Barabash, R., Liu, W.-J., and Ice, G.E. (2004) *TMS Lett.*, **1**, 5–6.

25 Zaiser, M., Grasset, F.M., Koutsos, V., and Aifantis, E.C. (2004) *Phys. Rev. Lett.*, **93**, 195507/1–195507/4.

26 Budai, J.D., Yang, W.G., Tamura, N., Chung, J.S., Tischler, J.Z., Larson, B.C., Ice, G.E., Park, C., and Norton, D.P. (2003) *Nat. Mater.*, **2**, 487–492.

27 Barabash, R.I., Barabash, O.M., Ice, G.E., Roder, C., Figge, S., Einfeldt, S., Hommel, D., Katona, T.M., Speck, J.S., DenBaars, S.P., and Davis, R.F. (2006) *Phys. Status Solidi A*, **203**, 142–148.

28 Barabash, R.I., Ice, G.E., Liu, W., Einfeldt, S., Roskovski, A.M., and Davis, R.F. (2005) *J. Appl. Phys.*, **97**, 013504-1–013504-5.

29 Budai, J.D. (2004) *Mater. Sci. Forum*, **467**, 1373–1378.

30 Sobiech, M., Welzel, U., Mittemeijer, E.J., Hugel, W., and Seekamp, A. (2008) *Appl. Phys. Lett.*, **93**, 011906.

31 Barabash, R.I., Ice, G.E., Tamura, N., Valek, B.C., Bravman, J.C., Spolenak, R., and Patel, J.R. (2003) *J. Appl. Phys.*, **93**, 5701–5706.

32 Miura, E., Ice, G.E., Specht, E.D., Pang, J.W.L., Kato, H., Hisatsune, K., and Inoue, A. (2006) X-ray study of Pd40Cu30Ni10P20 bulk metallic glass brazing filler for Ti-6Al-7Nb alloy. Proceedings THERMEC2006, International Conference on Processing & Manufacturing of Advanced Materials, Vancouver BC, Canada, July 4–8, 2006.

33 Ice, G.E., Barabash, R.I., and Liu, W. (2005) *Z. Kristallogr.*, **220**, 1076–1081.

Part IV

Applications

19

The Use of Neutron and Synchrotron Research for Aerospace and Automotive Materials and Components*

Wolfgang Kaysser, Jörg Eßlinger, Volker Abetz, Norbert Huber, Karl U. Kainer, Thomas Klassen, Florian Pyczak, Andreas Schreyer, and Peter Staron

19.1 Introduction

During the past century an increasing worldwide demand for individual and public mobility of people and freight has led to the construction of large automotive and aerospace industries, which is expected to grow at an even higher rate during the following decades [1–4]. This imposes a substantial impact on global economy, the environment, and individual and societal lifestyles [5]. More than 50 million passenger cars (amounting to about 87% of all motor vehicles) are produced annually; 600 million cars are estimated to be available at present worldwide [6]. The fleet of commercial aircrafts presumably will amount to more than 35 000 planes in 2020 [2]. The competition for customer satisfaction with respect to availability, performance, safety, and cost of transport has resulted in the manufacture of transportation vehicles with a high level of functionality and sophistication. Medium-term development of transportation vehicles may be strongly influenced by at least three upcoming "Grand Challenges": (i) the limited availability of affordable fossil fuel, (ii) the limitations of tolerated emissions, especially of greenhouse gases [2, 7], and (iii) on a longer term, upcoming competitive regional providers for currently growing or unexploited markets [4].

The limitation of affordable fossil fuel will necessitate the use of alternative fuels (biofuels, hydrogen, etc.) and alternative propulsion systems (electric, hybrid, etc.). The limitation of tolerated emissions (greenhouse gases, noise, etc.) will result in the need for vehicles with much lower emission values per transported person and kilometer, which can be achieved by reduced vehicle weight, alternative design, and the use of sustainable propulsion energy sources. Unexploited consumer markets, as, for example, in China, may vanish for the present manufacturers because of increasing production rates of cars and airplanes supplied by China, India, and other countries [4]. Due to the coupled pressure of the fuel and the emission challenge and, eventually, also the uncertainties in future exports, major changes in vehicle design and transportation philosophy are expected; this will require improvement of existing engineering materials, as well as development of new material concepts.

*This chapter is reproduced and updated from *Adv. Eng. Mater.* **13** (2011) 637–657 by permission of Wiley-VCH.

Neutrons and Synchrotron Radiation in Engineering Materials Science: From Fundamentals to Applications, Second Edition. Edited by Peter Staron, Andreas Schreyer, Helmut Clemens, and Svea Mayer.

Aircrafts and passenger cars, numerous materials, and components have been developed to their present state by using research performed with neutrons and synchrotron radiation [8]. The application of neutrons and synchrotron radiation in research and development can be expected to yield similar substantial support for future developments. This chapter focuses first on the impact of research with neutrons and synchrotron radiation on a number of materials and processing methods that are characteristic and essential for commercial passenger airplane (Section 19.2) and light duty car (Section 19.3) systems, such as lightweight construction and alternative propulsion systems. In Section 19.4, a few examples for research with neutrons and synchrotron radiation on other transportation vehicles and systems, for example, trains are given. Electronics, optics, and so on, are also important components of aeroplanes, cars, and other transport vehicles. The development of the respective materials and components is essentially driven by other branches of technology , such as information technology. Therefore, these materials will not be considered in this chapter. For information about research and development on these material groups using neutrons and synchrotron radiation see the published literature [7–11].

As will be seen from the selection of the examples given below, there are two important aspects in the scattering methods that make them so interesting for materials research: first, they can be used for evaluating numerical models nondestructively, and, secondly, *in situ* experiments can be performed that yield results that cannot be obtained by other experimental methods. Numerical methods, for example, based on finite-element models, are still growing in importance for materials research. Evaluation of such models with experiments is necessary, and scattering methods often have the advantage of being nondestructive, so that one sample can continue to be used for other tests. Examples are models for the prediction of residual stresses, textures, or precipitate distributions. *In situ* experiments are required for the study of transformations and processes when the examination of the conditions of a sample before and after a treatment does not suffice to answer all open questions about what happened in between. Examples are heat treatments, mechanical deformations, or a combination of both like, for example, friction stir welding (FSW).

As mentioned above, scattering techniques very often require models relating the scattering information to the structure (morphology, microstructure) of the material. Thus, in modern material science, scattering techniques are often used in combination with other methods (often real-space microscopic techniques) to investigate the structure [12]. Recent advances in transmission electron microscopy (TEM) tomography have shown that this technique has a high potential to obtain three-dimensional (3D) information from samples with thicknesses from the nanometer up to the micrometer range, depending on the sample material [13–15]. Another example is the use of the atom probe technique for metallic materials, which yields three-dimensional maps of atom positions (3DAP). Thus, this is a method that perfectly complements small-angle scattering techniques. An enormous progress for structural investigations of composite and nanocomposites will be the advancement of X-ray tomography, when spatial resolution in the range of a few nanometers will become possible.

19.2 Commercial Passenger Aircraft

Commercial passenger aircrafts, as shown in Figure 19.1, consist of an airframe and a propulsion system. The frame carries the crew and passengers, their belongings and travel supply, the cabin equipment, avionics, the landing gears, and so on, as well as the aero-engines and the fuel. The performance of the whole system is largely sensitive to the mass of its components [2]. Each kilogram of the aero-engine requires additional stiffness and strength of the frame and, thus, increases the weight of the system beyond the additional weight of the aero-engine itself. Similar arguments hold for components of the airframe. A heavier central wing box requires larger wings, heavier landing gears, and stronger, and thus usually heavier, aero-engines. An optimized set of airframe and aero-engines with affordable components, that is, having the highest possible performance-to-weight ratio, is a crucial precondition for present and future competitiveness. The Advisory Council for Aerospace Research in Europe (ACARE) defined ambitious targets for the year 2020 as compared to 2000 for the whole European aircraft industry [2]:

- Reduce fuel consumption and CO_2 emissions by 50% per passenger kilometer.
- Reduce NO_x emission by 80%.
- Reduce perceived external noise by 50%.
- Reduce production costs by 50%.

These ambitious goals can only be achieved by reduction in weight of the whole aircraft and an improved efficiency of the propulsion system.

What might future commercial airplane designs look like 25–30 years from now? Teams in the United States, triggered by NASA's N+3 program [16], generated visions of the "three generations beyond aircraft." Their concepts for the Subsonic Ultra Green Aircraft Research project included conventional reference configurations and versions of a new design high span, strut-braced wing aircraft, and a hybrid wing body configuration. Hybrid electric engine technology based on an electric battery gas turbine hybrid propulsion system was declared a clear winner, because it can reduce fuel burn

Figure 19.1 Commercial passenger aircrafts consist of an airframe and a propulsion system (Airbus A380). (Kaysser 2011 [281], reproduced with the permission of Wiley.)

by more than 70% and total energy use by 55%. Nevertheless, the hybrid electric concept to become competitive "needs to improve battery technology many, many times over what we have today." For conventional propulsion, a combination of improvements of air traffic management, airframe and propulsion was stated to reduce fuel burn by 44–58% [16].

19.2.1 Reduction of Airframe Weight of Commercial Passenger Aircrafts

The reduction of airframe weight is an ongoing development. It was initially driven by attempts to improve the technical performance and economics of the aircrafts, that is, to reduce fuel consumption and eventually manufacturing and maintenance costs. In the future, lowering CO_2 emission and the limited availability of kerosene will put additional pressure on weight-saving measures. At the same time, aircraft safety has to be increased [2]. In addition, passenger comfort and well-being cannot be neglected, nor can the practicability of loading and unloading passengers in both routine and emergency cases. The competition between the metal-based frame and frames consisting of a large fraction of fiber-reinforced polymers is still ongoing. In the future A350 Airbus family, about 70% of the airframe will constitute of composite material structures, titanium, and advanced aluminum alloys. Composites alone will account for about 53% [17]. The following three sections describe applications of neutron and synchrotron methods on metallic frame structures, while Section 19.2.1.4 will deal with polymeric materials and composites.

19.2.1.1 Welding Commercial Passenger Aircraft Frames: Reactions, Microstructure Development, and Mechanical Properties

Metal-based airframes consist of different Al-alloys, Ti-alloys, steels, and some further alloys of minor importance. For many decades, the Al-alloy components of the frame, such as the skin, stringers, and spars were joined by riveting. This technology requires considerable overlap of the components to be joined, resulting in a weight penalty. Furthermore, riveting is a very costly procedure. Since the mid-1990s laser beam welding (LBW) and, more recently, FSW (Figure 19.2) have started to replace riveting as a joining technology [18]. During conventional welding and LBW, the microstructure of precipitation-strengthened Al-alloys is drastically altered because of melting and subsequent solidification [19]. During the FSW processes, melting does not occur; however, there are changes to the microstructure due to growth, dissolution, and partially due to re-precipitation of intermetallic precipitates and the complete dynamic recrystallization of the stir zone. All these metallurgical phenomena are typically nonequilibrium phenomena and at present cannot be reasonably predicted from thermodynamics simulations. Small-angle neutron scattering (SANS) studies yield the required clarification of the precipitation kinetics and microstructure evolution. An example of this approach is the work carried out by Woo and co-workers [20] on *in situ* neutron diffraction measurements under transient temperature and stress fields in 6061-T6 thin plates. In this and other studies, the material's fast transient behavior (temperature, bulk stress/strain, microstructure changes) under complex conditions of "real-world" processing have been observed [21–27].

Welding processes not only strongly affect the microstructure and mechanical properties in the weld nugget itself, but also have a considerable influence on the mechanical

Figure 19.2 Part of a friction stir welded fuselage. (Kaysser 2011 [281], reproduced with the permission of Wiley.)

properties of the adjacent heat affected zone (HAZ). In welds of precipitation-hardened Al alloys, strength and hardness typically decrease significantly in and around the joint line compared to those in the base material. Mechanical properties could be measured either by tensile tests with samples of specific geometry (microtensile testing) [27, 28] or by hardness indentation measurements [29]. The results of the indentation measurements could be affected by residual welding stresses, an effect well understood from an investigation combining indentation measurements, neutron and synchrotron diffraction, and finite element simulations. Residual stress measurements are treated in Section 19.3.2 in more detail.

Novel approaches for the investigation of mechanisms of local deformation and fracture on the crystal scale combine FIB (focused ion beam) milling and nano-indentation with *in situ* measurements using SEM (scanning electron microscopy) or synchrotron X-ray techniques. Recent trends combine microcolumn testing with X-ray microLaue diffraction [29–34]. These methods have been developed mainly to improve understanding of deformation in small material volumes in general. Future application of these methods to welded materials offers a high potential to understand in detail the effects of the welding process on the local microstructure, deformation, and damage behavior of the material in the weld zones.

19.2.1.2 Welding Commercial Passenger Aircraft: Residual Stresses and Stress Modification

Advanced weight-saving joining technologies, which have recently found their application in welding of aerospace Al alloys, are FSW and LBW. Residual stresses in these welds depend on the parameters of the welding process, joint geometry, working piece clamping (i.e., constraints) as well as on the elastic moduli and the strength of the microstructure of the alloys. These residual stresses have been soundly investigated using neutrons, X-rays, and synchrotron radiation [25, 35–51]. In alloys without phase transformation, typically strong longitudinal residual tensile stresses develop in the weld, in the welding (longitudinal) direction. They are equilibrated by compressive residual stresses in the longitudinal direction in the heat-affected zones, thermo-mechanically affected zones, and the base material. Residual stresses often come along with geometry distortions, thus the joints are often straightened by thermal or mechanical processes. However, residual stresses still remain in the structure after that and their magnitude and distribution influences the fatigue crack propagation and damage tolerance behavior of the airframe. These phenomena have been investigated for a number of aerospace alloys in butt joint and T-joint configuration. One example of an interesting result of these investigations was the absence of any significant change of the residual stress distribution when a AA6056 material joint processed by welding in T4 condition underwent a post-weld heat treatment from T4 to T6 condition [43].

Methods like shot peening, mechanical tensioning, or roller tensioning can be used to modify the residual stress state in welds in order to minimize distortions and enhance the fatigue performance of welded structures. The working principle is to modify the induced plastic strains in the weld seam area during or directly after welding. The magnitude and distribution of residual stresses present in the as-welded state with and without modifications have been determined by synchrotron X-ray and neutron diffraction for different Al alloys. An example is shown by Figure 19.3. The residual stress change brought about by global mechanical tensioning is basically independent of the specific alloys; rather, it depends on the elastic pre-stretching stress during welding. The concept of global mechanical tensioning was recently further developed to residual stress engineering in friction stir welds by local mechanical tensioning approaches, such as roller tensioning and post-weld roller tensioning in FSW of aluminum alloy. In this case, a marked reduction in the longitudinal tensile residual stress was observed with increased loads imposed by the roller [52–58].

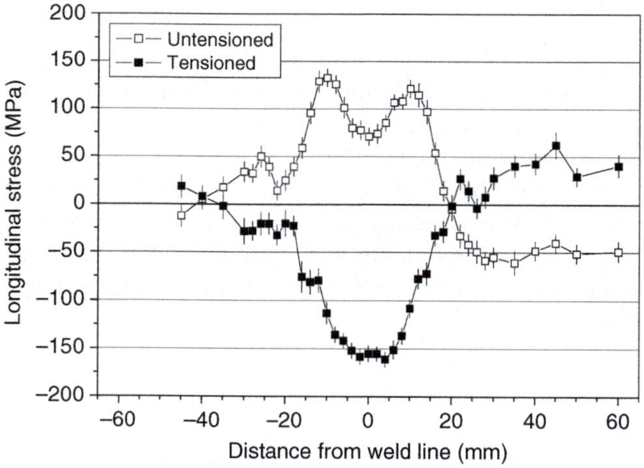

Figure 19.3 Residual stress distribution in an AA2024 Al sheet friction stir welded under mechanical tension. Stresses for a sheet welded without tensioning are shown for comparison. Measured by neutron diffraction. (Kaysser 2011 [281], reproduced with the permission of Wiley.)

19.2.1.3 Welding of Commercial Passenger Aircraft: Fatigue Crack Growth

Residual stresses strongly influence fatigue crack growth behavior. Here, besides knowing the initial residual stress field, it is also important to incorporate the effects of stress relaxation and redistribution due to crack growth. Neutron diffraction has been used to measure the evolution of the residual stresses in welded Al-2024 alloy with fatigue loading and subsequent crack growth. The finite element method, in conjunction with the modified virtual crack closure technique and the residual stress measurements, has turned out to be very efficient for the prediction of fatigue crack propagation in residual stress fields, showing very good agreement with experimental investigations [59–61].

The evolution of residual stresses near bonded crack retarders during fatigue crack growth has also been studied. The residual stress evolution has been measured *in situ* during crack growth using neutron diffraction, and modeled with a finite element approach [62, 63]. In this context, high-energy synchrotron X-ray diffraction (XRD) studies are particularly useful for the investigation of the stress fields around a fatigue crack and the observation of nonlinear effects caused, for example, by an overload. Recent evaluation of the overload effect on fatigue crack growth with the help of synchrotron XRD strain mapping show very good correlation with nonlinear elastic–plastic finite element simulations (Figure 20.3) [63]. For a better understanding of the mechanisms of microcracking, crack coalescence, and micromechanisms of fatigue crack closure, high-resolution synchrotron X-ray microtomography is a valuable tool to gain insight by 3D imaging of the location of the microcracks [63–70].

19.2.1.4 Weight Reduction of Aircraft by Polymers and Polymer Matrix-Based Composites

In aerospace applications, polymers and polymer composites play an increasingly important role and are replacing metals in many cases. Particularly in the case of fiber-reinforced composites, this is due to their lower densities as compared to metals, their high specific strength and stiffness, excellent fatigue resistance, the possibility to tailor directional properties, and their stability against natural metal-corroding agents, just to list some of the most important positive properties. Polymers can cover a broad range of applications in aircraft frames, including the fuselage, wing boxes, and so on, and less demanding interior parts of airplanes, like the floor, insulation, luggage compartments, inner part of windows, and so on. For all these various applications, polymers with different properties are required, which range from thermoplastics to thermosets. In fact, most polymers have to be combined with filler materials in order to improve their mechanical properties. To increase the tensile strength and stiffness of polymer composites most often glass fibers, carbon fibers, and aramide fibers are used.

Silica particles are often introduced in order to increase the toughness of glassy polymers. As in all these composites, the dispersion of fillers in the polymer matrix is highly important, especially in so-called nanocomposites, where at least one of the filler's dimensions is well below 100 nm. As a consequence, such fillers have an extremely high surface-to-volume ratio and, thus, they strongly tend to form agglomerates. In order to obtain the advantageous properties of a real nanocomposite by maximizing the effective interface between filler particles and polymer matrix, agglomerates of the filler have to be destroyed and this is achieved by a complete exfoliation of the filler particles.

In the case of layered silicates, the de-agglomeration of the layers occurs in several steps, which can be monitored by *in situ* wide angle X-ray scattering (WAXS or XRD) [71]. First, some intercalation is observed, which leads to a move of the typical scattering peaks of the filler particles to a lower angle, as the periodicity of the corresponding crystal plane is increased. When the different layers are separated completely (exfoliated), the scattering peak corresponding to the primary particles disappears. The key issue is the interaction between filler and matrix, as a good binding or adhesion is important for an effective transfer of mechanical load or impact.

X-ray scattering measurements under in situ or *ex situ* mechanical loading conditions can help to detect the orientation of domains (small-angle scattering) [72, 73] and also the change of crystallinity and crystallite orientations in semi-crystalline thermoplastics (wide-angle scattering) [73–75]. The occurrence of voids between filler particles and matrix due to delamination or between two different polymers can be easily detected by small-angle X-ray scattering (SAXS) [76, 77]. With static and dynamic neutron scattering techniques, namely SANS and neutron spin echo measurements, chain conformations around nanofillers [78], and the segmental dynamics of the polymer under confinements due to an interface [79] can be detected, thus helping to understand filler matrix interactions.

Polymer coatings of the aircraft skin are required as a corrosion-protective cover layer on modern aircraft fuselages. Development of polymer composites for the application as coatings to aircrafts and automotive vehicles is, therefore, another important field of research. The thickness and the homogeneity of the thickness (roughness) of polymer films can be investigated by X-ray reflectometry. Multilayered polymer coatings can be characterized by neutron reflectometry. The latter method also allows for the determination of the inter-diffusion profiles between different polymer layers [77, 80]. If polymer coatings possess a lateral nanostructuring, perpendicular to the substrate, grazing-incidence small-angle X-ray scattering (GISAXS), and grazing-incidence small-angle neutron scattering (GISANS) [81] are applicable to simultaneously monitor the perpendicular and the parallel to the surface structuring in films with thicknesses in the range of a few nanometers up to a few hundred nanometers.

Small-angle scattering techniques contribute to polymer science also in fundamental aspects that are often necessary pre-conditions for technical development. Some examples are mentioned here. With the help of SANS investigations on blends of deuterated and nondeuterated polymers, the conformational behavior of individual macromolecules could be investigated in the melt state [82–84]. In addition, thermodynamic properties as the segmental interaction parameter of multicomponent polymer systems are accessible by scattering techniques [80, 85, 86]. In phase-separating blends, the structural changes as a function of time can be followed by time-resolved SANS also under the influence of a shear field [87].

As mentioned in the introduction, scattering techniques always suffer from the fundamental problem that models are required to relate the scattering information with the structure (morphology) of the material. Thus, in modern polymer science, scattering techniques are often used in combination with other methods to investigate the structure (real-space microscopic techniques) [12]. Recent advances in TEM tomography show that this technique has a high potential to obtain 3D information from polymer samples with thicknesses up to the micrometer range [13]. A disadvantage of the investigations in the electron beams is the severe damage of the organic material by the electron bombardment as compared to the damage caused by X-rays or neutrons. An enormous progress for structural investigations of polymer composite and nanocomposites will be the advancement of X-ray tomography, when spatial resolution in the range of a few nanometers will become possible at low doses of irradiation.

Besides thermodynamics, also dynamic chain properties could be measured down to very short timescales. In polymer melts, there are different timescales relevant for the flow properties. For example, the so-called Rouse dynamics can be investigated by the neutron spin echo technique, a technique that takes advantage of the magnetic moment carried by neutrons [88, 89]. The nuclei of the sample atoms interact with this magnetic moment and depending on the motions of these atoms the polarizations of the neutrons in the beam are affected. These few examples show the strong contribution of scattering techniques to structural and dynamic aspects of polymer materials.

19.2.2 Aero-Engines

The thrust-to-weight ratio of aero-engines has been drastically increased during the past decades [8]. Major reasons are an improved design of components and the use of innovative materials [90]. An example of an improved system design is the introduction of an ever-increasing fan size in commercial plane aero-engines, shifting the major contribution of the propulsion from the central part of the aero-engine to the frontal fan and resulting in increased efficiency (Figure 19.4). A further example is improved inner cooling of the turbine blades, allowing higher gas temperatures entering the first rows of the turbine thus increasing the thermal efficiency. In addition, materials have been optimized in composition and microstructure design. Examples are Ni-based single crystal superalloy blades for the hot turbine section, which show a drastically improved creep resistance compared to polycrystalline blades of same composition with globular or directionally oriented grains [91]. Here, neutron diffraction techniques have

Figure 19.4 Design of a modern aero-engine for civil aircraft. The large fan at the left accelerates the incoming cold air pressing a major fraction through the bypass to the rear outlet of the engine. The minor fraction of the incoming air passes through compressor, combustion chamber, and turbine leaving the aero-engine as hot gas. (Kaysser 2011 [281], reproduced with the permission of Wiley.)

strongly contributed to alloy development and to optimizing the residual stress state (see Section 19.2.1).

Future improvements may follow along the same line with concepts such as counter-rotating fan blades, geared aero-engines that transmit the high rotation rates of the turbines into slower rotation rates of the fans [92]. New materials and material concepts may be used, such as intermetallics, fiber-reinforced polymers, and ceramics [93–95] or beyond band-aid ceramic thermal barrier coatings (TBCs) [96]. System and component design are also decisive for reduction of noise related to the propulsion system. Major noise reduction has already resulted from the introduction of the fan concept. Further reduction is feasible by introduction of geared and counter-rotating systems. The quantity of CO_2 emissions of an aero-engine is directly coupled to its fuel consumption. Therefore, increased aero-engine efficiency will reduce the emission of this greenhouse gas. The proportion of other emission gases critical for environment, for example, nitrous gases, depends critically on the design of the combustion process. The weight-to-thrust arguments also enters the discussion on the optimum future fuel. The energy density of kerosene presently used is considerably higher than the energy density of alternative fuels, for example, liquid hydrogen or other mobile energy storage systems [97]. It is, therefore, a common assumption for aircrafts that the shift from fossil to alternative fuels may occur much later than for other transport systems.

19.2.2.1 Metallic Materials to Improve the Thrust-to-Weight Ratio of Jet Aero-Engines

The performance of aero-engines increases with the "working temperature" of the combustion gases and the efficiency of the transition of the combustion energy into thrust. The hot gas stream passes through the aero-engine's turbine, starting at the high pressure section with maximum temperatures and pressures at the turbine inlet of approximately 1700 °C and 40–50 bar, respectively. The gas temperature and pressure decrease simultaneously during further flow toward the rear end of the turbine. At present, superalloys are used for the hottest sections [98–100]. Their creep and oxidation resistance are extraordinarily good but only up to temperatures of 1000–1100 °C. To keep the temperature of superalloy components in the hottest sections of the aero-engine below these values, a sophisticated air cooling system of the metallic components is required. Below 500 to 600 °C special steels and the lighter titanium alloys replace the heavier superalloys. Nevertheless, the only metallic materials that are promising contenders to replace Ni-based superalloys for mechanically loaded parts in the higher temperature regime of at least up to 800 °C are γ-TiAl alloys [101–104].

Superalloys and γ-TiAl alloys both are multiphase materials. They draw their mechanical strength at elevated temperatures from the presence of nanospaced internal interfaces that efficiently restrict the movement of dislocations. Additionally, both materials either consist fully of ordered intermetallic phases (γ-TiAl alloys) or have at least a significant amount of ordered intermetallic phases incorporated as precipitates into their microstructure (Ni-based superalloys) [98, 105]. Both alloy systems benefit from the inherent good mechanical properties of the ordered intermetallic phases at high temperatures. Therefore, the microstructure has a strong influence on the mechanical properties of both alloy systems and the microstructure evolution during processing and application needs close control. During processing, sophisticated heat or thermo-mechanical treatments are applied to achieve an optimum microstructure resulting in the desired property balance. Unfortunately, during service at elevated temperature the optimized initial microstructure can change rather fast leading to the degradation of the optimized mechanical properties [106–108]. Many of these microstructure alterations during processing and application at high temperatures are not directly accessible by classic microstructure characterization methods as TEM and SEM. Diffraction methods, however, are applicable up to very high material temperatures. Using either X-rays from synchrotron sources or neutrons the resolution of the diffraction methods is sufficient to measure the small changes in lattice parameters associated with internal stresses or deformation, which can occur during service or processing. In addition, large volumes of the specimens can be sampled with X-rays from synchrotron sources and especially with neutrons to determine global values effectively. On the other hand, the high brilliance of synchrotron sources allows for measurements with high spatial resolution, which yields diffraction information from defined small volumes. in addition, the recording of weak super-lattice reflections far beyond the possibilities of laboratory scale X-ray experiments is an additional benefit of these high brilliance X-ray sources as well as neutrons.

In Ni-based superalloys the precipitate shape and their coarsening are mostly influenced by coherency stresses in the nickel matrix and the coherently embedded precipitates of the intermetallic γ'-phase [98]. Measurements of these coherency stresses (or better to say the strains induced by them) are normally made by diffraction methods with lab-scale X-ray sources [109–111]. Here, the use of neutron and synchrotron diffraction [112–114] allows the measurements of higher order reflections where the peaks of the nickel matrix and γ'-phase are more widely separated than those of the lower order reflections accessible by lab-scale X-ray sources, as was demonstrated by Tiley and co-workers [115]. In addition, the possibility to focus X-rays from synchrotron sources can be used to an advantage. Typical examples are measurements of coherency stresses in coated Ni-based superalloy turbine blades after service by Biermann and co-workers [116, 117]. The same work also proved the possibility of analyzing the effects of sub-surface interdiffusion on the lattice parameters in the transition zone of the turbine blade and its Al-rich protective coating by synchrotron diffraction.

Important research topics in Ni-based superalloys and γ-TiAl alloys with considerable implications to their application are the deformation of the single phases as well as recovery and recrystallization phenomena during processing. Laboratory-scale X-ray diffraction methods [118–120], and in particular experiments with X-rays from synchrotron sources [46, 121] or neutrons [122] at service temperature provide a characterization of their microstructure development. In large-scale parts, such as turbine discs from Ni-based superalloys, high residual stresses in regions of millimeters to centimeters are induced during hot forging by inhomogeneous deformation and have to be considered in the design process. They hamper the reliability of these safety-sensitive parts. By neutron diffraction, these residual stresses could be determined precisely as shown, for example, by Staron and co-workers [46] (Figure 19.5).

SANS and SAXS are suitable methods to investigate microstructure evolution and degradation [122–127] during processing and application. In particular, changes in precipitate shape and size in Ni-base superalloys influence the mechanical properties. SANS and SAXS allow large specimen volumes to be sampled and changes to be monitored *in situ*. In a work by Zrník and co-workers [128], the change of the γ' microstructure in the Ni-based alloy CMSX-4 was investigated after creep deformation at different temperatures and loads. Modern high-strength γ-TiAl alloys are materials,

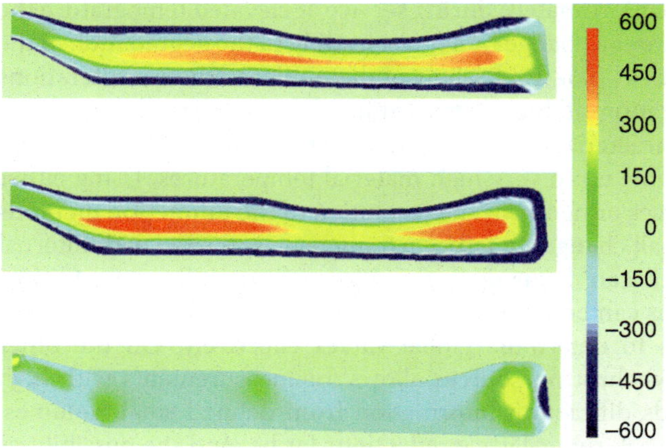

Figure 19.5 Residual stress distribution in water-quenched compressor disc (radial, tangential, and axial directions from top to bottom; stresses in MPa). The figure shows a result of a numerical simulation with a finite element model (DEFORM). By neutron diffraction these residual stresses could be determined precisely as shown by Staron and co-workers [46]. (Kaysser 2011 [281], reproduced with the permission of Wiley.)

which experience complex evolutions of phase constitution and microstructure during processing and service. These materials were frequently investigated by synchrotron diffraction in the past to monitor these changes [104, 111, 129, 130]. More details on this topic from an actual work can be found in a contribution of Schmoelzer and co-workers in this issue [131]. An example for the *in situ* observation of solid-state phase transitions in γ-TiAl including the formation of metastable phases from β is given in a paper of Stark and co-workers and is also included in this issue [132]. In a work of Liss and co-workers the split-up of initially overlapping 002 and 200 reflections in a strongly distorted so-called massively transformed (MT) microstructure of TiAl could be observed when internal stresses relaxed at higher temperatures, as shown in Figure 19.6 [133]. This work extensively benefits from the possibilities of specimen environments allowing heating and cool down of the sample during measurement and the recording of the diffraction information on area detectors. These advantages were frequently used to monitor the complex chain of solid state phase transformations in TiAl alloys with large additions of major alloying elements as Nb and Mo [134–139] or minor elements additions as boron or carbon [140, 141].

19.2.2.2 Thermal Barrier Coatings to Enhance the Thrust-to-Weight Increase of the Aero-Engine

TBCs used on advanced turbine blades can considerably increase the efficiency and the performance of the blades [142–143]. In state-of-the-art TBCs, often a MCrAlY (where M means Ni, Co, or both) bond coat and a top coat of yttria-partially stabilized zirconia (Y-PSZ) are deposited onto superalloy substrates either by plasma coating or by electron beam physical vapor deposition (EB-PVD). Zirconia is particularly suited as a thermal barrier layer due to its low thermal conductivity, its higher coefficient of thermal expansion than the superalloy substrate and its chemical inertness in kerosene combustion atmospheres. Zirconia is essentially transparent to oxygen at high temperatures and does not protect the superalloy from oxidation. Therefore, the superalloy blades

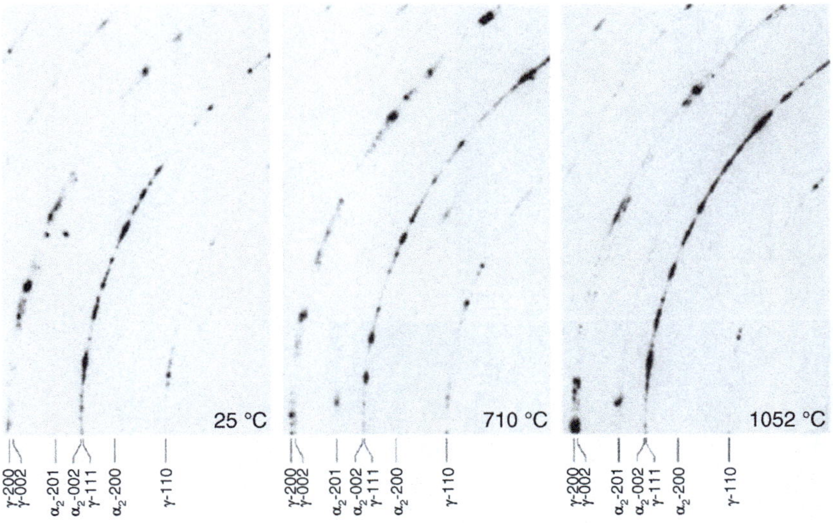

Figure 19.6 Ring patterns from the *in situ* observation of a split-up of 200 and 002 reflections of γ-phase when the internal stresses in a massively transformed microstructure of TiAl are relaxed at higher temperatures. (Kaysser 2011 [281], reproduced with the permission of Wiley.)

are coated by an oxidation-protective bond coat as mentioned above. The EB-PVD process offers the advantage of a superior strain and thermoshock tolerant behavior of the coatings due to their columnar microstructure (Figure 19.7) [144, 145].

A combination of ultra-small-angle X-ray scattering (USAXS) and SANS analysis enabled the determination of the anisotropic stereometric characteristics of the pores within the EB-PVD coatings. Morphological evolution from the initial state to different aging times at high temperatures could be followed. In particular, intercolumnar gaps between the columns, pores between feather arms of the columns, and isolated intra-columnar pores were investigated. These results could be related to the evolution of the thermal conductivity of the thermal barrier layers during long-term annealing [146–148].

During annealing at elevated temperatures so-called thermally grown oxide layers (TGO) form by oxidation of the metallic bond coat at the transition area to the Y-PSZ top coat. These layers are growing from a few hundred nanometers in the as manufactured state to several microns after long term temperature exposure (Figure 19.8) [149–153]. The TGOs are heavily loaded by growth stresses at temperature and thermal stresses during cooling. Crack formation and propagation due to these stresses are the major causes of the failure of TBCs by spallation. The stresses in the thin TGO layers have been determined by synchrotron radiation [154].

19.3 The Light-Duty Automotive Vehicle

The limitation of affordable fossil fuel and tolerated emissions will lead to vehicles with much lower net emission values per transported person and kilometer than present currently [3]. Higher efficiency by alternative design, innovative propulsion systems,

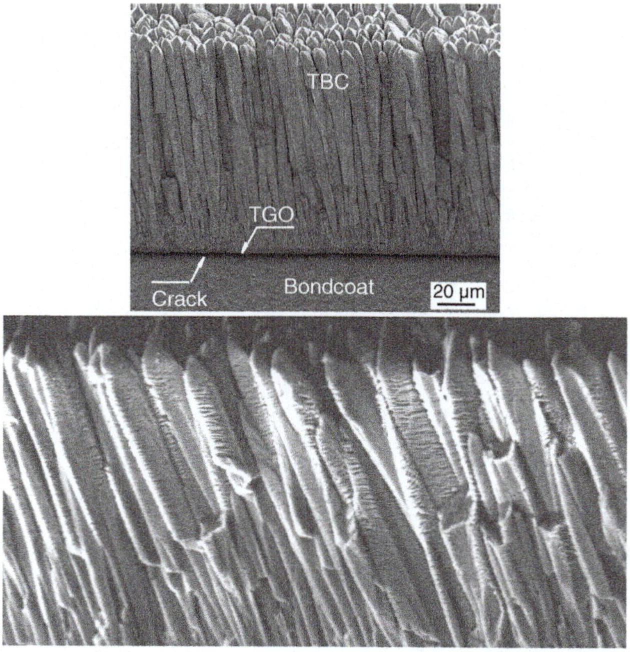

Figure 19.7 Thermal barrier coating (Y-PSZ, Zr02-4Y2O3) deposited by EB-PVD. The columnar morphology yields excellent strain and thermoshock tolerance. The site of failure is the TGO layer [139]. (Kaysser 2011 [281], reproduced with the permission of Wiley.)

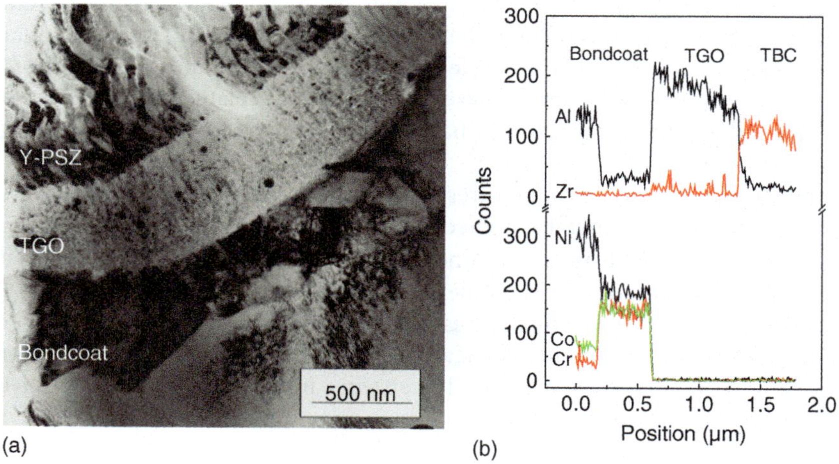

Figure 19.8 TGO (thermally grown oxide, essentially Al_2O_3) forming between the thermal barrier coating (Y-PSZ) and the bondcoat during coating and growing thicker at elevated temperatures. (a) TEM picture. (b) EDAX line scans of Al, Zr. (Kaysser 2011 [281], reproduced with the permission of Wiley.)

sustainable energy for propulsion, and a substantially reduced vehicle weight are cornerstones of future development. Fuel consumption of vehicles has been drastically reduced over the past decades, particularly if measured in liters per ton and per 100 km. Unfortunately, the average weight of the cars also increased at the same time, due to increasing car size and the addition of numerous parts and sub-systems providing safety and comfort to the drivers and additional passengers. As mentioned in the introduction, consequences for the future cars are obvious, although the timescale of the alterations is yet unclear and will depend on many essentially unpredictable upcoming events and developments. As medium- to longer-term consequences, extreme lightweight construction, the use of recyclable (sustainable lifecycle) materials, and the use of sustainable propulsion energies are obvious developments. Major concerns will be put on the very light car body comprising a combination of high functionality and inherent safety and on alternative propulsion systems including bio-generated fuels and the sub-products produced by zero-emission electricity [3]. In addition, changes in individual and public transportation philosophy are to be expected [5], requesting also an improvement of existing materials and the development of new materials concepts.

19.3.1 The Optimized Light-Duty Car Body

Applications for neutron and synchrotron diffraction to automotive applications cover the analysis of surface effects to investigate lubricants and wear as well as the investigation of deformation and fatigue behavior of materials, welds, and components, very often in connection with residual stress fields. These stress fields are produced from machining, surface heat treatment or nitriding, and joining and can have a strong effect on fatigue life of highly loaded components, such as gears and springs. Neutron and synchrotron radiation diffraction are nowadays widely used for the evaluation of residual stresses induced by thermal and mechanical treatments in materials and components for industrial applications. A review of the techniques used and some applications to materials for structural applications is presented by Fiori and co-workers [155]. In addition to steels and aluminum alloys, lightweight materials such as Mg alloys [156] and hybride components formed or joined from quite different materials [157] are expected to add less weight and more integrated functionality to the cars. Modern experimental techniques using neutrons and synchrotron radiation can help to provide an empirical understanding of the complex behavior of these materials. Moreover, they serve as excellent tools to verify materials models and to fit the parameters in simulation studies.

19.3.1.1 Lightweight Metallic Materials for Lightweight Car Bodies

The properties of semi- and final products of Mg alloys and most other alloys are strongly influenced by microstructure and texture, which are in turn attributed to thermomechanical processing and the specific alloying systems. Many researchers using neutrons and synchrotron radiation have opened up new opportunities of knowledge-based optimization of materials processing and performances of Mg alloys and other lightweight alloys with competitive properties. The deformation behavior of polycrystalline magnesium alloys has been investigated by *in situ* loading experiments [158]. Evaluation of the experimental results in combination with those from computer simulations, in terms of the time evolution of the crystallographic texture and the mechanical response under loading, revealed a correlation between the initial texture, the activities of deformation mechanisms and the mechanical anisotropy observed. *In situ* studies on twinning evolution during tensile and compressive loading of Mg alloys showed clearly that the strain hardening and asymmetry in mechanical response are related strongly to activation of mechanical twinning [159, 160]. Visualization of the recrystallization process, by which the microstructure of the heat treated product is determined, has been introduced using 3D X-ray diffraction microscopy [161]. The grain growth characteristics and kinetics, at the early recrystallization stage, were successfully examined by tracking a nucleus embedded in a deformed aluminum single crystal. By realizing a way of nondestructive, 3D mapping of grain shape, crystallographic orientation, and grain boundary geometry, it should be possible to study even more complex phenomena in materials experimentally. A study of the evolution of stress corrosion cracking in stainless steel using a grain-mapping technique combined with tomography is a successful example [162].

In the field of solidification of metals, the study of as-cast microstructures through conventional techniques provides some information about their characteristics and properties. However, more specific questions on the nucleation mechanisms leading to the growth of a primary solid phase, the microstructure evolution, the formation of intermetallic phases and their stability, and the occurrence of defects can only be answered through dynamic *in situ* studies. *In situ* characterization of solidification of aluminum alloys has been achieved through synchrotron X-ray diffraction, radiography, and tomography techniques. Most of these investigations have been focused in X-ray radiography [163, 164] and tomography [165, 166] after (*ex situ*) or during (*in situ*) solidification of Al alloys with respect to coarsening, fragmentation, columnar to equiaxed transition (CET), dendrite growth, and effects of convection in dendrite growth [165, 167–170]. Via real-time observation of the formation of individual grains during the solidification of aluminum alloys, the kinetics of nucleation, and growth have been studied using hard X-ray diffraction [171]. The results of this study demonstrated that grain nucleation is triggered by a meta-stable phase and thus confirmed the highly debated mechanism responsible for enhanced nucleation in a modified Al-based alloy. Microstructure evolution, partial remelting, microsegregation, branching mechanisms, coarsening, and behavior of the mushy zone in Al-Cu and Al-Mg-Si alloys are all aspects that have been addressed [166, 170]. Parallel to this research axis, neutron diffraction has also been employed to investigate the nucleation kinetics of Al alloys [171, 172]. For solidification of Mg alloys, comparable investigations are scarce due to the difficulties in handling Mg-melts. High-energy beams also allow analyzing the characteristics of hot cracks. This rather common defect is caused by tensile forces during cooling in an intermediate temperature range. The stresses occur in the vicinity of regions of high solid fractions situated in the mushy zone. Figure 19.9 shows transverse sections of volumes reconstructed after X-ray computer-based tomography of specimens that exhibit, in a particular region, a hot crack whose formation is a consequence of its geometry and the imposed solidification conditions.

Weight benefits for vehicles could also be realized by the use of high-performance steels. TRIP steels are characterized by high strength and extremely high-energy absorption during deformation to fracture due to deformation-induced martensitic deformation. The mechanical properties of the constituent phases of a TRIP 800 steel and the fitting parameters describing the martensite transformation kinetics were determined using the synchrotron-based *in situ* high-energy X-ray diffraction

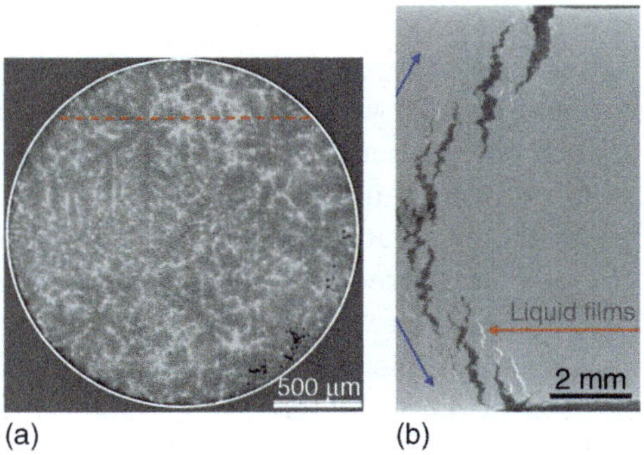

Figure 19.9 Crack formation in AZ61 alloys solidified in a sprue-rod mould. (a) 2D projection of a reconstructed volume region. (b) 2D section of reconstructed volume showing the formation of cracks. (Kaysser 2011 [281], reproduced with the permission of Wiley.)

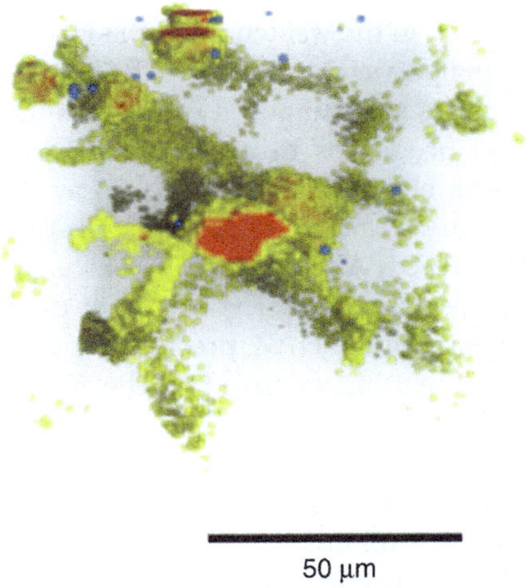

Figure 19.10 3D microtomography of an as-cast AZ91 alloy, segmented beta phase (yellow), micropores (blue), and Al–Mn phases (red), demonstrating the lamellar character of the beta phase enclosing the Al–Mn phases in a network on the grain boundary. (Kaysser 2011 [281], reproduced with the permission of Wiley.)

(HEXRD) experiments performed under a uniaxial tensile deformation [173]. The experimental results suggest that the HEXRD technique provides a powerful tool for characterizing the phase transformation behavior and the microstresses developed due to the phase-to-phase interaction of TRIP steels during deformation. In order to investigate formability, performance, and also to generate guidelines for the stamping process design of friction stir welded TWB (tailor welded blank) sheets, the hemispherical dome stretching test was experimentally performed and the results of the base and friction stir welded samples were compared [174].

Neutron as well as synchrotron radiation have also been used to investigate *in situ* the creep behavior, internal stresses and the influence of intermetallic phases on the properties of alloys, light metal matrix composites, and light metal foams [42, 79, 175–182]. Synchrotron X-ray diffraction and tomography at room temperature has been performed in order to characterize as-cast and extruded microstructures of AZ Magnesium alloys thoroughly (Figure 19.10) [183].

19.3.1.2 Optimized Joining Processes for Automotive Applications

Joining processes generally produce residual stress fields due to thermal shrinkage of the solidifying and/or strongly plasticized material in the weld area. Beyond neutron and X-ray diffraction [184], only a very few alternative approaches are available to determine these stresses, such as the full-field, experimental, optical technique of high sensitivity Moire interferometry [185]. Residual stress distributions have been determined, for example, in spot welds, which are generally used to join mild steel sheets [186]. Experimental and numerical studies were conducted to characterize laser and resistance spot welds to gain an understanding of load carrying capacity, temperature distributions, and residual stress states of different joint geometries. Different laser spot weld path geometries are compared with conventional resistance spot welds to find the residual stress distributions in each. Simulations showed good agreement with experimental temperature distributions, and were able to qualitatively predict the residual stress distributions in each of the weld geometries [186]. The determination of residual stresses in laser welds of Mg alloys was performed using the neutron diffraction technique. With this approach it was possible to determine the complete stress tensor field (in-depth and lateral) in the welded joint, the HAZ, and the base metal without matter removal [187].

In lightweight structures the joints are often exposed to fatigue loading, which creates stresses superimposed to the residual stress distribution. Recently, the microstructure and properties of a Mg AZ31B laser beam weld without filler have been studied using electron microscopy, X-ray diffraction, and mechanical tests. Microstructure, hardness, and residual stresses of the laser beam overlap welds between AZ31B sheets and AZ31, AZ61, and AZ80 extruded profiles were investigated using microscopy and X-ray diffraction. The results of the investigations revealed that weld microstructure, the size of the HAZ, precipitate density, and the maximum compressive residual stress values strongly depend on the Al content of the weld zone of two unequal Mg-alloys [188, 189].

In the past, the powerful techniques provided by neutron and synchrotron radiation were mainly used for fundamental investigations. Nowadays, materials scientists are rapidly exploiting the potential of neutrons and synchrotron X-rays in order to simulate engineering processes in environments close to industrial manufacturing reality. One example is the direct observation of a solidification process under pressure, simulating a continuous casting process. Another case is the *in situ* monitoring of the residual stresses and microstructures evolving at the front of the welding tool during FSW. It is envisaged that the approach of direct observation of simulated engineering processes will play an important role in future materials engineering. It provides a means to understand the materials behavior but beyond that also the process itself. Optimum processing conditions could be determined that result in better service performance of the final product.

19.3.2 The Automotive Power Train and the Propulsion System of Light-Duty Cars

As mentioned above, oil dependence and limitations of accepted CO_2 emissions may lead to a fundamental reshaping of the light-duty vehicle (LDV) power train during the next decades [190, 191]. Estimates on possible maximum reduction of CO_2 emissions from gasoline and diesel engines are 40% and 30%, respectively [190]. Most of this progress of optimizing the conventional internal combustion engine (ICE) powertrain is expected from downsizing, reduced friction, improved direct fuel or diesel injection, and by mild hybridization [192]. To reduce the CO_2 emissions further, the complete carbon cycle has to be considered. Biofuels and hydrogen are options to replace gasoline and diesel as fuels. Biofuels however, suffer from their competition with land use for food production and a CO_2 balance which is not as favorable as initially thought [193]. Full replacement of fossil fuels by hydrogen would require a huge number of sites for sustainable hydrogen production and a complex infrastructure for distribution and fuel stations. Full electric vehicles (FEVs) or plug-in hybrid electric vehicles (PHEVs) could achieve a 50% reduction in CO_2 emissions on paper, due to the highly efficient production of electricity in power plants. Recently cars with full hybrid power trains, combining ICE power train and a full electric power train with battery have entered the market. Due to their unattractive cost–benefit ratio, full hybrids may not be competitive on a longer term [192]. In contrast, FEVs with a driving range of more than 400 km or PHEVs with an electric driving range beyond 25 km would be very attractive. Unfortunately, even next generation Li-Ion batteries are expected to have an energy density of less than 400 $Wh\,l^{-1}$, a small value compared to gasoline at more than 9000 $Wh\,l^{-1}$. Much research is needed to achieve energy densities providing a driving range of 400 km with a battery weight less than 100 kg. In the next decades, PHEVs with additional downscaled ICE powered generators or fuel cells powered by stored hydrogen may be required to recharge the batteries during driving [192, 194]. In the following section, the application of neutrons and synchrotron radiation to measurements of residual stresses in automotive power train components and questions of wear and lubrication will be described. Some light will be shed on polymer membranes for fuel cells (PMFCs) and nanocrystalline metal hydrides for hydrogen storage.

19.3.2.1 Residual Stresses in Components

Automotive gears are mechanical components for which a hardening of the surface of the gear teeth is required to improve their wear behavior. This is generally obtained by applying specific thermal or mechanical treatments to the parts. However, these treatments usually also produce high residual stress fields, which in turn influence significantly the fatigue life of the gears. It is, therefore, particularly important to have a map of the residual stress fields in the component.

In an early work, neutron diffraction measurements were used for determination of residual stress on a UNI55Cr3 steel crown gear after austenitizing and tempering treatments and compared the results with X-ray measurements [195]. Results from neutron diffraction measurements of the triaxial stresses field in two steel automotive gears have been reported [196]. The strains were measured along the three principal directions at different depths from the surface of the throat between two adjacent teeth. These in depth results were compared to the very near surface stresses, determined by X-ray diffraction. Neutron scattering was also used to investigate the residual stress in two steel gears produced from extruded bars cut, machined, and then submitted to tempering and nitriding [197].

Nitriding is generating superficial stresses, influenced by nitriding conditions, composition of steel, and geometry of the part. The effect of shape on the residual stress profile obtained after nitriding on a gear tooth was determined using neutron diffraction techniques [198]. Indeed, the result of the diffusion process is different for a plane geometry and for a curvilinear one. Between two teeth of a gear, the determination of in-depth stress gradients by classical X-ray diffraction cannot be done with great accuracy. The spatial resolution is not sufficient because the irradiated area has the same dimension as the surface curvature of the component. Furthermore, it is very difficult to take into account the removal of matter that is required to determine in-depth stress profiles because of the particular geometry. The synchrotron diffraction technique is a well-adapted method to determine such stress gradients in strongly absorbing materials due to the capability of in-depth penetration of high energy X-rays [199]. The removal of matter is not required and it is possible to determine an in-depth map of the stress tensor. Another example of the beneficial use of neutron and synchrotron research methods for the power train is the crankshaft. The pistons moving linearly in the cylinder liners back and forth are connected by piston rods to the crankshaft, which translates the linear piston motion into rotation. The crankshafts require high hardness at the bearing surface for improved wear resistance, a high toughness at the center, and compressive residual stresses at the subsurface layer for improved fatigue behavior. In order to optimize surface hardness and the residual stress state, thermal and thermomechanical processes for transition hardening as well as mechanical processes for deformation hardening are employed [200]. The relationship of all the parameters that influence the residual stresses is highly complex, thus to cover and control the influencing effects of the hardening process requires numerical simulation. Neutron diffraction has been successfully used to determine the compressive stress at the surface and the transition to residual tensile stresses in the inner parts of the crankshaft [201]. Besides crankshafts, a number of other automotive parts were investigated in a similar way, such as impellers, pistons, and cylinder heads with equipment designed for high neutron flux and, therefore, high penetration depths [202]. A new X-ray stress measuring technique, namely the pseudo-ψ angle-changing method, has been developed to measure the circumferential stress on a round corner located at the bottom of a deep groove such as the tooth root of a gear or the cold-rolled fillet of a crankshaft. This is a multiple-point measurement technique applicable to a curved surface having a uniform stress state. An X-ray path is taken in the plane that does not intersect the sidewalls near the corner. Instead of ψ rotation, the irradiation position is changed along the circumference of the corner with the specimen's attitude against the X-ray path unchanged; diffraction angles at $\psi = 0°$ are measured for different positions of the corner. The position change causes a shift in the diffraction peak since the irradiated area forming a part of the round corner has a different angle of inclination to the incident X-rays. The circumferential stress on the corner is obtained from the relation between the irradiation position and the diffraction angle at $\psi = 0°$ [201].

Other applications, where residual stresses play an important role in automotive applications in the car body, are cold-coiled helical compression springs for automotive suspensions and cast iron brake disc rotors. Measurements using neutron diffraction for such components have been reported [203, 204].

19.3.2.2 Wear and Lubrication

The synthesis and study of oil-soluble metal carbonate colloids are of interest in the area of lubricant additives. These surfactant-stabilized nanoparticles are important components in marine and automotive engine oils. Recently introduced, environmentally driven legislation has focused on lowering of gaseous emissions by placing limits on the levels of phosphorous, sulfur, and ash allowed in engine oil systems. These chemical limits, coupled with improved engine performance and extended oil drainage intervals, have led to renewed interest in the production of stable, efficient nanodetergent systems. This has resulted in modifications of existing surfactant structures and the development of new generations of surfactants. Hudson and co-workers prepared a review on the state of research in the area of nanodetergents [205]. Synchrotron-based near-edge X-ray absorption fine structure (NEXAFS) has been used to investigate the chemical bonding mechanism of the Ti additive with the metal surface that affects the wear improvement mechanism. It is postulated that Ti provides antiwear enhancement through inclusion in the metal/metal oxide structure of the ferrous surface by forming $FeTiO_3$ [206].

19.3.2.3 Polymeric Membranes for Fuel Cells (PEMFCs)

Low- to medium-temperature polymeric fuels cells are one of the intensively discussed components of future propulsion systems [207–211]. The key component of a polymeric membranes for fuel cell (PEMFC) is the proton-conducting membrane that separates the cathodic and anodic compartments [212, 213]. Almost all current industrial PEMFC prototypes are built with Nafion™, a perfluorosulfonated polymer membrane developed in the 1960s and 1970s for the chlor-alkaline industry [214, 215]. Despite its impressive properties, Nafion™ and similar perfluorinated polymers are expensive to produce, considerably increasing the cost of fuel cells. Furthermore, their performance drastically drops at temperatures above 80 °C, limiting the operating temperature, and hence the efficiency, of the fuel cells [216, 217]. A precise knowledge of the microstructure of these polymers is essential both to understand the effect of operating conditions on the membrane, and to guide research on alternative materials [218–220] with the aim to both lower cost and enhance performance in fuel cell operating conditions.

X-ray and neutron scattering studies on Nafion™, and other ion-conducting polymers have provided extensive information on the complex microstructure, complementing the results of spectroscopic techniques and supplying quantitative data necessary for structural models [221–223]. The most salient feature of the scattering curves of these polymers is a broad maximum, the "ionomer peak" (Figure 19.11), generally attributed to the tendency of the ionic groups to form organized structures [224–231]. The distance between ionic groups along the polymer chain, the chain rigidity, and its steric hindrance determine both the dimension and shape of the organized structures. Structural models are then devised to relate the width, position, and intensity of the ionomer peak to the structural order, and its evolution with water content and stress in the membrane [232–234]. The proposed structures can be further correlated to membrane water percolation behavior, and consequently to the proton conductivity [235–238]. Ionomers

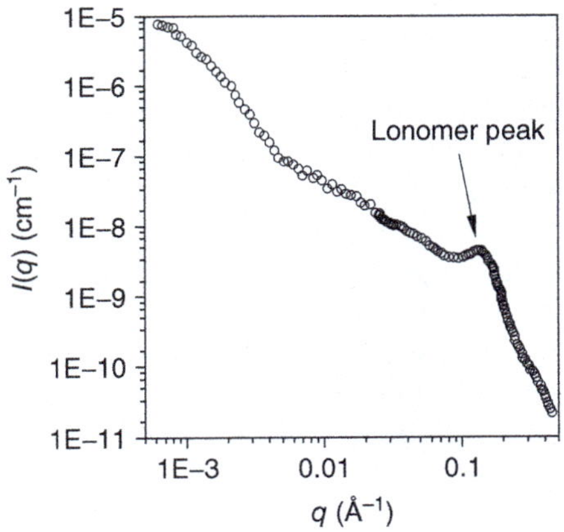

Figure 19.11 SAXS curve obtained for swollen Nafion. (Kaysser 2011 [281], reproduced with the permission of Wiley.)

with statistical distribution of ionic groups do not present a marked ionomer peak, but their composites with inorganic particles can be probed by anomalous scattering to determine average sizes and distribution of the particle aggregates [239–242]. The insights obtained over more than three decades of neutron and X-ray structural characterization of ionic polymers provide valuable direction to current fuel cell membrane development efforts.

Neutron and X-ray techniques are as well particularly powerful for characterization of other fuel cell components: radiography of H_2O and CO_2 of the active layers, tomography of the gas diffusion layers, and anomalous scattering of the catalyst layers. Examples of the structural information made available by these techniques are provided by Manke and co-workers [243].

19.3.2.4 Nanocrystalline Metal Hydrides for Hydrogen Storage

On-board storage of hydrogen is still one of the biggest challenges for hydrogen vehicles [244, 245]. Metal hydrides offer a safe alternative to hydrogen storage in pressurized tanks due to their long-term stability and low as well as constant pressures. In addition, they have a high volumetric energy density, which is up to 60% higher than that of liquid hydrogen. The main drawback of conventional metal hydrides is their poor gravimetric hydrogen storage density of about 2 m%, while modern high pressure composite 700 bar tank vessels have storage densities of 3.5 m% (1.2 kWh kg^{-1}) up to 5.5 m% (1.8 kWh kg^{-1}) [246]. Accordingly, the research on metal hydrides focuses on the exploitation and development of novel metal hydrides and especially complex hydrides with storage capacities higher than 6 m%. Most of these high-capacity hydrides are rather stable resulting in high working temperatures and large amounts of reaction heat to be dealt with [247]. Novel concepts and approaches have to be explored to favorably modify and tailor the enthalpy of the solid–hydrogen reaction [246, 248]. This will include the development of rather complex advanced systems like the reactive hydride composites [249–251]. These consist of a combination of two or more high-capacity hydrides, which react with each other during decomposition in an exothermic reaction. Thereby, the total value of reaction enthalpy required for the endothermic hydrogen release is lowered significantly. However, hydrogen absorption as well as desorption in such systems is a rather complex process involving several reaction steps and the formation of different intermediates and final product phases. A broad variety of diverse reaction pathways can be designed by changing stoichiometry, the type and amount of additives, and the applied pressures and temperatures. Understanding of the reaction mechanisms is essential to control the reaction pathway and to optimize the corresponding reaction kinetics. Synchrotron and neutron radiation studies [252] as well as neutron tomography and radiography have been employed for the investigation of the hydrogen uptake and release processes in the storage materials.

In situ synchrotron diffraction measurements performed during hydrogenation and dehydrogenation facilitate a detailed analysis of the reaction pathways [252–254]. Using *in situ* synchrotron XRD, the occurring phase transformations can be correlated with the corresponding hydrogen release steps by monitoring simultaneous pressure changes (Figure 19.12). These measurements gave a proof of concept, showing that the decomposition of $LiBH_4$ – which at this temperature is in the molten state and, therefore, not visible by XRD – and the formation of MgB_2 occur simultaneously [255]. Hence, the direct

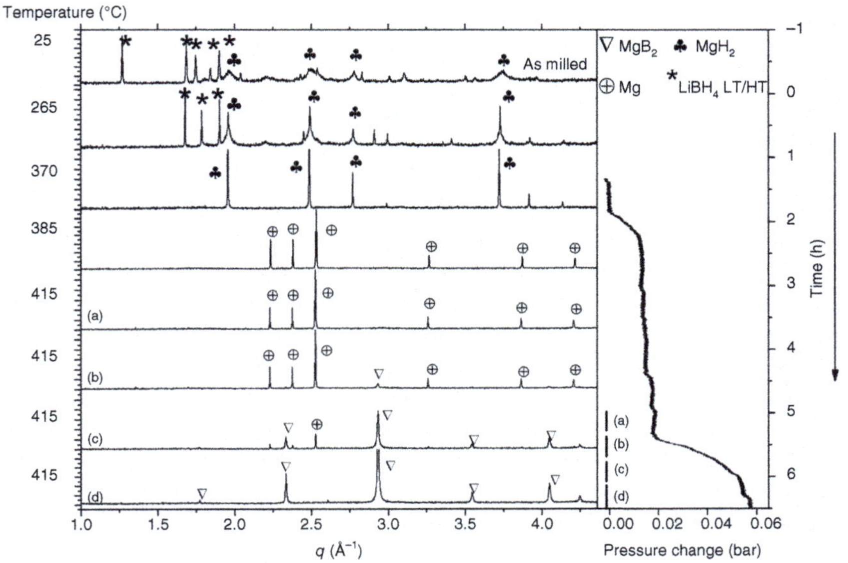

Figure 19.12 *In situ* XRD and pressure record of the desorption reaction of pure $LiBH_4$–MgH_2 composite under 5 bar hydrogen, measured at SNBL, ESRF. (Kaysser 2011 [281], reproduced with the permission of Wiley.)

formation of MgB_2 promotes the decomposition process of $LiBH_4$ under conditions, where the pure $LiBH_4$ itself would be stable.

Novel high-capacity lightweight metal hydrides and complex hydrides show rather sluggish reaction kinetics unless doped with suitable additives. These additives may have multiple different functionalities [256]. In novel complex metal hydrides, extended X-ray absorption fine structure (EXAFS) and X-ray absorption near-edge structure (XANES) measurements [257, 258], as well as anomalous small-angle scattering (ASAXS) [258, 259], have proven to be powerful tools to analyze the chemical state and distribution of additives. Based on these measurements, the function of the transition metal compound additives on improving sorption kinetics of reactive hydride composites could be explained. In addition, EXAFS [259] and *in situ* powder neutron diffraction (PND) [260] have been successfully applied to MgH_2 doped with Nb_2O_5. Previously, it was shown that up to a working temperature of 380 °C [261] a rather small amount of Nb_2O_5 [262] added to MgH_2 prior to ball-milling has a superior effect on reaction kinetics of MgH_2 [263], if compared to other additives. Using EXAFS and *in situ* PND, the catalytically active species was identified to be $Mg_xNb_yO_z$ [256, 264].

The measurement of EXAFS/XANES K-edge spectra also helps to identify the atomic surrounding of light atoms like boron or calcium [265] providing additional information on phases that cannot be detected by standard powder diffraction techniques. Photons (X-rays) are scattered by electrons, whereas neutrons are scattered by the nuclei and, therefore, PND can gather additional structural information like the exact position of the hydrogen atoms in the different hydrides [245]. Like Raman and infrared spectroscopy, inelastic neutron scattering (INS) is used to gather information about vibrational spectra. In comparison to the mentioned optical methods, INS has the advantage that it is sensitive to the vibrations of hydrogen atoms, whereas Raman and infrared are more sensitive to the motion of the heavier atoms. Furthermore, all vibrations contribute to INS, since it is not limited by optical selection rules. Using INS, hints for a so-far undetected volatile species in $NaAlH_4$ were found [266]. Together with Raman measurements and complementary computer modeling, the desorption mechanism could then be explained by a vacancy-mediated process [267].

Small-angle X-ray synchrotron scattering and (ultra) SANS have been employed to analyze grain and particle size as well as phase distributions in MgH_2 [268, 269]. Especially in the new field of nanoconfined hydrides [270], small-angle scattering offers useful information about both the pore-size distribution of the used nanoporous inert scaffold materials and the nanoconfined hydride particles [260].

In addition to these spectroscopic, diffraction and scattering methods, neutron computerized tomography (NCT) and neutron radiography are the only tools able to monitor hydrogen concentration profiles as well as to follow the hydrogen charging process of metal hydride hydrogen storage tanks. Pranzas and co-workers [268] describe NCT measurements of the hydrogen charging process of a $NaAlH_4$-based model tank. They were able to observe the formation of dendritic channels in the powder bed inside the storage tank caused by the applied hydrogen pressure of 80 bar. The use of neutron radiography gives additional information about the hydrogen distribution inside of the powder bed in the hydrogen storage tank [271]. On the basis of these results, metal hydrides and respective hydrogen storage tank designs are continuously optimized, yielding higher mass and higher volume-related capacities for the future zero-emission vehicles.

19.4 Other Transport Systems

Many of the neutron and synchrotron methods mentioned above are also used for other transport vehicles systems, such as ships, trains, or even bicycles, solving mostly same or similar materials and components questions as for the commercial passenger aircraft and the light duty vehicles. There are a number of phenomena, however, which are specific to particular transport systems. A prominent example is the stress and the degradation of railway tracks. Residual stresses and texture of rails have frequently been measured [272–278]. The increase of axle loads of the freight trains and the increase of speed of passenger trains within recent years – leading to integrated loads beyond 10 Mtons $year^{-1}$ – have led to an increase of wear of the rails that appeared, for example, in an increase in the frequency of structure modifications and defects in their surface layers [272]. Among structure modifications, the riffles and the white-etching layers are of practical as well as of theoretical interest due to their brittleness and due to the fact that their composition, crystallographic structure as well as their process of formation are still under vivid discussion. Investigations of these questions have been performed by a number of researchers [272, 279, 280].

All these examples demonstrate that neutron and synchrotron methods provide powerful tools for the basic understanding and systematic optimization of materials and components for aerospace, automotive, and other transportation systems.

References

1 Bugos, G.E. (2010) *The History of Aerospace Industry*, Economic History Services, http://eh.net/encyclopedia/the-history-of-the-aerospace-industry/ (accessed October 2016).
2 The European Commission (2001) European Aeronautics: A Vision for 2020.
3 Reed, J., Soble, J., Schaefer, D., Simon, B., and Jaggi, R. (2010) The Future of the Car. Financial Times Special Report (Oct. 4), http://media.ft.com/cms/082b8f82-cd16-11df-9bf0-00144feab49a.pdf (accessed June 2011).
4 Legget, D. (2010) Briefing Feature: What's the Prospect for China's Car Market, Just-Auto, http://www.just-auto.com/management-briefing/whats-in-prospect-for-chinas-car-market_id107052.aspx (accessed June 2011).
5 Polk, M. (2004) *J. Transp. Geogr.*, **12**, 185.
6 Worldometers (2010) http://www.worldometers.info/cars/ (accessed June 2011).
7 EUR-Lex (2007) EU Directive on Emission Standards 715/2007 of the European Parliament and the European Council, http://eur-lex.europa.eu/LexUriServe/LexUriServ.do?uri=OJ:L:2007:171:0001:0001:EN:PDF (accessed June 2011).
8 Dosch, H. and van de Voorde, M.H. (2009) *GENNESYS White Paper*, Max-Planck Institut für Metallforschung, Stuttgart.
9 Schobel, M., Altendorfer, W., Degischer, H.P., Vaucher, S., Buslaps, T., Di Michiel, M., and Hofmann, M. (2011) *Compos. Sci. Technol.*, **71**, 724.
10 Göbel, U., Lefranc, G., and Degischer, H.-P. (2003) *Materialwiss. Werkstofftech.*, **34**, 375.
11 Zschech, E., Huebner, R., Chumakov, D., Aubel, O., Friedrich, D., Guttmann, P., Heim, S., and Schneider, G. (2009) *J. Appl. Phys.*, **106**, 093711.
12 Nawani, P., Hsiao, B.S., Tsou, A.H., and Wenig, W.Q. (2010) *Polymer*, **51**, 5255.
13 Jinnai, H., Spontak, R.J., and Nishi, T. (2010) *Macromolecules*, **43**, 1675.
14 Weyland, M. (2001) Two and three dimensional nanoscale analysis: new techniques and applications. Dissertation. University of Cambridge, Cambridge.
15 Utess, D., Engelmann, H.J., and Zschech, E. (2008) Vortrag 42. Metallographie-Tagung, Jena 2008.
16 NASA (2008) N + 3, Three Generations Forward in Aviation Technology, www.aeronautics.nasa.gov/nra_awardees_10_06_08.htm (accessed June 2011).
17 Asian Aviation (2011) http://www.asianaviation.com/articles/97/Airbus-s-A350XWB-takes-wing (accessed June 2011).
18 Thomas, W.M., Nicholas, E.D., Needham, J.C., Murch, M.G., Temple-Smith, P., and Dawes, C.J. (1991) Improvements relating to friction welding. Patent GB 9125978; International Patent PCT/GB92/02203 (1991).
19 Thomas, W.M., Johnson, K.I., and Wiesner, C.S. (2003) *Adv. Eng. Mater.*, **5**, 485.
20 Woo, W., Feng, Z., Wang, X., Camden, K.A., Hubbard, R., David, S.A., and Choo, H. (2006) *Appl. Phys. Lett.*, **88**, 261903.
21 Martukanitz, R.P., Howell, P.R., Payzant, E.A., Spooner, S., and Hubbard, C.R. (1996) in *Nondestructive Evaluation and Materials Properties III* (eds P.K. Liaw, O. Buck, R.J. Arsenault, and R.E. Green), The Minerals, Metals and Materials Society, Warrendale, PA, p. 89.

22 Staron, P., Vaidya, W.V., and Kocak, M. (2009) *Mater. Sci. Eng., A*, **525**, 195.
23 Steuwer, A., Dumont, M., Altenkirch, J., Birosca, S., Deschamps, A., Prangnell, P.B., and Withers, P.J. (2011) *Acta Mater.*, **59**, 3002.
24 Lima, E.B.F., Wegener, J., Dalle Donne, C., Goerigk, G., Wroblewski, T., Buslaps, T., Pyzalla, A.R., and Reimers, W. (2003) *Z. Metallkd.*, **94**, 908.
25 Peel, M., Steuwer, A., Preuss, M., and Withers, P.J. (2003) *Acta Mater.*, **51**, 4791.
26 Woo, W., Feng, Z., Wang, X.-L., and David, S.A. (2011) *Sci. Technol. Weld. Joining*, **16**, 23.
27 Kraft, O. and Volkert, C.A. (2001) *Adv. Eng. Mater.*, **3**, 99.
28 Sano, S.H., Shono, T., Sonoda, H., Takatsu, T., Ciucchi, B., Carvalho, R., and Pashley, D.H. (1994) *Dent. Mater.*, **10**, 236.
29 Lee, Y., Kim, J.-Y., Lee, J.-S., Kim, K.-H., Koo, J.Y., and Kwon, D. (2006) *Philos. Mag.*, **86**, 5497.
30 Maass, R., Grolimund, D., van Petegem, S., Willimann, M., Jensen, M., van Swygenhoven, H., Lehnert, T., Gijs, M.A.M., Volkert, C.A., Lilleodden, E.T., and Schwaiger, R. (2006) *Appl. Phys. Lett.*, **88**, 151905.
31 van Swygenhoven, H., Schmitt, B., Derlet, P.M., van Petegem, S., Cervellino, A., Budrovic, Z., Brandstetter, S., Bollhalder, A., and Schild, M. (2006) *Rev. Sci. Instrum.*, **77**, 013902.
32 Holzapfel, C., Schäf, W., Marx, M., Vehoff, H., and Mücklich, F. (2007) *Scr. Mater.*, **56**, 697.
33 Huber, N. and Heerens, J. (2008) *Acta Mater.*, **56**, 6205.
34 Ice, G.E. and Pang, J.W.L. (2009) *Mater. Charact.*, **60**, 1191.
35 Winholtz, R.A. and Krawitz, A.D. (1996) in *Trends in Welding Research* (eds H.B. Smartt, J.A. Johnson, and S.A. David), ASM International, Materials Park, OH, p. 93.
36 Bruno, G., Carrado, A., Dunn, B., Fiori, F., Girardin, E., Pirling, T., and Rustichelli, F. (2000) *Mater. Sci. Forum*, **347**, 684.
37 Cho, J.R., Roberts, S.M., and Reed, R.C. (2003) *Proc. Soc. Photo-Opt. Instrum. Eng.*, **4831**, 192.
38 Owen, R.A., Preston, R.V., Withers, P.J., Shercliff, H.R., and Webster, P.J. (2003) *Mater. Sci. Eng., A*, **346**, 159.
39 Fiori, F., Albertini, G., Girardin, E., Giuliani, A., Manescu, A., and Rustichelli, F. (2004) *J. Alloys Compd.*, **382**, 39.
40 Bruno, G. and Dunn, B.D. (2004) *J. Pressure Vessel Technol.*, **126**, 284.
41 Bayraktar, F.S., Staron, P., Koçak, M., and Schreyer, A. (2006) *Mater. Sci. Forum*, **524**, 419.
42 Ganguly, S., Fitzpatrick, M.E., and Edwards, L. (2006) *Metall. Mater. Trans. A*, **37**, 411.
43 Staron, P., Vaidya, W.V., Koçak, M., Homeyer, J., and Hackius, J. (2006) *Mater. Sci. Forum*, **524**, 413.
44 Steuwer, A., Peel, M., and Buslaps, T. (2006) *Mater. Sci. Forum*, **524**, 267.
45 Ali, A., An, X., Rodopoulos, C.A., Brown, M.W., O'Hara, P., Levers, A., and Gardiner, S. (2007) *Int. J. Fatigue*, **29**, 1531.
46 Staron, P., Cihak, U., Clemens, H., Stockinger, M., Bayraktar, F.S., Koçak, M., and Schreyer, A. (2007) *Adv. Eng. Mater.*, **9**, 627.

47 James, M.N., Lombard, H., Hattingh, D.G., and Steuwer, A. (2007) *Key Eng. Mater.*, **348–349**, 561.

48 Bayraktar, F.S., Staron, P., Koçak, M., and Schreyer, A. (2008) *Mater. Sci. Forum*, **571–572**, 355.

49 Machold, W., Staron, P., Bayraktar, F., Riekehr, S., Koçak, M., and Schreyer, A. (2008) *Mater. Sci. Forum*, **571–572**, 375.

50 James, M.N., Hughes, D.J., Hattingh, D.G., Mills, G., and Webster, P.J. (2009) *Int. J. Fatigue*, **31**, 28.

51 Staron, P., Kocak, M., Williams, S., and Wescott, A. (2004) *Physica B*, **350**, e491.

52 Price, D.A., Williams, S.W., Wescott, A., Harrison, C.J.C., Rezai, A., Steuwer, A., Peel, M., Staron, P., and Kocak, M. (2007) *Sci. Technol. Weld. Joining*, **12**, 620.

53 Richards, D.G., Prangnell, P.B., Withers, P.J., Williams, S.W., Wescott, A., and Oliver, E.C. (2007) *Mater. Sci. Forum*, **539–543**, 4025.

54 Richards, D.G., Prangnell, P.B., Williams, S.W., and Withers, P.J. (2008) *Mater. Sci. Eng.*, **489**, 351.

55 Altenkirch, J., Steuwer, A., Peel, M.J., Withers, P.J., Williams, S.W., and Poad, M. (2008) *Metall. Mater. Trans. A*, **39**, 3246.

56 Altenkirch, J., Steuwer, A., Peel, M.J., Richards, D.G., and Withers, P.J. (2008) *Mater. Sci. Eng., A*, **488**, 16.

57 Altenkirch, J., Steuwer, A., Withers, P.J., Williams, S.W., Poad, M., and Wen, S.W. (2009) *Sci. Technol. Weld. Joining*, **14**, 185.

58 Richards, D.G., Prangneil, P.B., Withers, P.J., Williams, S.W., Wescot, A., and Oliver, E.C. (2006) *Mater. Sci. Forum*, **524–525**, 71.

59 Liljedahl, C.D.M., Tan, M.L., Zanellato, O., Ganguly, S., Fitzpatrick, M.E., and Edwards, L. (2008) *Eng. Fract. Mech.*, **75**, 3881.

60 Liljedahl, C.D.M., Zanellato, O., Fitzpatrick, M.E., Lin, J., and Edwards, L. (2010) *Int. J. Fatigue*, **32**, 735.

61 Servetti, G. and Zhang, X. (2009) *Eng. Fract. Mech.*, **76**, 1589.

62 Liljedahl, C.D.M., Fitzpatrick, M.E., and Edwards, L. (2009) *Mater. Sci. Eng., A*, **523**, 152.

63 Staron, P., Fischer, T., Lippmann, T., Stark, A., Daneshpour, S., Schnubel, D., Ulmann, E., Gerstenberger, R., Camin, B., Reimers, W., Eidenberger, E., Clemens, H., Huber, N., and Schreyer, A. (2011) *Adv. Eng. Mater.*, **13**, 658–663.

64 Toda, H., Sinclair, I., Buffiére, J.-Y., Maire, E., Khor, K.H., Gregson, P., and Kobayashi, T. (2004) *Acta Mater.*, **52**, 1305.

65 Khor, K.H., Buffiére, J.-Y., Ludwig, W., and Sinclair, I. (2006) *Scr. Mater.*, **55**, 47.

66 Ferrié, E., Buffiére, J.-Y., Ludwig, W., Gravouil, A., and Edwards, L. (2006) *Acta Mater.*, **54**, 1111.

67 Rahman, M., Fitzpatrick, M.E., Edwards, L., Pratihar, S., Peel, M., Steuwer, A., and Buslaps, T. (2008) *Mater. Sci. Forum*, **571–572**, 119.

68 Ferrié, E., Buffière, J.-Y., Ludwig, W., and Gravouil, A. (2008) *Mater. Sci. Forum*, **567–568**, 301.

69 Belnoue, J.P., Jun, T.-S., Hofmann, F., Abbey, B., and Korsunsky, A.M. (2010) *Eng. Fract. Mech.*, **77**, 3216.

70 James, M.N., Hughes, D.J., Hattingh, D.G., Bradley, G.R., Mills, G., and Webster, P.J. (2004) *Fatigue Fract. Eng. Mater. Struct.*, **27**, 187.

71 Reyna-Valencia, A., Deyrail, Y., and Bousmina, M. (2010) *Macromolecules*, **43**, 354.

72 Flores, A., Arribas, C., Fauth, F., Khariwala, D., Hiltner, A., Baer, E., Balta-Calleja, F., and Ania, F. (2010) *Polymer*, **51**, 4530.

73 Duan, Y.X., Thunga, M., Schlegel, R., Schneider, K., Rettler, E., Weidisch, R., Siesler, H.V., Stamm, M., Mays, J.W., and Hadjichristidis, N. (2009) *Macromolecules*, **42**, 4155.

74 Göschel, U., Deutscher, K., and Abetz, V. (1996) *Polymer*, **37**, 1.

75 Bartczak, Z. and Galeski, A. (2010) in *Layered Nanostructures – Polymers with Improved Properties* (eds G.H. Michler and S. Henning), Wiley-VCH Verlag GmbH, Weinheim, p. 67.

76 Wang, W., Murthy, N.S., Chae, H.G., and Kumar, S. (2009) *J. Polym. Sci., Part B: Polym. Phys.*, **47**, 2394.

77 Lorenz-Haas, C., Müller-Buschbaum, P., Ittner, T., Kraus, J., Mahltig, B., Cunis, S., Krosigk, G.V., Gehrke, R., Creton, C., and Stamm, M. (2003) *Phys. Chem. Chem. Phys.*, **5**, 1235.

78 Oberdisse, J., Hine, P., and Pyckhout-Hintzen, W. (2007) *Soft Matter*, **3**, 476.

79 Kusmin, A., Gruener, S., Henschel, A., de Souza, N., Richter, D., and Huber, P. (2010) *Macromolecules*, **43**, 8162.

80 Schubert, D.W., Abetz, V., Stamm, M., Hack, T., and Siol, W. (1995) *Macromolecules*, **28**, 2519.

81 Gutmann, J.S., Müller-Buschbaum, P., Wolkenhauer, M., Hermsdorf, N., and Stamm, M. (2004) *J. Macromol. Sci. Part B Phys.*, **43**, 207.

82 Schelten, J., Wignall, G.D., and Ballard, D.G.H. (1974) *Polymer*, **15**, 682.

83 Wu, L.F., Lodge, T.P., and Bates, F.S. (2006) *Macromolecules*, **39**, 294.

84 Guenet, J.M., Picot, C., and Benoit, H. (1979) *Macromolecules*, **12**, 86.

85 Nedoma, A.J., Robertson, M.L., Wanakule, N.S., and Balsra, N.P. (2008) *Macromolecules*, **41**, 5773.

86 White, R.P., Lipson, J.E.G., and Higgins, J.S. (2010) *Macromolecules*, **43**, 4287.

87 Zhang, R.Y., Cheng, H., Sun, T., Dong, X., and Han, C.C. (2008) *Macromolecules*, **41**, 6818.

88 Richter, D., Monkenbusch, M., Allgeier, J., Arbe, A., Colmenero, J., Farago, B., Cheol Bae, Y., and Faust, R. (1999) *J. Chem. Phys.*, **111**, 6107.

89 Ewen, B. and Richter, D. (1997) *Adv. Polym. Sci.*, **134**, 1.

90 Esslinger, J. (2009) *Landshuter Leichtbaukolloquium*, LC-Verlag, Landshut.

91 Williams, J.C. and Strake, E.A. (2003) *Acta Mater.*, **51**, 5775.

92 Esslinger, J. (2011) 10th Conference Gefüge und Bruch, Bochum, Germany, March/April, 2011.

93 Kumpfert, J., Peters, M., and Kaysser, W.A. (1998) NATO-RTO Symposium, Toulouse, France, May, 1998, NATO-RTO Report, p. 42-1.

94 Schmücker, M. and Schneider, H. (2004) in *Handbook of Ceramic Composites* (ed. N. Bansal), Kluwer Academic Publishers, Dodrecht, p. 423.

95 Dilba, D. (2010) MTU Report 1/2010, Munich, Germany.

96 Peters, M., Leyens, C., Schulz, U., and Kaysser, W.A. (2001) *Adv. Eng. Mater.*, **4**, 193.

97 Maus, S., Hapke, J., Ranong, C.N., Wüchner, E., Frielmeier, G., and Wenger, D. (2008) *Int. J. Hydrogen Energy*, **33**, 4612.

98 Reed, R. (2008) *The Superalloys: Fundamentals and Applications*, Cambridge University Press, Cambridge.

99 Ross, E.W. and Sims, C.T. (1987) in *Superalloys II* (eds C.T. Sims, N.S. Stoloff, and W.C. Hagel), John Wiley & Sons, Inc., New York, p. 97.
100 Seth, B.B. (2000) in *Proceedings of the 9th International Conference on Superalloys – Superalloys 2000* (ed. T.M. Pollock), The Minerals, Metals and Materials Society, Warrendale, PA, p. 3.
101 Kim, Y.-W. (1989) *J. Met.*, **41**, 24.
102 Kim, Y.-W. and Dimiduk, D.M. (1991) *J. Miner. Met. Mater. Soc.*, **43**, 40.
103 Kim, Y.-W. (1994) *J. Miner. Met. Mater. Soc.*, **46**, 30.
104 Appel, F., Oehring, M., and Wagner, R. (2000) *Intermetallics*, **8**, 1283.
105 Appel, F. and Wagner, R. (1998) *Mater. Sci. Eng.*, **22**, 187.
106 Hofmann, U. and Blum, W. (1999) *Intermetallics*, **7**, 351.
107 Mughrabi, H. (1996) in *The Johannes Wertmann Symposium* (ed. R.J. Arsenault), The Minerals, Metals and Materials Society, Warrendale, PA, p. 267.
108 Nabarro, F.R.N. (1996) *Metall. Mater. Trans. A*, **27**, 513.
109 Großmann, V., Biermann, B., Tetzlaff, H., Pyczak, F., and Mughrabi, H. (2000) *Scr. Mater.*, **43**, 859.
110 Kuhn, H.-A., Biermann, H., Ungár, T., and Mughrabi, H. (1991) *Acta Metall. Mater.*, **39**, 2783.
111 Guo, F., Ji, V., Francois, M., and Zhang, Y.G. (2003) *Acta Mater.*, **51**, 5349.
112 Bellet, D. and Bastie, P. (1991) *Philos. Mag.*, **64**, 135.
113 Glatzel, U. and Müller, A. (1994) *Scr. Metall. Mater.*, **31**, 285.
114 Glatzel, U. (1994) *Scr. Metall. Mater.*, **31**, 291.
115 Tiley, J., Srinivasan, R., Banerjee, R., Viswanathan, G.B., Toby, B., and Fraser, H.L. (2009) *Mater. Sci. Technol.*, **25**, 1369.
116 Biermann, H., von Grossmann, B., Mechsner, S., Mughrabi, H., Ungár, T., Snigirev, A., Snigireva, I., Souvorov, A., Kocsis, M., and Raven, C. (1997) *Scr. Mater.*, **37**, 1309.
117 Biermann, H., von Grossman, B., Ungár, T., Mechsner, S., Souvorov, A., Drakopolous, M., Snigirev, A., and Mughrabi, H. (2000) *Acta Mater.*, **48**, 2221.
118 Riemer, M., Jentsch, H.G., Biermann, H., and Mughrabi, H. (1999) *Intermetallics*, **7**, 241.
119 Guo, F.A., Ji, V., Zhang, Y.G., and Chen, C.Q. (2001) *Mater. Sci. Eng., A*, **315**, 195.
120 Guo, F.A., Ji, V., Francois, M., and Zhang, Y.G. (2003) *Mater. Sci. Eng., A*, **341**, 182.
121 Liss, K.D., Schmoelzer, T., Yan, K., Reid, M., Peel, M., Dippenaar, R., and Clemens, H. (2009) *J. Appl. Phys.*, **106**, 113526.
122 Ratel, N., Bruno, G., and Deme, B. (2005) *J. Phys. Condens. Matter*, **17**, 7061.
123 del Genovese, D., Strunz, P., Mukherjee, D., Gilles, R., and Rösler, J. (2005) *Metall. Mater. Trans. A*, **36**, 3439.
124 Zrnik, J., Strunz, P., Vrchovinsky, V., Muransky, O., Novy, Z., and Wiedenmann, A. (2004) *Mater. Sci. Eng., A*, **387–389**, 728.
125 Zickler, G.A., Schnitzer, R., Radis, R., Hochfellner, R., Schweins, R., Stockinger, M., and Leitner, H. (2009) *Mater. Sci. Eng., A*, **523**, 295.
126 Kelekanjeri, V., Moss, L.K., Gerhardt, R.A., and Ilavsky, J. (2009) *Acta Mater.*, **57**, 4658.
127 Dupke, R. and Reimers, W. (1995) *Z. Metallkd.*, **86**, 665.
128 Zrník, J., Strunz, P., Maldini, M., Wiedenmann, A., and Davydov, V. (2008) *J. Phys. Condens. Matter*, **20**, 104261.

129 Novoselova, T., Malinov, S., Sha, W., and Zhecheva, A. (2004) *Mater. Sci. Eng., A*, **371**, 103.

130 Brückner, U., Epishin, A., Link, T., and Dressel, K. (1998) *Mater. Sci. Eng., A*, **247**, 23.

131 Schmoelzer, T., Liss, K.-D., Staron, P., Mayer, S., and Clemens, H. (2011) *Adv. Eng. Mater.*, **13**, 685–699.

132 Stark, A., Oehring, M., Pyczak, F., and Schreyer, A. (2011) *Adv. Eng. Mater.*, **13**, 700–704.

133 Liss, K.D., Bartels, A., Clemens, H., Bystrzanowski, S., Stark, A., Buslaps, T., Schimansky, F.-P., Gerling, R., Scheu, C., and Schreyer, A. (2006) *Acta Mater.*, **54**, 3721.

134 Schmoelzer, T., Liss, K.D., Zickler, G.A., Watson, I.J., Droessler, L.M., Wallgram, W., Buslaps, T., Studer, T., and Clemens, H. (2010) *Intermetallics*, **218**, 1544.

135 Stark, A., Schwaighofer, E., Mayer, S., Clemens, H., Lippmann, T., Lottermoser, L., Schreyer, A., and Pyczak, F. (2013) *MRS Online Proc. Lib.*, **1516**. doi: 10.1557/opl.2012.1577

136 Stark, A., Oehring, M., Pyczak, F., Lippmann, T., Lottermoser, L., and Schreyer, A. (2014) *Mater. Sci. Forum*, **772**, 85.

137 Schmoelzer, T., Liss, K.-D., Kirchlechner, C., Mayer, S., Stark, A., Peel, M., and Clemens, H. (2013) *Intermetallics*, **39**, 25.

138 Schloffer, M., Rashkova, B., Schöberl, T., Schwaighofer, E., Zhang, Z., Clemens, H., and Mayer, S. (2014) *Acta Mater.*, **64**, 241.

139 Schwaighofer, E., Clemens, H., Mayer, S., Lindemann, J., Klose, J., Smarsly, W., and Güther, V. (2014) *Intermetallics*, **44**, 128.

140 Oehring, M., Stark, A., Paul, J.H.D., and Lippmann, T. (2013) *Intermetallics*, **32**, 12.

141 Schwaighofer, E., Rashkova, B., Clemens, H., Stark, A., and Mayer, S. (2014) *Intermetallics*, **46**, 173.

142 Peters, M., Fritscher, K., Staniek, G., Kaysser, W.A., and Schulz, U. (1997) *Materialwiss. Werkstofftech.*, **28**, 357.

143 Sheffler, D. and Gupta, D.K. (1988) *J. Eng. Gas Turbines Power*, **110**, 605.

144 Rhys-Jones, T.N. and Toriz, F.C. (1989) *High Temp. Technol.*, **7**, 73.

145 Schulz, U. and Schmücker, M. (2000) *Mater. Sci. Eng., A*, **276**, 1.

146 Renteria, F.A. (2007) A small-angle scattering analysis of the influence of manufacture and thermal induced morphological changes on the thermal conductivity of EB-PVD PYSZ Thermal Barrier Coatings. Thesis. RWTH Aachen.

147 Renteria, A.F., Saruhan, B., Ilavsky, J., and Allen, A.J. (2007) *Surf. Coat. Technol.*, **201**, 4781.

148 Renteria, A.F., Saruhan, B., Schulz, U., Raetzer-Scheibe, H.J., Haug, J., and Wiedenmann, A. (2006) *Surf. Coat. Technol.*, **201**, 2611.

149 Yang, Y.Q., Leyens, C., Schulz, U., Dudek, H.J., and Kaysser, W.A. (2000) Presentation for Materials Week 2000, Munich.

150 Leyens, C., Schulz, U., Braue, W., and Yang, Y.Q. (2001) in *High Temperature Corrosion and Materials Chemistry III* (eds M. McNallan and E. Opila), The Electrochemical Society, Pennington, NJ, p. 12.

151 Leyens, C., Schulz, U., Pint, B.A., and Wright, I.G. (1999) *Surf. Coat. Technol.*, **120/121**, 68.

152 Haynes, J.A., Ferber, M.K., Porter, W.D., and Rigney, E.D. (1999) *Oxid. Met.*, **52**, 31.

153 Tawancy, M., Sridhar, N., and Abbas, N.M. (1998) *J. Mater. Sci.*, **33**, 681.

154 Schmackers, T. and Reimers, W. (1999) in *Cyclic Oxidation of High Temperature Materials*, EFC Workshop No. 27, Frankfurt (eds M. Schütze and W.J. Quadakkers), Maney Publishing, p. 357.

155 Fiori, F., Girardin, E., Giuliani, A., Manescu, A., and Rustichelli, F. (2004) in *Metallic Materials with High Structural Efficiency*, NATO Sciences Series, Series II: Mathematics, Physics and Chemistry, vol. **146** (eds O.N. Senkov, D.B. Miracle, and S.A. Firstov), NATO, p. 425.

156 Jarfors, A.E.W., Kainer, K.U., Tan, M.-J., and Yong, J. (2009) *Cosmos*, **5**, 23.

157 Goede, M. (2009) *Proceedings of the International Conference Innovative Developments for Lightweight Vehicle Structures, Wolfsburg, May 26–27, 2009*, Volkswagen AG, Wolfsburg, p. 25.

158 Yi, S., Davies, C.H.J., Brokmeier, H.G., Bolmaro, R.E., Kainer, K.U., and Homeyer, J. (2006) *Acta Mater.*, **54**, 549.

159 Agnew, S.R., Tome, C.N., Brown, D.W., Holden, T.M., and Vogel, S.C. (2003) *Scr. Mater.*, **48**, 1003.

160 Davies, C.H.J., Yi, S., Bohlen, J., Kainer, K.U., Brokmeier, H.G., and Homeyer, J. (2005) *Mater. Sci. Forum*, **495-497**, 1633.

161 Schmidt, S., Nielsen, S.F., Gundlach, C., Margulies, L., Huang, X., and Jensen, D.J. (2004) *Science*, **305**, 229.

162 King, A., Johnson, G., Engelberg, D., Ludwig, W., and Marrow, J. (2008) *Science*, **321**, 382.

163 Mathiesen, R., Arnberg, L., Ramsoskar, K., Weitkamp, T., Rau, C., and Snigirev, A. (2002) *Metall. Mater. Trans. B*, **33**, 613.

164 Schenk, T., Thi, H.N., Gastaldi, J., Reinhart, G., Cristiglio, V., Mangelinck-Noel, N., Klein, H., Hartwig, J., Grushko, B., Billia, B., and Baruchel, J. (2005) *J. Cryst. Growth*, **275**, 201.

165 Gonzales, F. (2008) Etude des morphologies de solidification dans les alliages aluminium-zinc. PhD thesis 4260. Ecole Polytechnique Federale de Lausanne, Switzerland.

166 Limodin, N., Salvo, L., Boller, E., Suery, M., Felberbaum, M., Gailliegue, S., and Madi, K. (2009) *Acta Mater.*, **57**, 2300.

167 Mathiesen, R.H. and Arnberg, L. (2005) *Mater. Sci. Eng.*, **413–414**, 283.

168 Ruvalcalba, D., Mathiesen, R.D., Eskin, D.G., Arnberg, L., and Katgerman, L. (2007) *Acta Mater.*, **55**, 4287.

169 Baruchel, J., Buffiere, J.-Y., Cloetens, P., DiMichel, M., Ferrie, E., Ludwig, W., Maire, E., and Salvo, L. (2006) *Scr. Mater.*, **55**, 41.

170 Salgado-Ordorica, M.-A. (2009) Characterization and modeling of twinned dendrite growth. PhD thesis 4568. Ecole Polytechnique Federale de Lausanne, Switzerland.

171 Iqbal, N., van Dijk, N.H., Offerman, S.E., Moret, M.P., Katgerman, L., and Kearley, C.J. (2005) *Acta Mater.*, **53**, 2875.

172 Kasprzak, W., Sediako, D., Sahoo, M., Walker, M., and Swainson, I. (2010) *Supplemental Proceedings: Material Processing and Properties*, TMS, Warrendale, PA, p. 93.

173 Choi, K.S., Liu, W.N., Sun, X., Khaleel, M.A., Ren, Y., and Wang, Y.D. (2008) *Metall. Mater. Trans. A*, **39**, 3089.

174 Lee, W., Chung, K.H., Kim, D., Kim, J., Kim, C., Okamoto, K., Wagoner, R.H., and Chung, K. (2009) *Int. J. Plast.*, **25**, 1626.

175 Sediako, D. and Shook, S. (2009) in *Magnesium Technology 2009* (eds E.A. Nyberg, S.R. Agnew, N.R. Neelameggham, and M.O. Pekguleryuz), TMS, Warrendale, PA, p. 255.

176 Daymond, M.R. (1997) *Appl. Compos. Mater.*, **4**, 375.

177 Garces, G., Bruno, G., and Wanner, A. (2006) *Scr. Mater.*, **55**, 163.

178 Garces, G., Onorbe, E., Perez, P., Denks, I.A., and Adeva, P. (2009) *Mater. Sci. Eng., A*, **523**, 21.

179 Asghar, Z., Requena, G., Degischer, H.P., and Cloetans, P. (2009) *Acta Mater.*, **57**, 4125.

180 Maire, E., Grenier, J.C., Daniel, D., Baldacci, A., Klöcker, H., and Bigot, A. (2006) *Scr. Mater.*, **55**, 123.

181 García-Moreno, F., Rack, A., Helfen, L., Baumbach, T., Zabler, S., Babcsán, N., Banhart, J., Martin, T., Ponchut, C., and Di Michiel, M. (2008) *Appl. Phys. Lett.*, **92**, 134104.

182 Requena, G., Cloetens, P., Altendorfer, W., Poletti, C., Tolnai, D., Warchomicka, F., and Degischer, H.P. (2009) *Scr. Mater.*, **61**, 760.

183 Witte, F., Fischer, J., Beckmann, F., Störmer, M., and Hort, N. (2008) *Scr. Mater.*, **58**, 453.

184 Albertini, G., Bruno, G., Carrado, A., Fiori, F., Rogante, M., and Rustichelli, F. (1999) *Meas. Sci. Technol.*, **10**, R56.

185 Khanna, S.K., He, C., and Agrawal, H.N. (2001) *Trans. ASME*, **123**, 132.

186 Martinson, P., Daneshpour, S., Koçak, M., Riekehr, S., and Staron, P. (2009) *Mater. Des.*, **30**, 3351.

187 Barrallier, L., Fabre, A., Masse, J.E., and Ceretti, M. (2002) *Mater. Sci. Forum*, **404**, 399.

188 Coelho, R.S., Kostka, A., Pinto, H., Riekehr, S., Kocak, M., and Pyzalla, A.R. (2008) *Mater. Sci. Eng., A*, **485**, 20.

189 Coelho, R.S., Kostka, A., Pinto, H., Riekehr, S., Koçak, M., and Pyzalla, A.R. (2008) *Mater. Sci. Forum*, **571-572**, 361.

190 Valentine-Urbschat, M. and Bernhart, W. (2007) *Powertrain 2020 – The Future Drives Electric*, Roland Berger Strategy Consultants, http://www.rolandberger.com/expertise/publications/2007-07-20-rbsc-pub-Powertrain_Challenge.html and http://www.rolandberger.com/media/pdf/Roland_Berger_Powertrain_Voice_of_customer_20100504.pdf (accessed 30 June 2016).

191 Kirchsteiger, C. (2009) *Modelling and Cosimulation of an Automotive Power Management System for Hybrid-Electric Power Train Vehicles*, VDM Verlag, Saarbrücken.

192 Pischinger, S. (2005) Future Power Technology… Many Options, Even More Unknowns, SAE International, Automotive Engineering International Online (AEI), May 2005, http://www.sae.org/automag/features/futurelook/05-2005/1-113-5-86.pdf (accessed 30 June 2016).

193 Delucci, M.A. (2010) in *The Year of Ecology and Conservation Biology 2010*, vol. **1195** (ed. R.K. Stonjek), Annals of the New York Academy of Sciences, p. 28. doi: 10.1111/j.1749-6632.2010.05457.x

194 Fenton, J. (2005) *Handbook of Automotive Powertrain and Chassis Design*, Professional Engineering Publishing, London.

195 Albertini, G., Bruno, G., Fiori, F., Girardin, E., Giuliani, A., Quadrini, E., and Romani, F. (2000) *Physica B*, **276**, 925.

196 Ceretti, M., Magli, E., and Vangi, D. (2000) *Mater. Sci. Forum*, **321-323**, 732.

197 Annibali, G., Bruno, G., Fiori, F., Giuliani, A., Manescu, A., Marcantoni, M., and Turquier, F. (2002) *Appl. Phys. A*, **74**, 1698.

198 Fabre, A. and Barrallier, L. (2005) *Mater. Sci. Forum*, **490-491**, 251.

199 Goret, V., Fabre, A., Barrallier, L., and Vardon, P. (2006) *Mater. Sci. Forum*, **524-525**, 285.

200 Mayer, H.M., Pyzalla, A., Reimers, W., and Achmus, C. (2000) *Mater. Sci. Forum*, **347–349**, 340.

201 Oguri, T., Murata, K., and Yamaguchi, K. (2008) *J. Strain Anal. Eng. Des.*, **43**, 395.

202 Poeste, T., Wimpory, R.C., and Schneider, R. (2006) *Mater. Sci. Forum*, **524–525**, 223.

203 Matejicek, J., Brand, P.C., Drews, A.R., Krause, A., and Lowe-Ma, C. (2004) *Mater. Sci. Eng., A*, **367**, 306.

204 Ripley, M.I. and Kirstein, O. (2005) *Physica B*, **385**, 604.

205 Hudson, L.K., Eastoe, J., and Dowding, P.J. (2006) *Adv. Colloid Interface Sci.*, **123**, 425.

206 Guevremont, J.M., Guinther, G.H., Szemenyei, D., Devlin, M.T., Jao, T.C., Jaye, C., Woicik, J., and Fischer, D.A. (2008) *Tribol. Lubr. Technol.*, **64**, 41.

207 Hoffert, M.I., Caldeira, K., Benford, G., Criswell, D.R., Green, C., Herzog, H., Jain, A.,.K., Kheshgi, H.S., Lackner, K.,.S., Lewis, J.S., Lightfoot, H.D., Manheimer, W., Mankins, J.C., Mauel, M.E., Perkins, L.J., Schlesinger, M.E., Volk, T., and Wigley, T.M.L. (2002) *Science*, **298**, 981.

208 Jacobson, M.Z. (2009) *Energy Environ. Sci.*, **2**, 148.

209 Bartels, J.R., Pate, M.B., and Olson, N.K. (2010) *Int. J. Hydrogen Energy*, **35**, 8371.

210 Züttel, A., Borgschulte, A., and Schlapbach, L. (2008) *Hydrogen as a Future Energy Carrier*, Wiley-VCH Verlag GmbH, Weinheim.

211 Haug, M. and Neef, H. (2010) in *Proceedings of the 18th World Hydrogen Energy Conference 2010– WHEC 2010, Essen, May 16–21, 2010*, Schriften des Forschungszentrums Jülich: Energy and Environment, vol. **78-5** (eds D. Stolten and T. Grube), Forschungszentrum Jülich.

212 Vielstich, W., Lamm, A., and Gasteiger, H.A. (2003) *Handbook of Fuel Cells: Fundamentals, Technology, Applications*, Wiley-VCH Verlag GmbH, Weinheim.

213 Peinemann, K.-V. and Pereira-Nunes, S. (2008) *Membrane Technology: Membranes for Energy Conversion*, vol. **2**, Wiley-VCH Verlag GmbH, Weinheim.

214 Mehta, V. and Cooper, J.S. (2003) *J. Power Sources*, **114**, 32.

215 Mauritz, K.A. and Moore, R.B. (2004) *Chem. Rev.*, **104**, 4535.

216 de Bruijn, F.A., Dam, V.A.T., and Janssen, G.J.M. (2008) *Fuel Cells*, **8**, 3.

217 Wannek, C., Glüsen, A., and Stolten, D. (2010) *Desalination*, **250**, 1038.

218 Rozière, J. and Jones, D.J. (2003) *Annu. Rev. Mater. Res.*, **33**, 503.

219 Hickner, M.A., Ghassemi, H., Kim, Y.S., Einsla, B.R., and McGrath, J.E. (2004) *Chem. Rev.*, **104**, 4587.

220 Smitha, B., Sridhar, S., and Khan, A.A. (2005) *J. Membr. Sci.*, **259**, 10.

221 Tant, M.R., Mauritz, K.A., and Wilkes, G.L. (1997) *Ionomers: Synthesis, Structure, Properties and Applications*, Blackie Academic and Professional and Springer, Netherlands.
222 Eisenberg, A. and Kim, J. (1998) *Introduction to Ionomers*, Wiley-VCH Verlag GmbH, New York.
223 Gebel, G. and Diat, O. (2005) *Fuel Cells*, **5**, 261.
224 Gierke, T.D., Munn, G.E., and Wilson, F.C. (1981) *J. Polym. Sci., Part B: Polym. Phys.*, **19**, 1687.
225 Yarusso, D.J. and Cooper, S.L. (1983) *Macromolecules*, **16**, 1871.
226 Yarusso, D.J. and Cooper, S.L. (1985) *Polymer*, **26**, 371.
227 Gebel, G. and Lambard, J. (1997) *Macromolecules*, **30**, 7914.
228 Haubold, H., Vad, T., Jungbluth, H., and Hiller, P. (2001) *Electrochim. Acta*, **46**, 1559.
229 Kreuer, K.D. (2001) *J. Membr. Sci.*, **185**, 29.
230 Rubatat, L., Rollet, A.L., Gebel, G., and Diat, O. (2002) *Macromolecules*, **35**, 4050.
231 Rubatat, L., Gebel, G., and Diat, O. (2004) *Macromolecules*, **37**, 7772.
232 Kim, M., Glinka, C.J., Grot, S.A., and Grot, W.G. (2006) *Macromolecules*, **39**, 4775.
233 Page, K.A., Landis, F.A., Phillips, A.K., and Moore, R.B. (2006) *Macromolecules*, **39**, 3939.
234 Schmidt-Rohr, K. and Chen, Q. (2008) *Nat. Mater.*, **7**, 75.
235 Knox, C.K. and Voth, G.A. (2010) *J. Phys. Chem. B*, **114**, 3205.
236 Devanathan, R., Venkatnathan, A., Rousseau, R., Dupuis, M., Frigato, T., Gu, W., and Helms, V. (2010) *J. Phys. Chem. B*, **114**, 13681.
237 Liu, J., Suraweera, N., Keffer, D.J., Cui, S., and Paddison, S.J. (2010) *J. Phys. Chem. C*, **114**, 11279.
238 Wu, D., Paddison, S.J., Elliott, J.A., and Hamrock, S.J. (2010) *Langmuir*, **26**, 14308.
239 Prado, L.A.S.de.A., Wittich, H., Schulte, K., Goerigk, G., Garamus, V.M., Willumeit, R., Vetter, S., Ruffmann, B., and Nunes, S.P. (2004) *J. Polym. Sci., Part B: Polym. Phys.*, **42**, 567.
240 Takimoto, N., Wu, L., Ohira, A., Takeoka, Y., and Rikukawa, M. (2009) *Polymer*, **50**, 534.
241 Prado, L.A.S. de A., Ponce, M.L., Goerigk, G., Funari, S.S., Garamus, V.M., Willumeit, R., Schulte, K., and Nunes, S.P. (2009) *J. Non-Cryst. Solids*, **355**, 6.
242 Komarov, P.V., Veselov, I.N., Chu, P.P., and Khalatur, P.G. (2010) *Soft Matter*, **6**, 3939.
243 Manke, I., Markötter, H., Tötzke, C., Kardjilov, N., Grothausmann, R., Dawson, M., Hartnig, C., Haas, S., Thomas, D., Hoell, A., Genzel, C., and Banhart, J. (2011) *Adv. Eng. Mater.*, **13**, 712–729.
244 Felderhoff, M., Weidenthaler, C., Helmolt, R., and Eberle, U. (2007) *Phys. Chem. Chem. Phys.*, **9**, 2643.
245 Sakintuna, B., Lamari-Darkrim, F., and Hirscher, M. (2007) *Int. J. Hydrogen Energy*, **32**, 1121.
246 Klell, M. (2010) in *Handbook of Hydrogen Storage* (ed. M. Hirscher), Wiley-VCH Verlag GmbH, Weinheim, p. 1.
247 Dornheim, M. and Klassen, T. (2009) *Encyclopedia of Electrochemical Power Sources*, p. 459.

248 Dornheim, M. (2010) in *Handbook of Hydrogen Storage* (ed. M. Hirscher), Wiley-VCH Verlag GmbH, p. 187.

249 Barkhordarian, G., Klassen, T., Dornheim, M., and Bormann, R. (2007) *J. Alloys Compd.*, **440**, L18.

250 Xiong, Z., Hu, J., Wu, G., Chen, P., Luo, W., Gross, K., and Wang, J. (2005) *J. Alloys Compd.*, **398**, 235.

251 Vajo, J.J. and Olson, G.L. (2007) *Scr. Mater.*, **56**, 829.

252 Pranzas, P.K., Boesenberg, U., Karimi, F., Münning, M., Metz, O., Bonatto-Minella, C., Schmitz, H.-W., Beckmann, F., Vainio, U., Zajac, D., Welter, E., Jensen, T.R., Cerenius, Y., Bormann, R., Klassen, T., Dornheim, M., and Schreyer, A. (2011) *Adv. Eng. Mater.*, **13**, 730–736.

253 Nakamura, Y., Nakamura, J., Iwase, K., and Akiba, E. (2009) *Nucl. Instrum. Methods Phys. Res., Sect. A*, **600**, 297.

254 Ravnsbaek, D., Mosegaard, L., Joergensen, J.E., and Jensen, T.R. (2008) *Proceedings of the 29th Riso International Symposium on Materials Science: Energy Materials – Advances in Characterization, Modelling and Application*, Risö National Laboratory for Sustainable Energy, Technical University of Denmark, p. 349.

255 Boesenberg, U., Ravnsbaek, D.B., Hagemann, H., d'Anna, V., Bonatto-Minella, C., Pistidda, C., van Beek, W., Jensen, T.R., Bormann, R., and Dornheim, M. (2010) *J. Phys. Chem. C*, **114**, 15212.

256 Dornheim, M., Eigen, N., Barkhordarian, G., Klassen, T., and Bormann, R. (2006) *Adv. Eng. Mater.*, **8**, 377.

257 Deprez, E., Munoz-Marquez, M.A., Roldan, M.A., Prestipino, C., Palomares, J., Bonatto-Minella, C., Bösenberg, U., Dornheim, M., Bormann, R., and Fernandez, A. (2010) *J. Phys. Chem. C*, **114**, 3309.

258 Boesenberg, U., Vainio, U., Pranzas, P.K., Bellosta von Colbe, J.M., Goerigk, G., Welter, E., Dornheim, M., Schreyer, A., and Bormann, R. (2009) *Nanotechnology*, **20**, 204003-1.

259 Friedrichs, O., Martinez-Martinez, D., Guilera, G., Sanchez-Lopez, J.C., and Fernandez, A. (2007) *J. Phys. Chem.*, **111**, 10700.

260 Sartori, S., Knudsen, K.D., Zhao-Karger, Z., Bardaji, E.G., Muller, J., Fichtner, M., and Hauback, B.C. (2010) *J. Phys. Chem. C*, **114**, 18785.

261 Huhn, P.A., Dornheim, M., Klassen, T., and Bormann, R. (2005) *J. Alloys Compd.*, **404–406**, 499.

262 Barkhordarian, G., Klassen, T., and Bormann, R. (2004) *J. Alloys Compd.*, **364**, 242.

263 Barkhordarian, G., Klassen, T., and Bormann, R. (2003) *Scr. Mater.*, **49**, 213.

264 Schimmel, H.G., Huot, J., Chapon, L.C., Tichelaar, F.D., and Mulder, F.M. (2005) *J. Am. Chem. Soc.*, **127**, 14348.

265 Suarez-Alcantara, K., Ramallo Lopez, J.M., Bösenberg, U., Saldan, I., Pistidda, C., Requejo, F.G., Jensen, T., Cerenius, Y., Soerby, M., Avila, J., Bellosta von Colbe, J., Taube, K., Klassen, T., and Dornheim, M. (2012) *J. Phys. Chem.*, **116**, 7207–7212.

266 Fu, Q.J., Ramirez-Cuesta, A.J., and Tsang, S.C. (2006) *J. Phys. Chem. B*, **110**, 711.

267 Borgschulte, A., Züttel, A., Hug, P., Barkhordarian, G., Eigen, N., Dornheim, M., Bormann, R., and Ramirez-Cuesta, A.J. (2008) *Phys. Chem. Chem. Phys.*, **10**, 4045.

268 Pranzas, P.K., Dornheim, M., Bellmann, D., Aguey-Zinsou, K.-F., Klassen, T., and Schreyer, A. (2006) *Physica B*, **385–386**, 630.

269 Pranzas, P.K., Dornheim, M., Bösenberg, U., Ares-Fernandez, J.R., Goerigk, G., Roth, S.V., Gehrke, R., and Schreyer, A. (2007) *J. Appl. Crystallogr.*, **40**, 262.

270 Nielsen, T.K., Bösenberg, U., Gosalawit, R., Dornheim, M., Cerenius, Y., Besenbacher, F., and Jensen, T. (2010) *ACS Nano*, **4**, 3903.

271 Bellosta von Colbe, J.M., Metz, O., Lozano, G.A., Pranzas, P.K., Schmitz, H.W., Beckmann, F., Schreyer, A., Klassen, T., and Dornheim, M. (2012) *Int. J. Hydrogen Energy*, **37**, 2807–2811.

272 Wroblewski, T., Wild, E., Poeste, T., and Pyzalla, A. (2000) *Mater. Sci. Lett.*, **19**, 975.

273 Song, X. and Korsunsky, A.M. (2011) *J. Appl. Mech.*, **78**, 031019.

274 Sasaki, T., Yaguchi, O., and Suzuki, H. (2010) *Mater. Sci. Forum*, **652**, 260.

275 Jun, T.-S., Hofmann, F., Belnoue, J., Song, X., Hofmann, M., and Korsunsky, A.M. (2009) *J. Strain Anal. Eng. Des.*, **44**, 563.

276 Webster, P.J., James, M.N., Ting, S.-P., Chen, Z., Hughes, D.J., Steuwer, A., Malard, B., and Ratel, N. (2004) *Applications of X-Rays in Mechanical Engineering*, ASME, Anaheim, CA, pp. 63–69.

277 Webster, P.J., Mills, G., Wang, X.D., Kang, W.P., and Holden, T.M. (1997) *J. Strain Anal. Eng. Des.*, **32**, 389.

278 Webster, P.J., Wang, X., Mills, G., and Webster, G.A. (1992) *Physica B*, **180**, 1029.

279 Österle, W., Pyzalla, A., Wang, L., and Rooch, H. (2001) *Mater. Sci. Eng., A*, **303**, 150.

280 Pyzalla, A., Wang, L., Wild, E., and Wroblewski, T. (2001) *Wear*, **251**, 901.

281 Kaysser, W., Esslinger, J., Abetz, V., Huber, N., Kainer, K.U., Klassen, T., Pyczak, F., Schreyer, A., and Staron, P. (2011) *Adv. Eng. Mater.*, **13**, 637–657.

20

In situ Experiments with Synchrotron High-Energy X-rays and Neutrons*

Peter Staron, Torben Fischer, Thomas Lippmann, Andreas Stark, Shahrokh Daneshpour, Dirk Schnubel, Eckart Uhlmann, Robert Gerstenberger, Bettina Camin, Walter Reimers, Elisabeth Eidenberger-Schober, Helmut Clemens, Norbert Huber, and Andreas Schreyer

20.1 Introduction

Engineering materials science strives for both, understanding the basic principles behind materials properties and performance, and development and optimization of the processing steps required for production. In both fields, not only static but also dynamic properties are of interest. An example is the cold or hot deformation of materials, where not only the degree of deformation but also the deformation speed and temperature determines the microstructure and, thus, the properties of the material. As a consequence, often static material studies do not give a sufficient answer to crucial questions. Here, "static studies" means studies that are performed after the material treatment is finished ("post mortem") and the sample, or in most cases only parts of it, can be prepared for studies like, for example, transmission electron microscopy (TEM) observations.

Instead of static studies, dynamic studies would often be more appropriate for answering specific questions. Among these questions are, for example, how exactly a material has evolved from state A to state B, which often cannot be answered adequately from studies of state A and state B alone. However, the change of a material from state A to state B, for example, in metal processing, requires some sort of treatment, like, for example, high temperatures, or mechanical forces, or a combination of both. In many cases, the process environment hinders *in situ* studies because it cannot be combined with the probe to be used for the study. Nevertheless, a variety of *in situ* experiments already exists, among which furnaces and mechanical testing units are probably the most common ones used in diffractometers or microscopes.

However, there are also *in situ* experiments that do not yet exist because of the difficulties and costs of the required instrumentation. In addition, available spatial and time resolutions are often limiting such studies. With the advent of third-generation synchrotron sources like PETRA III at DESY, Hamburg, some of these restrictions belong

*This chapter is reproduced and updated from *Adv. Eng. Mater.* **13** (2011) 658–663 by permission of Wiley-VCH.

Neutrons and Synchrotron Radiation in Engineering Materials Science: From Fundamentals to Applications, Second Edition. Edited by Peter Staron, Andreas Schreyer, Helmut Clemens, and Svea Mayer.

to the past. Therefore, HZG has started realizing a number of *in situ* experiments, making use of the new opportunities. In the following sections, one of the first experiments at the *in situ* quenching and deformation dilatometer with differential scanning dilatometry (DSC) unit called FlexiTherm will be described. Moreover, first results of an *in situ* tensile test and an *in situ* cutting experiment will be introduced. Finally, an *in situ* study of precipitation kinetics using small-angle neutron scattering (SANS) will be presented. The presented synchrotron experiments were performed at the former HZG beamline HARWI II, which had been at the second-generation synchrotron source DORIS III at DESY, Hamburg [1]. Today, these experiments benefit from the largely improved brilliance at the new HZG high-energy materials science beamline HEMS at PETRA III at DESY, Hamburg.

20.2 *In situ* Dilatometry

20.2.1 Motivation

Intermetallic γ-TiAl-based alloys are promising candidates as high temperature lightweight materials for turbines and engines at temperatures up to 800 °C because of their excellent strength and low density ($4\,g\,cm^{-3}$) [2]. Suitable production processes like casting or forging have to be developed for the application of such alloys. This development requires a detailed knowledge of the phases present at the processing temperatures. However, investigating the constitution at high temperatures is difficult by microstructural characterization and phase analysis at room temperature. The reason is that different phase transformations might have occurred on cooling even if high cooling rates have been applied. Furthermore, it is also important to study the kinetics of phase transformations at temperatures relevant for processing. The latter can be studied, for example, by diffraction, by dilatometry, or by calorimetry. Usually, diffraction is not performed in combination with these two techniques of thermal analysis. The lack of such simultaneous measurements, however, made a coherent interpretation of the results difficult, especially in complex multiphase materials such as γ-TiAl-based alloys.

To overcome these problems, a commercial quenching and deformation dilatometer with an additional DSC unit was installed in the X-ray beam at the HZG beamlines at DESY for simultaneous measurement of heat flux or thermal expansion curves and diffraction patterns.

20.2.2 FlexiTherm

A dilatometer DIL 805A/D from Bähr (now TA Instruments) was equipped with windows for the X-ray beam (Figure 20.1). The X-ray beam with a cross-section of up to 1 mm × 1 mm passes through the heating coil. Samples with 4 or 5 mm diameter can be used at photon energies of 100 keV, which is the standard dimension for the used dilatometer. The samples can be heated inductively and quenched by blowing on with water-cooled gas. A maximum heating rate of $4000\,K\,s^{-1}$ and a maximum cooling rate of $2500\,K\,s^{-1}$ (employing hollow samples) can be achieved.

Figure 20.1 Sketch of the dilatometer induction coil in the X-ray beam. (Staron 2011 [3]. Reproduced with the permission of Wiley.)

20.2.3 Results

Conventional γ-TiAl-based alloys consist of tetragonal γ-TiAl ($L1_0$ structure; *P4/mmm*) and small amounts of hexagonal α_2-Ti_3Al ($D0_{19}$ structure; $P6_3/mmc$). In order to improve their workability as well as their mechanical properties, various elements can be alloyed. These additional alloying elements can also promote the formation of ternary intermetallic phases. However, up to now the exact pathways of formation and transformation of these phases are not fully understood [2, 4].

Here, we present the results of a first test illustrating the capabilities of the combined measuring techniques. An as-cast Ti-43Al (in at%) sample was inductively heated in Ar atmosphere up to 1200 °C and subsequently cooled with a rate of $40\ K\ min^{-1}$. The diffraction rings were recorded with a Mar555 flat panel detector every 2 min.

In the beginning of the measurement, only reflections of α_2-Ti_3Al occur, as shown in Figure 20.2. This indicates that the single-phase high-temperature state was frozen due to a high cooling rate after casting. At about 800 °C additional reflections of γ-TiAl appear, which can be attributed to the onset of thermally activated diffusion processes. Thus, the metastable nonequilibrium α_2 decomposes to $\alpha_2 + \gamma$ lamellar colonies [5]. At the same time, the thermal expansion significantly changes. Obviously, the smaller volume of the precipitated γ compensates for the thermal expansion of the sample. During further heating, the ordered hexagonal α_2-Ti_3Al transforms to disordered hexagonal α-Ti(Al) (A3 structure; $P6_3/mmc$), which is indicated by the vanishing superlattice reflections. Subsequently, γ starts to transform to α and, as a consequence, the thermal expansion rate increases. This behavior proceeds reversely during cooling. However, due to the slow cooling rate, Ti-43Al remains two-phase down to room temperature.

These results illustrate the correlation of phase transformations with changes in the thermal expansion rate. Discontinuous volume changes, as observed during decomposition of the metastable α_2-phase, can cause high internal stresses and lead to damage. However, the results will help designing suitable heat treatments in order to minimize these stresses. Other examples for experiments performed with FlexiTherm can be found in the literature [6–9].

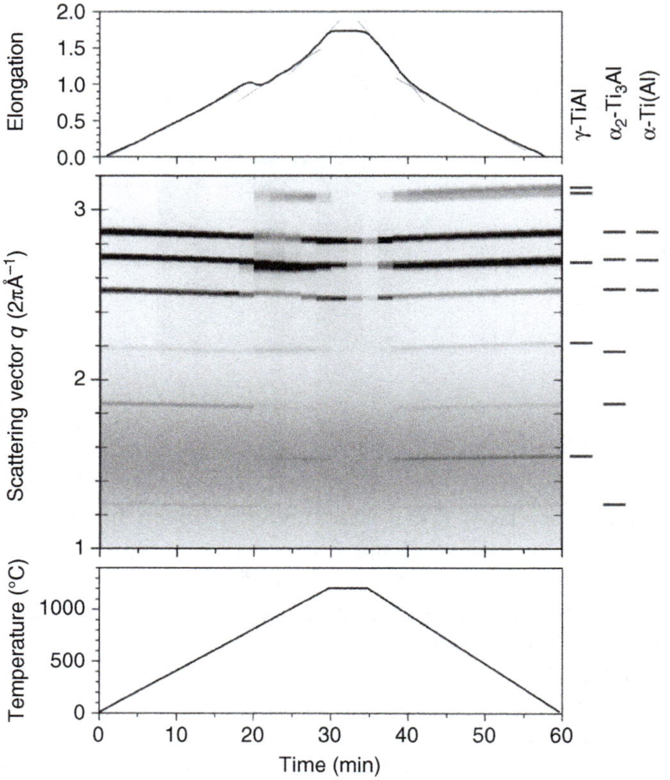

Figure 20.2 Development of the elongation and diffraction pattern of a Ti-43 at.% Al sample with temperature and time. (Staron 2011 [3]. Reproduced with the permission of Wiley.)

20.3 *In situ* Study on Single Overload of Fatigue-Cracked Specimens

20.3.1 Motivation

Reliable prediction of fatigue life has become a central topic in the design of aerospace components and assemblies ever since the 1970s [10, 11]. Since fatigue damage occurs locally at the crack tip, the knowledge of the stress–strain fields around the crack tip is essential for the development of reliable life prediction models. Due to a highly nonlinear evolution of deformation and damage, the local stress–strain field depends on the previous loading history. As a consequence, fatigue life predictions are mainly based on a number of simplifications and assumptions reducing the reliability of the life prediction method. In this context, X-ray diffraction at a synchrotron source provides direct insight into the stress–strain fields in the vicinity of the crack tip, which is essential for a deeper understanding of the mechanisms of fatigue damage. In addition, such experiments can be used to quantitatively investigate load interaction effects during fatigue crack growth, which is an important problem that can occur in the damage tolerant design and is the focus of current research.

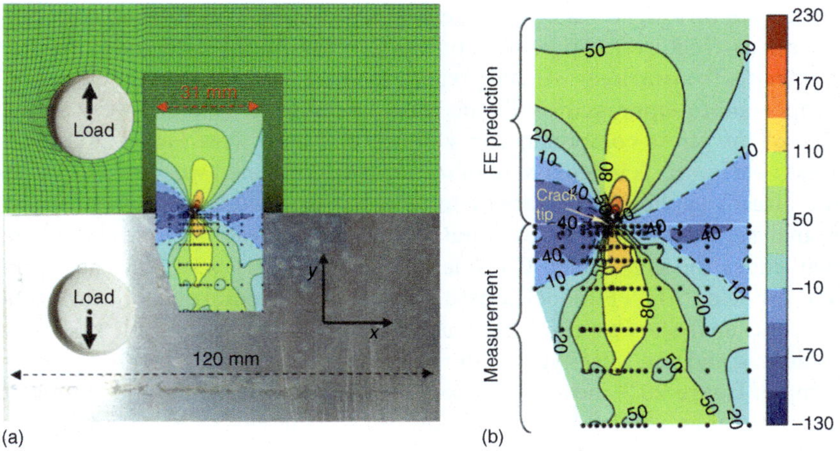

Figure 20.3 Stress field results $\sigma_y - \sigma_x$ (in MPa) close to the crack tip at an overload of 10.2 kN taken from simulation (top part) and diffraction patterns (bottom part). (a) Overview and (b) area close to the crack tip. The dots mark the measurement positions. (Staron 2011 [3]. Reproduced with the permission of Wiley.)

The occurrence of an overload (high/low load sequence) can lead to strong fatigue crack growth retardation [12]. To provide a deeper understanding of the mechanisms of retardation of the fatigue crack tip due to an overload, stress fields were determined in specimens containing a fatigue crack under external load. In conjunction with the finite element (FE) simulation, such experiments will be used in subsequent work to enhance existing life prediction methods.

20.3.2 Experimental

Compact-Tension (CT) specimens made of Al alloy AA6056-T6 sheet material with 6.2 mm thickness were used (Figure 20.3a). To investigate load interaction effects on fatigue crack propagation, different overloads during cyclic loading with a frequency of 10 Hz were applied. The maximum cyclic load was 5.1 kN and the load ratio *R* was 0.1. Overloads ranged from 7.65 to 17.15 kN.

Diffraction measurements under static loading condition were conducted on the CT specimens at the HZG beamline HARWI II at DESY using a hydraulic tensile testing machine (INSTRON 8800) mounted in the beam. In total, 160 diffraction patterns have been measured along the crack path by a biased area scan within an area of 31 mm × 32 mm (Figure 20.3a). The experiments were performed at a photon energy of 70 keV. The beam cross-section was 1 mm × 1 mm to reach a good lateral resolution. A Mar555 area detector was used to record several complete diffraction rings.

20.3.3 Results

The small beam cross-section in combination with the relatively large grain size of the AA6056-T6 sheet leads to noisy results when using the normal data evaluation procedure based on only four small sectors of the measured diffraction rings. To overcome this difficulty, a data evaluation routine was programmed to fit the strains in the global coordinate axis directions using the information of the whole diffraction rings.

In future work, our investigations will be extended to welded specimens. Since welding changes both microstructure and precipitation state, it also alters locally the stress-free lattice parameter d_0. To avoid long-winded preparation of stress-relieved reference specimens, it is important to suppress the influence of d_0 on the stress results. By calculating the difference of the stress components in y and x directions (see Figure 20.3a) the resulting stress field $\sigma_y - \sigma_x$ is independent of d_0. In view of planned work on welds, this approach was already applied to the base material specimen used in the present study.

A 3D FE analysis was performed to simulate crack tip stress–strain fields in the CT-specimen under the given load. Considering the symmetries, only a half of the specimen was modeled. The FE model consists of linear hexahedral elements available in ABAQUS. The analysis was conducted using a plasticity model with nonlinear isotropic hardening and von Mises yield criterion.

As an example, Figure 20.3 shows a comparison between measured and simulated stress fields $\sigma_y - \sigma_x$ for an applied load of 10.2 kN. The dots in Figure 20.3b indicate the actual measurement positions. The contour color plot was created as a natural neighbor interpolation based on these points. The simulation results were extracted in a similar way from the fine-meshed region as shown in the upper half of Figure 20.3a. In general, a good agreement between simulation and experimental results is observed, especially for the areas a bit further away from the crack tip. As expected, Figure 20.3b reveals high-gradient regions of stresses confined to the crack tip for both simulation and experiment. A lack of data points at the vicinity of the crack tip in combination with noisy raw data limits the capability of capturing stresses at this highly stressed region. However, experimental data are essential for the calibration and validation of the simulation approach. To obtain higher resolution within the region of interest and to provide the additionally needed strain field, the simulation results can be used to extrapolate the experimental data near the crack tip. Furthermore, the 3D simulation provides the depth-dependent distribution of the stress and strain fields, which is integrated through the thickness of the specimen by the synchrotron beam. Such 3D information is very useful to predict the different crack propagation behaviors at the specimen surface and inside the specimen.

20.4 *In situ* Cutting Experiment

20.4.1 Motivation

The cutting process with a geometrically defined cutting edge is characterized by a highly complex chip formation process, which is influenced by many factors. Numerous analytical models have been developed to describe the material flow through the chip formation zone, enabling the calculation of mechanical and thermal state variables, such as strain, stress, and temperature. The validity of these models relies on a number of restricting preconditions, such as the assumption of a plane strain state and a homogeneous stress state within the entire shear plane. These postulations represent a significant simplification of the general cutting problem. However, experimental observations are difficult because of the restricted accessibility to the chip formation zone, the high process speed, and the nonstationary conditions of the chip formation process.

Therefore, a first experiment for the evaluation of the applicability of X-ray diffraction for the study of the cutting process was performed at the HZG beamline HARWI II at

DESY [13]. The experimentally determined strains and stresses were compared with results from a FEM cutting simulation.

20.4.2 Experiment

The cutting experiments were conducted with orthogonal kinematics as well as a cutting depth of $h = 0.3$ mm and a chip width of $b = 1.6$ mm. The setup was integrated in a tension/compression test machine. The work piece moved with a cutting speed of $v_c = 0.01$ m min^{-1}. Since practically used cutting speeds are significantly higher, the cutting speed that was applied in this experiment has been termed *quasistatic*. However, a long exposure time of 45 s was required to increase the signal-to-noise ratio. The steel S235JR was used as work piece material because of its stationary chip formation. Complete diffraction rings were recorded on a Mar345 image plate detector using a photon wavelength of 100 keV. The beam cross-section was 100 μm × 100 μm. The setup and measurement positions are sketched in Figure 20.4a.

20.4.3 Results

The strain tensor components $\varepsilon_{11}(hkl)$, $\varepsilon_{22}(hkl)$, and $\varepsilon_{12}(hkl)$ were calculated from the diffractograms for all reflections and measuring positions. The calculation of the stress tensor components was carried out under the assumption of both plane strain and plane

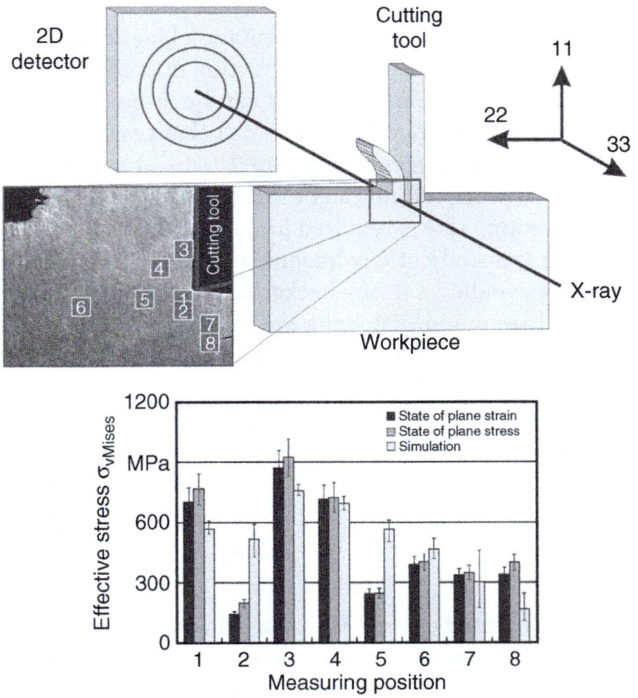

Figure 20.4 (a) Sketch of the experimental setup for *in situ* strain measurements during orthogonal cutting. (b) Effective stresses determined from the measured strain data at the eight positions indicated in (a), compared with simulation results. (Staron 2011 [3]. Reproduced with the permission of Wiley.)

stress states. FEM cutting simulations were conducted with the software DEFORM 3D in order to evaluate the experimental results. The geometric model of the cutting wedge and the work piece as well as the cutting parameters complied with the experimental conditions. Coulomb friction with $\mu = 0.2$ was assumed between the work piece and the rigid cutting tool. The results show a very good qualitative and, partially, also quantitative consistency of experimental and simulation data [13, 14]. As an example, the effective stresses after von Mises at the eight measurement positions are shown in Figure 20.4b.

The results, however, give evidence that this first experiment suffered from insufficient spatial resolution (beam cross-section) to capture the high stress gradients. In addition, the number of data points was found to be insufficient. Thus, these experiments were continued at the HEMS beamline at the PETRA III ring of DESY, which is a third-generation synchrotron providing an intensity increased by a factor of about 10 000. Further experiments with a cutting speed of $v_c = 0.03\,\mathrm{m\,min^{-1}}$ and a decreased beam size of 20 μm × 20 μm were carried out [15]. Cutting speeds up to $0.1\,\mathrm{m\,min^{-1}}$ are expected to be possible with the presented experimental setup.

20.5 *In situ* Study of Precipitation Kinetics Using Neutrons

20.5.1 Motivation

For decades, SANS has been a standard tool for studying precipitation kinetics [16, 17]. It enables the determination of size distribution and volume fraction of precipitates as a function of time and temperature. Typical sample volumes can be around $0.1\,\mathrm{cm^3}$, thus leading to good statistics as well as to high sensitivity to small changes in the precipitate dispersion. This makes SANS a complementary tool to TEM and atom probe tomography (3DAP). The early stages of precipitation can be investigated by isothermal *in situ* heating experiments. A high heating rate is required for reaching the treatment temperature within a short time. For the study of ferromagnetic material, the sample has to be magnetized to saturation for avoiding a strong background scattering of magnetic domains. Therefore, a furnace is required that can be used in a magnet. A new compact furnace with a relatively high heating rate for use in a 2 T electromagnet was developed at HZG. The prototype furnace was tested by studying precipitation in an Fe-25 at.% Co-9 at.% Mo alloy strengthened by precipitates of the intermetallic μ-phase ($(Fe,Co)_7Mo_6$) with sizes of a few nanometer [18]. Only little was known about details regarding the early stage of precipitate formation in this alloy system before these investigations were conducted.

20.5.2 Experimental Details

The furnace uses ceramic heating elements between which a coin-shaped sample is fixed. Its water-cooled Al housing had Al windows for the neutron beam. The evacuated furnace was placed between the pole shoes of an electromagnet at the SANS-2 instrument at HZG and the sample was magnetized to saturation. The temperature was measured with a thermocouple placed into a hole drilled into the sample. The heating rate of the prototype furnace was $5\,\mathrm{K\,s^{-1}}$; thus, the temperature of 493 °C was reached in 100 s (later versions of the furnace achieved significantly higher heating rates). A neutron

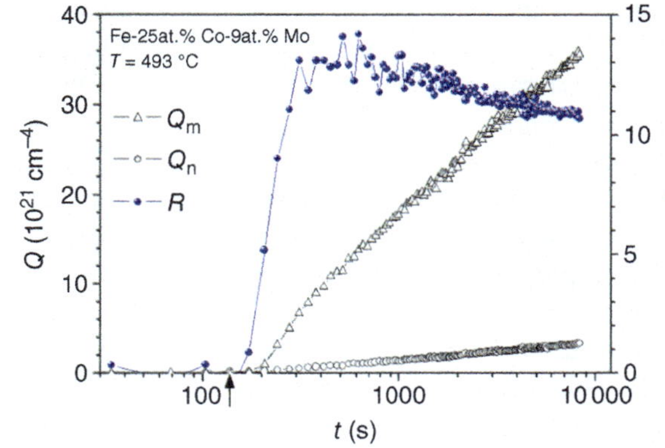

Figure 20.5 Integrated SANS intensities Q_m and Q_n after baseline subtraction and ratio $R = Q_m/Q_n$ for Fe-25 at.% Co-9 at.% Mo. The arrow marks the time when 493 °C are almost reached. (Staron 2011 [3]. Reproduced with the permission of Wiley.)

wavelength of 0.58 nm and a detector distance of 1 m were used. The data acquisition time at the beginning of the reaction was 30 s.

20.5.3 Results

Since the statistics of the SANS signal during the early stages was weak, the integrated small-angle scattering intensity Q was calculated from the experimental scattering curves [19]. Q is a measure for the product of the precipitate volume fraction f and the square of the scattering contrast $\Delta\eta$. The ratio R of integrated intensities calculated from magnetic and nuclear scattering curves, $R = Q_m/Q_n$, thus reflects the ratio of magnetic and nuclear scattering contrasts squared, since both signals originate in the same precipitate structure. From the fact that Q_m initially has a higher slope than in the later stages, although Q_n does not show this change in slope it was concluded that strong compositional changes in the newly formed precipitates lead to building up of the magnetic contrast during the first 150 s (Figure 20.5).

In the literature, the formation of precipitates in Fe–Co–Mo alloys was often described to proceed via spinodal decomposition [20]. From the present results it is concluded that, should spinodal decomposition play a role in this alloy at 493 °C, it is most likely during the first 150 s where the strong increase of R indicates strong compositional changes. After 150 s, the precipitates should be close to their final composition and only smaller compositional changes seem to occur as indicated by the small changes in R.

Additional *in situ* SANS experiments at different temperatures help to understand the early decomposition stages in the investigated material [21]. Although small-angle X-ray scattering experiments at a synchrotron source can give much better time resolution due to the high intensity of the source, using the magnetic properties for analysis, for example, of the ratio R is not possible. The combination of SANS with 3DAP and TEM as well as suitable decomposition models leads to a comprehensive description of the precipitation reaction in the investigated material [22].

20.6 Conclusions

Besides the experiments reported in this paper, other *in situ* experiments were developed. FlexiStir is an *in situ* friction stir welding (FSW) instrument, a transportable unit equipped with a state-of-the-art welding head, which is not used for *in situ* experiments alone. FlexiLas is an *in situ* laser beam welding experiment using a fiber laser and a furnace for sample preheating up to 800 °C. Other *in situ* experiments are still under design or under construction. Among them is a nano-indentation experiment (FlexiIndent) that is planned for use with the 3DXRD and diffraction contrast tomography techniques. These experiments will be performed at the new grain mapper at the HEMS beamline at PETRA III. In addition, *in situ* tomography experiments can be performed, for example, with a gradient furnace, which will mainly be limited by the speed at which a sample can be rotated.

In future, the high brilliance of the third-generation synchrotron source PETRA III will enable further *in situ* experiments in the field of engineering materials science, making use of the high penetration power of high-energy photons.

References

1 Staron, P., Schell, N., Haibel, A., Beckmann, F., Lippmann, T., Lottermoser, L., Herzen, J., Fischer, T., Koçak, M., and Schreyer, A. (2010) *Mater. Sci. Forum*, **638**, 2470.

2 Appel, F., Paul, J.D.H., and Oehring, M. (2011) *Gamma Titanium Aluminide Alloys – Science and Technology*, Wiley-VCH Verlag GmbH, Weinheim.

3 Staron, P., Fischer, T., Lippmann, T., Stark, A., Daneshpour, S., Schnubel, D., Uhlmann, E., Gerstenberger, R., Camin, B., Reimers, W., Eidenberger, E., Clemens, H., Huber, N., and Schreyer, A. (2011) *Adv. Eng. Mater.*, **13**, 658–663.

4 Schmoelzer, T., Stark, A., Schwaighofer, E., Lippmann, T., Mayer, S., and Clemens, H. (2012) *Adv. Eng. Mater.*, **14**, 445–448.

5 Liss, K.-D., Stark, A., Bartels, A., Clemens, H., Buslaps, T., Phelan, D., and Yeoh, L.A. (2008) *Adv. Eng. Mater.*, **10**, 389.

6 Stark, A., Schwaighofer, E., Mayer, S., Clemens, H., Lippmann, T., Lottermoser, L., Schreyer, A., and Pyczak, F. (2013) *MRS Proc.*, **1516**, 71–76.

7 Tolnai, D., Mendis, C.L., Stark, A., Szakács, G., Wiese, B., Kainer, K.U., and Hort, N. (2013) *Mater. Lett.*, **102–103**, 62–64.

8 Suwanpinij, P., Stark, A., Li, X., Römer, F., Herrmann, K., Lippmann, T., and Bleck, W. (2014) *Adv. Eng. Mater.*, **16**, 1044–1051.

9 Stark, A., Oehring, M., Pyczak, F., and Schreyer, A. (2011) *Adv. Eng. Mater.*, **13**, 700–704.

10 Codrington, J. (2009) *Int. J. Fract.*, **155**, 93.

11 Korsunsky, A.M., Song, X., Belnoue, J., Jun, T., Hofmann, F., De Matos, P.F.P., Nowell, D., Dini, D., Aparicio-Blanco, O., and Walsh, M.J. (2009) *Int. J. Fatigue*, **31**, 1771.

12 Daneshpour, S., Koçak, M., Langlade, S., and Horstmann, M. (2009) *Int. J. Fatigue*, **31**, 1603.

13 Uhlmann, E., Gerstenberger, R., Herter, S., Hoghé, T., Reimers, W., Camin, B., Martins, R.V., Schreyer, A., and Fischer, T. (2011) *Prod. Eng. Res. Dev.*, **5**, 1–8.

14 Uhlmann, E., Henze, S., Gerstenberger, E., Brömmelhoff, K., Reimers, W., Fischer, T., and Schell, N. (2013) *Prod. Eng. Res. Dev.*, **7**, 401–408.

15 Brömmelhoff, K., Henze, S., Gerstenberger, R., Fischer, T., Schell, N., Uhlmann, E., and Reimers, W. (2013) *J. Mater. Process. Technol.*, **213**, 2211–2216.

16 Guinier, A. and Fournet, G. (1955) *Small-Angle Scattering of X-Rays*, John Wiley & Sons, Inc., New York, Chapman & Hall, London.

17 Kostorz, G. (1979) in *Treatise on Materials Science and Technology – Neutron Scattering*, vol. **15** (eds G. Kostorz and H. Herman), Academic Press, New York, p. 227.

18 Eidenberger, E., Stergar, E., Leitner, H., Staron, P., Spitaler, J., Ambrosch-Draxl, C., and Clemens, H. (2009) *Appl. Phys. A*, **97**, 331.

19 Staron, P., Eidenberger, E., Schober, M., Sharp, M., Leitner, H., Schreyer, A., and Clemens, H. (2010) *J. Phys. Conf. Ser.*, **247**, 012038.

20 Doi, M., Tanabe, H., and Miyazaki, T. (1987) *J. Mater. Sci.*, **22**, 1328.

21 Eidenberger, E., Schober, M., Staron, P., Casliskanoglu, D., Leitner, H., and Clemens, H. (2010) *Intermetallics*, **18**, 2128.

22 Eidenberger, E., Schober, M., Stergar, E., Leitner, H., Staron, P., and Clemens, H. (2010) *Metall. Mater. Trans. A*, **41**, 1230.

21

Application of Photons and Neutrons for the Characterization and Development of Advanced Steels*

Elisabeth Eidenberger-Schober, Ronald Schnitzer, Gerald A. Zickler, Michael Eidenberger-Schober, Michael Bischof, Peter Staron, Harald Leitner, Andreas Schreyer, and Helmut Clemens

21.1 Introduction

Following the request of tailor-made high-performance properties, it is essential that, for the development of modern steels, detailed knowledge is gained on all aspects of their intrinsic structure and the evolution of their microstructural constituents, especially precipitates, during processing and heat treatments. In this chapter, a short review is given on how synchrotron radiation and neutrons have been used to study the microstructure of advanced steels. Special emphasis is laid on the measurement details, the employed equipment and the developments therein. In addition, the drawn conclusions on the microstructural development of the investigated alloys are highlighted and possible ways to combine the described scattering techniques with complementary experimental methods are pointed out.

For the development of advanced steels, it is of vital importance to understand the prevailing precipitation reactions. Because modern alloys are complex mult-iphase and multi-component materials, their comprehensive characterization is a demanding task and the analysis of microstructural components in such materials down to the nanometer-scale requires exceptional experimental expertise [1]. Direct imaging techniques like scanning electron microscopy (SEM), transmission electron microscopy (TEM), and also atom probe (AP) are powerful tools widely used for the mentioned task, but often exhibit the drawback of a small sample volume accompanied by time-consuming specimen preparation. Also, the set up of *in situ* experiments, for example, upon heating and/or under mechanical loading is prevalently difficult to establish [2]. By using indirect methods, such as scattering and diffraction, samples of macroscopic dimensions can be probed, thereby providing enhanced statistical relevance. In addition, the experimental environment is flexible, allowing for a wide variety of possible measurement setups. Laboratory-size X-ray sources, for example, Cu-Kα, suffer from the disadvantage of a high photon absorption and fluorescence radiation in iron-based alloys. The use of synchrotron radiation or neutrons provided in large-scale facilities satisfies the need of the experimentalists for short measurement

*This chapter is reproduced and updated from *Adv. Eng. Mater.* **13** (2011) 664–673 by permission of Wiley-VCH.

Neutrons and Synchrotron Radiation in Engineering Materials Science: From Fundamentals to Applications, Second Edition. Edited by Peter Staron, Andreas Schreyer, Helmut Clemens, and Svea Mayer.

times and large sample volumes. Short exposure times also entail the possibility to perform *in situ* experiments. To give a complete overview of the use of synchrotron radiation and neutrons in the field of material science [3] would go beyond the scope of this work; hence, the following pages will concentrate on a compendium of recent developments and investigations in the field of advanced steels performed in the framework of a collaboration between the Montanuniversität Leoben, Leoben, Austria, and the Helmholtz-Zentrum Geesthacht, Germany. For detailed information, the interested reader is referred to the corresponding references.

21.2 Characterization Using Synchrotron Radiation

21.2.1 *Ex situ* and *In situ* High-Energy X-ray Diffraction (HE-XRD) during Heating

The class of high-strength steels, so-called maraging steels, characterized by very low carbon contents, transforms to a soft martensite, exhibiting high dislocation densities after quenching from the solution treatment [4]. In this condition, machining and cold-working of the material is easy and only after subsequent aging between 500 and 600 °C the required strength for the respective application is achieved. The outstanding mechanical properties of these steels are due to a dispersion of nanometer-sized intermetallic precipitates formed during aging situated in the soft martensitic matrix. Maraging steels are used in many high-performance fields like aerospace and military as well as for tools and dies. To further improve the performance of modern steels it is essential to understand the underlying mechanisms of phase transformations as well as the influence of composition and heat treatment on the microstructural development. Due to the fact that the amount of austenite in steels is of great scientific and technological relevance, literature reports several experimental techniques for quantifying the austenitic phase in steels: for example, [5–8] X-ray diffraction (XRD) is commonly used because crystallographic phases, volume fractions, crystallite sizes, and lattice parameters can be detected quite easily. The use of synchrotron radiation overcomes the above mentioned deficiencies of the probed materials' high photon absorption and enables a superior potential of experimental studies. High-energy X-rays, such as provided by modern synchrotron radiation sources, bear the advantage of high photon brilliance and low absorption in most materials [3, 9], enabling time-resolved *in situ* measurements in transmission geometry of bulk materials of several millimeters in thickness [10, 11]. *In situ* high-energy X-ray diffraction (HE-XRD) experiments allow for the determination of the entire path of phase transformation on one single specimen. Therefore, influences from different chemical compositions and microstructures in various samples are avoided. To be able to obtain information of the early stages, it is important to apply fast heating rates during *in situ* experiments. Therefore, a new *in situ* induction furnace was used, where the sample is positioned in an induction coil and can be heated up to aging temperature of 400–700 °C within 50 s [12]. The temperature is measured directly on the sample via thermocouples. In combination with *ex situ* measurements on longer aged samples, a wide range of aging times can be easily investigated.

In the presented experiment, the investigated maraging steel (chemical composition listed in Table 21.1, alloy no. 1) was characterized in a time range of 120 s

Table 21.1 Chemical composition of all investigated alloys (all values in at.%).

Alloy number in text	Precipitate type	Fe	C	Cr	W	Mo	V	Si	Mn	Ni	Co	Ti	Al	Cu	O	N	References	Complementary techniques
No. 1	NiAl	Bal.	0.14	13.52	—	1.27	—	—	—	7.73	—	—	2.26	—	—	—	[13–15]	AP
No. 2	Carbides: MC, M_2C	Bal.	4.30	4.50	2.00	3.00	2.00	0.55	0.28	0.18	0.82	—	—	—	—	—	[16, 17]	EFTEM, AP
No. 3	Carbides: MC, M_2C, M_6C	Bal.	1.71	5.10	—	1.59	0.62	0.79	0.33	0.17	—	—	—	—	—	—	[18–20]	EFTEM, XRD, TEM
No. 4	Ni_3Ti, $Ti_6Si_7Ni_{16}$	Bal.	0.05	13.03	—	0.56	—	1.08	0.11	8.34	—	0.95	1.16	—	—	—	[21–23]	EFTEM
No. 5	Carbides and NiAl	Bal.	1.40	2.60	—	1.40	0.30	—	—	6.00	—	—	5.00	—	—	—	[24–27]	TEM, AP
No. 6	$(Fe,Co)_7Mo_6$	Bal.	0.10	—	—	9.30	—	0.40	—	—	25.10	—	—	—	—	—	[28–33]	TEM, AP, DSC
No. 7	bcc-Cu	Bal.	0.01	0.01	—	—	—	0.02	0.03	0.02	—	—	—	0.99	0.34	0.08	[34–36]	AP
No. 8	NiAl	Bal.	—	—	—	—	—	—	—	4.05	—	—	6.55	—	—	—	[37]	AP

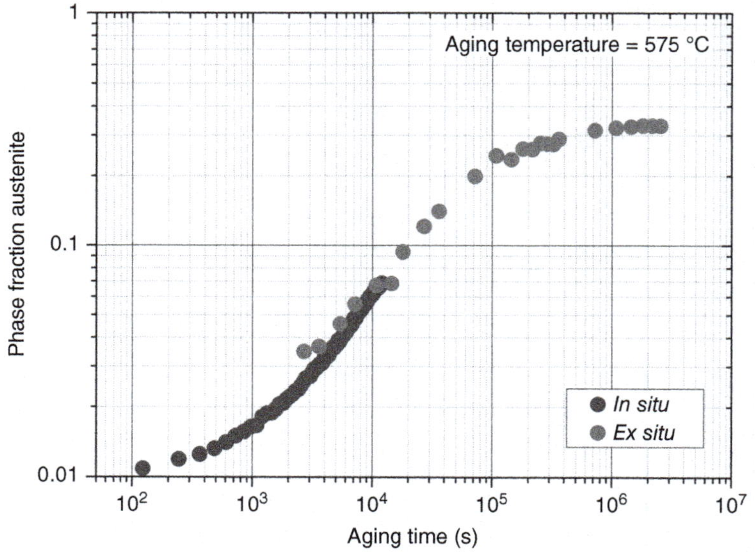

Figure 21.1 Phase fraction of austenite as a function of aging temperature as determined from the diffraction patterns for a maraging steel (see Table 21.1, alloy no. 1). (Eidenberger 2011 [39]. Reproduced with permission of Wiley.)

to700 h of aging. XRD experiments using synchrotron radiation were conducted at the Engineering Materials Science Beamline HARWI-II [12] of Helmholtz-Zentrum Geesthacht, located at DESY, Hamburg, Germany. Synchrotron radiation with a nominal energy of 100 keV was used for measuring diffraction patterns. The volume fractions of austenite and martensite were determined by using a method that directly compares the integrated intensities of martensite and austenite diffraction peaks [38]. The results are shown in Figure 21.1, where an increase of the phase fraction of austenite with increasing aging time can be seen. After 100 h of aging, the austenite phase fraction reaches a plateau at about 32 vol%, which is in the range of the equilibrium phase fraction at that aging temperature determined by thermodynamic simulations [13].

21.2.2 Small-Angle X-ray Scattering (SAXS)

In order to reveal the changes of the phase fraction of the intermetallic precipitates during aging, small-angle X-Ray scattering (SAXS) was performed on the same maraging steel samples as used for XRD. In Figure 21.2 the integrated intensity of the scattering signal, which is proportional to the volume fraction of the intermetallic precipitates, is shown above the aging time. It can be seen that the phase fraction of the precipitates increases drastically after short aging times, slightly during further aging, and remains almost constant after 20 000 s. Therefore, it can be concluded that with increasing amounts of reverted austenite (see Figure 21.1) the phase fraction of the precipitates remains almost constant during aging. This is consistent with AP measurements, thermodynamic, and kinetic simulations on the same maraging steel in a previous study [13]. Furthermore, it supports the assumption of an independent formation of reverted austenite and intermetallic precipitates as both are formed simultaneously from the beginning of aging.

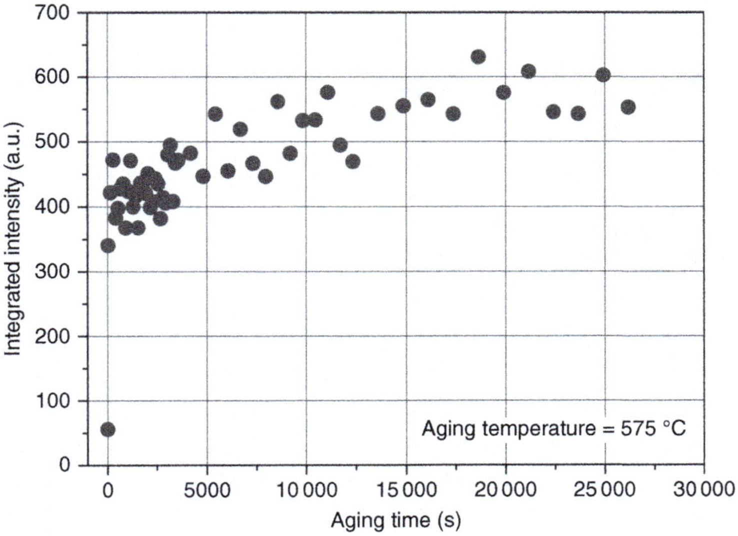

Figure 21.2 Integrated intensity of the SAXS signal as a function of aging temperature of the maraging steel investigated (see Table 21.1, alloy no. 1). (Eidenberger 2011 [39]. Reproduced with permission of Wiley.)

In conclusion, these measurements are examples of how XRD measurements and SAXS using synchrotron radiation can be combined for advanced maraging steels to get information of the microstructural evolution during aging.

Another example for applying *in situ* small-angle scattering for the characterization of the microstructural evolution during aging in steels can be found in [28]. In this study, information about the characteristic length scale and the volume fraction of precipitates in a novel steel alloyed with 25 at.% Co and 9 at.% Mo were gained, for the exact chemical composition see Table 21.1, alloy no. 6. A scaling model was used for the evaluation of the SAXS patterns. This model describes different morphologies of the precipitates and in case of the steel investigated a transition from spherical toward rod-shaped precipitates during aging was proposed.

21.2.3 *In situ* High-Energy X-ray Diffraction under Tensile Loading

For studying martensitic phase transformations in steels, *in situ* XRD using synchrotron radiation is valuable to get information on the onset and early stages of transformation. In maraging steels containing reverted austenite, it is known from literature that the amount of reverted austenite significantly influences the mechanical properties [14, 40]. The mechanical stability of reverted austenite in maraging steels under external loading was examined by *in situ* XRD using high-energy synchrotron radiation for the first time [15]. For the experiments, a tensile testing machine was mounted at the high-energy X-ray beamline.

For all samples, subjected to different aging times, a decrease of the phase fraction of austenite with increasing strain was observed. In [13] it is shown that this transformation behavior can be described by a model introduced by Ludwigson and Berger [41]. During that course, an increase of the stability of reverted austenite was shown with increasing

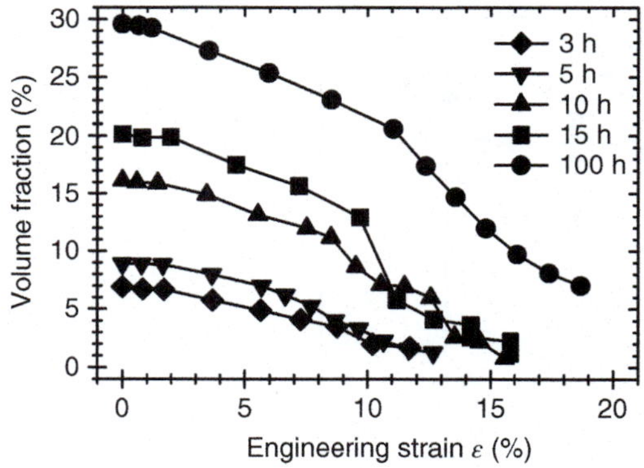

Figure 21.3 Volume fraction of austenite as a function of engineering strain after aging treatments at 575 °C. (Eidenberger 2011 [39]. Reproduced with permission of Wiley.)

aging time. Further evaluation of the diffraction patterns are given in [15], where the domain size of the crystallites and the lattice parameters were determined as a function of strain for differently aged samples. Exemplarily, the volume fraction of austenite versus the engineering strain in heat-treated samples of the investigated maraging steel is shown in Figure 21.3 [15].

21.3 Characterization Using Small-Angle Neutron Scattering (SANS)

All investigations conducted to characterize modern materials described in the following were small-angle neutron scattering (SANS) experiments performed at the SANS-2 beamline of the Helmholtz-Zentrum Geesthacht in Germany.

21.3.1 Use of SANS to Study Precipitates in Steels

21.3.1.1 Analysis of Secondary Hardening Carbides

During the complex heat treatments of modern steels, different precipitation reactions occur. Such steels, for example, high-speed or hot-work tool steels, contain a small volume fraction of large blocky primary carbides and different types of extremely fine secondary hardening carbides formed during tempering (e.g., [42, 43]). These nanometer-sized secondary-hardening carbides are of the MC or M_2C type and are responsible for a balanced combination of high strength and ductility. Their volume fractions strongly depend on the chemical composition and the heat-treatment of the steel [44]. During application of the steel, the secondary hardening carbides are prone to coarsening and the unstable M_2C-type may also dissolve, leading to the formation of the stable M_6C and $M_{23}C_6$ carbides. This over-aging behavior of complex steels has been excessively studied in a variety of publications, for example, [42, 43, 45] but due to the smallness of the investigated microstructure it is difficult to obtain representative size distributions, even with TEM or AP. Because the modeling of precipitation reactions necessary for a prediction of mechanical properties requires information on the type, the volume fractions, and the size distributions of the prevailing precipitates, often indirect complementary methods have to be applied. To gain these mentioned parameters, SANS is a suitable technique. Especially for ferromagnetic samples,

SANS has the advantage of yielding information on the chemical composition of the precipitates, even in their very early stages, for example, [46–48]. Our group has used this powerful method in recent years and selected results are summarized in the following text.

In different studies concentrating on the over-aging behavior of secondary hardening carbides in two different tool steels (for chemical composition see Table 21.1, alloys no. 2 and 3), SANS was used to gain the size distributions of the prevailing precipitate fractions [16–20]. The size distributions were calculated using the two-phase model [49] and compared to data gained by TEM, energy-filtered transmission electron microscopy (EFTEM), and high-energy XRD measurements. The combination of these methods has been proven successful in characterizing the types of secondary hardening carbides that occur as well as their development during the applied heat treatments [18]. This data has been further compared to a simulation of the precipitate evolution using a novel thermo-kinetic approach. The results of these calculations agree well with the experimental data, although an improvement of the databases is stated to be inevitable to improve the quality of the simulations [19, 20].

The above-mentioned chemical information yielded by SANS has been used to cross-check the chemistry of carbides gained by TEM and AP [16, 17]. Therefore, the so-called A-value has been calculated from the measured SANS curves as well as from chemical compositions gained by the complementary methods used. The A-value describes the ratio of magnetic to nuclear scattering contrast and can be directly related to the measured particles' magnetism and chemical composition. In a further theoretical approach, the A-value as a function of the Fe-content of the occurring carbides has been estimated. It could be shown that the Fe-content in the precipitates decreases during tempering and a hint to an overestimation of the Fe-content in AP data was found [17].

21.3.1.2 Analysis of Intermetallic Precipitates

To gain knowledge on the size and volume fractions of nanometer-sized precipitates is a demanding task, especially if statistically relevant data is required. As already mentioned above, the method of SANS is designed for yielding size distributions of very small particles dispersed in a homogeneous matrix and when probing ferromagnetic samples also provides information on the chemistry. SANS was first used for the study of maraging steels by Servant and Bouzid [50]. Complementary methods like TEM, electron energy loss spectroscopy (EELS), or AP have been supportively used to study the precipitation behavior, for example, [51, 52] but details on the precipitation sequence often remain unclear. Hence, various studies employing SANS concentrating on the description of the precipitation behavior of different maraging steels were carried out by our group and are shortly summarized in the following text.

Staron *et al.* [21] and Jamnig *et al.* [22] determined the precipitate size distributions in a cobalt-free corrosion resistant maraging steel (for chemical composition see Table 21.1, alloy no. 4), first investigated by Gemperle *et al.* [53]. They fitted the measured scattering cross-section using the two-phase model [49] accounting for interparticle interference by employing a monodisperse approximation described by Pedersen [54]. Qualitative information on the precipitates' chemistry gained by additional EFTEM measurements was compared to SANS data evaluated with regard to A-value of the precipitates. It could be shown, that at an aging temperature of

475 °C, the investigated maraging steel exhibits rather fast precipitation kinetics and the precipitates were concluded to be of the $Ti_6Si_7Ni_{16}$ G-phase type. In a comment to [22], Guo and Sha [55] pointed out the problem of a straightforward particle analysis if more than one type of precipitate is present. They claimed a probable coexistence of $Ti_6Si_7Ni_{16}$ G-phase and Ni_3Ti precipitates in the investigated steel. Our groups' successive AP investigations on the material performed by Schober *et al.* [23] showed that the precipitation of a precursor phase precedes the formation of both a spherically shaped $Ti_6Si_7Ni_{16}$ G-phase and a coarse rod-shaped Ni_3(Ti,Al) η-phase.

Another maraging-type alloy has also been a subject of SANS investigations due to its interesting precipitation sequence and its extraordinary mechanical properties (see [28, 29]). The alloy Fe-25 at.%Co-9 at.%Mo (see also Table 21.1, alloy no. 6) follows a long tradition of Co-free and Co-containing age-hardenable Fe-Mo and Fe-W alloys, see, for example, [30, 56–59]. First SANS investigations by Eidenberger *et al.* [60] on a solution annealed, quenched, and age-hardened sample exhibited scattering curves with a distinct maximum due to a high number density of precipitates resulting in interparticle interference. The size distribution for the peak-hardened state was obtained from the magnetic scattering curves assuming spherical precipitates. However, additional TEM and AP investigations showed a percolated precipitate structure, hence putting the gained size-distribution into question and asking for an adapted analysis model. Nevertheless, it could be shown that the phase responsible for the age-hardening of the alloy is of the type μ-$(Fe,Co)_7Mo_6$, thus proving the findings of preliminary AP studies of Leitner *et al.* [29] SANS curves of two different aging states have been further analyzed in a comprehensive study, combined with *ab initio* calculations performed in the framework of density functional theory and AP investigations [61]. The *ab initio* calculations were implemented in the WIEN2k code [62] using the augmented plane wave + local orbital (apw + lo) basis set [63] and yielded the magnetic moment per unit cell of the equilibrium precipitate $(Fe,Co)_7Mo_6$ μ-phase. Although no direct information on the chemistry of the precipitates can be gained by SANS, hypothetic concentrations, for example, from AP measurements can be checked using the already mentioned A-value. It could be shown, that the equilibrium $(Fe,Co)_7Mo_6$ precipitates indeed carry a magnetic moment agreeing with the one calculated *ab initio*. Discrepancies in A-value at the lower aging state lead to the conclusion that metastable Mo-rich precursor phases, as were mentioned in [29], probably precede the formation of the μ-phase. Especially at these early stages, an overestimation of the Fe-content is probable, leading to false precipitate compositions (see also [17]). In order to investigate this matter, the dependence of the A-value on the Fe-content has been calculated, yielding the conclusion that in addition to the mentioned Fe-overestimation, the magnetic moment of possible precursor phases has to be higher than that of the equilibrium μ-phase [61]. Another benefit of the use of the A-value calculated from SANS curves is the possibility of critically considering the parameters for the AP data analysis. In the referred study [61], this has been successfully accomplished in finding the most reasonable isoconcentration threshold for the analyzed precipitates (for further reading on isoconcentration surfaces see, e.g., [64, 65]).

Another complex maraging steel, which hardens not only by carbides but additionally due to the formation of an intermetallic NiAl-phase has also been subject of SANS investigations performed by our group in recent years [24–26]. For the alloy's chemical composition see Table 21.1, alloy no. 5. At least four different precipitate types occur simultaneously after a standard heat treatment and their size distributions and chemical compositions change with ongoing annealing. To characterize these precipitates, SANS has been combined with AP and TEM investigations, both methods yielding only limited statistical data but quantitative compositional analyses or crystallographic information, respectively. In addition, TEM analyses become difficult for particles in the size range of a few nanometers. Furthermore, if ferromagnetic samples are investigated, a strong deflection of the electron beam complicates the experiment [66]. For the analysis of the SANS data gained in the mentioned publications [24, 25] all particles were assumed to be of spherical shape and to be nonmagnetic. Deviations from a homogeneous magnetization due to magnetoelastic coupling of magnetic moments with strain fields of dislocations [67] or micromagnetic effects [68] have been neglected. From the magnetic scattering curves the size distributions were determined by performing an indirect Fourier transformation (IFT), using a method described by Glatter [69], calculating a curve based on the sum of 20 cubic B-splines. In addition, another fit was performed using four independent Gaussian distributions defined in its limits by the findings of the complementary TEM and AP measurements. By comparing these two fits, it was possible to decide on the accurateness of the found particle classes. Generally, a good approximation of the particle volume fractions could be accomplished, because the size distributions only weakly overlap. Only when dealing with the smallest particle fractions, problems arose because of technical limitations of both TEM and AP. The results of these studies also indicated that scattering from structures in the size range of MC carbides (radius $\approx$ 40 nm) might not only be attributed to particle scattering, but possibly to the occurrence of micromagnetic effects as well.

In [26], the microstructural analysis of the investigated complex steel was extended with simulations of the observed precipitation behavior using a novel model developed by Svoboda *et al.* [70] and Kozeschnik *et al.* [71] employed into the software "Mat-Calc," see, for example, [72]. The performed simulations were successful and the experimental results of the combined methods SANS, TEM, and AP could be consistently reproduced. It was shown that the computer simulations can support the interpretation of the experimental findings, whereas the experiments lead to an improvement of the quality of thermodynamic databases and the significance of kinetic simulations.

As already mentioned above, for the analysis of scattering curves with the two-phase model [49], precipitates in steels are commonly regarded as nonmagnetic holes in a ferromagnetic matrix [48]. Especially in the early stages of precipitation, this assumption is often imprecise and leads to errors in the interpretation of SANS data. In a study performed on a model Fe-alloy containing 1 at.%Cu (for the exact composition see Table 21.1, alloy no. 7) Schober *et al.* [34] emphasized the question of Fe-content in nanometer-sized bcc Cu clusters in bcc Fe. This topic had been an "unsolved mystery" [73] and subject of various publications comparing SANS and AP measurements of Fe-Cu alloys, see, for example, [74, 75]. To solve that "mystery," the R-value, which is the ratio of integrated magnetic and nuclear scattering intensities, and directly related to the above-mentioned A-value, was calculated from the magnetic and nuclear scattering curves of differently aged Fe-1 at.%Cu samples. These values were compared to R-values that had been derived from the chemical compositions of Cu-rich precipitates measured by means of AP, combined with *ab initio* calculations of the magnetic moment of Fe-containing bcc-Cu clusters. The comparison of measured and calculated R-values versus aging time is shown in Figure 21.4. It could be shown, that even at aging times

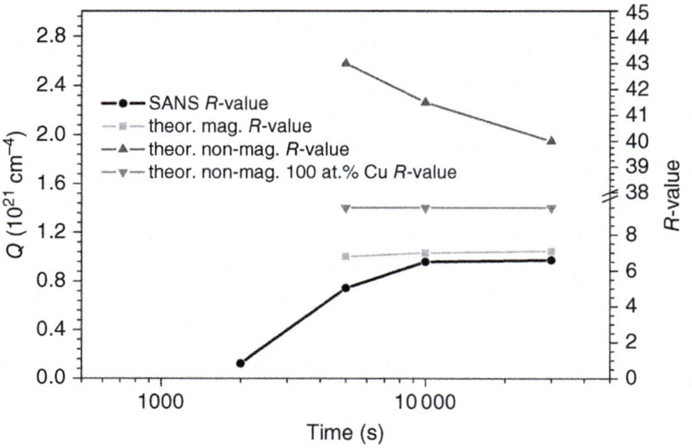

Figure 21.4 Measured and calculated *R*-values versus aging time for an Fe-1 at.%Cu alloy (see Table 21.1, alloy no. 7). (Eidenberger 2011 [39]. Reproduced with permission of Wiley.)

of 30 000 s at 500 °C, bcc-Cu clusters contain almost 50 at.% of Fe and hence carry a considerable magnetic moment. Only by taking this magnetic moment into account, AP and SANS yielded results that were in accordance and the former "mystery" was solved [34]. In a continuative work of Schober *et al.* [35], the method of comparing AP and SANS data via the yielded chemical compositions during data analysis is further described. Thereby, the conducted AP measurements were generally improved regarding the size of the analyzed volume compared to former works of other groups (see, e.g., [76, 77]). It was shown that the combination of SANS and AP opens up new possibilities for the study of precipitation kinetics.

21.3.2 *In situ* SANS during Continuous and Isothermal Aging

During the course of the investigation of the Fe-Co-Mo alloy mentioned above [28, 29, 60, 61] it became clear that knowledge on the nature of the precipitation sequence is of utmost importance for the understanding of the development of the prevailing phases. In studies on the precipitation behavior of Fe-Mo and Fe-Co-Mo alloys there was a discussion on the possibility of spinodal decomposition being the underlying mechanism of the occurring phase separation during the applied heat treatments [59, 78, 79]. Another group investigating the precipitation behavior of Fe-Mo alloys distinguished two stages of the decomposition process and assigned the first stage to a nucleation and growth-type formation of Guinier–Preston zones along the crystallographic ⟨100⟩ directions due to large coherency strains [58, 80]. To shed light on the issue, a series of *in situ* aging experiments have been performed by our group, employing a newly developed *in situ* furnace [31–33]. This furnace was placed between the pole shoes of the electromagnet of the SANS-2 beamline of the Helmholtz-Zentrum Geesthacht, Germany, and has been elaborately described by Staron *et al.* [36] In this work, the differences in the precipitation behavior of the Fe-Co-Mo alloy and the Fe-Cu alloy, also described in [34, 35], have been compared. It was shown that the two alloys exhibit clearly distinguishable precipitation sequences. In [31] the *R*-value development of the measured Fe-Co-Mo alloy during continuous aging with two different heating rates was described,

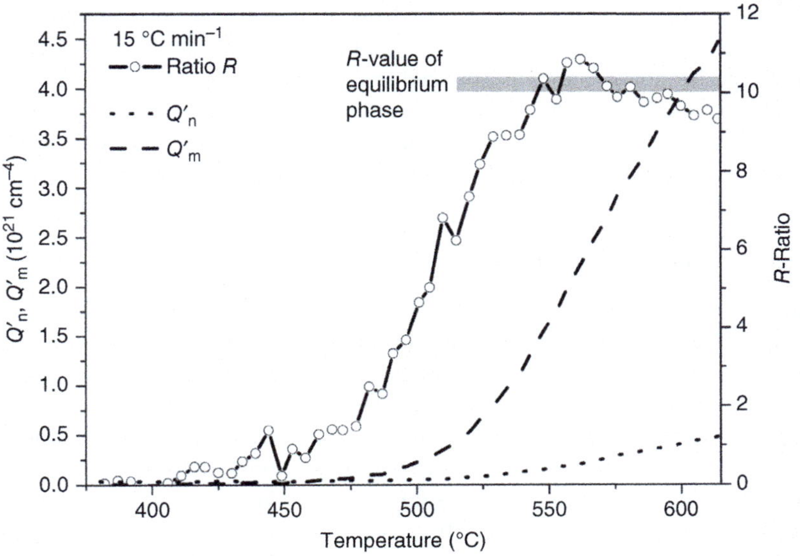

Figure 21.5 Development of the *R*-value, and hence chemical composition of the Fe-Co-Mo alloy, see Table 21.1, alloy no. 6. (Eidenberger 2011 [39]. Reproduced with permission of Wiley.)

yielding the onset temperature of the precipitation reaction as well as a first insight into the developing chemistry of precipitates. By a comparison with AP and DSC measurements, the three-step precipitation reaction mentioned in [29] was further investigated. With additional DSC and high-resolution TEM, as well as high-energy XRD measurements during continuous heating presented in [32] a thorough understanding of the precipitation sequence could be reached. With the evaluation of the *in situ* SANS data via the calculation of the *R*-value it was possible to show a gradual change in chemistry in the early stages of the precipitation reaction. This development of the *R*-value is presented in Figure 21.5. Supported by the *in situ* high-energy XRD and DSC measurements it could be shown that a bcc-precursor phase formed by the diffusion of Mo precedes the formation of the equilibrium $(Fe,Co)_7Mo_6$ μ-phase. During the first stage of the precipitation process, rectangular Mo-rich features are formed, which could be visualized by AP measurements [32]. The question of spinodal decomposition was dealt with in [33], by performing isothermal *in situ* SANS experiments at different temperatures. In a first attempt to differentiate the behavior of the material during aging at higher and lower temperatures, the peak height and peak position of the measured scattering curves were described with power law fits and the *R*-value development was compared. The chemistry of the developing precipitates was monitored with additional AP measurements. Because the Fe-25 at.%Co-9 at.%Mo alloy is situated close to the calculated spinodal presented in [78] an unambiguous determination between spinodal decomposition and nucleation is complicated. However, by the combination of the *R*-value characteristics with information gained by AP a gradual change in decomposition mechanism from spinodal decomposition toward nucleation with decreasing supersaturation could be ascertained [33].

The monitoring of the *R*-value measured in an isothermal *in situ* SANS experiment has been also successfully adopted by Schober *et al.* [37] for an investigation of the precipitation behavior of intermetallic NiAl particles in an Fe-Ni-Al model alloy (for composition see Table 21.1, alloy no. 8). In this study, the comparability of the complementary methods AP and SANS was assessed and compared to hardness values by using the Friedel formula [81]. It could be shown that the precipitated NiAl particles carry a magnetic moment due to their considerable Fe-content in the early stages. The combination of *in situ* SANS and AP yielded unique information about the prevailing precipitation kinetics [37].

21.3.3 SANS with Variable Magnetic Field

As demonstrated above, SANS is a powerful technique to study even nanometer-sized precipitates in modern steels. By exposing the sample to an external field of normally 1.3–2 T the measured material is supposed to consist of a single domain, magnetized to saturation, for example, [16, 25, 82] SANS data gained by such measurements are then interpreted by using the two-phase model [49] in terms of particles in a homogeneously saturated matrix. In reality, for example, due to the magnetocrystalline anisotropy of individual crystallites [83] or the anisotropies due to strain fields around dislocations [67], SANS intensities not stemming from dispersed particles may arise. The magnitude of this contribution depends on the microstructure of the steel. These deviations from the homogeneously magnetized state are usually neglected in the analysis of precipitates, see also the above-mentioned investigations on the precipitation behavior of a complex maraging steel [24–26], which already supposed the incidence of micromagnetic effects. In a subsequent work conducted by Bischof *et al.* [27] SANS measurements at different external magnetic fields were performed. To identify the effect of spin-misalignment scattering is very important because such micromagnetic effects are of the same length scale as the resolution range of the SANS technique and the investigated precipitates [82]. In our study [27], the ferromagnetic steel samples containing small amounts of carbide forming elements as well as Ni and Al have been analyzed with magnetic field-dependent SANS after being exposed to different heat treatments. A schematic illustration of a small detail of a typical steel microstructure subjected to an external magnetic field is shown in Figure 21.6a,b. An algorithm allowing for the correction of the contribution of spin-misalignment scattering to the SANS cross-sections has been developed thereby eliminating large errors due to micromagnetic effects. Using the proposed correction method scattering curves measured in an off-saturation regime can be nevertheless analyzed successfully. Especially for nuclear scattering curves or for the investigation of magnetically anisotropic materials this correction is recommended [27]. Figure 21.6c,d show an example of the successful use of the correction for SANS curves of the maraging steel investigated [27].

21.4 Conclusions

The use of synchrotron radiation and neutrons for the application in materials science, especially for the characterization and development of modern steels, offers a wide range of possible experiments. Because of the high X-ray and neutron intensities provided by large-scale facilities, short exposure times and extraordinary statistics are possible, leading to a large amount of valuable data. Additionally, these facts entail the feasibility of *in situ* experiments, for example, during heating or mechanical loading, in which

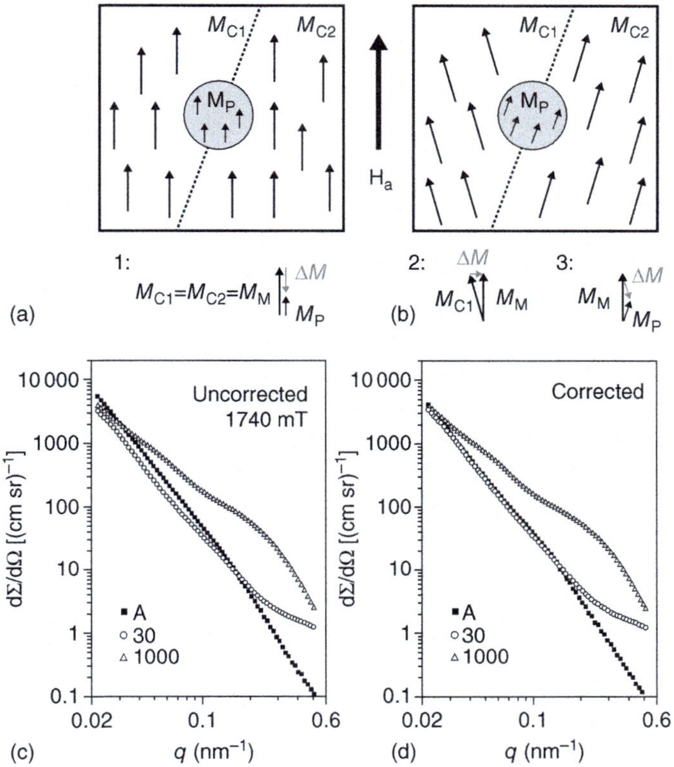

Figure 21.6 Schematic illustration of a small detail of a typical steel microstructure containing two matrix crystals and a particle located at the boundary in between (dotted line). By applying an external field, the sample is either magnetized to saturation (a), or not (b). In case (b), spin-misalignment scattering occurs, leading to a distortion of the measured SANS intensities; nuclear scattering curves for an as-quenched sample (A) and samples aged at 610 °C for 30 min (30) and 1000 min (1000) of alloy no. 5. (c) Experimental and (d) corrected as proposed. (Eidenberger 2011 [39]. Reproduced with permission of Wiley.)

phase transformations can be readily observed. Large and flexible sample environments allow for the mounting of macroscopic samples with arbitrary dimensions or even components, which in turn stand for a large sample volume and hence excellent statistical relevance of the gained data. This is a huge advantage compared to direct measurement techniques, such as SEM, TEM, or AP, which often require complicated sample preparation and only allow the mounting of very small specimens. On the other hand, the information obtained by these methods is of great use when it comes to the interpretation of data gained by scattering techniques, for example, SANS scattering curves. Here, for instance, the inputs regarding particle shape and chemical composition are important factors that can guarantee correct data evaluation. As it has been shown in the above-mentioned text, the combination of scattering techniques with complementary methods has proved successful for the analysis of precipitates in steels. In addition, the use of *ab initio* simulations and thermo-kinetic calculations has emerged to be capable of further improving our knowledge of the ongoing reactions in the investigated materials. To summarize, it can be stated that scattering and diffraction with synchrotron radiation and neutrons have made an important impact in the characterization and development of modern steels in the past 20 years and will surely continue to yield interesting results in years to come.

References

1 Clemens, H. and Scheu, C. (2008) in *Neutrons and Synchrotron Radiation in Engineering Materials Science* (eds W. Reimers, A.R. Pyzalla, A. Schreyer, and H. Clemens), Wiley-VCH Verlag GmbH, Weinheim, p. 3.

2 Zickler, G.A., Clemens, H., and Schreyer, A. (2007) *BHM*, **152**, 282.

3 Reimers, W., Pyzalla, A.R., Schreyer, A., and Clemens, H. (2008) in *Neutrons and Synchrotron Radiation in Engineering Materials Science* (eds W. Reimers, A.R. Pyzalla, A. Schreyer, and H. Clemens), Wiley-VCH Verlag GmbH, Weinheim.

4 Honeycombe, R.W.K. and Bhadeshia, H.K.D.H. (1995) *Steels-Microstructure and Properties*, Butterworth Heineman, Oxford, p. 197.

5 Sato, A., Sunaga, Y., and Mori, T. (1977) *Acta Metall.*, **25**, 627.

6 Katz, Y., Mathias, H., and Nadiv, S. (1983) *Metall. Trans. A*, **14**, 801.

7 Fanninger, G. and Hartmann, U. (1972) *HTM*, **27**, 233.

8 Taran, Y.V., Daymond, M.R., Eifler, D., and Schreiber, J. (2002) *Appl. Phys. A*, **74**, 1391.

9 Liss, K.-D., Bartels, A., Schreyer, A., and Clemens, H. (2003) *Texture Microstruct.*, **35**, 219.

10 Offerman, S.E., van Dijk, N.H., Sietsma, J., Grigull, S., Lauridsen, E.M., Margulies, L., Poulsen, H.F., Rekveldt, M.T., and van der Zwaag, S. (2002) *Science*, **298**, 1003.

11 Pyzalla, A., Camin, B., Buslaps, T., Di Michiel, M., Kaminski, H., Kottar, A., Pernack, A., and Reimers, W. (2005) *Science*, **308**, 92.

12 Lippmann, T., Lottermoser, L., Beckmann, F., Martins, R.V., Dose, T., Kirchhof, R., and Schreyer, A. (2007) in *HASYLAB Annual Report* (eds W. Caliebe, W. Drube, K. Rickers, and J.R. Schneider), HASYLAB/DESY, Hamburg, p. 113.

13 Schnitzer, R., Radis, R., Noehrer, M., Schober, M., Hochfellner, R., Zinner, S., Povoden-Karadeniz, E., Kozeschnik, E., and Leitner, H. (2010) *Mater. Chem. Phys.*, **122**, 138.

14 Nakagawa, H. and Miyazaki, T. (1999) *J. Mater. Sci.*, **34**, 3901.

15 Zickler, G.A., Schnitzer, R., Hochfellner, R., Lippmann, T., Zinner, S., and Leitner, H. (2009) *Int. J. Mater. Res.*, **100**, 1566.

16 Leitner, H., Staron, P., Clemens, H., Marsoner, S., and Warbichler, P. (2005) *Mater. Sci. Eng., A*, **398**, 323.

17 Leitner, H., Clemens, H., Akre, J., Danoix, F., and Staron, P. (2007) *Int. J. Mater. Res.*, **98**, 11.

18 Bischof, M., Staron, P., Caliskanoglu, D., Leitner, H., Scheu, C., and Clemens, H. (2008) *Mater. Sci. Eng., A*, **472**, 148.

19 Leitner, H., Staron, P., Bischof, M., Clemens, H., Kozeschnik, E., Svoboda, J., and Fischer, F.D. (2004) Proceedings of *Mater. Sci. Technol.* 2004, New Orleans, Louisiana, p. 623.

20 Sonderegger, B., Kozeschnik, E., Bischof, M., Leitner, H., Clemens, H., Svoboda, J., and Fischer, F.D. (2006) Proceedings of 7th Tooling Conference, Tool 06, Torino, Italy, p. 533.
21 Jamnig, B., Ebner, R., Staron, P., Clemens, H., Leitner, H., Jeglitsch, F., and Warbichler, P. (2002) Proceedings of 6th International Tooling Conference, Karlstad, Sweden, p. 549.
22 Staron, P., Jamnig, B., Leitner, H., Ebner, R., and Clemens, H. (2003) *J. Appl. Crystallogr.*, **36**, 415.
23 Schober, M., Schnitzer, R., and Leitner, H. (2009) *Ultramicroscopy*, **109**, 553.
24 Bischof, M., Erlach, S., Leitner, H., Staron, P., and Clemens, H. (2005) Proceedings of Solid–Solid Phase Transformations in Inorganic Materials, PTM, Phoenix, AZ, p. 529.
25 Bischof, M., Erlach, S., Staron, P., Leitner, H., Scheu, C., and Clemens, H. (2005) *Z. Metallkunde*, **96**, 9.
26 Leitner, H., Bischof, M., Clemens, H., Erlach, S., Sonderegger, B., Kozeschnik, E., Svoboda, J., and Fischer, F.D. (2006) *AEM*, **8** (**11**), 1066.
27 Bischof, M., Staron, P., Michels, A., Granitzer, P., Rumpf, K., Leitner, H., Scheu, C., and Clemens, H. (2007) *Acta Mater.*, **55**, 2637.
28 Zickler, G.A., Eidenberger, E., Leitner, H., Stergar, E., Clemens, H., Staron, P., Lippmann, T., and Schreyer, A. (2008) *Mater. Charact.*, **59**, 1809.
29 Leitner, H., Schober, M., Clemens, H., Caliskanoglu, D., and Danoix, F. (2008) *Int. J. Mater. Res.*, **99**, 4.
30 Danninger, H., Rouzbahani, F., Harold, C., Ponemayr, H., Daxelmüller, M., Simancik, F., and Izdinsky, K. (2005) *Powder Metall. Prog.*, **5** (**2**), 92.
31 Eidenberger, E., Schober, M., Stergar, E., Leitner, H., Staron, P., and Clemens, H. (2010) *Met. Mater. Trans. A*, **41** (**5**), 1230.
32 Eidenberger, E., Schober, M., Schmoelzer, T., Stergar, E., Staron, P., Leitner, H., and Clemens, H. (2010) *Phys. Status Solidi A*, **207**, 2238.
33 Eidenberger, E., Schober, M., Staron, P., Caliskanoglu, D., Leitner, H., and Clemens, H. (2010) *Intermetallics*, **18**, 2128.
34 Schober, M., Eidenberger, E., Leitner, H., Staron, P., Reith, D., and Podloucky, R. (2010) *Appl. Phys. A*, **99**, 697.
35 Schober, M., Eidenberger, E., Staron, P., and Leitner, H. (2011) *Microsc. Microanal.*, **17**, 26.
36 Staron, P., Eidenberger, E., Schober, M., Sharp, M., Leitner, H., Schreyer, A., and Clemens, H. (2009) *J. Phys. Conf. Ser.*, Proceedings of the SAS, **247**, 1.
37 Schober, M., Lerchbacher, C., Eidenberger, E., Staron, P., Clemens, H., and Leitner, H. (2010) *Intermetallics*, **18**, 1553.
38 Faninger, G. and Hartmann, U. (1972) *J. Heat Treat. Mater.*, **27**, 233.
39 Eidenberger, E., Schnitzer, R., Zickler, G.A., Schober, M., Bischof, M., Staron, P., Leitner, H., Schreyer, A., and Clemens, H. (2011) *Adv. Eng. Mater.*, **13**, 664–673.
40 Schnitzer, R., Zickler, G.A., Lach, E., Clemens, H., Zinner, S., Lippmann, T., and Leitner, H. (2010) *Mater. Sci. Eng., A*, **527**, 2065.
41 Ludwigson, D.C. and Berger, J.A. (1969) *J. Iron Steel Inst.*, **207**, 63.
42 Rong, W., Andren, H.-O., Wisell, H., and Dunlop, G.L. (1992) *Acta Metall. Mater.*, **40**, 1727.
43 Fischmeister, H.F., Karagöz, S., and Andren, H.-O. (1988) *Acta Metall.*, **36**, 817.

44 Ebner, R., Leitner, H., Caliskanoglu, D., Marsoner, S., and Jeglitsch, F. (2001) *Z. Metallkunde*, **92**, 820.

45 Karagöz, S., Fischmeister, H.F., Andren, H.-O., and Cai, G.-J. (1992) *Metall. Trans. A*, **23**, 1631.

46 Große, M., Gokhman, A., and Böhmert, J. (2000) *Nucl. Instrum. Methods Phys. Res. B*, **160**, 515.

47 Mathon, M.H., de Carlan, Y., Geoffroy, G., Averty, X., Alame, A., and de Novion, C.H. (2003) *J. Nucl. Mater.*, **312**, 236.

48 Coppola, R., Fiori, F., Little, E.A., and Magnani, M. (1997) *J. Nucl. Mater.*, **205**, 131.

49 Kostorz, G. (1979) in *Treatise on Materials Science and Technology: Neutron Scattering*, vol. **15** (eds G. Kostorz and H. Herman), Academic Press, New York, p. 227.

50 Servant, C. and Bouzid, N. (1989) *Philos. Mag. B*, **60**, 659.

51 Vasudevan, V.K., Kim, S.J., and Wayman, C.M. (1990) *Metall. Trans. A*, **21**, 2655.

52 Sha, W., Cerezo, A., and Smith, G.D.W. (1993) *Metall. Trans. A*, **24**, 1221.

53 Gemperle, A., Gemperlova, J., Sha, W., and Smith, G.D.W. (1992) *Mater. Sci. Technol.*, **8**, 546.

54 Pedersen, J.S. (1994) *J. Appl. Crystallogr.*, **27**, 595.

55 Guo, Z. and Sha, W. (2004) *J. Appl. Crystallogr.*, **37**, 325.

56 Geller, Y. (1983) *Instrumentalniye Staly*, Metallurgiya, Moscow, p. 500.

57 Karpov, M.I. (1998) Proceedings of PM World Congress (EPMA), Granada, Spain, p. 519.

58 Isheim, D. (2000) *Acta Mater.*, **48**, 2873.

59 Miyazaki, T., Takagishi, S., Mori, H., and Kozakai, T. (1980) *Acta Metall.*, **28**, 1143.

60 Eidenberger, E., Stergar, E., Leitner, H., Scheu, C., Staron, P., and Clemens, H. (2008) *BHM*, **153** (**7**), 247.

61 Eidenberger, E., Stergar, E., Leitner, H., Staron, P., Spitaler, J., Ambrosch-Draxl, C., and Clemens, H. (2009) *Appl. Phys. A*, **97**, 331–340.

62 Blaha, P., Schwarz, K., Madsen, G., Kvasnicka, D., and Luitz, J. (2001) in *Program Package WIEN2k* (ed. K. Schwarz), TU Wien, Wien.

63 Sjöstedt, E., Nordström, L., and Singh, D.J. (2000) *Solid State Commun.*, **114**, 15.

64 Hellman, O.C., Blatz du Rivage, J., and Seidman, D.N. (2003) *Ultramicroscopy*, **95**, 215.

65 Cerezo, A. and Davin, L. (2007) *Surf. Interface Anal.*, **39**, 184.

66 Williams, D.B. and Carter, C.B. (1996) *Transmission Electron Microscopy II (Diffraction)*, Plenum Press, New York, p. 253.

67 Göltz, G., Kronmüller, H., Seeger, A., Scheuer, H., and Schmatz, W. (1986) *Philos. Mag. A*, **54**, 213.

68 Weissmüller, J., Michels, A., Barker, J.G., Wiedenmann, A., Erb, U., and Shull, R.D. (2001) *Phys. Rev. B*, **63**, 214414.

69 Glatter, O. (1980) *J. Appl. Crystallogr.*, **13**, 7.

70 Svoboda, J., Fischer, F.D., Fratzl, P., and Kozeschnik, E. (2004) *Mater. Sci. Eng., A*, **385**, 166.

71 Kozeschnik, E., Svoboda, J., Fratzl, P., and Fischer, F.D. (2004) *Mater. Sci. Eng., A*, **385**, 157.

72 Kozeschnik, E. and Buchmayr, B. (2001) *Mathematical Modelling of Weld Phenomena*, vol. **5**, Book 738, Institute of Materials, London, p. 349.

73 Fine, M.E., Liu, J.Z., and Asta, M.D. (2007) *Mater. Sci. Eng., A*, **463**, 271.

74 Osamura, K., Okuda, H., Takashima, M., Asano, K., and Furusaka, M. (1993) *Metall. Trans. JIM*, **34**, 205.

75 Miller, M.K., Wirth, B.D., and Odette, G.R. (2003) *Mater. Sci. Eng., A*, **353**, 133.

76 Worrall, G.M., Buswell, J.T., English, C.A., Hetherington, M.G., and Smith, G.D.W. (1987) *J. Nucl. Mater.*, **148** (**1**), 107.

77 Pareige, P.J., Russell, K.F., and Miller, M.K. (1996) *Appl. Surf. Sci.*, **94–95**, 362.

78 Kozakai, T., Aihara, H., Doi, M., and Miyazaki, T. (1985) *Trans. Iron Steel Inst. Jpn.*, **25**, 159.

79 Doi, M., Tanabe, H., and Miyazaki, T. (1987) *J. Mater. Sci.*, **22**, 1328.

80 Isheim, D., Hellman, O.C., Seidman, D.N., Danoix, F., and Blavette, D. (2000) *Scr. Mater.*, **42**, 645.

81 Nembach, E. (1997) *Particle Strengthening of Metals and Alloys*, John Wiley & Sons Inc., New York.

82 Wagner, W. and Kohlbrecher, J. (2005) in *Modern Techniques for Characterizing Magnetic Materials* (ed. Y. Zhu), Kluwer Academic Publishers, Boston, MA, p. 65.

83 Löffler, J.F., Braun, H.B., Wagner, W., Kostorz, G., and Wiedenmann, A. (2005) *Phys. Rev. B*, **71**, 134410.

22

The Contribution of High-Energy X-rays and Neutrons to Characterization and Development of Intermetallic Titanium Aluminides*

Thomas Schmoelzer, Klaus-Dieter Liss, Peter Staron, Andreas Stark, Emanuel Schwaighofer, Thomas Lippmann, Helmut Clemens, and Svea Mayer

22.1 Introduction

Ever increasing economic and environmental pressure promotes the development of more fuel-efficient combustion engines. One way of decreasing the fuel consumption of vehicles and airplanes is to reduce component weights. In addition to careful design, new structural materials are necessary to achieve substantial weight reductions. Intermetallic γ-TiAl-based alloys have low densities while exhibiting excellent high-temperature strength [1–8]. Therefore, this material class is expected to replace conventional high-temperature materials with high densities such as Ni-baed super-alloys in applications where temperatures range up to 750 °C. Possible applications are turbocharger wheels for reciprocating combustion engines, valves in sports and racing cars, and low pressure turbine blades in jet-engines [1, 3–5, 9–18].

During the past 20 years, considerable effort has been made to develop γ-TiAl-based alloys that are suited for service in these demanding areas while being economically competitive at the same time. To achieve this goal, different processing routes and alloy compositions have been intensively studied [9, 19, 20]. The multitude of phase transformations that occur in the Ti-Al binary system along with the pronounced dependency of transformation temperatures on chemical composition complicates the development of alloys that can be produced in a robust industrial production process [21]. Additionally, TiAl alloys frequently show rather coarse-grained microstructures after casting. Efforts have been undertaken to develop heat treatments that allow adjusting a microstructure with the desired mechanical properties in cast TiAl parts [20]. An additional approach was chosen for wrought alloys where cast ingots are forged and heat treated under conditions that result in the desired properties [19, 22]. In both cases a profound knowledge of the phase diagram and the kinetics of the occurring phase transformations are prerequisites for the design of a reliable production route.

Conventional *ex situ* characterization methods suffer from certain limitations in the case of the TiAl system. That the ordering reactions α-Ti(Al) (A3, *P6$_3$/mmc*) → α_2-Ti_3Al ($D0_{19}$, *P6$_3$/mmc*) and β (A2, $Im\bar{3}m$) → β_o (B2, $Pm\bar{3}m$) are too fast to preserve the

*This chapter is reproduced and updated from *Adv. Eng. Mater.* **13** (2011) 685–699 and *Adv. Eng. Mater.* **14** (2012) 445–448 by permission of Wiley-VCH

Neutrons and Synchrotron Radiation in Engineering Materials Science: From Fundamentals to Applications, Second Edition. Edited by Peter Staron, Andreas Schreyer, Helmut Clemens, and Svea Mayer.

disordered state by rapid quenching is only one example [23]. Complications might arise by the possible change in phase fractions if the phases present at high temperatures are determined by microscopical methods on heat-treated and quenched specimens. Additional problems may occur due to the fine γ-TiAl phase ($L1_0$, *P4/mmm*) lamellae that precipitate from supersaturated α/α_2 grains. With their width being as small as 10 nm, light-optical microscopy (LOM) or scanning electron microscopy (SEM) have insufficient resolution for determining phase fractions within lamellar regions [24]. Transmission electron microscopy (TEM), on the other hand, suffers from its restriction to small sample volumes.

As a tool for materials characterization, conventional diffraction methods have been used since the early beginnings of research activities on TiAl alloys [25–29]. Phase fractions can be determined independently of grain sizes and ordering phenomena are observed through the occurrence of superstructure reflections [30, 31]. Thus, in many cases, these methods offer considerable advantages compared to microscopical methods. Due to the widespread availability, most diffraction experiments performed on TiAl alloys have been conducted by means of laboratory X-ray sources. *In situ* studies were performed with a high-temperature diffractometer with all the ensuing limitations discussed in Section 22.3.1. An alternative that offers higher penetration depth, as compared to laboratory sources, is the use of synchrotron radiation. The high brilliance of the beam in combination with the availability of fast area detectors allow highly time resolved *in situ* studies [32, 33]. Neutrons are also used for diffraction experiments on TiAl and offer even higher penetration depth than synchrotron radiation [32]. A more detailed introduction to the different types of radiation and their properties is given in Section 22.2.

In this chapter, methods used for characterizing TiAl alloys are illustrated by several examples. Advantages and limitations of different methods and experimental setups are discussed and possible pitfalls highlighted.

22.2 High-Energy X-rays and Neutrons

Despite their different nature, high-energy X-rays and neutrons are both suited for performing diffraction as well as scattering experiments in materials science [34]. While neutrons are particle waves, X-rays are electromagnetic waves that cause the modes of interaction with the investigated material to differ correspondingly. For the case of nonmagnetic materials, neutrons interact with the nucleus of the atoms, while X-rays are scattered by their electrons. This leads to different contrasts and atomic form factors. While the scattering length in the case of X-rays is the classical electron radius r_e times the number of electrons, which corresponds in general to the atomic number of the investigated element, no clear dependency between atomic isotope and scattering length exists for neutrons [35]. Further, in the case of X-rays, the scattering centers in the electron cloud are extended, as compared to the wavelength, in contrast to point centers for neutrons. This causes the form factors to decrease with increasing momentum transfer $\mathbf{q}$ in the case of X-rays, while they remain constant for neutrons. Apart from a few highly absorbing isotopes, neutrons have generally a larger attenuation length than X-rays, which can range from centimeters to meters and micrometers to centimeters, respectively [32].

Another important difference is imposed by the corresponding radiation sources. X-rays can be generated by standard laboratory sources that are easy to handle and are comparatively cheap. If higher brilliance is required, synchrotron sources are used although at greatly increased costs and limited availability [36]. Neutrons, on the other hand, are only available at large-scale facilities such as research reactors and spallation sources [37]. Compared to synchrotrons, neutron sources are of low brilliance, which entails correspondingly larger beam cross-sections and longer acquisition times for obtaining diffraction patterns. Despite these limitations, neutron diffraction experiments can yield important information on the behavior of materials in a variety of conditions.

The scattering behavior of atoms is characterized by b_c for neutrons and $r_e \cdot f(q)$ for X-rays. Here, b_c is the bound coherent neutron scattering length, while r_e is the classical electron radius and $f(q)$ is the so-called atomic X-ray form factor. The spatial arrangement of the atoms is accounted for by the phase factor, which is dependent on the position of an individual atom within the unit cell. Summing up the atomic scattering lengths weighted by the respective phase factors for all atoms within the unit cell yields the structure factor F_{hkl}, which finally determines the intensity of the diffraction peaks [38, 39].

If the elements are distributed randomly in the crystal lattice, a virtual atom with a mean scattering length is calculated. Intensities of single reflections are derived from the structure factors, which in turn are calculated by summing up the mean atomic scattering lengths of the atoms in one unit cell and weighting them by an individual phase factor that depends on the position of the respective atom in the unit cell. For the main reflections of a phase, the phase factors are close to one and the scattering lengths of the individual atoms add up directly. Therefore, high intensities are usually observed for these reflections. Superstructure reflections do not occur in disordered crystals because they exhibit phase factors that are equal but of opposite sign, which results in a structure factor that is zero and causes the corresponding reflections to vanish.

In the case of ordered structures, no average scattering length is calculated. The scattering lengths from the individual atomic sites are multiplied by the phase factors for the corresponding reflection. Summing up these values gives the structure factor, which determines the intensity of the reflection. If the different elements exhibit different scattering lengths, the structure factor for superstructure reflections is no longer zero. Consequently, the intensity of the reflections then attains a finite value, which indicates that some degree of order exists in the investigated phase.

In the case of γ-TiAl, which exhibits $L1_0$ structure, this behavior has a peculiar consequence: Since for neutrons the scattering lengths of Ti (−3.370 fm) and Al (3.449 fm) [40] are very similar but of opposite sign, the structure factor for the main reflections is close to zero and the observed intensities are, therefore, rather small. On the contrary, if this calculation is performed for superstructure reflections, large structure factors and, therefore, high intensities are obtained.

When X-rays are used to probe the material, superstructure reflections have rather low intensities while the main reflections exhibit high intensities. Since these statements are valid for α_2, β_o, and γ phase, neutron and X-ray diffraction (XRD) are in fact complementary techniques if applied to TiAl alloys [30, 31]. For neutron and high-energy X-ray diffraction (HEXRD) experiments, the diffraction pattern is recorded in transmission

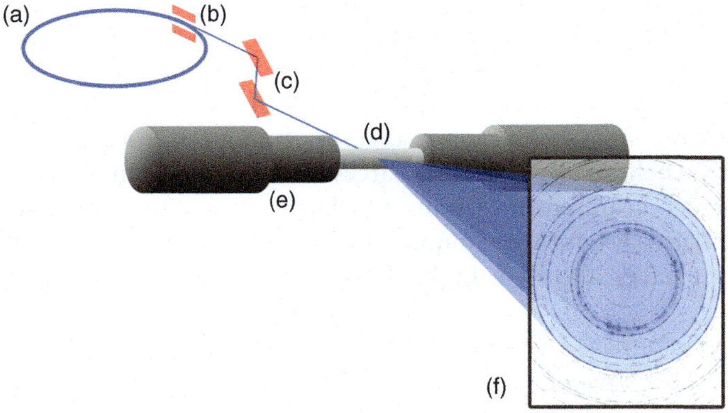

Figure 22.1 A typical experimental setup for a HEXRD experiment in transmission geometry is shown. Electrons or positrons in the storage ring (a) are deflected by an insertion device (b), which causes them to emit a white X-ray beam. After the monochromator (c), a beam with a narrow energy bandwidth penetrates the sample (d). Different environments can be adjusted by additional devices such as furnaces or load-frames (e). The diffracted intensity on the Debye–Scherrer rings is then recorded by a flat-panel detector (f). (Schmoelzer 2011 [41], reproduced with the permission of Wiley.)

geometry which is easily possible due to their high penetration depth. A typical example for a HEXRD experimental setup is shown in Figure 22.1.

A white X-ray beam is produced by the deflection of charged particles (usually electrons or positrons) in a storage ring. In order to deflect the particles, an insertion device such as a bending magnet, wiggler, or undulator is used [36]. A narrow energy range is selected from the white beam by the monochromator [32]. After that, the monochromatic beam is collimated and interacts with the specimen. The diffracted intensity is recorded by a suitable detector. Common recording devices for high-energy X-rays are fluorescent screens coupled to a charge-coupled Device (CCD) detectors, amorphous silicon pixel detectors, or online image plates. Further information on detectors can be found in [32, 33]. For *in situ* diffraction experiments, different environments can be adjusted at the sample position. A variety of furnaces and load-frames are available for heating the specimen in a controlled atmosphere or performing *in situ* deformation experiments.

22.3 *In situ* Investigation of Phase Evolution

22.3.1 General Aspects

A common feature of structural materials, including TiAl alloys, is the dependency of their properties on phase composition and the arrangement of the constituting phases. Therefore, which phases are present to what amount is a decisive factor for materials behavior under service conditions as well as during processing.

There is a plethora of methods by which a quantitative evaluation of phase-fractions is possible. LOM on etched specimens for instance is usually readily available and a very cost-effective method for determining phase-fractions. In the case of advanced TiAl alloys, however, up to three different phases have to be distinguished. It is, therefore, difficult to find etchants that provide reproducible results and allow determining and quantifying the present phases unambiguously. Additionally, γ and α_2-phase frequently occur in the form of lamellar colonies. The thickness of the γ-lamellae depends on the thermal history of the specimen and can be as small as 10 nm [24]. Therefore, the evaluation of the overall phase fractions in lamellar microstructures by means of LOM is not possible. Even in the case of SEM investigations, those two phases can usually not be properly distinguished. For the investigation of phase fractions, SEM images are recorded in back-scattered electron (BSE) mode. Due to the large volume from which BSEs emerge and the low contrast between α and γ-phase, the determination of phase fractions with SEM is not feasible if lamellar colonies are present. TEM is a method that would allow determining phase fractions in lamellar grains, although at the expense of time-consuming sample preparation and the restriction to small sample volumes.

By diffraction methods, the phase fractions can be determined with high accuracy even in the instance that lamellar colonies are present in the material. Additionally, it is possible to adjust special sample environments that allow conducting these experiments at elevated temperatures or during deformation. Especially at synchrotrons, highly time-resolved *in situ* experiments can be performed due to the high brilliance of the source [33].

Due to the numerous advantages that the determination of phase fractions with diffraction methods offers, compared to competing techniques in the case of TiAl alloys, laboratory X-ray sources have been used since the beginning of research activities on TiAl alloys [25, 38, 42, 43]. One way to investigate phase fractions present at elevated temperatures is to perform conventional XRD on heat-treated and quenched specimens. A drawback of this method is that changes in the microstructure might occur during cooling that could potentially lead to erroneous results.

To circumvent this problem, *in situ* experiments have been performed by means of high-temperature X-ray diffractometry [25, 38, 42]. In this case a different complication arises: If exposed to temperatures >1000 °C for a prolonged time span, the surface-near regions of TiAl samples deplete in Al and a so-called α-case forms, even in controlled atmospheres or vacuum. This surface layer can be (depending on temperature and holding time) several 100 μm thick and consists almost exclusively of α/α_2-phase [31]. Due to the low penetration depth of laboratory X-ray sources (e.g., 50% absorbed by a 80 μm thick Al-plate for CuKα), high-temperature XRD experiments are limited to temperatures where no α-case formation is observed. At dedicated high-energy beamlines high penetration depths can be attained (e.g., 50% absorbed by 12.8 mm Al at 100 keV) so that α-case formation affects the experiments less severely [33].

22.3.2 Phase Evolution in β/γ-Alloys

It has been shown that the presence of disordered β-phase at elevated temperatures improves the hot-workability of TiAl alloys significantly [19, 21, 44]. Although room temperature ductility and creep resistance at service temperature are negatively influenced by high β_o volume fractions [45], the concept of so-called β/γ-alloys has been conceived to facilitate the cost-effective production of TiAl parts for applications in which lower temperature and mechanical loads are imposed. In this type of alloys, the β/β_o-phase is stabilized by elements such as Mo and Nb [20, 46–48]. One of the challenges to be faced is that the commercially available TiAl thermodynamic database [49]

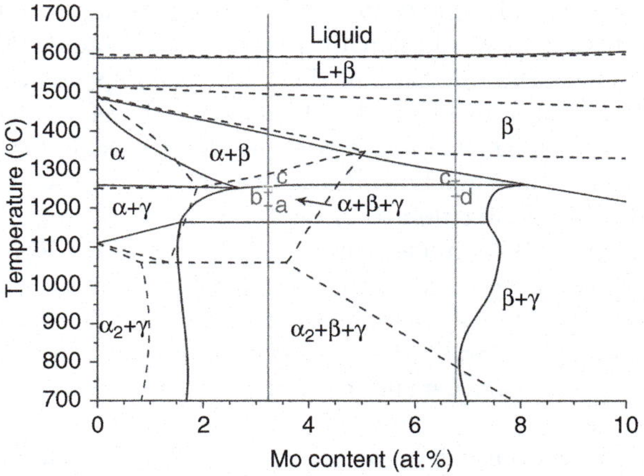

Figure 22.2 Two super-imposed Ti-44Al-(0–10) Mo phase diagrams derived from literature data (continuous) and obtained by thermodynamic simulations (dashed) are shown. Substantial differences exist in terms of transformation temperatures and in the shape and position of the occurring phase fields. The compositions of the two investigated alloys are given by vertical lines and their experimentally determined transformation temperatures by short horizontal dashes. The eutectoid temperature is indicated by the letter a, whereas b marks the disordering temperature of β_o and at c the γ phase dissolves. At the temperature indicated by d the onset of α-phase formation was observed. (Schmoelzer 2011 [41], reproduced with the permission of Wiley.)(Schmoelzer 2011 [41], reproduced with the permission of Wiley.)

is insufficient for describing binary phase diagrams with high contents of β/β_o stabilizing elements properly [22, 50]. Therefore, the predictions of the thermodynamic calculations are inaccurate and the dependence of the phase fractions and transition temperatures on the amount of β/β_o stabilizing elements is not well known.

In Figure 22.2a, thermodynamically calculated Ti-Al-Mo phase diagram and one derived from data published in the open literature [51–53] are presented. Both diagrams are shown for an Al content of 44 at.% and Mo contents in the range from 0 to 10 at.%. It is obvious that significant differences between the two phase diagrams exist.

For validating the accuracy of calculated and literature data, alloys with nominal compositions of Ti-44Al-3Mo-0.1B and Ti-44Al-7Mo-0.1B were produced by GfE Materials and Metals GmbH (Nuremberg, Germany) and investigated by means of *in situ* HEXRD at the HARWI II beamline of Helmholtz-Zentrum Geesthacht (HZG) at DESY in Hamburg (Germany) [41, 54]. Note that all alloy compositions in this chapter are stated in atomic percentage unless indicated otherwise. In these experiments, cast and hot-isostatically pressed (HIP) specimens were heated rapidly to 1000 °C, held for 2 min and then continuously heated to 1350 °C at a rate of 0.033 K s^{-1}. Integration of the patterns acquired with a mar555 flat panel detector (Marresearch GmbH, Norderstedt, Germany) was performed by the software fit2D [55]. Phase fractions present were evaluated by means of the intensity ratio method where the weighting factors were derived from the results of Rietveld analysis performed at different temperatures. Rietveld fitting was carried out with the commercial software package TOPAS (Brucker AXS, Madison, USA). The SEM images were obtained from samples heated to the desired temperature, annealed for 1 h and subsequently water quenched.

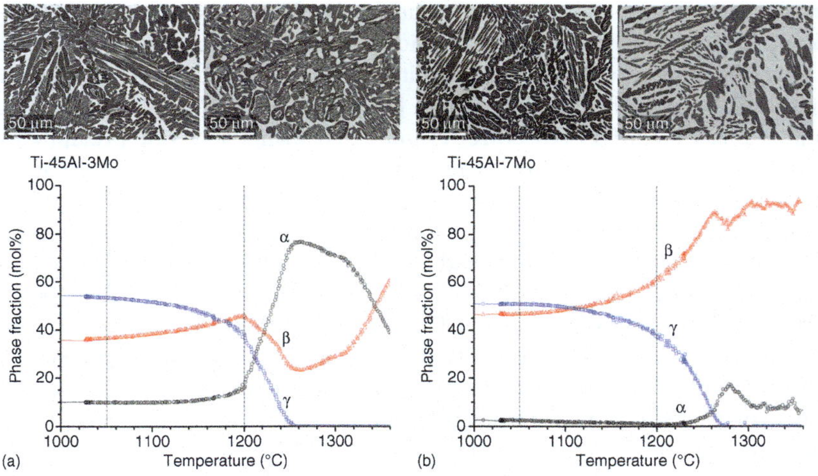

Figure 22.3 The course of phase fractions is shown for two alloys with a nominal composition of (a) Ti-44Al-3Mo-0.1B and (b) Ti-44Al-7Mo-0.1B. At the temperatures indicated by vertical lines, additional micrographs are provided which were obtained in the SEM in BSE mode. (Schmoelzer 2011 [41], reproduced with the permission of Wiley.)

The results of the phase fraction analyses are depicted in Figure 22.3. All microstructural images were acquired in BSE mode, in which the β/β_0 phase appears brightest, the γ-phase is dark and the α/α_2 phase presents with an intermediate contrast. It is obvious that with increasing Mo content, the amount of α/α_2-phase diminishes. Consequently, the fraction of β/β_0-phase is increasing with increasing Mo content. Concerning the dissolution temperature of γ, the so-called α-transus temperature, a slight increase is observed with increasing Mo content.

A comparison of the results in Figure 22.3 with the phase diagrams derived from literature and thermodynamic calculations shows that a significant deviation between experimental results and the phase diagrams exists. While the calculated phase diagram and the course of phase fractions in Figure 22.3 show only a faint resemblance, both in terms of transition temperatures and phases present, the phase diagram derived from literature fits the experimental data better. Due to deviations of the actual from the nominal chemical composition of the specimens, slight differences in phase fractions or transformation temperatures could be explained. However, if the deviations are as large as in the case of the experimental results and the calculated phase diagram, it must be concluded that the calculation is erroneous.

In a next step, the database used for the calculation has to be modified on the basis of the experiments conducted in order to be able to correctly predict the phase diagram. Thus, the HEXRD experiment not only allowed to verify the accuracy of the prediction but also provided data that allow improvement of the thermodynamical model.

22.3.3 Formation and Identification of a Transition Phase

In technological processes high heating and cooling rates are frequently used. Therefore, the materials behavior under nonequilibrium conditions is of paramount technological interest. As shown in Figure 22.2 for Ti-44Al-3Mo-0.1B in equilibrium conditions the phases α_2, β_o, and γ co-exist at room temperature. At high temperatures, α_2 and β_o disorder to α and β, respectively [54, 56]. In addition to these main constituents, several phases with lower symmetry are mentioned in literature. Among these are hexagonal ω-related phases [57] and orthorhombic phases [23, 58]. Abe *et al.* [23] reported on an orthorhombic phase with B19 structure (*Pmma*) that forms very fine precipitates upon rapid cooling in a Ti-48 Al alloy. Tanimura *et al.* [59] observed the formation of B19-phase as a metastable transitional phase in the $A3 \rightarrow D0_{19} + L1_0$ precipitation sequence. Here also small regions of B19 were formed in the α/α_2 matrix upon quenching of a Ti-40 Al alloy from 1250 °C which acted as nucleation sites for γ-phase formation during a subsequent annealing step at 1000 °C. Appel *et al.* [58] observed B19 phase in Ti-(40–44) Al-8.5Nb alloys after extrusion at 1250 °C and a subsequent heat treatment at 1030 °C. Although the existence of the B19 phase in the TiAl system has been well established, no detailed information on the chemical or temperature regime in which the B19 phase forms in these alloys is available as yet.

It should be noted that the B19-phase is structurally closely related to the α/α_2, γ, and β/β_o phases [60]. Figure 22.4a shows a hexagonal closest packed plane with atoms of only one kind, which corresponds to the {0001} plane of the disordered α-Ti(Al) phase. Figure 22.4b,c show closest packed planes containing two atomic species with compositional ratios of 3 : 1 or 1 : 1, similar to the {0001} plane in α_2-Ti_3Al and the {100} plane in B19-TiAl, respectively. All three phases exhibit the stacking sequence ...ABAB... corresponding to a hexagonal closest packed structure. Nonetheless, the B19 structure possesses an orthorhombic symmetry due to the fact that the two atomic species are arranged in alternating lines (Figure 22.4c). It should be mentioned, however, that the B19 lattice is only slightly distorted as compared with an ideal hexagonal lattice.

It is interesting also to note that the atoms in the closest packed plane of γ-TiAl (i.e., the {111} plane) are arranged in the same way as in the {100} plane of the B19 structure (Figure 22.4c), but contrary to B19, the $L1_0$ structure exhibits the stacking sequence ...ABCABC... Also in β_o-TiAl, the atomic arrangement in the closest packed plane (which is the {110} plane) corresponds to that of the {100} plane of B19 (Figure 22.4c). The structural relationship between B19 and β_o is similar to that between α and β-Ti.

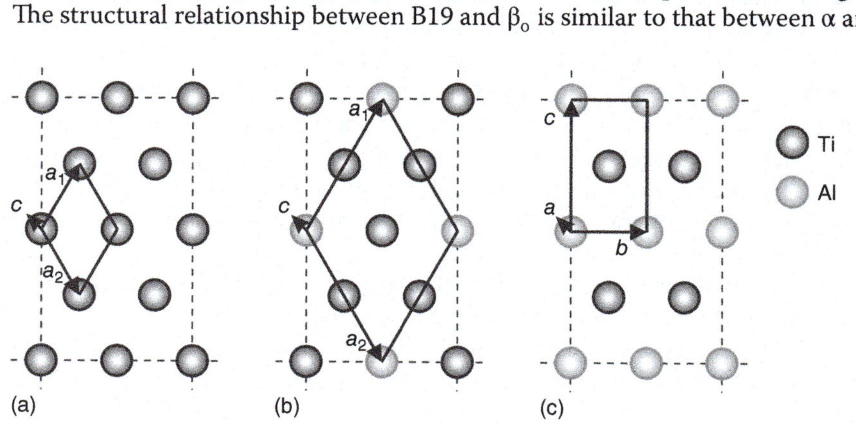

Figure 22.4 Structural relations between the individual phases in TiAl alloys illustrated by sections of their closest packed planes. (a) {0001} α-Ti(Al), (b) {0001} α_2-Ti_3Al, and (c) {100} B19-TiAl. The dimensions of the unit cells of α-Ti(Al), α_2-Ti_3Al, and B19-TiAl are marked by solid lines. It should be noted that (a) also corresponds to {110} β-Ti(Al) and (c) also corresponds to {111} γ-TiAl as well as {110} β_o-TiAl. (Schmoelzer 2011 [41, 54], reproduced with the permission of Wiley.)

The phases occurring under nonequilibrium conditions in a Ti-44Al-3Mo-0.1B alloy were studied by *in situ* HEXRD experiments [61]. The investigations were conducted at the HARWI II beamline of the DESY synchrotron in Hamburg, Germany [62]. The starting microstructure is shown in Figure 22.3a. A beam of monochromatic synchrotron radiation with a mean energy of 100 keV and a cross-section of $1 \times 1\,mm^2$ was used for diffraction experiments. Heating of the cylindrical specimens with a diameter of 5 mm and a length of 15 mm was performed by a modified Bähr DIL 805A/D dilatometer (see Chapter 20). Temperature was controlled by an S-type thermocouple. Note that the temperatures given in this chapter are rounded to 5 °C steps. A mar555 detector (Marresearch GmbH, Norderstedt, Germany) recorded the diffraction patterns at a distance of 1837 mm from the specimen position. Integration of the diffraction patterns was performed within the fit2D software package [55]. The software PowderCell was employed for Rietveld fitting.

Prior to the *in situ* experiment, the specimens were heat treated at 1450 °C for 30 min and subsequently water quenched. This resulted in a $\alpha_2 + \beta_o$ microstructure where the α_2-phase formed rapidly during cooling and it is in a chemical nonequilibrium. Due to the suppression of γ-phase formation, the α_2-phase is highly supersaturated in Al. Heating during the HEXRD experiments was performed continuously at 25 and 2 °C min^{-1}.

Employing HEXRD allowed for *in situ* monitoring of the phases evolution. A series of selected diffraction patterns recorded during heating at 2 °C min^{-1} is presented in Figure 22.5. The first pattern acquired at 565 °C features single peaks of the α_2 and β_o-phase as indexed in Figure 22.2. At this temperature, the phase fractions as determined by means of Rietveld analysis are approximately 80 vol.% α_2 and 20 vol.% β_o. As temperature rises, the $(20\bar{2}l)$ and $(22\bar{4}l)$ α_2 peaks ($l = 0, 1, 2, \ldots$) begin to split. This is most obvious for the $(20\bar{2}0)$ and $(22\bar{4}0)$ peaks at 640 °C where the separation

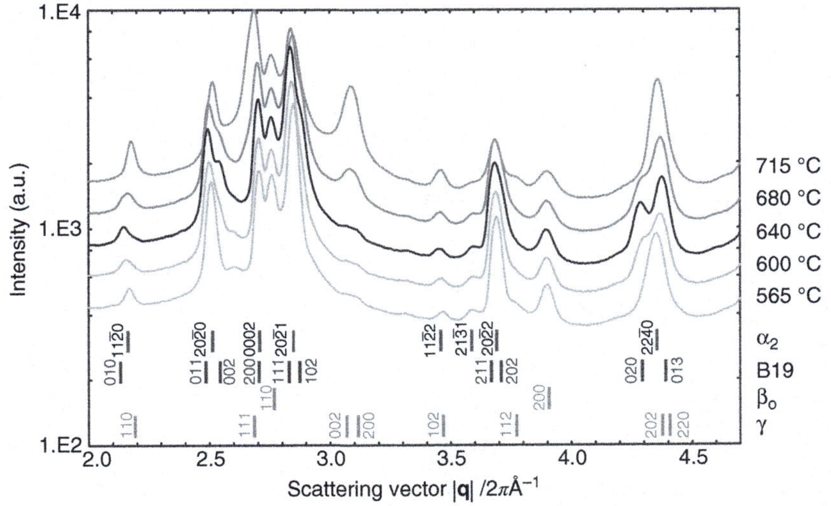

Figure 22.5 Diffractograms obtained by HEXRD during continuous heating. Temperatures at which the patterns were obtained are indicated on the right. It is obvious that peaks of α_2 reflections split due to the formation of the orthorhombic B19 phase. All reflections are indexed at the bottom. (Schmoelzer 2011 [41], reproduced with the permission of Wiley.)

between the original and the newly emerged reflection reaches its maximum. As the temperature is increased further, the peak splitting gradually diminishes until a single peak is visible again at 715 °C. Within this temperature interval, also the intensity of the γ-peaks changes significantly. Starting with low intensities below 640 °C, a gradual gain in intensity with increasing temperature is observed. Simultaneously, the intensity of the α_2 reflections is abating. A phase composition of about 35 vol.% α_2, 20 vol.% β_o, and 45 vol.% γ is attained at the highest temperature of 715 °C.

When a higher heating rate (25 °C min^{-1}) is used, the diffraction patterns show similar changes, only shifted to higher temperatures. The splitting of α_2-phase reflections starts at about 650 °C and the maximum peak separation is reached at approximately 710 °C. The appearance of the γ reflections and the reduction in α_2-peak intensity occurs at 740 °C.

In order to investigate if any microstructural changes occurred during heating to 640 °C, one specimen was heated to this temperature (at a rate of 2 °C min^{-1}) and subsequently quenched. A second specimen was only subjected to the primary heat treatment at 1450 °C and quenched. Using standard metallographic methods [63], both samples were prepared for scanning electron microscopic (SEM) investigations using a Zeiss Evo 50 microscope in the BSE mode. Comparison of the obtained micrographs revealed no apparent changes (see Figure 22.6 with the specimen heated to 640 °C). Only two phases (α_2 and β_o) can be distinguished, which exhibit morphologies that strongly resemble martensitic microstructures observed in commercial Ti alloys (e.g., [48, 64]). Although a small γ peak is observed in the diffraction pattern obtained at 640 °C, no indication for γ formation was found in the micrograph. This might be due to the fact that γ forms ultrafine lamellae during its early stages of precipitation [65] which cannot be resolved from SEM investigations.

From the sequence of events, it is tempting to speculate that a precursor phase forms in supersaturated α_2 and facilitates the precipitation of γ-phase. The splitting of the hexagonal α_2 reflections (20−2l) and (22−4l) (l = 0, 1, 2, ...) is an indication of an orthorhombic distortion of the lattice. All former α_2 reflections and the newly split double peaks can clearly be attributed to a transient orthorhombic phase with B19 structure (*Pmma*) as shown in Figure 22.5. At the temperature where the peak separation is most clearly visible (T = 640 °C), the lattice parameters of B19 were evaluated to a = 4.65 Å, b = 2.93 Å, and c = 4.95 Å. At this stage, all of the α_2 phase had transformed to B19, and the phase fractions are approximately 80 vol.% B19 and

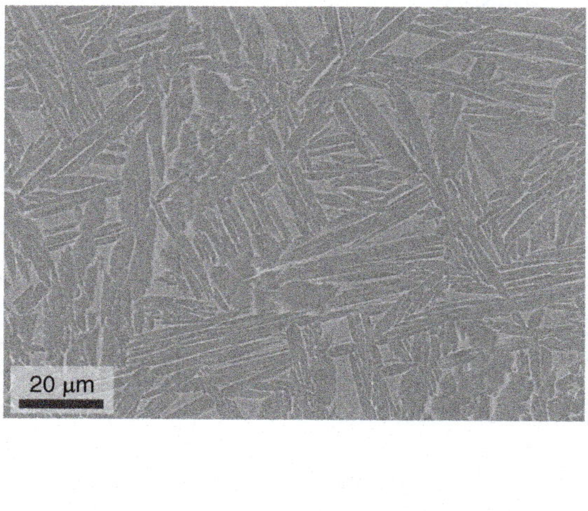

Figure 22.6 SEM image of the specimen after heating to 640 °C at a rate of 2 °C min^{-1}. The image was obtained in BSE mode. β_o appears in light gray and B19 in dark gray. (Schmoelzer 2011 [41], reproduced with the permission of Wiley.)

20 vol.% β_o. The finding that all of α_2 transforms to B19 before γ precipitation starts is in contrast to the observations of Abe *et al.* [23] and Tanimura *et al.* [59]. However, this apparent disagreement may be a consequence of the different alloy compositions used in these studies.

Concerning the formation of B19 it should be noted that the splitting of α_2 reflections increases continuously as the temperature rises. One example for the development of double peaks is the former 22−40-α_2 reflection that gradually splits into two clearly separated peaks, 020-B19 and 013-B19. This indicates that the B19 formation is associated with a diffusional process and a continuous change in the chemical composition occurs. This conclusion is supported by the observation that the temperature of formation of B19 strongly depends on the heating rate.

In this study, the transformation processes were observed *in situ* whereas former studies depended on post-mortem analyzed samples. It was, however, not possible to identify any changes between the as-quenched and the annealed to 640 °C samples by means of SEM in BSE contrast. Since the chemical compositions and the crystal structures of α_2 and B19 are quite similar, it is not unlikely that the microstructural changes induced by the phase transformation are too subtle to be detected in the SEM. Unfortunately, no information on the stability of the B19 phase within the observed temperature range was obtained and further experiments are needed to clarify this matter. This study, however, is the first report on the formation of B19 phase in TiAl alloys observed *in situ*. It was found that B19 forms as a transient phase facilitating the precipitation of γ from α_2. Furthermore, it was possible to determine the temperature range in which the B19 phase occurs in the investigated alloy. Since industrial processes are frequently conducted at high heating and cooling rates, the observation of B19 phase formation during processes far from thermal equilibrium has a strong technological relevance.

22.3.4 Formation of Lamellar Microstructure

In addition to the course of phase fractions over temperature, the processes during the formation or dissolution of phases are frequently of particular interest. In a recent study by Shuleshova *et al.* [66], the solidification behavior of Ti-Al-Nb ternary alloys was investigated by means of HEXRD, which allowed a reassessment of the thermodynamic database used for phase calculations. Senkov [67] investigated the crystallization behavior of an amorphous TiAl film produced by physical vapor deposition and was able to show that this process occurred in a two-step fashion. Studies concerning the formation and coarsening of Perovskite-type precipitates in carbon containing TiAl alloys were conducted by means of small angle neutron scattering [68].

In TiAl alloys, the γ-phase dissolves at the α-transus temperature. If the specimen is cooled slowly afterward, the formation of γ-lamellae occurs within the supersaturated, disordered α-grains [2, 69, 70]. If the alloy is cooled from the $(\alpha + \beta)$ two-phase region at a sufficiently high rate, the two-phase microstructure can be preserved down to room temperature and no γ-phase is formed. The only phase transformation occurring is the ordering of the supersaturated α-phase. By a subsequent heat treatment, γ-lamellae are formed in the supersaturated α_2-grains [21, 71]. The formation of fine γ-lamellae strongly increases the performance of the material in terms of strength, plastic fracture strain, fracture toughness, and creep resistance at service temperature [24, 26, 70, 72]. If the γ-lamellae are formed upon cooling, the cooling rate is the decisive parameter for the interlamellar spacing [69, 73]. In fast-cooled material, in which the formation of γ-phase was suppressed, γ-lamellae can form during an ensuing heat-treatment step [74]. The lamellar spacing then depends either on the heating rate and maximum temperature reached if continuously heated or solely on the temperature if the heat-treatment is performed isothermally [24, 75].

To determine the onset of γ-lamellae precipitation during continuous heating in a Ti-45Al-7.5Nb alloy, experiments were conducted at the ESRF in Grenoble, France [74]. Diffraction data acquired with a two-dimensional detector allowed investigating the diffuse and inter-Bragg-peak scattering. Streaking of the reflections in reciprocal space has been observed, which is correlated to the precipitation and formation of ultrafine γ-lamellae within globular α_2-grains. The obtained results were verified by repeating the experiment in a high temperature laser scanning confocal microscope [74].

In an ensuing experiment, a two-step heat treatment was conducted on a Ti-43.5Al-4Nb-1Mo-0.1B specimen to quantify the development of phase fractions. At first, the sample was heated to 1230 °C, which is slightly below the dissolution temperature of the γ phase and held for 300 s at this temperature to reduce the amount of γ-phase. Subsequently, the specimen was quenched to room temperature in order to produce supersaturated α_2 grains. During the second heat treatment step a continuous heating ramp with a constant heating rate of $2\,\mathrm{K\,s^{-1}}$ was adjusted and γ lamellae precipitate in the supersaturated α_2 grains [21, 71].

Figure 22.7a superimposes the γ(200) and γ(002) peak intensities with the temperature program of the conducted heat treatment as a function of time. Upon heating, a drop in peak intensity is observed, which indicates that the volume fraction of γ-phase decreases. At the same time, the peak positions plotted in Figure 22.7b shift toward lower q-values due to the thermal expansion of the material. Furthermore, the tetragonality c/a of the γ unit cell initially decreases but as the γ-phase fraction diminishes, this trend is reversed and the c/a ratio increases rapidly. While the temperature is kept constant at 1230 °C, the γ-phase fraction decreases slightly while peak positions and c/a ratio remain approximately constant. The peak width does not change significantly during the holding time, while it decreases slightly upon heating to 1230 °C. As the sample is quenched, the peak positions move toward larger scattering vectors and the c/a ratio rises further. Small increases in peak intensity and width are also observed on quenching, which indicates a small increase in the γ-phase volume fraction.

After quenching, the specimen was heated at a rate of $2\,\mathrm{K\,s^{-1}}$ as indicated in Figure 22.7a. Apart from the thermal expansion of the sample, no major changes are observed until 1000 s (765 °C) are reached. At this time, the c/a ratio starts to decrease. At $t > 1050$ s (867 °C) dramatic changes in all parameters become visible. The γ-phase fraction starts to increase rapidly, while the c/a ratio is diminishing. Additionally, the peak width of both γ reflections (γ(200) and γ(002)) increases. After reaching a maximum at 1090 s (947 °C), the peak width decreases while the intensity of the γ-peaks continues to rise. The c/a ratio also keeps following its initial trend andgets smaller until, at 1170 s the specimen is cooled.

It is obvious from Figure 22.7 that only little γ-phase forms during quenching. As the sample is reheated the γ-phase fraction increases, which indicates the precipitation of γ-lamellae within the supersaturated α_2 grains. Although more detailed investigations would be necessary to make any definitive statements, it is tempting to speculate that the changes in c/a ratio and in peak width are caused by coherency strains as well as by the

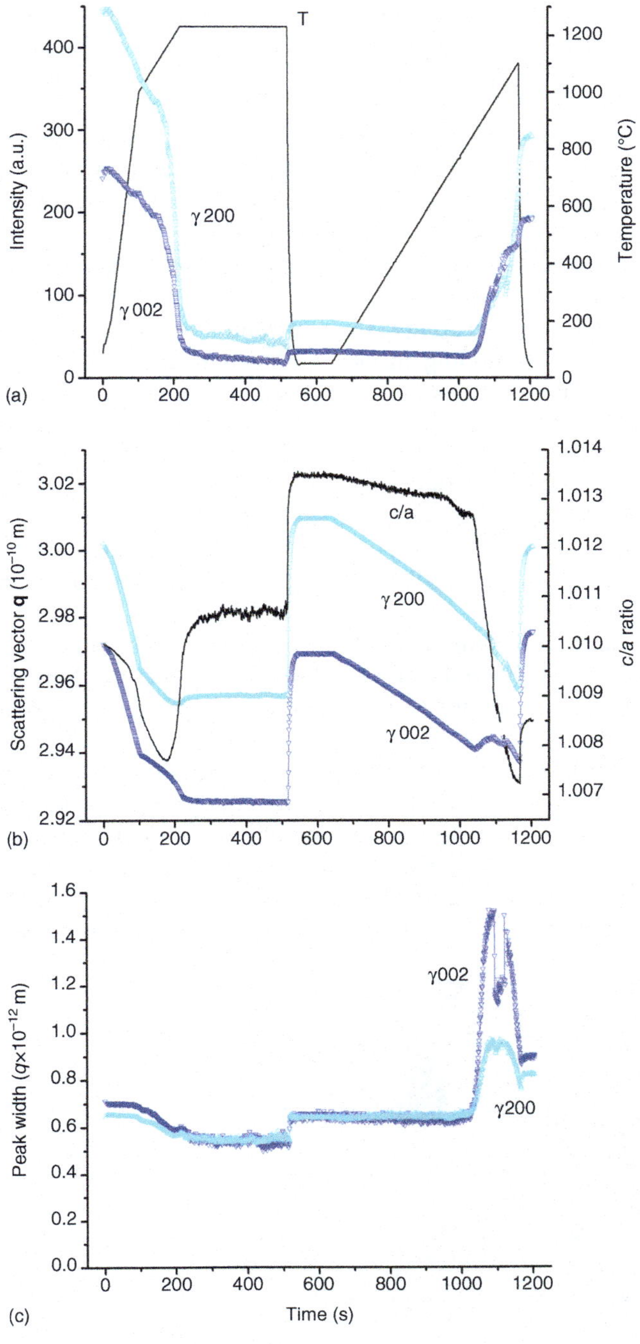

Figure 22.7 Parameters of the γ200 and γ002 peaks as well as the temperature are plotted versus time for a Ti-43.5Al-4Nb-1Mo-0.1B alloy. In (a) the peak intensity and the temperature are shown. Panel (b) indicates the peak position and the tetragonality of the γ unit cell. Diagram (c) represents the peak width. After dissolution of most of the γ-phase at 1230 °C, the specimen is quenched. Upon reheating with a rate of 2 K s^{-1}, the γ-phase fraction starts to increase. Simultaneously, the tetragonality of the unit cell is changing and the peak widths are increasing. (Schmoelzer 2011 [41], reproduced with the permission of Wiley.)

chemical composition being far from thermodynamic equilibrium. As the precipitation process progresses, which is indicated by an increase in intensity, the γ-lamellae grow in thickness, causing the coherency strains to diminish. This effect could explain the decrease in width of the γ peaks.

Peak height, width, and intensity were determined by fitting two Gaussians to the double-peak of γ-002/200. As the peak width increases and the *c*/*a* ratio decreases, the separation of the two peaks becomes smaller, which causes the fitting algorithm to produce incoherent results. This explains the discontinuities in the data plots of Figure 22.7c. Despite the poor fitting in Figure 22.7c, there is no doubt that the reflections broaden significantly.

The precipitation of γ lamellae during reheating of the specimen was also observed in TEM investigations [21, 71]. The corresponding images are shown in Figure 22.8.

After annealing at 1230 °C and subsequent water quenching, only few γ lamellae are visible in the TEM. Within the lamellar colonies, the orientation of γ and α_2-lamellae is described by the so-called Blackburn orientation relationship where γ(111) || α_2(0001) and γ[110] || α_2[11−20] [77]. In contrast to this, a large number of γ lamellae can be observed when the material is isothermally annealed at 850 °C for 6 h subsequently to quenching. It is, therefore, concluded that after annealing at 1230 °C and water quenching the α_2-grains are supersaturated and γ-lamellae precipitate during an ensuing heat treatment.

The *in situ* HEXRD experiments complement the TEM investigations because they allow determining the temperature at which the precipitation process starts. Additionally, the data show that the *c*/*a* ratio changes with temperature and during the formation of γ-lamellae.

As shown in this section, *in situ* HEXRD experiments yield a multitude of data on phase transformations. Not only phase fractions are extracted, but also the evolution of lattice parameters and the width of the reflections are simultaneously obtained. The lattice parameter may be of interest since the formation of a new phase from a supersaturated parent phase usually entails a change in the chemical composition and, therefore, according to Vegards law, a change in the lattice parameter of the corresponding phases

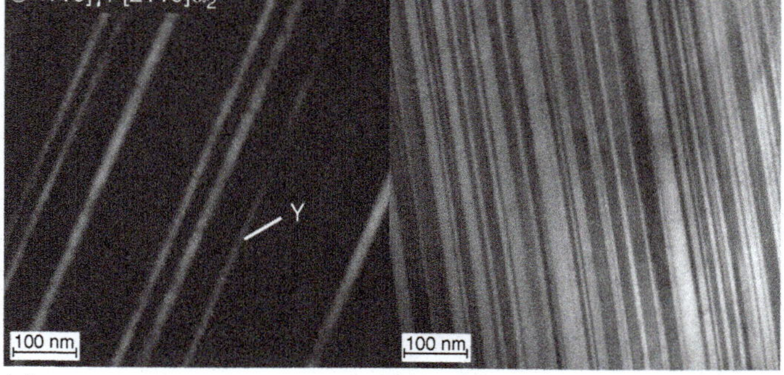

(a) (b)

Figure 22.8 TEM images of lamellar colonies in a Ti-43.5Al-4Nb-1Mo-0.1B alloy. An α_2-grain is shown after quenching from 1230 °C (a) in which only few γ-lamellae have formed. Lamellar colonies exist after annealing at 850 °C as shown in (b). It is obvious, that the number and volume of γ-lamellae has increased upon annealing. (Schmoelzer 2011 [76], reproduced with the permission of Wiley.)

[78–80]. From the measured peak widths, information on the defect density and the internal strains can be derived [81].

22.4 Atomic Order and Disorder in TiAl Alloys

Ordering phenomena are important in TiAl alloys, because the mechanical behavior of the occurring phases changes upon ordering. The disordered β-phase, for instance, improves the hot-workability of TiAl alloys at elevated temperatures [82–84] whereas the ordered β_o phase affects the room-temperature ductility negatively [45]. An additional motivation for determining ordering temperatures is that the accuracy of thermodynamic calculations can be validated by comparing ordering temperatures obtained from experiments with calculated ones.

Ex situ studies have been performed with laboratory X-ray sources in order to study the site occupancy in different phases [27, 85, 86]. Complementary investigations concerning the ordering behavior were conducted by means of TEM [23]. HEXRD was used to investigate the site occupancy in the α-phase by means of Rietveld analysis and thereby derive the disordering temperature of α_2 in a Ti-45Al-7.5Nb-0.5C alloy [87]. Due to the fast kinetics of the ordering reactions, it is not possible to preserve the disordered states of α and β-phase down to room temperature even if the specimens are water-quenched [23, 88]. It is, therefore, inevitable to investigate the ordering behavior at those temperatures where the ordering reactions occur.

As described in Section 22.2, neutron diffraction is particularly well suited for investigating the ordering behavior of TiAl. Experiments for determining the ordering/disordering temperatures of α/α_2 and β/β_o-phase in a Ti-43.9Al-4Nb-1Mo-0.1B alloy were performed with the structure powder diffractometer SPODI [89, 90] at the neutron source FRM II of the Heinz Meier-Leibnitz Zentrum (MLZ) in Garching, Germany. Since the acquisition time for one diffraction pattern was set to 3 h, the specimen was heated in a step-wise fashion where the temperature remained constant during pattern acquisition. In Figure 22.9 the patterns acquired at four different temperatures are shown exemplarily.

While at 1000 and 1100 °C superstructure reflections of all three phases are visible, the α_2 reflections vanish at 1200 °C. Between 1200 and 1250 °C, the β_o reflections also attain zero intensity. The presence of α and β phase above 1300 °C has been proven by earlier work [22] and the vanishing of the superstructure reflections indicates the loss of order in the respective phase due to the average scattering length of the disordered material being close to zero.

During the experiment at SPODI, only few data-points were generated due to the long acquisition times for the individual patterns. To be able to determine the order-to-disorder transition temperatures more accurately, neutron diffraction experiments were performed at the WOMBAT facility at the OPAL reactor of the Australian Nuclear Science and Technology Organization (ANSTO) in Lucas Heights, Australia [91]. The facility is optimized for high intensity rather than resolution and hosts a true two-dimensional, position sensitive detector covering 120 in-plane and 15 out-of-plane. The area is binned into 968 × 128 pixels, each with intrinsic time response (microseconds), which makes the acquisition fast and only intensity limited. With the two-dimensional detector installed, a single pattern with sufficient counting

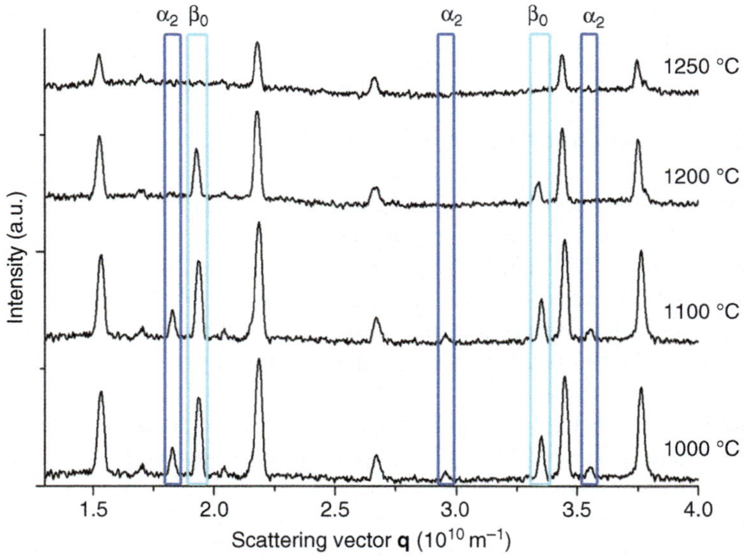

Figure 22.9 Neutron diffraction patterns of a Ti-43.5Al-4Nb-1Mo-0.1B alloy obtained with the powder diffraction spectrometer SPODI at the FRM II research reactor (Garching, Germany). Peaks of the α_2 and β_o-phases are indicated by rectangles and labeled at the top. The temperatures at which the patterns were recorded are given on the right hand side of the image. Between 1100 and 1200 °C, the α_2 reflections vanish, whereas the β_o phase disorders between 1200 and 1250 °C. (Schmoelzer 2011 [41], reproduced with the permission of Wiley.)

statistics was obtained within an exposure time of 18.2 s [30]. At a heating, respectively cooling rate of 0.166 K s^{-1}, this frame rate is sufficiently high for determining the order-to-disorder transition temperatures with adequate accuracy. The results of these experiments are shown in Figure 22.10.

In Figure 22.10, the intensity of the corresponding reflection is plotted versus temperature for a heating and a cooling ramp. It is clearly visible that different transition temperatures are observed upon heating and cooling, which is attributed to the kinetics of the corresponding material. As reported earlier, the $\alpha_2 \leftrightarrow \alpha$ and $\beta_o \leftrightarrow \beta$ reactions progress relatively fast, whereas the dissolution and formation of γ phase occurs more reluctantly [30].

Although the ordered phases in TiAl alloys show only weak intensities for superstructure peaks in X-ray diffraction, they are usually sufficiently high for determining the order/disorder transition temperatures. It is important to note that the alloying elements (e.g., Mo and Nb) have large scattering lengths and can, therefore, influence the structure factors significantly. In Figure 22.11, the intensities are plotted for superstructure reflections of the ordered phases in a Ti-44Al-3Mo alloy. The data for this plot were obtained from the same experiment described in Section 22.3.2.

In Figure 22.11, the loss of order in the α_2 and β_o phase upon heating is indicated by vanishing of the intensity, while the disappearance of the γ-phase reflection indicates its dissolution at the α-transus temperature. The results were obtained from fitting a Gaussian to the corresponding peak after performing a background correction. As soon as the intensity reached zero values, the fitting routine was disrupted.

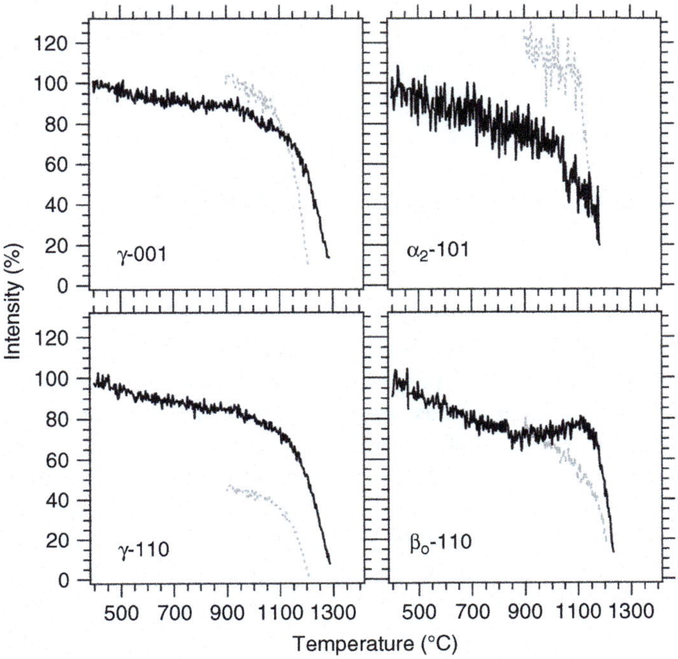

Figure 22.10 Ordering-disordering temperatures of α/α_2 and β/β_o in a Ti-43.27Al-3.84Nb-1.01Mo-0.11B alloy. Additionally, the dissolution and formation temperatures of the γ phase are shown. Data obtained on heating are shown as solid lines whereas intensity values recorded during cooling are represented by dashed lines. The intensities of superstructure peaks are plotted as a function of temperature. Intensity values are normalized to the intensity observed at 400 °C. Disordering or dissolution of a phase is marked by a sharp decrease in intensity of the corresponding peak. (Schmoelzer 2011 [76], reproduced with the permission of Wiley.)

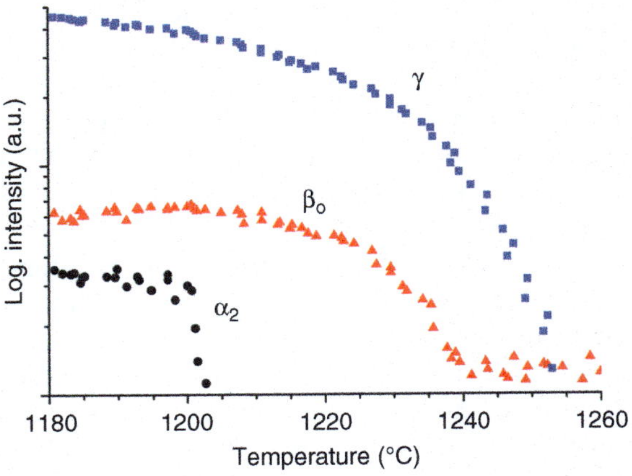

Figure 22.11 Disordering of α_2 and β_o as well as dissolution of the γ-phase in Ti-44Al-3Mo during continuous heating with a rate of $0.033\ \mathrm{K\,s^{-1}}$. The intensity of superstructure peaks is plotted as a function of temperature. Intensity values were obtained by fitting a Gaussian to the corresponding peak of the azimuthally integrated HEXRD pattern. (Schmoelzer 2011 [41], reproduced with the permission of Wiley.)

From these results, it is obvious that the ordering behavior of intermetallic TiAl alloys can be investigated elegantly by means of *in situ* diffraction methods. While the interpretation of neutron diffraction data is comparatively simple, smaller intensities of superstructure reflections may complicate the interpretation in HEXRD experiments.

22.5 Recovery and Recrystallization during Deformation of TiAl

22.5.1 General Aspects

A range of methods exist for performing X-ray diffraction experiments with the scope to investigate structural changes in crystalline materials. Conventional angle-dispersive or energy-dispersive powder diffraction methods, which average over a large number of crystallites, allow obtaining information on phase fractions, texture, and lattice parameters. If, however, a small number of crystallites are illuminated by a monochromatic beam, information on individual crystallites can be gained from their reflection spots on the Debye–Scherrer rings [83]. Alternatively, white beam Laue diffraction can be used to perform diffraction experiments in which only a few grains are illuminated. In principle, there are two different methods of performing diffraction experiments in materials science. If a monochromatic beam is used, a sample with many crystallites is illuminated, which creates Debye–Scherrer cones that image as rings or sections of rings on a two-dimensional area detector. If experiments with a white X-ray beam are performed, usually only one or very few crystallites are illuminated during the experiment. This results in a Laue pattern, which is characteristic for the crystal lattice and its orientation with respect to the incident beam. A synchrotron-based method of microbeam Laue diffraction has recently evolved large impact in the materials science and engineering community [92–94].

One classical field of application of the white beam Laue technique is the adjustment of the orientation of TiAl single crystals as performed by Inui *et al.* [95]. In contrast to powder diffraction experiments, the reflections of the lattice planes stay visible in Laue experiments if the crystal is rotated about an arbitrary axis. Therefore, the Laue technique is very well suited for investigating the deformation and recrystallization behavior of single crystals. With this method, the occurrence of dynamic recrystallization in high purity Al single crystals has been shown [96, 97]. While this method offers new insights in the behavior of one or a few crystals, it is less suited for investigating a fine-grained polycrystalline sample.

In contrast to this, a large number of grains are usually illuminated in powder diffraction experiments. Investigations with monochromatic radiation were originally performed in such a way that only a small section of the Debye–Scherrer rings was recorded on a photographic film or with a one-dimensional detector. In these cases, it is necessary to use texture-free samples with a small grain size in order to obtain an intensity profile that is representative for the whole specimen. These requirements can be met by using a powder sample, which is the reason why the term *powder diffraction* has been established for this type of investigations, even if solid polycrystalline samples are concerned. As the electronic detector technology evolved, first one-dimensional and later on two-dimensional detectors became available. With modern CCD detectors used at synchrotron sources, it is possible to acquire a complete two-dimensional powder diffraction pattern in fractions of a second [98].

However, for most applications such as the evaluation of phase fractions, third order internal stresses, and so on, the two-dimensional detector images are usually integrated over the azimuthal angle ψ in order to obtain a pattern in which the intensity is plotted versus the scattering vector $\mathbf{q}$. Thereby, the influence of fiber textures averages out for the condition that the fiber is perpendicular to the incident beam and the requirement of illuminating a large number of grains is much less stringent compared to powder diffraction experiments with a point detector. If investigations are targeted toward directionally dependent properties, integration is performed only over azimuthal angle segments [99]. With this method it is possible to determine for instance the macroscopic strain perpendicular to the beam direction from one single detector image.

One drawback of the classical powder diffraction approach is that no information on individual grains can be retrieved from such an experiment. A new technique has emerged during the past years by which it is possible to track the reflections of individual grains over time. This is achieved by rotating the specimen in such a way that reflections from all grains in the investigated volume are obtained. For further information on this so-called three dimensional X-ray diffraction microscopy (3DXRD) technique the reader is referred to the corresponding literature [100–102] and Chapter 17. The result of a 3DXRD experiment is a reconstructed volume in which the shape and orientation of the individual grains are indicated. Drawbacks of this technique are that only a limited number of grains can be tracked at the same time and that a large number of detector images have to be acquired at different detector positions to be able to reconstruct the illuminated sample volume. Since this is time consuming, 3DXRD is only suited for *in situ* studies if the processes of interest are occurring at a slow rate. In addition to that it has to be borne in mind that the reconstruction process is mathematically demanding and requires the use of fast algorithms [102].

A different approach to study processes such as recovery, recrystallization, and grain growth is to monitor the intensity distribution of one or more Debye–Scherrer rings over time. If a large number of grains are illuminated by the incident beam, so that the reflection spots of single crystallites are no longer discernable, only information concerning texture and strain can be derived from the intensity distribution as a function of ψ. However, if reflections from single grains can be distinguished, additional information is contained in the data [33]. The patterns can be evaluated with respect to a variety of parameters, such as number of spots, morphology of individual reflection spots, arrangement of different spots, and their evolution with time.

A simplified model of the development of individual reflection spots dependent on the different processes that occur in the material during hot-working is presented in Figure 22.12.

In Figure 22.12a, a perfect grain is depicted, which produces a reflection spot whose width is only determined by the instrumental parameters and the grain size. If the grain undergoes plastic deformation, defects are generated. Frequently, the deformation in metals is facilitated by dislocation motion, which was also assumed for Figure 22.12. If the temperature during deformation is sufficiently high, dynamic recovery might occur, which is characterized by the arrangement of dislocations in low-energy configurations (Figure 22.12b). Newly formed low-angle grain boundaries tilt different regions of the grain by a small angle. This causes the corresponding reflection spot to broaden in the

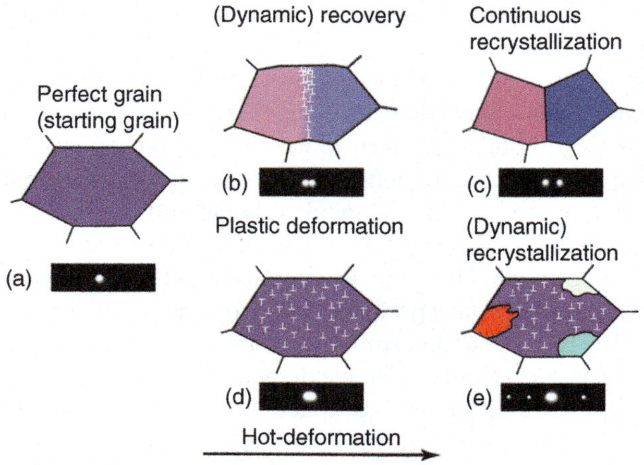

Figure 22.12 Schematic model that illustrates the development of reflection spots dependent on the different processes that might occur during hot-deformation of TiAl alloys. Starting with a perfect grain, plastic deformation increases from left to right. The development of the corresponding reflection spot is indicated below. In (a) a sharp reflection spot of a perfect grain is visible. Images (b) and (c) show how dislocations are introduced by plastic deformation, which form a subgrain boundary in (b) through dynamic recovery processes and eventually a high-angle grain boundary in (c). In contrast to this, images (d) and (e) show a case where the defect density increases until the onset of recrystallization, which is illustrated by the nucleation of new grains. The sketch reflects the development if the incident beam is monochromatic. (Schmoelzer 2011 [41], reproduced with the permission of Wiley.)

direction normal to the tilt axis. Recovery can occur to such an extent that high-angle grain boundaries are formed in which case this process is referred to as *continuous recrystallization* [103]. The reflection spot then splits into two, each one corresponding to an individual crystallite (Figure 22.12c).

An alternative method for decreasing the defect density is illustrated in the lower section of Figure 22.12. In Figure 22.12d, the dislocations are randomly distributed, which causes the reflection to broaden with respect to the azimuthal angle and the scattering vector distribution (i.e., the Bragg angle). If a certain defect density is attained and the temperature for dynamic recrystallization reached, the grain begins to recrystallize (Figure 22.12e). The newly nucleated grains have low defect density and, therefore, produce sharp reflection spots. They also have a different orientation than the parent grain, which causes reflection spots to appear at different azimuthal angles if they fulfill the Bragg condition (Figure 22.12e). Depending on whether the material or phase of interest predominantly shows recovery or recrystallization under the conditions of the experiment, the diffraction patterns are affected in different ways. By the development of the patterns during deformation it is, therefore, possible to distinguish these two processes.

22.5.2 Analysis of Diffraction Data

By evaluating the relative positions of reflection spots on different Debye–Scherrer rings, which is referred to as *reciprocal space mapping*, it was possible to investigate how differently oriented grains in a massively transformed TiAl sample rearrange into single grains upon heating [88, 104]. Careful analysis of single diffraction patterns even allowed tracing the phase transformation of single grains by the observation of diffuse streaking between parent and newly emerging grains in a TiAl alloy [74]. To be able to observe streaking or the grouping of reflection spots of different phases at similar azimuthal angles, individual detector images have to be analyzed manually. With modern flat panel detectors operated at several frames per second this is tedious and time consuming even for rather short *in situ* experiments. Additionally, the tracking of individual reflections over time is difficult to achieve. A simple and elegant method for visualizing the development of spots of the same crystallographic plane is to integrate over a small range of the Bragg angle θ in which the corresponding reflections occur and to plot the thereby acquired intensity as a function of azimuthal angle ψ and time which results in a so-called azimuthal angle-time (AT) plot [33].

This kind of data representation was used by Yan *et al.* to illustrate the formation of texture during deformation of an initially coarse grained Cu specimen [105]. In this study, the occurrence of microstructural processes, such as dislocation activity and subgrain formation is mentioned, but not elaborated on. In contrast to this, recovery and recrystallization processes were found to occur in a Zr alloy during hot compression in a similar experimental setup [106]. The angular spread of single reflection spots allowed deriving the mosaicity of the corresponding grains and the formation of subgrains was observed by the development of single spots with large angular spread into many sharp reflections separated only by a small angle. Recrystallization was identified by the rapid appearance and disappearance of reflections during a later stage of the experiment. Austenitic steel, which can deform by a twinning induced plasticity mechanism, has also been subject to investigation by this newly developed method [107]. In this case, it was shown that the material deforms by dislocation slip as well as twinning and the texture development during deformation was compared with the results of an elasto-plastic self-consistent model.

22.5.3 Hot Deformation of a Multi-phase Alloy

With the above-described approach, the hot-deformation behavior of a Ti-43.5Al-4Nb-1Mo-0.1B was investigated at the ESRF in Grenoble, France [83]. This is of particular interest since the difficult hot-working characteristics of TiAl alloys are one of the reasons why this material class has found only limited applications so far. Therefore, the aim of these investigations was to obtain a better understanding of the materials behavior during hot-forming operations. To this end, specimens were resistively heated between the anvils of an electro thermo-mechanical tester from Instron, Norwood (MA), USA, to 1300 °C and subsequently compressed. Details on the experimental procedure are given in [83]. Specimens were cylindrical with a diameter of 4 mm and a length of 8 mm. These dimensions were chosen because they enable characterizing the materials behavior in the bulk of the sample rather than its properties close to the surface. For recording the diffraction patterns, a Pixium 4700 (Thales Group, Neuilly-sur-Seine, France) flat panel detector [98] was used and operated at a frame rate of approximately 2 Hz. Throughout the experiment, the temperature was controlled with a type S thermocouple spot welded to the sample surface.

In Figure 22.13 AT-plots of two selected reflections (a,b) and graphs of applied temperature, true stress, and true strain (c) are presented. Lines produced by the evolution of individual spots with time are designated as timelines. Continuous timelines during heating are observed for both phases, α and β. As the specimen is held within the

($\alpha + \beta$) phase field at 1300 °C, fluctuations in the β-phase reflections are visible while the α-reflections remain static. When deformation starts, immediate changes in the appearance of the intensity distribution are apparent in the insets of Figure 22.13a,b. While the α-phase exhibits inclined, continuous timelines, the β-phase reflections broaden, and become dotty. Another distinctive attribute is the blurred appearance of the α-phase timelines, while those of the β-phase remain as sharp dots in the AT-plot. At $t = 215$ s, the sample was unloaded for 30 s to investigate the materials response. It can be seen in Figure 22.13a that the α-timelines remain continuous during this period, but still appear to be blurred. In contrast to this, the β-phase reflections fluctuate fast and the timelines are sharp (Figure 22.13b).

Inclined timelines are observed if the crystal lattice of a grain rotates about an axis that is parallel to the incident beam, whereas broadening of a reflection is indicative of an increasing defect density within the corresponding grain [81, 108]. The formation of many sharp dots stemming from one original reflection and occurring within a small angular range is caused by the formation of subgrains through a dynamic recovery process [83, 106]. Since peak broadening immediately followed by the formation of reflection dots was observed in Figure 22.13b it was derived that fast dynamic recovery processes are dominating in the β-phase. In the α-phase, however, the reflections are inclined and blurred indicating the deformation by dislocation slip and the presence of high defect densities. The absence of reflections with low angular spread points toward a slow recovery process instead of a discontinuous recrystallization process.

To validate the results from the *in situ* HEXRD experiment, electron-backscatter diffraction (EBSD) experiments were performed on the deformed specimens. The images shown in Figure 22.14a,b were acquired with a LEO 1525 field emission SEM equipped with an EDAX EBSD system. Orientations are indicated in color-code, the legends for the corresponding phases are located above the EBSD images. Figure 22.14c shows a micrograph of the same sample area obtained in BSE mode. From this image, it is obvious that the deformation experiment was conducted in the ($\alpha + \beta$) phase region.

From Figure 22.14 it can be seen that the α-phase formed elongated grains during deformation that are divided into a small number of equiaxed subgrains, whereas practically no small angle grain boundaries are visible in the β/β_0-grains. It should also be noted that the β_0-grains exhibit complex shapes, which can be attributed to the fact that at high temperatures, the β-phase accommodates a disproportionately large fraction of the plastic strain [58, 109, 110]. In both phases, no indication for dynamic recrystallization was found. Consequently, the conclusions drawn from the results of the AT plots could be verified by EBSD.

With the method described above, it is possible to derive the deformation, recovery, and recrystallization behavior of the individual phases during *in situ* experiments conducted on multi-phase alloys. It is, however, not possible to monitor and track a single grain throughout the whole experiment or to determine the shape and position of individual grains. Nonetheless, valuable information about the microstructural development can be derived by applying this method.

A complementary approach to investigate the materials evolution during production processes is to measure the texture before and after forming or heat-treatment steps of interest. Orientation distribution functions can be determined by means of neutron diffraction [111, 112] as well as by XRD by using laboratory [113–115] or synchrotron sources [32, 105]. Neutrons and high-energy X-rays offer the benefit of high penetration

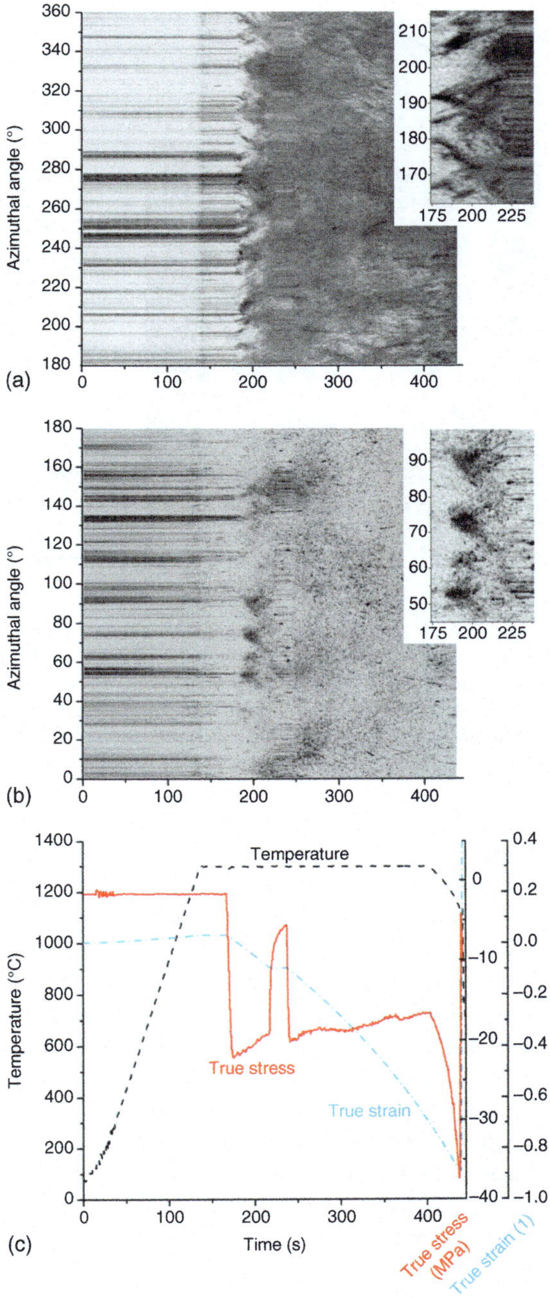

Figure 22.13 *In situ* hot-deformation experiment conducted on a Ti-43.5Al-4Nb-1Mo-0.1B alloy (nominal). Azimuthal angle-time plots of the (a) α(012) reflection and the (b) β(002) reflection. The experimental parameters such as temperature (dashed line), true stress (solid line), and true strain (dash-dotted line) are shown in (c). The insets visible at the upper right corner of (a) and (b) indicate the development of the reflections at the beginning of the hot-compression test at higher magnification. (Schmoelzer 2011 [76], reproduced with the permission of Wiley.)

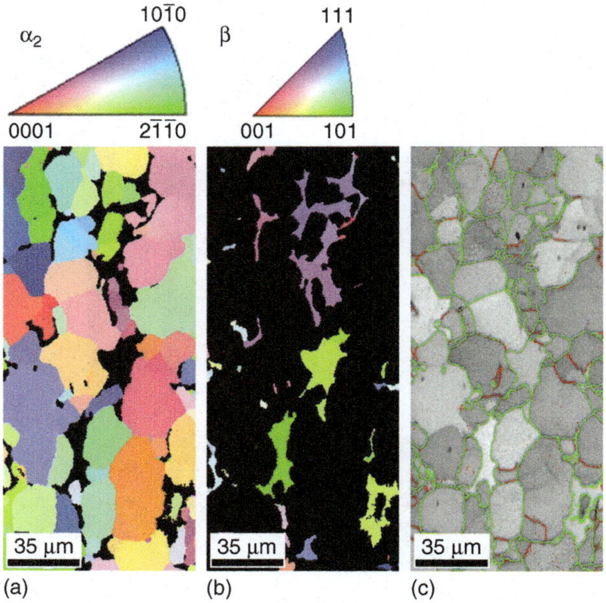

Figure 22.14 Micrographs obtained after the *in situ* hot-compression experiment on a Ti-43.5Al-4Nb-1Mo-0.1B alloy (nominal) by means of EBSD (a,b) and SEM in BSE mode (c). Image (a) shows elongated grains of the α_2-phase, whereas the intricately shaped β_o-phase is visible in (b). In image (c), high-angle grain boundaries (misorientation >15°) are indicated by green lines, whereas small-angle grain boundaries are red. The loading direction was horizontal. (Schmoelzer 2011 [41], reproduced with the permission of Wiley.)

depths and, therefore, allow investigating large sample volumes which leads to excellent grain statistics. Another advantage of neutron diffraction in case of TiAl alloys is that only a reduced number of Bragg-reflections are visible which minimizes problems that arise from overlapping α and γ reflections [111].

Brokmeier *et al.* [111] used this method for investigating the texture of TiAl alloys after extrusion. It was found that the sharpness of the texture is dependent on extrusion temperature as well as on alloy composition. The influence of carbon additions to a TiAl alloy on texture formation during hot-rolling was investigated by Stark *et al.* [114], while Schillinger *et al.* [115] derived the primary deformation mechanisms during rolling from texture measurements. Texture development during hot compression of Cu, Ni, and a γ-TiAl alloy was investigated by Hasegawa *et al.* which compared the dynamic recrystallization behavior of these three metals [113]. From the results of these experiments, these authors were able to derive the primary deformation mechanisms at different temperatures and deformation conditions.

22.6 Lattice Parameter and Thermal Expansion

For a material class that is intended for use in high-temperature applications, the response of the material on exposure to high-temperatures is of vital interest. Lattice parameters in all three dimensions can be tracked with relative ease in diffraction

Figure 22.15 Ratio of the lattice parameters in crystallographic c and a axis of the γ-phase in Ti-43.5Al-4Nb-1Mo-0.1B as a function of temperature. (Schmoelzer 2011 [41], reproduced with the permission of Wiley.)

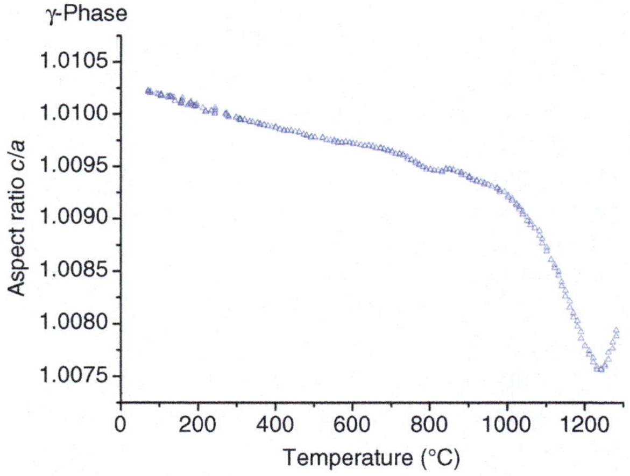

experiments. If the volume of the unit cell is plotted as a function of temperature, the gradient of the resulting curve at a certain temperature is the linear coefficient of thermal expansion (CTE) at this temperature [116].

With this method Novoselova *et al.* [116] have determined the CTE of a Ti-46Al-1.9Cr-3Nb alloy. It has to be stressed that the values measured in diffraction experiments only reflect the increase in lattice parameter, whereas the CTE is known to be affected also by the formation of thermal vacancies. Additionally, the lattice parameter may be influenced by the formation of elastic strains between adjacent phases due to differences in the CTEs [81]. In other publications only the development of lattice parameters was investigated and related to changes in the microstructure and phase composition [87, 88]. Figure 22.15 shows the temperature dependence of the *c*/*a* ratio of the γ-phase in a Ti-43.5Al-4Nb-1Mo-0.1B alloy. In this experiment, a heating rate of 10 K s^{-1} was used. Yeoh *et al.* [87] investigated a Ti-45Al-7.5Nb-0.5C alloy and related the development of the *c*/*a* ratio of the γ-phase similar to the one observed in Figure 22.15 to the decrease in its Al content at temperatures higher than 900 °C.

At 1230 °C, the γ-phase exhibits the lowest *c*/*a* ratio before it starts to increase again. From this behavior it is concluded that the Al content is lowest at that temperature. This shows exemplarily how information on the materials behavior can be derived from the development of the lattice parameter. In many applications, knowledge on the development of the lattice parameters with temperature is important, that is, when the use of protective coatings is considered or the impact of phase transitions on internal stresses is of relevance. Diffraction experiments allow obtaining information on the lattice parameters of individual phases in a multi-phase material. Additionally, they reveal effects such as the one shown in Figure 22.15 that strongly influence the thermal behavior.

22.7 Conclusions

Diffraction methods using neutrons and/or high-energy X-rays offer unique opportunities for investigating various aspects of importance for the development of intermetallic TiAl alloys. The possibility of performing experiments *in situ* is a great advantage especially for investigating fast phase transformations, such as the $\alpha \leftrightarrow \alpha_2$ and $\beta \leftrightarrow \beta_o$ ordering/disordering reactions. Additionally, diffraction methods are capable of incorporating small regions of an individual phase if phase fractions are determined. This is especially important in this alloy system, since lamellar grains that consist of fine γ and α_2-lamellae are a common microstructural feature. The different contrast formation for neutrons and X-rays in TiAl alloys is a peculiarity that makes these quite similar techniques complementary ones, which was emphasized in this review with selected examples. Progresses in detector technology and the construction of dedicated HEXRD beamlines have enabled conducting diffraction experiments at high frame rates, offering the opportunity to directly observe processes that occur comparatively fast. Parallel to the hardware development, new techniques have emerged that comprise valuable tools for various applications in materials science. Experiments with nano-beams and the three-dimensional X-ray diffraction microscopy are examples for methods that have great potential for future investigations. Without doubt, during the past 20 years, the use of diffraction methods has significantly contributed to the development of TiAl alloys and production routes and will continue to provide interesting results in the near future.

References

1 Yamaguchi, M., Inui, H., and Ito, K. (2000) *Acta Mater.*, **48**, 307.
2 Appel, F. and Wagner, R. (1998) *Mater. Sci. Eng., R*, **22**, 187.
3 Kim, Y., Clemens, H., and Rosenberger, A. (2003) *Gamma Titanium Aluminides 2003*, The Minerals, Metals and Materials Society (TMS), Warrendale, PA.
4 Kim, Y. (2008) *Structural Aluminides for Elevated Temperature Applications* (eds Y.-W. Kim, D. Morris, R. Yang, and C. Leyens), TMS, Warrendale, PA, p. 215.
5 Peters, M. and Leyens, C. (2003) *Titanium and Titanium Alloys*, Wiley-VCH Verlag GmbH, Weinheim.
6 Kim, Y. (1994) *J. Appl. Meteorol.*, **46**, 30.
7 Kesler, M.S., Goyel, S., Rios, O., Cupid, D.M., Seifert, H.J., and Ebrahimi, F. (2010) *Mater. Sci. Eng., A*, **527**, 2857.
8 Wu, X. (2006) *Intermetallics*, **14**, 1114.
9 Appel, F., Oehring, M., and Wagner, R. (2000) *Intermetallics*, **8**, 1283.
10 Cui, W.F. (2007) *Intermetallics*, **15**, 675.
11 Baur, H., Joos, R., Smarsly, W., and Clemens, H. (2000) *Intermetallics and Superalloys*, Wiley-VCH Verlag GmbH, Weinheim, p. 384.
12 Tetsui, T. (1999) *Gamma Titanium Aluminides 1999*, The Minerals, Metals and Materials Society (TMS), Warrendale, PA, p. 15.
13 Smarsly, W., Baur, H., Glitz, G., Clemens, H., Khan, T., and Thomas, M. (2001) *Structural Intermetallics 2001*, The Minerals, Metals and Materials Society (TMS), Warrendale, PA, p. 25.
14 Sommer, A. and Keijzers, G. (2003), *Gamma Titanium Aluminides 2003*, The Minerals, Metals and Materials Society (TMS) (eds Y.-W. Kim, H. Clemens, and A.H. Rosenberger), Warrendale, PA, p. 3.
15 Lasalmonie, A. (2006) *Intermetallics*, **14**, 1123.
16 Clemens, H., Bartels, A., Bystrzanowski, S., Chladil, H., Leitner, H., Dehm, G., Gerling, R., and Schimansky, F.P. (2006) *Intermetallics*, **14**, 1380.

17 Clemens, H. and Mayer, S. (2013) *Adv. Eng. Mater.*, **15**, 191.

18 Appel, F., Paul, J.D.H., and Oehring, M. (2011) *Gamma Titanium Aluminide Alloys – Science and Technology*, Wiley-VCH Verlag GmbH, Weinheim.

19 Clemens, H., Wallgram, W., Kremmer, S., Güther, V., Otto, A., and Bartels, A. (2008) *Adv. Eng. Mater.*, **10**, 707.

20 Imayev, R.M., Imayev, V.M., Oehring, M., and Appel, F. (2007) *Intermetallics*, **15**, 451.

21 Wallgram, W., Schmoelzer, T., Cha, L., Das, G., Güther, V., and Clemens, H. (2009) *Int. J. Mater. Res.*, **100**, 1021.

22 Clemens, H., Chladil, H.F., Wallgram, W., Zickler, G.A., Gerling, R., Liss, K.D., Kremmer, S., Güther, V., and Smarsly, W. (2008) *Intermetallics*, **16**, 827.

23 Abe, E., Kumagai, T., and Nakamura, M. (1996) *Intermetallics*, **4**, 327.

24 Cha, L., Scheu, C., Clemens, H., Chladil, H., Dehm, G., Gerling, R., and Bartels, A. (2008) *Intermetallics*, **16**, 868.

25 Shull, R.D. (1988) *High Temp. Sci.*, **26**, 95.

26 Kim, Y.W. (1995) *Mater. Sci. Eng., A*, **192**, 519.

27 Swaminathan, S. (1996) *Philos. Mag. Lett.*, **73**, 319.

28 Kim, Y.W. (1992) *Acta Metall. Mater.*, **40**, 1121.

29 Grinfeld, M.A. (1998) *Metall. Mater. Trans. A*, **29**, 937.

30 Schmoelzer, T., Liss, K.D., Zickler, G.A., Watson, I.J., Droessler, L.M., Wallgram, W., Buslaps, T., Studer, A.J., and Clemens, H. (2010) *Intermetallics*, **18**, 1544.

31 Watson, I.J. (2009) *Adv. Eng. Mater.*, **11**, 932.

32 Liss, K., Bartels, A., Schreyer, A., and Clemens, H. (2010) *Textures Microstruct.*, **35**, 219.

33 Liss, K. and Yan, K. (2010) *Mater. Sci. Eng., A*, **528**, 11.

34 Reimers, W., Pyzalla, A., Schreyer, A., and Clemens, H. (2008) *Diffraction Methods in Engineering Materials Science* (eds W. Reimers, A.R. Pyzalla, A. Schreyer, and H. Clemens), Wiley-VCH Verlag GmbH, Weinheim.

35 Schreyer, A. (2008) *Diffraction Methods in Engineering Materials Science* (eds W. Reimers, A.R. Pyzalla, A. Schreyer, and H. Clemens), Wiley-VCH Verlag GmbH, Weinheim, p. 79.

36 Treusch, R. (2008) *Diffraction Methods in Engineering Materials Science* (eds W. Reimers, A.R. Pyzalla, A. Schreyer, and H. Clemens), Wiley-VCH Verlag GmbH, Weinheim, p. 97.

37 Knop, W., Pranzas, P., and Schreiner, P. (2008) *Diffraction Methods in Engineering Materials Science* (eds W. Reimers, A.R. Pyzalla, A. Schreyer, and H. Clemens), Wiley-VCH Verlag GmbH, Weinheim, p. 91.

38 Hashimoto, K. (1998) *Intermetallics*, **6**, 667.

39 Ashcroft, N. (1976) *Solid State Physics*, Holt Rinehart and Winston, New York.

40 Rauch, H. and Waschowski, W. (2003) *Neutron Data Booklet* (eds A.J. Dianoux and G. Lander), Old City Publishing, Philadelphia, PA, p. 1.

41 Schmoelzer, T., Mayer, S., Sailer, C., Haupt, F., Güther, V., Staron, P., Liss, K., and Clemens, H. (2011) *Adv. Eng. Mater.*, **13**, 306.

42 Kimura, M. (1999) *J. Phase Equilib.*, **20**, 224.

43 Qin, G.W. (2001) *Metall. Mater. Trans. A*, **32**, 1927, doi:10.1007/s11661-001-0005-1.

44 Kremmer, S., Chladil, H., Clemens, H., Otto, A., and Güther, V. (2008) *Ti-2007 Science and Technology*, The Japan Institute Of Metals (JIM), Sendai, p. 989.

45 Sun, F.S., Cao, C.X., Kim, S.E., Lee, Y.T., and Yan, M.G. (2001) *Metall. Mater. Trans. A*, **32**, 1573.

46 Kainuma, R., Fujita, Y., Mitsui, H., Ohnuma, I., and Ishida, K. (2000) *Intermetallics*, **8**, 855.

47 Chladil, H.F., Clemens, H., Zickler, G.A., Takeyama, M., Kozeschnik, E., Bartels, A., Bulaps, T., Gerling, R., Kremmer, S., Yeoh, L., and Liss, K.-D. (2007) *Int. J. Mater. Res.*, **98**, 1131.

48 Takeyama, M. and Kobayashi, S. (2005) *Intermetallics*, **13**, 993.

49 Saunders, N. (1999) *Gamma Titanium Aluminides* (eds Y.-W. Kim, D.M. Dimiduk, and M.H. Loretto), TMS, Warrendale, PA, p. 183.

50 Clemens, H., Boeck, B., Wallgram, W., Schmoelzer, T., Droessler, L.M., Zickler, G.A., Leitner, H., and Otto, A. (2008) *Materials Research Society Symposium Proceedings*, MRS, Warrendale, PA, p. 115.

51 Hamajima, T., Luetjering, G., and Weissmann, S. (1972) *Red Books*, vol. **XVIII**, Kombinat Viniti, Moskau (Moskva).

52 Hansen, R. and Aravamudham, R. (1970) *Metallkunde*, **61**, 115.

53 Singh, A.K. and Banerjee, D. (1997) *Metall. Mater. Trans. A*, **28**, 1745.

54 Mayer, S., Sailer, C., Nakashima, H., Schmoelzer, T., Lippmann, T., Staron, P., Liss, K.-D., Clemens, H., and Takeyama, M. (2011) *Mater. Res. Soc. Symp. Proc.*, **1295**, 113.

55 Hammersley, A.P., Svensson, S.O., Hanfland, M., Fitch, A.N., and Häusermann, D. (1996) *High Pressure Res.*, **14**, 235.

56 Kabra, S., Yan, K., Mayer, S., Schmoelzer, T., Reid, M., Dippenaar, R., Clemens, H., and Liss, K.-D. (2011) *Int. J. Mater. Res.*, **102**, 697.

57 Stark, A., Oehring, M., Pyczak, F., and Schreyer, A. (2011) *Adv. Eng. Mater.*, **13**, 700.

58 Appel, F., Oehring, M., and Paul, J.D.H. (2006) *Adv. Eng. Mater.*, **8**, 371.

59 Tanimura, M., Inoue, Y., and Koyama, Y. (2001) *Scr. Mater.*, **44**, 365.

60 Stark, A. (2010) *Textur- und Gefügeentwicklung bei der thermomechanischen Umformung Nb-reicher γ-TiAl-Basislegierungen*, Shaker Verlag, Aachen.

61 Schmoelzer, T., Stark, A., Schwaighofer, E., Lippmann, T., Mayer, S., and Clemens, H. (2012) *Adv. Eng. Mater.*, **14**, 445.

62 Lippmann, T., Lottermoser, L., Beckmann, F., Martins, R.V., Dose, T., Kirchhof, R., and Schreyer, A. (2007) in *Hasylab Annual Report* (eds W. Caliebe, W. Drube, K. Rickers, and J.R. Schneider), HASYLAB/DESY, Hamburg, p. 113.

63 Schnitzer, R., Chladil, H.F., Scheu, C., Clemens, H., Bystrzanowski, S., Bartels, A., and Kremmer, S. (2007) *Prakt. Metallogr.*, **44**, 430.

64 Matsumoto, H., Yoneda, H., Sato, K., Kurosu, S., Maire, E., Fabregue, D., Konno, T.J., and Chiba, A. (2011) *Mater. Sci. Eng., A*, **528**, 1512.

65 Cha, L., Clemens, H., and Dehm, G. (2011) *Int. J. Mater. Res.*, **102**, 703.

66 Shuleshova, O., Holland-Moritz, D., Löser, W., Voss, A., Hartmann, H., Hecht, U., Witusiewicz, V.T., Herlach, D.M., and Büchner, B. (2010) *Acta Mater.*, **58**, 2408.

67 Senkov, O.N. (2002) *Scr. Mater.*, **46**, 187.

68 Staron, P., Christoph, U., Appel, F., and Clemens, H. (2002) *Appl. Phys. A*, **74**, 1163.

69 Beschliesser, M., Chatterjee, A., Lorich, A., Knabl, W., Kestler, H., Dehm, G., and Clemens, H. (2002) *Mater. Sci. Eng., A*, **329**, 124.

70 Zhu, H.A. (2005) *Scr. Mater.*, **52**, 45.

71 Droessler, L.M., Schmoelzer, T., Wallgram, W., Cha, L., Das, G., and Clemens, H. (2009) Advanced Intermetallic-Based Alloys for Extreme Environment and Energy Applications. *Material Research Society Symposium Proceedings* (eds M. Palm, B.P. Bewley, Y.-H. He, M. Takeyama, and J.M.K. Wiezorek), MRS, Warrendale, PA, vol. 1128 p. 121–126.
72 Dehm, G.A.B. (2006) *Adv. Eng. Mater.*, **8**, 1033.
73 Tang, J. (2001) *Mater. Res. Bull.*, **36**, 1737.
74 Liss, K.-D., Stark, A., Bartels, A., Clemens, H., Buslaps, T., Phelan, D., and Yeoh, L.A. (2008) *Adv. Eng. Mater.*, **10**, 389.
75 Zhu, H. (2005) *Mater. Sci. Eng., A*, **397**, 58.
76 Schmoelzer, T., Liss, K.-D., Staron, P., Mayer, S., and Clemens, H. (2011) *Adv. Eng. Mater.*, **13**, 685–699.
77 Blackburn, M. (1970) *The Science, Technology and Application of Titanium*, Pergamon Press Ltd., Oxford, p. 633.
78 Eidenberger, E., Schober, M., Schmoelzer, T., Stergar, E., Staron, P., Leitner, H., and Clemens, H. (2010) *Phys. Status Solidi A*, **207**, 2238.
79 Liss, K.-D., Whitfield, R.E., Xu, W., Buslaps, T., Yeoh, L.A., Wu, X., Zhang, D., and Xia, K. (2009) *J. Synchrotron Radiat.*, **16**, 825.
80 Yan, K., Carr, D.G., Kabra, S., Reid, M., Studer, A., Harrison, R., Dippenaar, R., and Liss, K.-D. (2010) *World J. Eng.*, **7** (**Suppl. 2**), P422.
81 Warren, B. (1990) *X-Ray Diffraction*, Dover Publications, New York.
82 Grujicic, M. and Zhang, Y. (1999) *J. Mater. Sci.*, **34**, 1419.
83 Liss, K., Schmoelzer, T., Yan, K., Reid, M., Peel, M., Dippenaar, R., and Clemens, H. (2009) *J. Appl. Phys.*, **106**, 113526.
84 Tetsui, T., Shindo, K., Kaji, S., Kobayashi, S., and Takeyama, M. (2005) *Intermetallics*, **13**, 971.
85 Kim, S. (1998) *Mater. Sci. Eng., A*, **250**, 77.
86 Kim, S. (1997) *Mater. Sci. Eng., A*, **239**, 229.
87 Yeoh, L.A., Liss, K., Bartels, A., Chladil, H., Avdeev, M., Clemens, H., Gerling, R., and Buslaps, T. (2007) *Scr. Mater.*, **57**, 1145.
88 Liss, K.D., Bartels, A., Clemens, H., Bystrzanowski, S., Stark, A., Buslaps, T., Schimansky, F.P., Gerling, R., Scheu, C., and Schreyer, A. (2006) *Acta Mater.*, **54**, 3721.
89 Hoelzel, M. (2004) *Phys. Rev. B: Condens. Matter*, **350**, E671.
90 Gilles, R. (2006) *Z. Kristallogr.*, **221**, 183.
91 Studer, A.J., Hagen, M.E., and Noakes, T.J. (2006) *Phys. Rev. B: Condens. Matter*, **385**, 1013.
92 Barabash, R.I., Ice, G., and Walker, F. (2003) *J. Appl. Phys.*, **93**, 1457.
93 Larson, B.C., Yang, W., Ice, G., Budai, J., and Tischler, J. (2002) *Nat. Geosci.*, **415**, 887.
94 Kirchlechner, C., Kiener, D., Motz, C., Labat, S., Vaxelaire, N., Perroud, O., Micha, J., Ulrich, O., Thomas, O., Dehm, G., and Keckes, J. (2012) *Philos. Mag.*, **91**, 1256.
95 Inui, H., Matsumuro, M., Wu, D., and Yamaguchi, M. (1997) *Philos. Mag. A*, **75**, 395.
96 Ihara, K. and Miura, Y. (2004) *Mater. Sci. Eng., A*, **387**, 651.
97 Miura, Y., Ihara, K., and Fukaura, K. (2000) *Mater. Sci. Eng., A*, **280**, 134.

98 Daniels, J.E. and Drakopoulos, M. (2009) *J. Synchrotron Radiat.*, **16**, 463.
99 Daniels, J.E. (2008) *J. Appl. Crystallogr.*, **41**, 1109.
100 Margulies, L., Winther, G., and Poulsen, H.F. (2001) *Science*, **291**, 2392.
101 Margulies, L., Lorentzen, T., Poulsen, H.F., and Leffers, T. (2002) *Acta Mater.*, **50**, 1771.
102 Poulsen, H. (2004) *Three-Dimensional X-Ray Diffraction Microscopy: Mapping Polycrystals and Their Dynamics*, Springer, Berlin, New York.
103 Doherty, R.D. (1997) *Mater. Sci. Eng., A*, **238**, 219.
104 Liss, K.D., Clemens, H., Bartels, A., Stark, A., and Buslaps, T. (2006) *Materials Research Society Symposium Proceedings*, MRS, Warrendale, PA, p. 425.
105 Yan, K., Liss, K.D., Garbe, U., Daniels, J., Kirstein, O., Li, H., and Dippenaar, R. (2009) *Adv. Eng. Mater.*, **11**, 771.
106 Liss, K.D., Garbe, U., Li, H., Schambron, T., Almer, J.D., and Yan, K. (2009) *Adv. Eng. Mater.*, **11**, 637.
107 Yan, K., Carr, D.G., Callaghan, M.D., Liss, K., and Li, H. (2010) *Scr. Mater.*, **62**, 246.
108 Sahu, P. (2006) *Intermetallics*, **14**, 180.
109 Tetsui, T.A. (2002) *Scr. Mater.*, **47**, 399.
110 Tetsui, T.A. (2003) *Intermetallics*, **11**, 299.
111 Brokmeier, H.-G., Oehring, M., Appel, F., and Clemens, H. (2002) *Appl. Phys. A*, **74**, S1222.
112 Brokmeier, H.-G. (2004) *Metall. Mater. Trans. A*, **35**, 3563.
113 Hasegawa, M., Yamamoto, M., and Fukutomi, H. (2003) *Acta Mater.*, **51**, 3939.
114 Stark, A., Bartels, A., Gerling, R., Schimansky, F.P., and Clemens, H. (2006) *Adv. Eng. Mater.*, **8**, 1101.
115 Schillinger, W., Bartels, A., Gerling, R., Schimansky, F.P., and Clemens, H. (2006) *Intermetallics*, **14**, 336.
116 Novoselova, T., Malinov, S., Sha, W., and Zhecheva, A. (2004) *Mater. Sci. Eng., A*, **371**, 103.

23

In situ μLaue: Instrumental Setup for the Deformation of Micron Sized Samples*

Christoph Kirchlechner, Jozef Keckes, Jean S. Micha, and Gerhard Dehm

23.1 Introduction

Small dimensions influence the mechanical properties of all materials. Examples are pores in ceramics, precipitates in metals, the Hall–Petch effect due to grain size, or the inevitable small geometrical dimensions of materials in miniaturized systems. It is nowadays a well-documented fact in materials science [1, 2] that small single crystals exhibit a higher flow stress compared to their macroscopic counter parts. This fact is neither fully analyzed nor completely understood, although it is not only of scientific interest but also of paramount concern as in for example, microelectronic applications to understand the plasticity of metallic domains in the micron and submicron regime.

Uchic and co-workers were the first ones who performed uniaxial compression tests on micron-sized samples [3] and inspired scientists worldwide to perform similar microcompression [4, 5], tension [6, 7], or bending [8, 9] experiments. During the past decade, experimental setups where considerably improved and efforts have been put into sample preparation [10, 11] or into improving the test conditions [12]. Nevertheless, interlinking the mechanical properties in these dimensions with the underlying microstructure is still hindered by the lack of appropriate characterization methods. Scanning electron microscopy (SEM) can, for instance, guaranty a precise sample alignment and, therefore, enables to test samples well below 1 μm in diameter. Nevertheless, SEM just probes the sample surface and, therefore, is only able to monitor glide steps formed by dislocations escaping at the sample surface. These glide steps imply that the activation of discrete dislocation sources are responsible for the plastic deformation of metallic structures. Nevertheless, dislocations in the sample volume cannot be observed with SEM. In contrast, transmission electron microscopy (TEM) is able to monitor dislocation sources during operation and dislocation movements [13, 14], but the use of TEM is limited due to electron transparency, which is achieved for most metals only at thicknesses well below 0.5 μm. Furthermore, it has to be noted, that focused ion beam (FIB) milling introduces a surface damage layer [15] whose fraction increases with decreasing sample size. Even though the damage layer is not fully investigated, it is clear that it consists of stacking fault tetrahedrons, dislocation loop-like structures, and probably also of a huge

*This chapter is reproduced and updated from *Adv. Eng. Mater.* **13** (2011) 837–844 by permission of Wiley-VCH.

Neutrons and Synchrotron Radiation in Engineering Materials Science: From Fundamentals to Applications, Second Edition. Edited by Peter Staron, Andreas Schreyer, Helmut Clemens, and Svea Mayer.

number of interstitials and vacancies – all well known to hinder dislocation movement and, therefore, to increase the strength of metallic structures.

Laue diffraction [16] is partly able to overcome the above-mentioned challenges of *in situ* SEM and TEM testing of micron- and submicron-sized specimens. Up to now, Laue experiments on miniaturized specimens were performed *ex situ* [16–19] and *in situ* [20–25]. The data of the *ex situ* experiments revealed the presence of strain gradients, subgrains [17], and FIB damage [19]. Furthermore, it was shown that the diffraction peak width of submicron-sized pillars did not change during deformation. Hence, the authors concluded that the dislocation density did not change [18]. In contrast, micron-sized copper tensile samples showed pronounced peak streaking caused by the storage of excess dislocations [16]. These *ex situ* experiments as well as some *in situ* observations [20, 23] showed the early activation of an unpredicted slip system. Without going into detail, the mentioned publications impressively demonstrate the need for and the advantages of Laue diffraction. Here, a new experimental approach for micromechanical testing during Laue experiments with a micron-sized white X-ray beam (μLaue) is presented. A detailed description of the instrumentation and the measurement is shown with a representative example of a compressed micron-sized pillar. In addition, challenges and limitations during the measurement and the data evaluation are discussed.

23.1.1 μLaue Diffraction, a Short Introduction

For Laue diffraction experiments, a white X-ray beam consisting of a broad energy band pass is used. Thus, Bragg's condition is satisfied for several crystallographic planes – all for different and automatically selected wavelengths. Advantages of using the white X-ray beam are that (i) a huge number of reflections are obtained simultaneously and (ii) no sample rotation is necessary to analyze single crystalline materials. Both, (i) and (ii) guaranty fast measurements and, therefore, favor Laue diffraction for *in situ* experiments. In addition, a constant illuminated volume is assured as no sample rotation or tilt is required [26, 27]. Another consequence is that all crystals in the illuminated volume contribute to the Laue pattern and, hence, the presence and/or the formation of subgrains will definitely be noticed. This point is very important in case of micromechanical tests where most of the investigated samples are expected to be single crystalline.

Due to the overlap of all diffracting grains a major disadvantage occurs: If more than approximately 10 [27] grains contribute to the Laue pattern it is impossible to assign the observed reflections to each grain and an analysis of the full Laue pattern is prevented. Thus, μLaue experiments are suitable for materials where the grain size is in the range of the primary beam size [28]. Another disadvantage of Laue diffraction is the fact that the energy responsible for a reflection is unknown. Thus, the absolute lattice spacing cannot be derived from white beam experiments. Consequently, Laue diffraction provides the shape of the unit cell (i.e., the deviatoric strains) but not its size (hydrostatic strains). Complementary fine energy scans can be performed by inserting a tunable monochromator or a multi-colored rainbow filter [29] in the white beam path in order to determine the energy of selected reflections and to further analyze the full strain tensor [26].

Besides determining the strain tensor, Laue diffraction also enables to explore the defect structure of imperfect crystals [30]. Thereby, two kinds of defects can be discriminated: The first kind is known to keep a sharp but weaker Bragg reflection accompanied by some redistributed broad diffusive intensity. Defects of the first kind are, for instance,

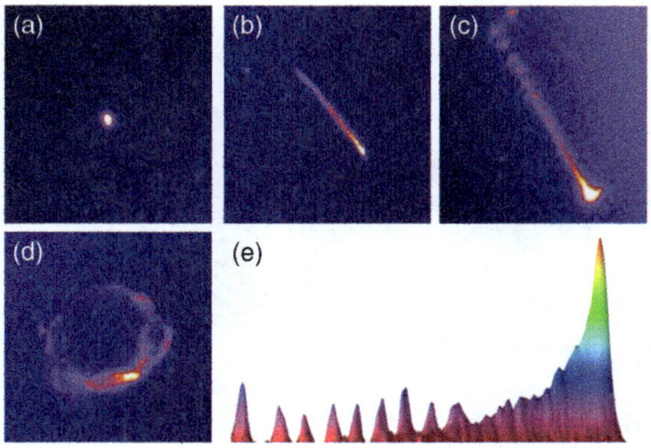

Figure 23.1 Different (111) Laue peak shapes: (a) single crystalline, focused ion beam milled Cu; (b) GNDs on the primary system in a tensile sample; (c) "spikes" formed in a bending beam evidencing the formation of subgrains; (d) more complex structure of GNDs in a compression pillar; and (e) 3D profile of the spikes in Figure 23.1c.

point defects or coherent precipitates [30]. In this chapter, we focus on defects of the second kind, such as dislocations, stacking faults, or grain boundaries [30]. Defects of the second kind are known to cause diffraction peak broadening. For instance, statistical stored dislocations (SSDs), where the total Burgers vector equals zero, cause peak broadening in every direction and the diffraction peak exhibits a symmetric (circular) peak shape (Figure 23.1a). In contrast, introducing geometrically necessary dislocations (GNDs) – also known as *unpaired dislocations* – with a nonzero net Burgers vector, induces a local lattice curvature as described by Nye [31]. Thus, the orientation of the lattice planes – and, therefore, also the diffraction vector – slightly changes the orientation. This orientation gradient leads to streaking of diffraction peaks (Figure 23.1b), with a natural streaking axes ξ [30] defined by Eq. (23.1):

$$\xi = \frac{l \times G_{hkl}}{|l \times G_{hkl}|} \tag{23.1}$$

with l being the direction of the dislocation line and G_{hkl} the reciprocal lattice vector. In the case of edge type dislocations, the direction of the dislocation line is given by the cross product of the slip plane normal vector $\mathbf{n}$ with the Burgers vector $\mathbf{b}$. Thus, by analyzing the orientation of peak streaking directions the type of GNDs can – at least in simple cases – be deduced. However, if dislocations with different Burgers vectors, on different slip systems or an unknown screw character are present in the crystal, the evaluation of activated slip systems is ambiguous (Figure 23.1d).

Dislocations are forming cells and boundaries to minimize the internal strain energy. Without going into detail, it should be mentioned, that subgrains separated by such boundaries can be observed with Laue diffraction and are evidenced by split intensities. An example of a split diffraction spot is presented in Figure 23.1c,e. For a thorough review on white beam diffraction experiments on imperfect crystals the reader is referred to the work of Ice and Barabash [30].

23.2 Experimental Instrumentation

23.2.1 The Straining Device

The straining device shown in Figure 23.2a,b is able to perform compression, tensile, or bending experiments. Loading is performed by using a closed loop piezo (PU-38,

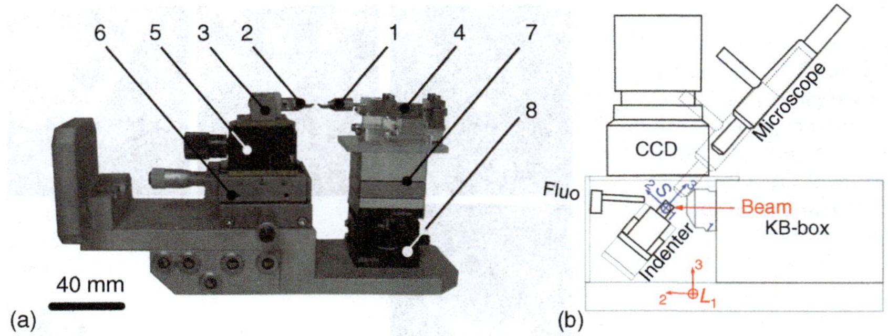

Figure 23.2 (a) Image of the straining device. (1) sample holder with sample; (2) counterbody; (3) closed loop piezo for straining; (4) load cell; (5) rotation stage; (6,8) coarse; and (7) fine translation stage. (b) Schematic drawing of the instrumental setup at BM32, ESRF (European Synchrotron Radiation Facility). For clarity, only the CCD camera, the Kirckpatrick-Beaz (KB)-box, the fluorescence detector (Fluo), the optical microscope and the straining device are shown. In addition, the sample (*S*) and laboratory (*L*) coordinate system are plotted.

provided by piezosystems Jena, Germany) with a traveling range of 32 μm and a typical reproducibility in the nanometer range. The displacement is measured by a resistance strain gauge at the fixed body joint of the piezo. Forces F are measured by using the eigenfrequency of a 30 μm thick tungsten wire according to Eq. (23.2), where L is the length of the wire, f the frequency, ρ the density of tungsten, and A the cross-section area.

$$F = 4 \cdot L^2 \cdot f^2 \cdot \rho \cdot A \tag{23.2}$$

The wire is prestrained by a calibration weight of 39 g leading to a typical initial frequency of 5000 Hz. The frequency of the wire is then evaluated by counting the time necessary for 1024 periods. Based on these settings, the load cell provided by Kammrath and Weiss (Germany) is able to measure the forces with a precision of 10 μN up to approximately 2*N* (fracture of the wire). Thereby, the load cell has typically a sampling rate of $0.2\,s^{-1}$ which is sufficient in case of displacement controlled micro mechanical experiments in these dimensions. Some of the advantages of this setup are that the wire can be renewed and calibrated within minutes and, therefore, the setup is very damage tolerant. Furthermore, by varying the wire diameter and/or the standard weight for prestraining the maximum load, the precision, and the sampling rate of the load cell can easily be modified.

Additionally, the straining device has to enable an independent alignment of the sample and the corresponding counter body. Therefore, the device consists of translation stages for coarse alignment, which are free from backlash and a piezo stage for positioning in x, y, and z direction in the nanometer regime, as well as a rotation stage enabling rotations with a precision of approximately 0.01°. The device is mounted on the translation stage of the beamline in order to perform scans with the entire device in S_1, S_2, and S_3 direction.

23.2.2 The Synchrotron Beamline

Our experiments are performed at French beamline CRG-IF [32] at the exit of the ESRF BM32 bending magnet synchrotron source. The white X-ray beam, with an energy range of 5–25 keV, is focused by a set of Kirkpatrick–Baez (KB) mirrors to a size typically less than 1 μm in diameter. The exact peak shape is measured using a gold wire as a sharp edge and the fluorescence detector system. The center position of the charge couple device (CCD, MAR165) is at a 2θ position of 90° and a sample detector distance of approximately 70 mm, as can be seen in Figure 23.2b. In addition, an optical microscope and a fluorescence detector are used to find the sample in the beam. Unfortunately, due to the restricted dimensions, either the optical microscope or the CCD camera can be used. Hence, during alignment of sample and counter body, the CCD has to be brought in a parking position 150 mm above the measuring position and during Laue experiments the optical microscope cannot be used (Figure 23.2b). Sample stages and focusing KBs are fixed on a massive frame and passive pads to reduce vibrations coming from the ground. Besides, the CCD camera is mechanically decoupled from sample stages. The temperature in the hutch is controlled by an air conditioner and is kept constant within 0.2° h^{-1}, as monitored during our experiments. Nevertheless, to avoid any air-stream onto the straining device caused by the air conditioner, the setup was covered.

23.2.3 The Experiment

In contrast to other experimental setups [20] we are using a single optical microscope with the optical axes being perpendicular to the S_1-S_2 plane (Figure 23.2b) of the sample coordinate system. Using the different translation and rotation stages the sample can be aligned in this plane with a precision in the submicrometer regime. Due to the image sharpness in the focal plane of the microscope we are also able to adjust the sample height (S_3) with a precision of less than 100 nm, which is sufficient for our measurements. Subsequently, the entire setup has to be brought in the beam. This is assisted by a fluorescence detector system while a stage scan with the entire straining device is performed. This scan is performed as a mesh scan using the S_1 and S_2 sample coordinates and typically leads to a fluorescence map as presented in Figure 23.3a,b.

Before the *in situ* experiments, the setup is calibrated using a Germanium wafer glued to the sample base. Subsequently, the sample is mesh scanned with step sizes in the range of 1 or 2 μm. The exposure time varies with the sample diameter (irradiated volume) and is typically 0.2–2 s, followed by a read out time of the CCD (6 s). Hence, typically it takes 6–8 s to record a Laue pattern. During straining, neither the stage nor the primary beam are moved in order to avoid any vibrations of the setup. Unfortunately, due to straining, the irradiated volume slightly changes. The relevance of this problem increases with

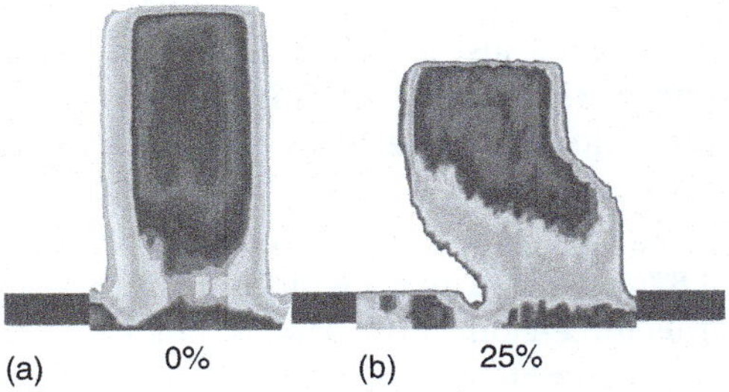

Figure 23.3 Fluorescence scans of a Cu micropillar with a rectangular cross-section used for analyzing the sample shape and measure the position of the pillar with respect to the X-ray beam. (a) Undeformed and (b) deformed to 25% engineering strain.

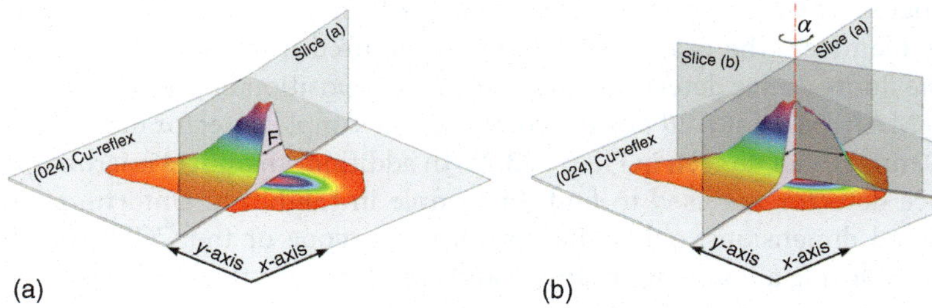

Figure 23.4 The analysis of the angular distribution of diffraction peak width (FWHM = F) is presented for two different slices (a) and (b).

increased sample length, applied strain, smaller primary beam width, and the position of the primary beam with respect to the sample. Consequently, these conditions have to be taken into account when performing the data evaluation, which will be discussed later.

23.2.4 Data Analysis

The diffraction patterns are evaluated by the software package XMAS developed by Tamura and co-workers [33]. Hereby, the crystal orientation and also the deviatoric strains can be evaluated. Unfortunately, due to the shape of streaked diffraction peaks, the evaluation of deviatoric strains becomes inaccurate and is not within the focus of our work. Nevertheless, the indexed Laue patterns lead to the exact orientation of the sample and, hence, are of paramount interest for subsequent data processing. As a next step, the peak full width at half maximum (FWHM) of the Laue spots is analyzed by cutting the Laue spot into various slices, perpendicular to the detector plane (Figure 23.4a) so that the slice always intersects the maximum peak intensity. Thus, by varying the angle of intersection α (Figure 23.4b) the angular distribution of the FWHM can be analyzed. As mentioned before, the storage of excess dislocations (GNDs) leads to a streaking with a natural streaking axis given by (1). Here, we superimpose the natural streaking axes of edge type dislocations on the diffraction images and so – for simple cases – the slip system, which stored GNDs can be evaluated.

23.2.5 Example: *In situ* Deformation of a Copper Pillar

Figure 23.5a shows the stress strain curve of a single slip oriented copper pillar with the dimensions of $7 \times 7 \times 21\,\mu m^3$. The red points indicate positions where Laue patterns were recorded, whereby the sample was not moved and, hence, the primary beam illuminated just an almost constant volume of the sample. In addition to this local information during straining, mesh scans covering the entire sample with a lateral resolution of 1 µm where performed at positions marked with "M_n" in the stress strain curve.

The sample was loaded sequentially in three steps with a total engineering strain of 0.8%, 8.4%, and 16.8% with comparable elastic slops with a stiffness of 3980 ± 120 MPa. The force-elongation data is not corrected for machine stiffness and though, the slope cannot be compared with the elastic modulus of copper (in the $\langle 234 \rangle$ direction).

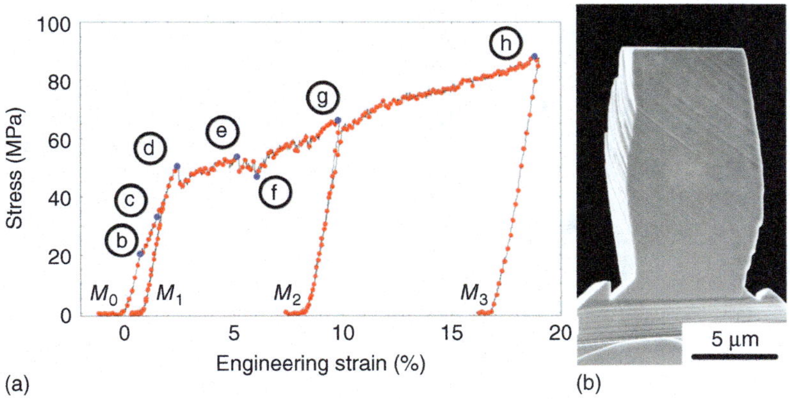

Figure 23.5 (a) Stress–strain curve of a single slip oriented Cu pillar. The red points superimposed to the stress– strain curve indicate positions where Laue patterns were recorded. Some selected Laue patterns marked in the stress–strain curve are plotted in Figure 23.6b–h. In addition, at positions marked with M_n, mesh scans covering the entire sample were performed. (b) SEM image showing the deformed pillar after M_3.

The metamorphosis of the (−2−24) spot during loading is depicted in Figure 23.6. Initially, all Laue spots (see for instance Figure 23.6b) are circular shaped and there is no evidence for a strain gradient as observed by Maaß *et al.* [16]. This shape does not change during elastic loading. Directly after the kink at 0.7% strain (point b in Figure 23.5a) the diffraction peak streaks, as can be seen by the elongated diffraction peak (Figure 23.6c). By superimposing the streaking axes of several slip systems as presented in Figure 23.6a with the number representing the rank in Schmid factor (1 highest, 12 lowest), one can notice, that slip system number 5 – an unpredicted slip system – matches the measured peak best. Upon further straining, the peak width in the described streaking direction increases (Figure 23.6d,e). Nevertheless, at 5% engineering strain the diffraction peak is still elongated in one direction and shows just one maximum. This evidences the presence of a dislocation structure with randomly distributed dislocations of one distinct type – most likely edge type dislocations on system 5 being the (111)[0−11] slip system. After the load drop at 6% strain, the diffraction spot splits. In other words, one can observe two separated maxima indicating the formation of two subgrains – divided by a dislocation wall. Furthermore, from now on, the diffraction peak shows the contribution of a second slip system and the continuous formation of new subgrains (Figure 23.6g) being responsible for the observed hardening in the stress–strain curve. At the maximum applied plastic strain of 16.8% the intensities of several (−2−24) Laue peaks caused by numerous subpeaks are very weak, nevertheless seven subgrains can be observed in this pattern.

The mesh scans performed at positions M_n can be analyzed in the same way. In addition to the qualitative analysis of activated slip system, the diffraction peak width – proportional to the dislocation density – and the streaking path, being the direction containing the maximum FWHM can be evaluated. Note that the streaking path does not necessarily coincide with a natural streaking axes of one distinct dislocation type. The maps presented in Figure 23.7a,b show the local evolution of the FWHM and the streaking path of the (111) Laue spot in the pillar. In the case

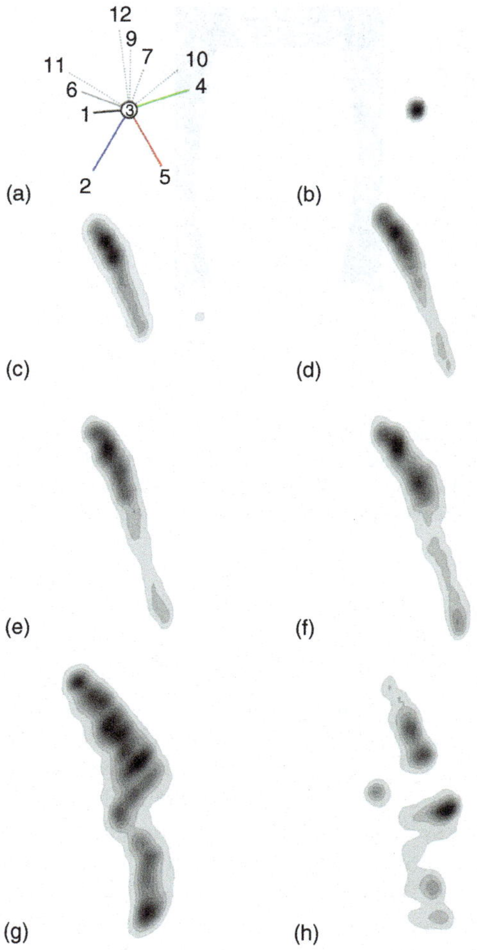

Figure 23.6 (a) Streaking directions of pure edge type dislocations calculated for the (−2−24) Laue spot. The number next to each streak indicates the rank in Schmid factor (1 = highest, 12 = lowest) of the slip system which stored GNDs. (b–h) The alteration of diffraction peaks with the corresponding position in the stress–strain curve (Figure 23.5) is shown.

of the undeformed pillar (mesh scan M_0, Figure 23.7a), the FWHM is constant over the entire sample and no streaking occurs. After 0.8% engineering strain the FWHM, and hence also the dislocation density, in the entire sample increased (mesh scan M_1, Figure 23.7b). Note that, typically, we do not observe any slip steps at this point of the stress strain curve during *in situ* SEM experiments and one can assume, that all multiplied dislocations were stored in the crystal. However, the FWHM is not homogenous over the entire specimen and a localization of the increased FWHM can be made out at the upper left third, accompanied by pronounced peak streaking in this region.

After the deformation to 16.8%, engineering strain the shape of individual Laue spots is more complicated. At these high strains, the shape of the individual Laue spots is very complicated and, therefore, the above-described analysis cannot be performed. Several streaking directions and split peaks are observed, all well known to increase the strength of metallic domains and, therefore, responsible for the observed hardening from a flow stress of 45 MPa to a stress of more than 90 MPa at 16.8%.

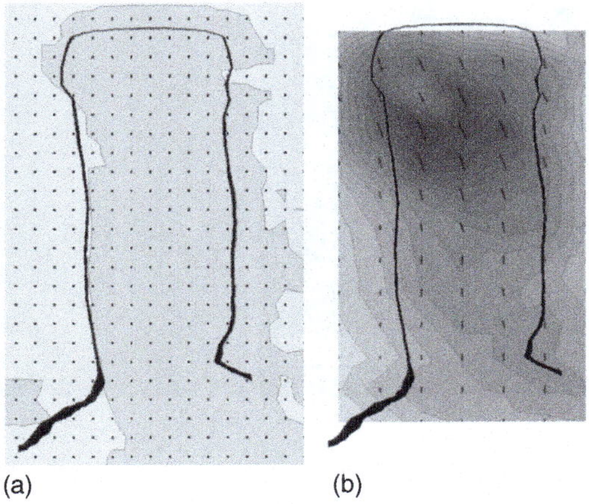

Figure 23.7 The presented contour plots show the maximum FWHM of the (111) Laue spot. In addition, the streaking path being the direction in which the maximum FWHM was measured is plotted. (a) The initial pillar exhibits a uniform maximum diffraction peak width over the entire pillar without any streaking. (b) After the deformation to 0.8% the FWHM increased over the entire pillar and a localized storage of GNDs in the upper left part can be observed.

23.3 Discussion

23.3.1 Deformation Behavior of the Pillar

During the first stage of deformation, the unpredicted slip system (111)[0−11] with a comparably low Schmid factor of 0.22 was activated, which is unexpected. Nevertheless, the observed streaking axes was maintained during the entire experiment and even after 16% strain no visible slip steps of this system were formed on the sample surface. Therefore, most likely, this system was activated in order to accommodate a misalignment of the pillar and the flat punch. Due to misalignment, the stress state in the pillar is not uniaxial – but unfortunately also unknown. Consequently, the system with the highest Schmid factor, the expected primary slip system, does not necessarily be the one with the highest shear stress and therefore, deformation does not automatically start on the primary slip system. However, the presence of dislocations on the (111)[0−11] slip system hinders the dislocation motion of all intersecting slip systems (for instance, the one with the highest Schmid factor which later formed the slip steps observed in the SEM, Figure 23.5b). As one possible consequence of the hindered dislocation motion, a dislocation pile up can form, which may cause the evolution of subgrains, such as those observed in Figure 23.6d. Both, the activation of an unpredicted slip system especially in combination with slip on a second system – most likely on the primary one – and the formation of subgrains are well known to lead to hardening as observed in the stress-strain curve (Figure 23.5a).

23.3.2 Tails of the Primary Beam

The primary beam typically exhibits a beam diameter in FWHM of less than 1 μm. Nevertheless, due the Gaussian like distribution, the intensities far of the beam center are not zero and this "tail" also contributes to the Laue pattern. In case of a homogenous diffracting volume, with a comparably small X-ray beam, the lateral resolution is not seriously influenced. However, if the sample size is reduced, the tail might hit the sample base – typically having a larger volume. Thus, the contribution of the sample base to the Laue pattern increases with respect to the contribution of the gauge length. This might lead to a misinterpretation of Laue patterns in case of small sample dimensions and

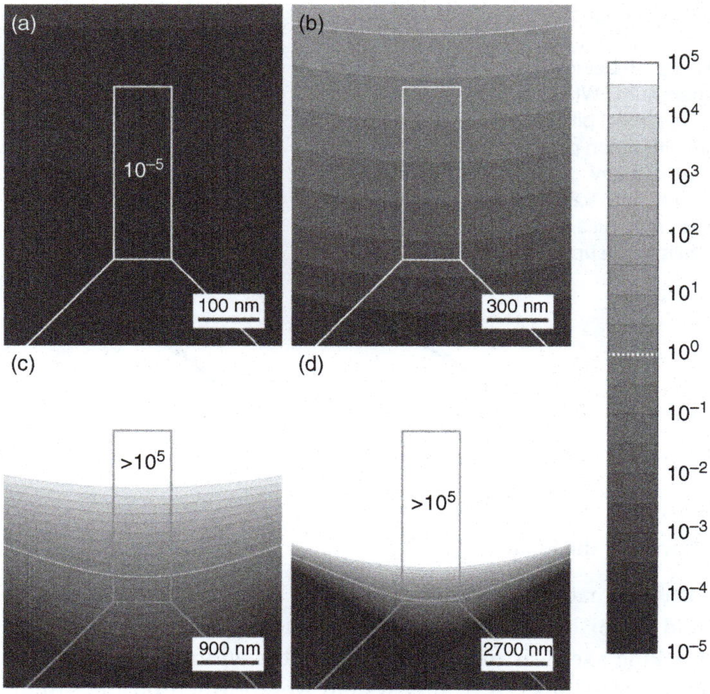

Figure 23.8 Influence of the "tail" of the primary beam on the lateral resolution of μLaue diffraction. The ratio of the scattered intensity of the gauge length over the scattered intensity of the sample base is plotted for four pillar sizes ($100 \times 300\,nm^2$, $300 \times 900\,nm^2$, $900 \times 2400\,nm^2$, and $2700 \times 8400\,nm^2$). The white dashed line indicates where the sample base and the gauge length similarly contribute to the Laue pattern.

needs to be discussed in detail. Therefore, the quantitative study presented in Figure 23.8 shows the ratio of the scattered intensity of the gauge length to the scattered intensity of the sample base.

Four different pillar sizes ranging from 100 nm up to 2.7 μm in diameter are simulated. Herby, the intensity distribution of the primary beam was weighted by the diffracting volume of the gauge length or the sample base respectively, leading to a measure for the total diffracted intensity. Note that the simulation does not correct for mass absorption. The pillar cross-section is square shaped and the pillar base is formed by a pyramid with an opening angle of 45°. In case of the largest and also the second largest pillar (Figure 23.8c,d), the contribution of the sample base to the Laue pattern can be neglected. But smaller samples such as those presented in Figure 23.8b with a diameter of 300 nm show major contributions of the sample base. If the sample length is comparable to the beam size, useful quantitative information cannot be obtained (Figure 23.8b).

Note, that the entire diffracted intensity is plotted. Real micromechanical samples store GNDs within the gauge length, whereas the sample base keeps being undeformed in most cases. Thus, the maximum peak intensity obtained inside the gauge length decreases (due to the streaking) and keeps constant in the sample base. Therefore, with ongoing deformation, the contribution of the sample base to the maximum peak intensity (caused by the tail of the X-ray beam) increases. This might lead to misinterpretations during the evaluation of Laue patterns – except the sample base is made out of a different material with different crystal structure or a different crystallographic orientation.

23.3.3 Sample Movements during Deformation

During deformation, the sample might move in the beam. Assuming that the sample base and the counter body (flat punch or gripper) are infinite stiff – which is a good assumption for the used testing setup – the entire displacement is applied to the gauge length. Since only the counter body is moved with respect to the primary beam, the vicinity of the sample base does not move due to an applied strain and the volume illuminated by the X-ray beam is constant. In contrast, at the upper part of the sample in the vicinity of the flat punch or gripper, the initially illuminated volume can move out of the beam.

In case of pure elastic deformation, this effect can be neglected because of the comparable large beam size (1 μm) and the small sample length (typically 15–20 μm). However, if the sample deforms plastically, the applied strain is much higher and the illuminated volume changes. To minimize this effect, our samples are loaded sequentially and the primary beam is placed either in the middle of the sample or in the vicinity of the sample base, but never in the upper half of the sample. Although typically our samples are 5–20 μm long and a strain of 10% is applied, the illuminated volume does not change significantly.

23.3.4 Streaking of Laue Patterns

Streaking is caused by the storage of GNDs, also known as *unpaired* or *excess dislocations*. It is worth mentioning that dislocations that left the crystal cannot contribute to a lattice curvature and, therefore, do not cause streaking. Therefore, perfect unconstrained slip, where no dislocations are expected to be stored, does not cause any peak streaking. In our case, almost the entire plastic deformation is governed by dislocations of the primary (−111)[101] slip system with a Schmid factor of 0.423, evidenced by the slip steps on the sample surface detected by subsequent SEM studies (Figure 23.5b). Dislocations of other slip systems than the primary one were not able to break through the surface in a huge number and formed no visible slip steps on the sample surface. In contrast to this observation obtained by SEM images, μLaue patterns reveal that – in each subvolume of the pillar – at least one additional slip system was activated and stored GNDs. Thus, SEM and μLaue diffraction have to be combined in order to understand the plasticity of micron and submicron sized specimens.

In addition, the sensitivity of a Laue spot to a distinct dislocation type has to be taken into account when quantitative analyses are made. This is described by the contrast factor of Laue diffraction as shown by Ice and Barabash [30, 34] being a number between 0 and 1. An example is given in Figure 23.6a, where edge type dislocations of the (111)[−110] slip system (number 3) do not cause streaking in case of the displayed (−2−24) Laue spot. This is simply because the crystal rotation axes of edge type dislocations off the (111)[−110] slip system is the [−1−12] crystal axes. The contrast factor of the (−2−24) diffraction vector and the [−1−12] dislocation line equals zero and though, the (−2−24) Laue spot is not sensitive to edge type dislocations of system

3. To avoid any misinterpretations, the streaking direction of all available Laue spots needs to be analyzed.

Considering the above-mentioned pit fall, the analysis of streaking directions might – at least in the simple case of edge type dislocations of one slip system – lead to the activated slip system. Generally, one is able to calculate the lattice curvature tensor based on μLaue data. Nevertheless, due to possible combinations of all 12 slip systems with different dislocation types (edge, screw) it is impossible to split the curvature tensor in the contributions of distinct slip systems without additional assumptions. Thus, in case of more than one slip system it is impossible to calculate the density of GNDs on each slip system unambiguously.

23.4 Conclusion

Interlinking the mechanical properties with the microstructure is the key to understanding the material behavior in macroscopic and microscopic dimensions. Regrettably, methods for the investigation of the microstructure of micron-sized samples are limited. μLaue diffraction is able to provide statistical information on stored defects during deformation with a very high angular resolution. Using μLaue and, complemented, for instance by SEM, new insights into the microstructure of deformed volumes can be obtained, which definitely will contribute to the understanding of the plasticity in small dimensions.

Acknowledgments

The authors thank O. Ulrich and O. Robach from BM32/ESRF for the support during the design and installation of the straining device as well as for the support during the beamtime. Assistance during the experiment and the inspiring discussions by and with W. Grosinger, P.J. Imrich, M. Kapp, J. Kreith, and C. Motz are gratefully acknowledged.

References

1 Uchic, M.D., Shade, P.A., and Dimiduk, D.M. (2009) Plasticity of micrometer-scale single crystals in compression. *Annu. Rev. Mater. Res.*, **39**, 361–386.

2 Dehm, G. (2009) Miniaturized single-crystalline fcc metals deformed in tension: new insights in size-dependent plasticity. *Prog. Mater Sci.*, **54** (**6**), 664–688.

3 Uchic, M.D., Dimiduk, D.M., Florando, J.N., and Nix, W.D. (2004) Sample dimensions influence strength and crystal plasticity. *Science*, **305** (**5686**), 986–989.

4 Volkert, C.A. and Lilleodden, E.T. (2006) Size effects in the deformation of sub-micron Au columns. *Philos. Mag.*, **86** (33–35 Spec. Issue), 5567–5579.

5 Greer, J.R. and Nix, W.D. (2006) Nanoscale gold pillars strengthened through dislocation starvation. *Phys. Rev. B – Condens. Matter Mater. Phys.*, **73** (**24**), 245410.

6 Kiener, D., Grosinger, W., Dehm, G., and Pippan, R. (2008) A further step towards an understanding of size-dependent crystal plasticity: in situ tension experiments of miniaturized single-crystal copper samples. *Acta Mater.*, **56** (**3**), 580–592.

7 Kim, J.-Y., Jang, D., and Greer, J.R. (2009) Tensile and compressive behavior of tungsten, molybdenum, tantalum and niobium at the nanoscale. *Acta Mater.*, **58** (**7**), 2355–2363.

8 Motz, C., Schöberl, T., and Pippan, R. (2005) Mechanical properties of micro-sized copper bending beams machined by the focused ion beam technique. *Acta Mater.*, **53** (**15**), 4269–4279.

9 Demir, E., Raabe, D., and Roters, F. (2010) The mechanical size effect as a mean-field breakdown phenomenon: example of microscale single crystal beam bending. *Acta Mater.*, **58** (**5**), 1876–1886.

10 Bei, H., Shim, S., George, E.P., Miller, M.K., Herbert, E.G., and Pharr, G.M. (2007) Compressive strengths of molybdenum alloy micro-pillars prepared using a new technique. *Scr. Mater.*, **57** (**5**), 397–400.

11 Buzzi, S., Dietiker, M., Kunze, K., Spolenak, R., and Loffler, J.F. (2009) Deformation behavior of silver submicrometer-pillars prepared by nanoimprinting. *Philos. Mag.*, **89** (**10**), 869–884.

12 Korte, S. and Clegg, W.J. (2009) Micropillar compression of ceramics at elevated temperatures. *Scr. Mater.*, **60** (**9**), 807–810.

13 Oh, S.H., Legros, M., Kiener, D., Gruber, P., and Dehm, G. (2007) In situ TEM straining of single crystal Au films on polyimide: change of deformation mechanisms at the nanoscale. *Acta Mater.*, **55** (**16**), 5558–5571.

14 Oh, S.H., Legros, M., Kiener, D., and Dehm, G. (2009) In situ observation of dislocation nucleation and escape in a submicrometre aluminium single crystal. *Nat. Mater.*, **8** (**2**), 95–100.

15 Kiener, D., Motz, C., Rester, M., Jenko, M., and Dehm, G. (2007) FIB damage of Cu and possible consequences for miniaturized mechanical tests. *Mater. Sci. Eng., A*, **459** (**1-2**), 262–272.

16 Kirchlechner, C., Kiener, D., Motz, C., Labat, S., Vaxelaire, N., Perroud, O., Micha, J.S., Ulrich, O., Thomas, O., Dehm, G., and Keckes, J. (2011) Dislocation storage in single slip-oriented Cu micro-tensile samples: new insights via X-ray microdiffraction. *Philos. Mag.*, **91** (**7–9**), 1256–1264.

17 Maaß, R., Grolimund, D., Van Petegem, S., Willimann, M., Jensen, M., Van Swygenhoven, H., Lehnert, T., Gijs, M.A.M., Volkert, C.A., Lilleodden, E.T., and Schwaiger, R. (2006) Defect structure in micropillars using x-ray microdiffraction. *Appl. Phys. Lett.*, **89** (**15**), 151905.

18 Budiman, A.S., Han, S.M., Greer, J.R., Tamura, N., Patel, J.R., and Nix, W.D. (2008) A search for evidence of strain gradient hardening in Au submicron pillars under uniaxial compression using synchrotron X-ray microdiffraction. *Acta Mater.*, **56** (**3**), 602–608.

19 Zimmermann, J., Van Petegem, S., Bei, H., Grolimund, D., George, E.P., and Van Swygenhoven, H. (2010) Effects of focused ion beam milling and pre-straining on the microstructure of directionally solidified molybdenum pillars: a Laue diffraction analysis. *Scr. Mater.*, **62** (**10**), 746–749.

20 Maaß, R., Van Petegem, S., Van Swygenhoven, H., Derlet, P.M., Volkert, C.A., and Grolimund, D. (2007) Time-resolved laue diffraction of deforming micropillars. *Phys. Rev. Lett.*, **99** (**14**), 145505.

21 Maaß, R., Van Petegem, S., Borca, C.N., and Van Swygenhoven, H. (2009) In situ Laue diffraction of metallic micropillars. *Mater. Sci. Eng., A*, **524** (**1–2**), 40–45.

22 Kirchlechner, C., Keckes, J., Motz, C., Grosinger, W., Kapp, M.W., Micha, J.S., Ulrich, O., and Dehm, G. (2011) Impact of instrumental constraints and imperfections on the dislocation structure in micron-sized Cu compression pillars. *Acta Mater.*, **59** (**14**), 5618–5626.

23 Kirchlechner, C., Imrich, P.J., Grosinger, W., Kapp, M.W., Keckes, J., Micha, J.S., Ulrich, O., Thomas, O., Labat, S., Motz, C., and Dehm, G. (2012) Expected and unexpected plastic behavior at the micron scale: an in situ μLaue tensile study. *Acta Mater.*, **60** (**3**), 1252–1258.

24 Kirchlechner, C., Grosinger, W., Kapp, M.W., Imrich, P.J., Micha, J.S., Ulrich, O., Keckes, J., Dehm, G., and Motz, C. (2012) Investigation of reversible plasticity in a micron-sized, single crystalline copper bending beam by X-ray mu Laue diffraction. *Philos. Mag.*, **92** (**25–27**), 3231–3242.

25 Barabash, R.I., Kirchlechner, C., Robach, O., Ulrich, O., Micha, J.-S., Sozinov, A., and Barabash, O.M. (2013) In-situ observation of stress-induced stochastic twin boundary motion in off stoichiometric NiMnGa single crystals. *Appl. Phys. Lett.*, **103** (**2**), 021909-5.

26 Chung, J.-S. and Ice, G.E. (1999) Automated indexing for texture and strain measurement with broad-bandpass x-ray microbeams. *J. Appl. Phys.*, **86** (**9**), 5249–5255.

27 Macdowell, A.A., Celestre, R.S., Tamura, N., Spolenak, R., Valek, B., Brown, W.L., Bravman, J.C., Padmore, H.A., Batterman, B.W., and Patel, J.R. (2001) Submicron X-ray diffraction. *Nucl. Instrum. Methods Phys. Res., Sect. A*, **467–468** (PART II), 936–943.

28 Tamura, N., Padmore, H.A., and Patel, J.R. (2005) High spatial resolution stress measurements using synchrotron based scanning X-ray microdiffraction with white or monochromatic beam. *Mater. Sci. Eng., A*, **399** (**1–2**), 92–98.

29 Robach, O., Micha, J.S., Ulrich, O., Geaymond, O., Sicardy, O., Härtwig, J., and Rieutord, F. (2013) A tunable multicolour 'rainbow' filter for improved stress and dislocation density field mapping in polycrystals using X-ray Laue microdiffraction. *Acta Crystallogr., Sect. A: Found. Crystallogr.*, **69** (**2**), 164–170.

30 Ice, G.E. and Barabash, R.I. (2007) White beam microdiffraction and dislocations gradients, in *Dislocations in Solids* (eds F. Nabarro and J. Hirth), Elsevier, pp. 499–601.

31 Nye, J.F. (1953) Some geometrical relations in dislocated crystals. *Acta Metall.*, **1** (**2**), 153–162.

32 Ulrich, O., Biquard, X., Bleuet, P., Geaymond, O., Gergaud, P., Micha, J.S., Robach, O., and Rieutord, F. (2011) A new white beam X-ray microdiffraction setup on the BM32 beamline at the European Synchrotron Radiation Facility. *Rev. Sci. Instrum.*, **82** (**3**), 033908-6.

33 Tamura, N., Celestre, R.S., Macdowell, A.A., Padmore, H.A., Spolenak, R., Valek, B.C., Chang, N.M., Manceau, A., and Patel, J.R. (2002) Submicron x-ray diffraction and its applications to problems in materials and environmental science. *Rev. Sci. Instrum.*, **73** (3 II), 1369.

34 Barabash, R.I., Ice, G.E., Kumar, M., Ilavsky, J., and Belak, J. (2009) Polychromatic microdiffraction analysis of defect self-organization in shock deformed single crystals. *Int. J. Plast.*, **25** (**11**), 2081–2093.

24

Residual Stresses in Thin Films and Coated Tools: Challenges and Strategies for Their Nondestructive Analysis by X-ray Diffraction Methods

Manuela Klaus and Christoph Genzel

24.1 Introduction

Thin films and coatings nowadays fulfill many important functions in science and technology as well as in industry and our daily life. The applications range from protective layers for cutting tools or drills, thin films in micro- and optoelectronic devices, films for energy conversion in solar cells up to layers used for decorative purposes. Irrespective of the various areas of application they are used for, thin films in many cases have a polycrystalline structure with a more or less preferred orientation (i.e., crystallographic texture) of the crystallites. The mechanical, physical, and chemical properties of the films depend strongly on the materials' microstructure, its texture as well as on the residual stress state of the film [1]. In general, compressive residual stresses in the highly stressed near surface zone (i.e., also in surface layers of technical components, if they are coated) are known to have a positive effect on the material properties [2]. For example, cutting tools made of cemented carbide (WC-Co) are coated by chemical vapor deposition (CVD) with Al_2O_3 and $TiC_{1-x}N_x$ ($x = 0, \ldots, 1$) or combinations of these materials to improve wear resistance. Due to the high deposition temperature >800 °C and the differences in the thermal expansion coefficients, the as-grown CVD coating systems show unfavorable tensile residual stresses. To generate beneficial compressive residual stresses that improve the toughness behavior of the tools, mechanical post-treatment processing like dry or wet blasting is introduced after coating deposition. By variation of the blasting parameters as well as of the coating design and the structural coating properties, it is possible to tailor the residual stresses with respect to both amount and depth distribution [3].

An efficient assessment of this "residual stress engineering" step is possible by employing X-ray methods that allow for nondestructive, phase-selective, and depth-resolved materials characterization. X-ray stress analysis (XSA) in thin films and coating systems, however, involves a series of the following challenges that X-ray analysts are faced with:

- The thinner the film the smaller the effectively scattering volume becomes, which results in (very) weak diffraction signals.
- Similar crystal structure and lattice parameters may lead to a strong overlap of diffraction lines originating from the film and the substrate, respectively.
- Thin polycrystalline films are usually strongly textured that reduces the available measuring directions to small angular ranges around the intensity poles.

Neutrons and Synchrotron Radiation in Engineering Materials Science: From Fundamentals to Applications, Second Edition. Edited by Peter Staron, Andreas Schreyer, Helmut Clemens, and Svea Mayer.

- To achieve optimum properties, thin films are often deposited as functionally graded layers with gradients of the chemical composition and, therefore, also of the strain-free lattice parameter, which superimpose the internal film stress distribution.
- Multi-layered coatings consist of alternating layers of different crystal structure, which leads from the diffraction perspective to a situation, where diffracting layers are separated by only absorbing layers.

To meet the challenges outlined above, a series of measuring techniques and data evaluation concepts have been developed in the past, which may be classified with respect to different criteria [4] such as their applicability to films with random or preferred crystallographic texture, see Figure 24.1.

Common features of all methods listed in Figure 24.1 are: (i) They were developed for angle-dispersive (AD) diffraction using monochromatic X-rays. (ii) The corresponding measurements are performed in reflection geometry. (iii) The methods can be applied under laboratory conditions. (iv) They are nondestructive. Concerning the criteria (ii)–(iv), these methods, therefore, differ from the so-called micro- and nanobeam methods which are strongly linked with the increasing availability of high-flux synchrotron beamlines [5]. With respect to film stress analysis, strain depth profiling employing micro- or nanobeam techniques is performed in transmission geometry with highly focused synchrotron X-ray beams of very small cross-section on thin penetrable cross-sectional sample slices [6, 7]. Therefore, these methods enable direct access to the real space residual stress depth profiles $\sigma_{||}(z)$ within the films; however, they are not nondestructive and require time-consuming sample preparation as well as careful corrections for the residual macro stresses released by the preparation.

The aim of this chapter is to demonstrate with selected examples how some of the issues in thin film stress analysis outlined above can be solved. The focus is on nondestructive methods that do not require any sample preparation. Besides AD diffraction

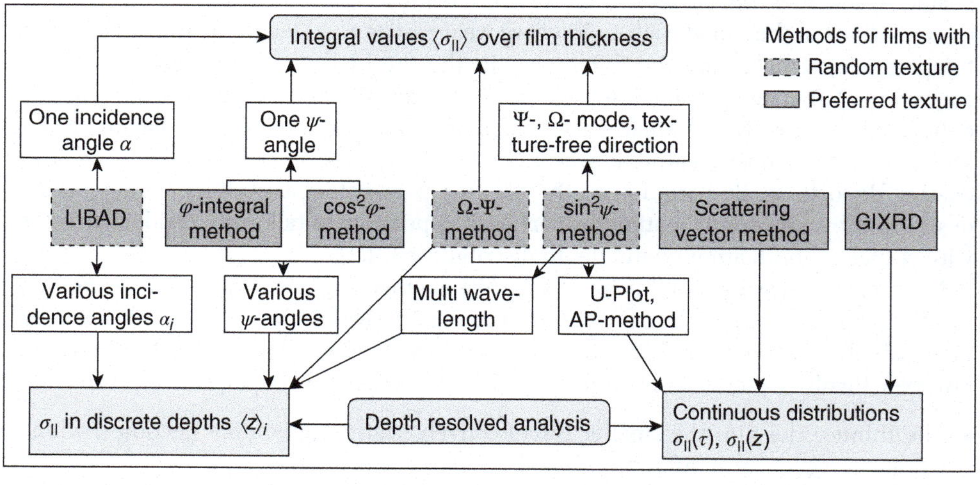

Figure 24.1 X-ray diffraction techniques in thin film (residual) stress analysis. LIBAD and GIXRD stand for *L*ow *I*ncident *B*eam *A*ngle *D*iffraction and *G*razing *I*ncidence *X-ray D*iffraction, respectively. The dark-gray boxes mark the methods, while the white boxes denote the concepts used for data acquisition and/or stress evaluation. (Taken from [4].)

techniques we will also discuss the application of energy-dispersive (ED) diffraction to thin film stress analysis. It will be shown that the special features of the ED mode can be used conveniently to study various cases of thin film stress even under difficult conditions (see also Chapter 9). In order to present the examples in a compact form, we have deliberately refrained from describing the fundamentals of X-ray residual stress analysis, which are necessary for an improved understanding of the matter. Some important relationships such as the fundamental equation of XSA and the basic principles of residual strain and stress depth profiling have been already introduced in Chapter 9. For a comprehensive survey of the "state of the art" of residual stress analysis by diffraction methods, the reader is referred to the textbooks 8–11, monographs, and topical reviews [12–14] given at the end of this chapter.

24.2 Compilation of Approaches to Meet the Challenges in Thin Film X-ray Stress Analysis (XSA)

24.2.1 Stress Analysis under Grazing and Glancing Diffraction Conditions

With regard to thin film XSA there are different reasons why the information depth τ of the X-rays should be confined to the film thickness D or values even (much) smaller than D. First of all, a small incidence angle α between the incoming beam and the sample surface increases the effective diffracting film volume and thus, improves the peak to noise ratio of the film reflections in the diffraction pattern. Figure 24.2 demonstrates with the example of stress evolution in a thin Al_2O_3 oxide scale grown on a single crystal α-Fe(Al) substrate how ED diffraction can be applied to thin film XSA. Stress analysis was performed by means of the $\sin^2\psi$-method under grazing incidence conditions keeping the angle α between the incident beam and the surface fixed at 2°. In this way, diffraction patterns with evaluable diffraction lines became available even in the early stages of oxidation when the total film thickness reached values between 50 and 100 nm. The example in Figure 24.2 reveals a further advantage of ED-XSA on thin films: Because the ED method provides complete diffraction patterns with a multitude of diffraction lines for each ψ-angle, which can be used for stress analysis (Figure 24.2b), statistically better verified results may be achieved (Figure 24.2c). For more experimental details the reader is referred to [15, 16].

A further reason for adapting the X-ray information depth τ to the film thickness can be seen from the example in Figure 24.3, which also shows another feature of the ED method using rather high energy photons. Residual stress analysis was performed by means of the $\sin^2\psi$-method in the symmetrical Ψ-mode on a mechanically surface blasted CVD multilayer system [17]. Since the white synchrotron beam penetrates the coating completely, the ED diffraction patterns contain reflection lines from all layers (here Al_2O_3 and TiCN) as well as from the substrate (WC). Therefore, only *one* $\sin^2\psi$-measurement is necessary to evaluate the in-plane residual stress state in *all* material phases. However, what is more important in terms of the issue discussed in this section is the fact that the very steep intralayer residual stress gradient generated by surface blasting in the Al_2O_3 layer on top is only detectable if the information depth τ matches with the layer thickness and the steepness of the gradient to be analyzed, respectively. Comparing the $d_\psi^{012} - \sin^2\psi$-plots obtained for a large ($E^{012} = 34.1$ keV)

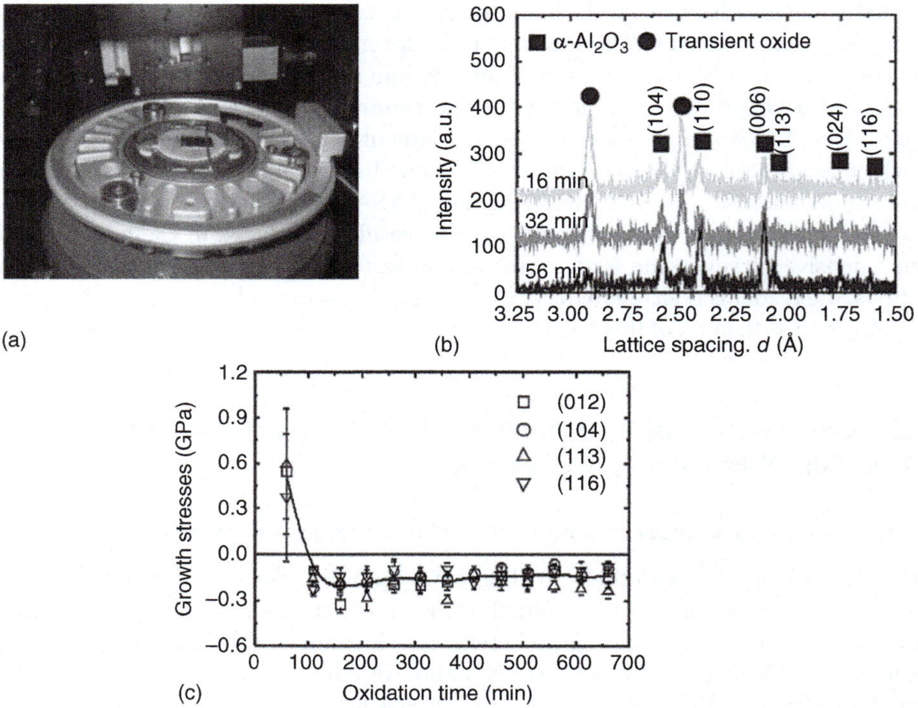

Figure 24.2 *In situ* monitoring of phase and stress evolution during the oxidation of an α-Fe(Al) (111) single crystal at 1100 °C. (a) Substrate in the heating station DHS 1100 (Anton Paar) mounted on the five-axes sample positioning unit (EDDI (Energy Dispersive Diffraction) beamline @ BESSY II). (b) ED diffraction patterns showing the development of the oxide phases during the early stages of oxidation. (c) Evolution of growth stresses in the α-Al_2O_3 layer grown during oxidation. (Panels (b,c): Brito 2011 [15]. Reproduced with the permission of Elsevier.)

and a smaller ($E^{012} = 18.6$ keV) photon energy, it can be realized that the stress gradient becomes "visible" in the form of a strong concave curvature at large ψ-angles (i.e., at grazing diffraction (GD) conditions) only in the latter case. Thus, making use of the feature of ED diffraction that the position of the diffraction lines can be shifted to smaller or larger energies by changing the Bragg angle θ (cf. Figure 9.1), it becomes possible to tune τ and, therefore, to adapt it to the particular conditions defined by the sample/coating design and the residual stress depth distribution expected within the near surface materials zone, respectively.

A third reason for employing GD techniques in thin film XSA is given, if film and substrate diffraction lines superimpose each other more or less completely. An important practical case where this phenomenon occurs is illustrated in Figure 24.4. When cutting tools made of WC-Co are coated with diamond films, one must be aware that diamond (D) and cobalt (β-Co) have not only a similar cubic crystal structure (diamond and fcc (face centered cubic), respectively), but that they also nearly coincide with respect to their lattice parameters ($a_0^{D} = 0.3566$ nm, $a_0^{\beta\text{-Co}} = 0.3544$ nm). Consequently, *all* diffraction lines of the diamond film are superimposed by those of the β-Co phase in the substrate, and the only free-standing β-Co reflection (within the covered 2θ-range) is the 200 diffraction line (see Figure 24.4a). Thus, the only way to extract the

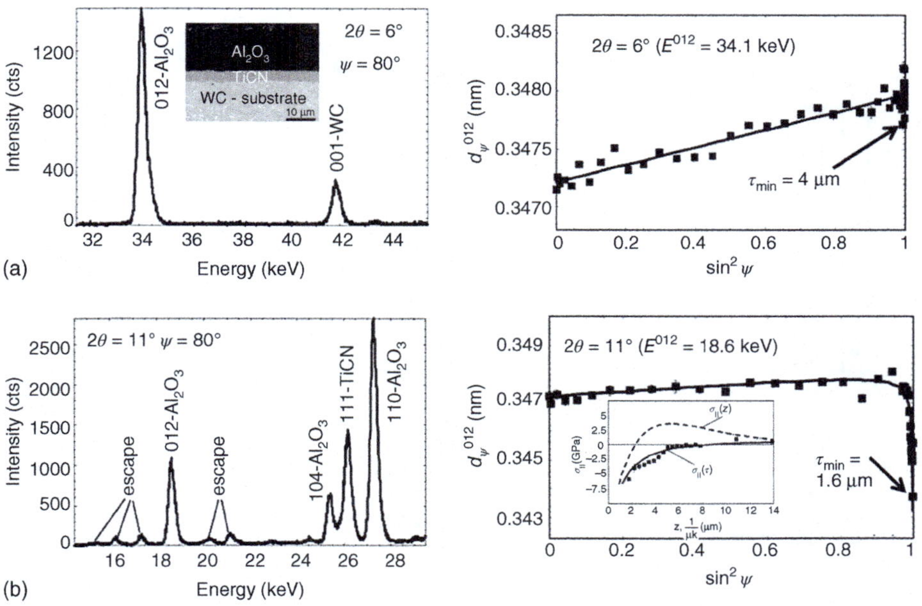

Figure 24.3 ED-XSA performed for different diffraction angles 2θ on the Al_2O_3 top layer of a CVD multilayer system. See inset in the left in (a). The strong line asymmetry of the 012-Al_2O_3 diffraction line is due to the intralayer stress gradient in the Al_2O_3 top layer. See inset at the right in (b). $\tau_{min} = \sin\theta\cos\psi_{max}/[2\mu(E)]$ denotes the minimum information depth that is achieved at the maximum inclination angle $\psi_{max} = 88°$ (μ – linear absorption coefficient). (Klaus 2009 [17]. Reproduced with the permission of Cambridge University Press.) See text for further details.

information originating from the film is to confine the X-ray information depth τ to the film thickness t. Figure 24.4b shows the situation for diamond. Due to the very weak absorption of X-rays by carbon values of $\tau < 8$ μm (the thickness of the film considered here) are achieved for $\psi > 86.3°$ which corresponds to incidence angles $\alpha < 1.38°$ that are in the range of the critical angle of total external reflection, $\alpha_c = 0.27°$. Taking into consideration that the lattice planes detected in the symmetrical Ψ-mode at these large ψ-angles are nearly perpendicular to the film plane and, therefore, are directly affected by the in-plane film stress, an average value for $\sigma_{\parallel}$ can be evaluated using the following relation:

$$\langle \sigma_{\parallel}^{D} \rangle = \frac{E^{111,D}}{(1 - \nu^{111,D})} \langle \varepsilon_{\parallel}^{111,D} \rangle \tag{24.1}$$

with $\langle \varepsilon_{\parallel}^{111,D} \rangle = \left[\frac{1}{N}\sum_{n=1}^{N}(d_{\perp,n}^{111,D} - d_0^{111,D})\right]/d_0^{111,D}$ (N – number of data points, $E^{111,D}$, $\nu^{111,D}$ – direction-dependent Young's modulus and Poisson ratio, respectively, $d_0^{111,D}$ – strain-free diamond lattice parameter). Making use of Eq. (24.1) with the reduced data set shown in the diagram in Figure 24.4d yields $\langle \sigma_{\parallel}^{D} \rangle = -1.6 \pm 0.6$ GPa which is in good agreement with the film thermal stress of about −2 GPa expected from the difference of the thermal expansion coefficients of diamond and tungsten carbide. Finally, it has to be stressed that the application of the classical $\sin^2\psi$-method in the present case leads to wrong results. This is to be seen from Figure 24.4c: Because the

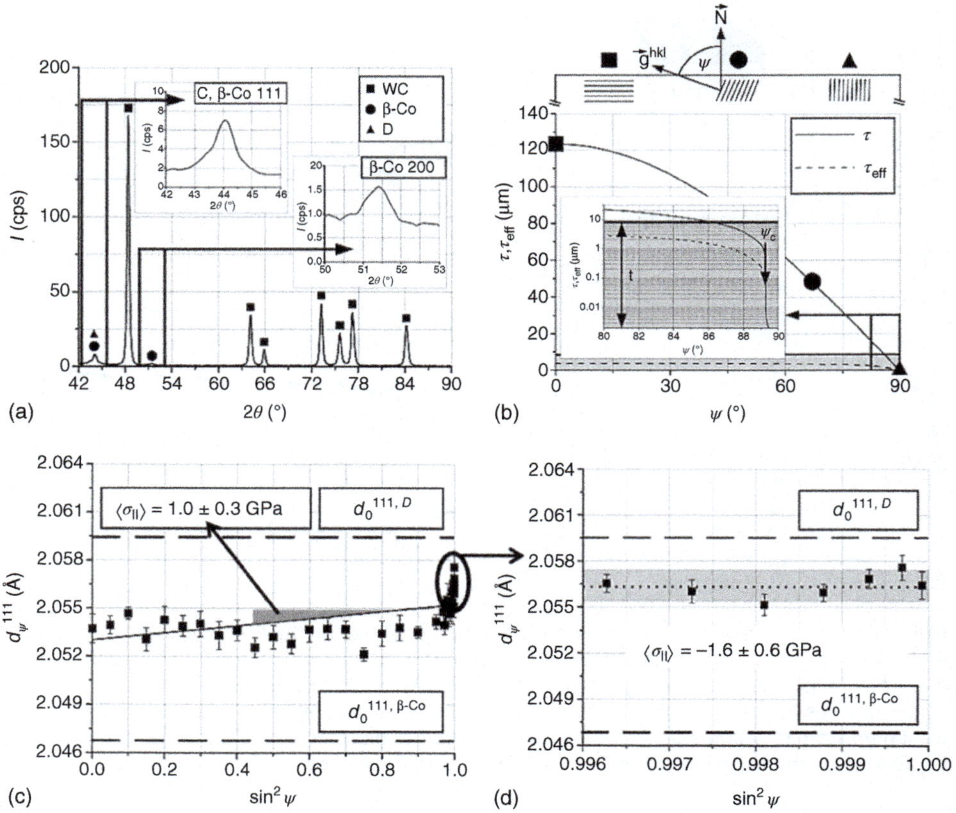

Figure 24.4 X-ray residual stress analysis performed on a diamond (D) coating (thickness $t = 8\,\mu m$) deposited by hot-filament CVD on a WC-Co cutting tool. (a) X-ray diffractogram (CuKα) showing the positions of the substrate and film peaks. (b) Effective and actual information depth for different inclination angles ψ in diamond under nongrazing and grazing (inset) diffraction conditions. (c) $d_\psi^{111} - \sin^2\psi$-distribution for the superimposed 111-D/Co diffraction line. The circle marks the angular range $\psi > 86.3°$ that was taken into account for the coating residual stress evaluation. (d) Zoom into the region of interest. (Meixner 2013 [18]. Reproduced with the permission of International Union of Crystallography.) See text for further details.

d_ψ^{111}-data for $\psi < 86.3°$ (i.e., outside of the circle) were obtained from superimposed diffraction lines, they cannot be assigned solely to the diamond lattice. The contribution of the β-Co phase in the substrate (smaller lattice parameter compared with diamond) becomes smaller with increasing inclination angle ψ. A least squares fit of the complete $d_\psi^{111} - \sin^2\psi$-data set without paying attention to this fact would result in a positive slope of the regression line and, therefore, lead to the wrong conclusion that there are (high) tensile residual stresses in the film. For a more detailed discussion the reader is referred to [18].

24.2.2 Separation of Residual Stress and Composition Gradients

Thin films often are not uniform over their thickness, but occur in the form of gradients with respect to the composition, or, they are composed of layers with uniform but slightly different composition. In these cases, the strain-free lattice parameter a_0 becomes a function of depth, which superimposes possible residual stress gradients, $\sigma_{\|}(z)$. In $d_{\psi}^{hkl} - \sin^2\psi$-diagrams a $a_0(z)$ depth gradient even in the absence of any stress gradient would lead to a nonlinear (curved) distribution similar to that obtained for a stress depth gradient. Hence, it becomes clear that a nonconsideration of composition depth gradients in the residual stress evaluation would lead to erroneous results. In order to solve this problem various methods have been developed in the past, which are based on lattice strain analysis in constant, predefined information depths within the film, such as the LIBAD (low incident beam angle diffraction) method or the $\Omega - \Psi$ ("mixed mode") method (cf. Figure 24.1). Summaries of these and other special approaches for residual stress gradient analysis in thin films are found, for example, in the topical reviews [12–14].

In this section, we present two alternative experimental techniques to separate the impact of residual stress from that of the composition on the lattice parameter. The first approach shown in Figure 24.5 is based on ED diffraction and uses a combination of the scattering vector method [20] and the $\sin^2\psi$-method to evaluate depth profiles of the in-plane residual stress and the materials composition, respectively. The investigations were performed on a layer of expanded austenite that was generated by nitriding of stainless steel 316L (for experimental details see [19]). The diffraction pattern in Figure 24.5a reveals that the lattice parameter of the nitrided layer ($\gamma - N$) differs completely from that of the substrate (γ). Moreover, while the diffraction lines of the γ base material have a small width, the corresponding $\gamma - N$ line profiles show significant broadening which is due to the strong variation of the lattice parameter with depth. Lattice strain depth profiling was performed for different inclination angles ψ ($\sin^2\psi = 0.1, \ldots, 0.9$) in the scattering vector mode, that is, by stepwise rotation η of the sample around the diffraction vector [20]. Since the $E_{\psi=\text{const.}}^{hkl}[\tau(\eta)]$ profiles overlap with respect to the covered depth ranges (see Figure 24.5b), it becomes possible to draw sections through the obtained data sets and to construct $\sin^2\psi$-plots for predefined depths τ (inset in Figure 24.5b). In this way, the in-plane residual stress for the respective depths can be evaluated from the slope of the regression line using the $\sin^2\psi$-method (Figure 24.5c). Furthermore, keeping in mind that the lattice spacing $d_{\psi^*}^{hkl}$ obtained from the $d_{\psi}^{hkl} - \sin^2\psi$-plots in the strain-free direction of a biaxial stress state is given by $\psi^{*,hkl} = \sin^{-1}\sqrt{-2S_1^{hkl}/\frac{1}{2}S_2^{hkl}}$, the strain-free lattice parameter a_0 becomes available as a function of depth and enables the evaluation of composition gradient (Figure 24.5d).

The second approach that should be discussed here exploits the abrupt change of X-ray absorption and, therefore, of the information depth τ, when diffraction experiments are performed using monochromatic synchrotron X-rays with photon energies in the immediate vicinity of absorption edges [21]. Figure 24.6 shows a multilayer system with a rather sophisticated layer composition from the point of view of XSA. Using conventional CoKα-radiation available in the lab, the two buried TiC_xN_{1-x} layers of similar composition yield diffraction line profiles that are close together and thus, very difficult to separate.

On the other hand, using highly parallel monochromatic synchrotron radiation with photon energies closely below and above the TiKβ-absorption edge at 4.9 keV, respectively, the contribution of the thin upper $TiC_{0.3}N_{0.7}$ layer is either suppressed or increased (Figure 24.7a). The reason for this behavior becomes clear from Figure 24.7b.

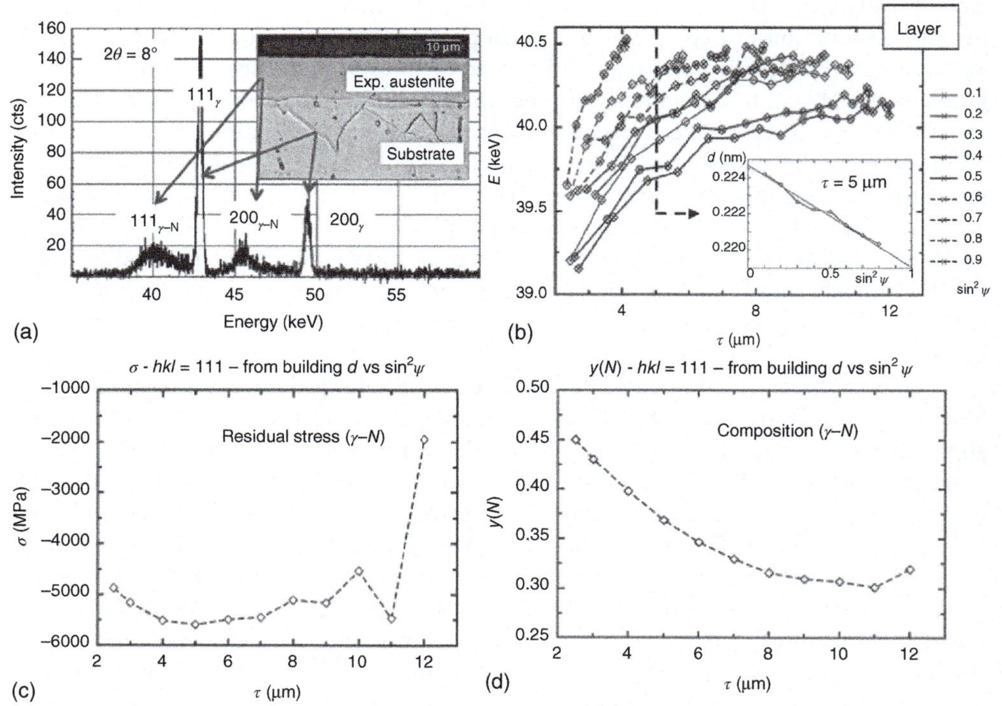

Figure 24.5 Energy-dispersive XSA performed on an expanded austenite layer. (After [19].) (a) Diffraction pattern of nitrided 316L stainless steel. Expanded austenite and austenite are denoted by $\gamma - N$ and γ, respectively. (b) Energy line positions E^{111} versus information depth τ profiles of the expanded austenite layer (see inset in (a)) measured in the scattering vector mode for different inclination angles ψ. The inset shows a reconstructed $d_{\psi}^{111,\gamma-N} - \sin^2\psi$-distribution in a predefined depth $\tau = 5\ \mu$m. (c) Depth profile of the in-plane residual stress component in the expanded austenite layer. The average error is about $\pm$250 MPa. (d) Composition depth profile showing the nitrogen content in the expanded austenite layer. See text for further details.

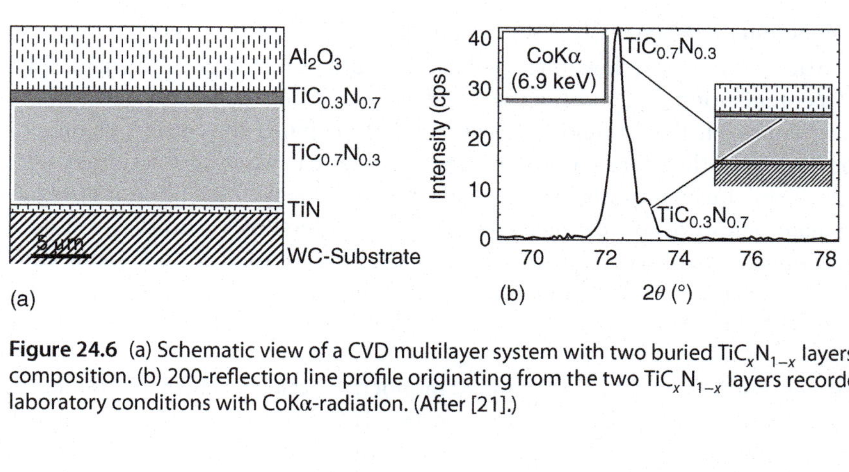

Figure 24.6 (a) Schematic view of a CVD multilayer system with two buried TiC_xN_{1-x} layers of different composition. (b) 200-reflection line profile originating from the two TiC_xN_{1-x} layers recorded under laboratory conditions with CoKα-radiation. (After [21].)

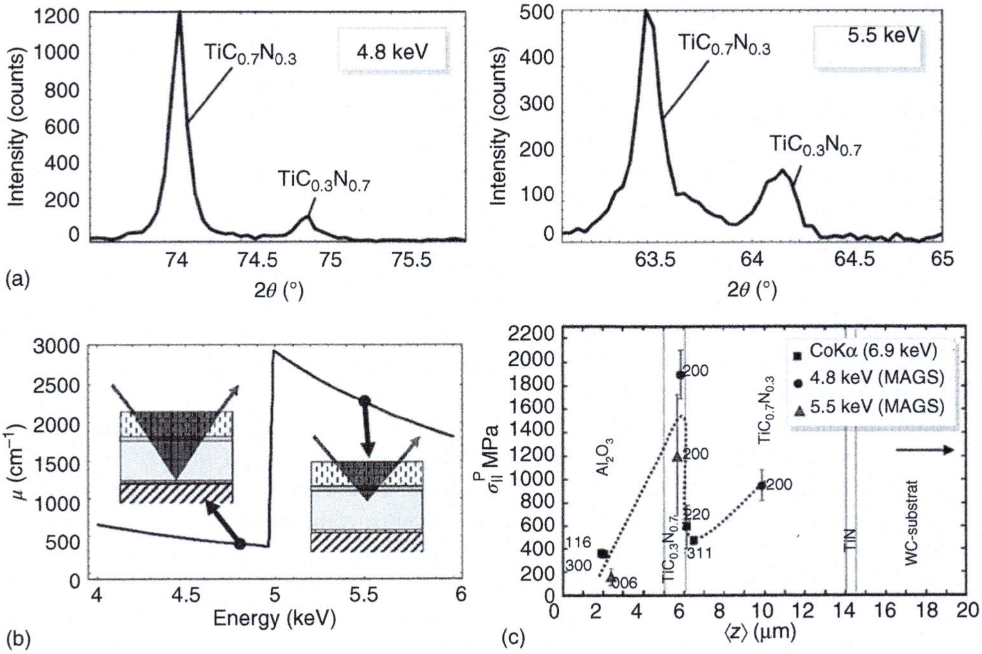

Figure 24.7 (a) X-ray diffractograms of the multilayer system in Figure 24.6a recorded using monochromatic synchrotron radiation with photon energies below and above the TiK β-absorption edge at 4.9 keV, respectively. (b) Linear absorption coefficient and schematic view of the X-ray information depth below and above the TiK β-absorption edge. (c) Results of the $\sin^2\psi$-based XSA in the individual layers of the multilayer system. (After [21].)

Below the absorption edge the X-ray beam penetrates the thin layer more or less completely and the main contribution to the diffraction line originates from the thick $TiC_{0.7}N_{0.3}$ layer beneath. Above the absorption edge the situation is the opposite: Here the thin layer absorbs most of the incoming photons, which yields a stronger peak for this layer in relation to that originating from the thick layer beneath. Thus, making use of the behavior of X-rays in the immediate vicinity of absorption edges may be helpful for analyzing residual stresses in neighboring layers of similar composition (see Figure 24.7c).

24.3 Final Remarks and Recommendations

X-ray residual stress analysis conducted on thin polycrystalline films and coating systems is a challenging task, since it involves a series of demands, problems, and pitfalls that have to be taken into account. To meet these demands, various experimental methods have been developed in the past years and decades. However, there is no method that would be able to answer all the questions arising in thin film XSA at once.

In this chapter, some recently developed approaches for analyzing residual stresses in thin films are presented, which have in common that they are based on reflection mode measurements and, therefore, do not require any sample preparation and correction for released macro stresses. However, this advantage is achieved at the cost of resolution: On the one hand, especially under GD conditions, the irradiated spot on the sample surface may become very large, which reduces the lateral resolution. On the other hand, under nongrazing conditions, the incoming X-ray beam in most cases penetrates the film completely and reduces the sensitivity to detect stress depth gradients within the film. As shown very recently in [22, 23] a possible way to improve the spatial resolution in reflection mode XSA experiments consists in the reduction of the gage volume size that is used to probe the near surface region.

It should be emphasized that many issues concerning thin film XSA can be solved with conventional X-ray sources and diffraction equipment available in the lab. Grazing and even glancing diffraction conditions can be realized, for example, with comparatively little effort by means of a GD unit consisting of an equatorial soller slit and a single crystal monochromator, both to be placed in the diffracted beam path. Thus, before applying for expensive synchrotron beamtime, it is strongly recommended to perform some preliminary investigations with laboratory equipment on the films to be analyzed, in order to figure out the problems involved in their characterization (cf. Figure 24.6).

Synchrotron radiation for residual stress analysis in thin films should be reserved for issues that require a high photon flux. Its application is recommended, for example, for fast *in situ* measurements to study the stress evolution during film growth (cf. Figure 24.2), for measurements that are based on tuning the photon energy (cf. Figure 24.7), or for investigations that require very small beam cross-sections.

References

1 Nix, W.D. (1989) Mechanical properties of thin films. *Metall. Trans. A*, **20**, 2217–2245.

2 Scholtes, B. (1991) *Eigenspannungen in mechanisch randschichtverformten Werkstoffen. Ursachen, Ermittlung und Bewertung*, DGM Informationsgesellschaft, Oberursel.

3 Klaus, M. and Genzel, C. (2011) Multilayer systems for cutting tools: on the relationship between coating design, surface processing and residual stress. *Adv. Eng. Mater.*, **13**, 845–850.

4 Klaus, M. and Genzel, C. (2013) X-ray residual stress analysis on multilayer systems: an approach for depth-resolved data evaluation. *J. Appl. Crystallogr.*, **46**, 1266–1276.

5 Ice, G.E., Budai, J.D., and Pang, J.W. (2011) The race to X-ray microbeam and nanobeam science. *Science*, **334**, 1234–1239.

6 Keckes, J., Bartosik, M., Daniel, R., Mitterer, C., Maier, G., Ecker, W., Vila-Comamala, J., David, C., Schoeder, S., and Burghammer, M. (2012) X-ray nanodiffraction reveals strain and microstructure evolution in nanocrystalline thin films. *Scr. Mater.*, **67**, 748–751.

7 Krywka, C., Neubauer, H., Priebe, M., Salditt, T., Keckes, J., Buffet, A., Roth, S.V., Doehrmann, R., and Mueller, M. (2012) A two-dimensional waveguide beam for X-ray nanodiffraction. *J. Appl. Crystallogr.*, **45**, 85–92.

8 Noyan, I.C. and Cohen, J.B. (1987) *Residual Stress Measurement by Diffraction and Interpretation*, Springer, New York.

9 Hauk, V. (1997) *Structural and Residual Stress Analysis by Nondestructive Methods* (contributions by H. Behnken, Ch. Genzel, W. Pfeiffer, L. Pintschovius, W. Reimers, E. Schneider, B. Scholtes, and W. A. Theiner), Elsevier, Amsterdam.

10 Birkholz, M. (2006) *Thin Film Analysis by X-Ray Scattering*, Wiley-VCH Verlag GmbH, Weinheim.

11 Spieß, L., Teichert, G., Schwarzer, R., Behnken, H., and Genzel, C. (2009) *Moderne Röntgenbeugung*, 2. Aufl. edn, Vieweg + Teubner, Wiesbaden.

12 Genzel, C. (2004) in *Diffraction Analysis of the Microstructure of Materials*, Springer Series in Materials Science, vol. **68** (eds E.J. Mittemeijer and P. Scardi), Springer, Berlin, Heidelberg, pp. 473–503.

13 Welzel, U., Ligot, J., Lamparter, P., Vermeulen, A.C., and Mittemeijer, E.J. (2004) Stress analysis of polycrystalline thin films and surface regions by X-ray diffraction. *J. Appl. Crystallogr.*, **38**, 1–29.

14 Noyan, I.C., Huang, T.C., and York, B.R. (1995) Residual stress/strain analysis in thin films by X-Ray diffraction. *Crit. Rev. Solid State Mater. Sci.*, **20**, 125–177.

15 Brito, P., Pinto, H., Klaus, M., Genzel, C., and Kaysser-Pyzalla, A.R. (2011) Impact of transition oxides on growth stresses and texture of alumina scales formed during oxidation of iron metals. *Scr. Mater.*, **65**, 312–315.

16 Brito, P., Pinto, H., Genzel, C., Klaus, M., and Kaysser-Pyzalla, A. (2012) Epitaxial stress and texture in thin oxide layers grown on Fe–Al alloys. *Acta Mater.*, **60**, 1230–1237.

17 Klaus, M., Reimers, W., and Genzel, C. (2009) Application of energy-dispersive diffraction to the analysis of highly inhomogeneous residual stress fields in thin film structures. *Powder Diffr.*, **24** (Suppl.), S82–S86.

18 Meixner, M., Klaus, M., Genzel, C., and Reimers, W. (2013) Residual stress analysis of diamond coated WC-Co cutting tools – Separation of film and substrate information by grazing X-ray diffraction. *J. Appl. Crystallogr.*, **46**, 1323–1330.

19 Jegou, S., Christiansen, T.L., Klaus, M., Genzel, C., and Somers, M.A.J. (2013) Determination of composition, residual stress and stacking fault depth profiles in expanded austenite with energy-dispersive diffraction. *Thin Solid Films*, **530**, 71–76.

20 Genzel, C. (1999) A self-consistent method for X-ray diffraction analysis of multiaxial residual stress fields in the near surface region of polycrystalline materials. I. Theoretical concept. *J. Appl. Crystallogr.*, **32**, 770–778.

21 Klaus, M., Denks, I., and Genzel, C. (2006) X-ray diffraction analysis of nonuniform residual stress fields $\sigma_{ij}(\tau)$ under difficult conditions. *Mater. Sci. Forum*, **524–525**, 601–606.

22 Meixner, M., Klaus, M., and Genzel, C. (2013) $\mathrm{Sin}^2\psi$-based residual stress gradient analysis by energy-dispersive synchrotron diffraction constrained by small gauge volumes. I. Theoretical concept. *J. Appl. Crystallogr.*, **46**, 610–618.

23 Meixner, M., Klaus, M., and Genzel, C. (2013) $\mathrm{Sin}^2\psi$-based residual stress gradient analysis by energy-dispersive synchrotron diffraction constrained by small gauge volumes. II. Experimental implementation. *J. Appl. Crystallogr.*, **46**, 619–627.

Index

Neutrons and Synchrotron Radiation in Engineering Materials Science: From Fundamentals to Applications, Second Edition. Edited by Peter Staron, Andreas Schreyer, Helmut Clemens, and Svea Mayer.

f

t

y

z